Springer Textbooks in Earth Sciences, Geography and Environment

The Springer Textbooks series publishes a broad portfolio of textbooks on Earth Sciences, Geography and Environmental Science. Springer textbooks provide comprehensive introductions as well as in-depth knowledge for advanced studies. A clear, reader-friendly layout and features such as end-of-chapter summaries, work examples, exercises, and glossaries help the reader to access the subject. Springer textbooks are essential for students, researchers and applied scientists.

Jeff B. Chaumba

Introductory Physical Geology

Springer

Jeff B. Chaumba
Department of Geological Sciences
Texas Christian University
Fort Worth, TX, USA

ISSN 2510-1307 ISSN 2510-1315 (electronic)
Springer Textbooks in Earth Sciences, Geography and Environment
ISBN 978-3-032-20230-7 ISBN 978-3-032-20231-4 (eBook)
https://doi.org/10.1007/978-3-032-20231-4

This Springer imprint is published by the registered company Springer Nature Switzerland AG
The registered company address is: Gewerbestrasse 11, 6330 Cham, Switzerland

To Mr. A. P. K. and the late Mrs. S. T. Chaumba for their sacrifice in sending us all their children to elite boarding schools

Foreword

Introductory Physical Geology is designed to accompany an initial college-level, physical geology course for geology majors; however, the book could also serve as the text in a survey geology course, in which the students are not necessarily geology majors. Oftentimes these courses are popular and serve to satisfy a natural science or lab science requirement. The author, Dr. Jeff B. Chaumba, has taught in-person and online physical geology courses at a number of universities including Reinhardt University, the University of Georgia, Georgia State University, the University of North Carolina at Pembroke, and Texas Christian University. Consequently he has taught a variety of student populations and used this experience to develop this introductory text.

Each chapter begins with a bulleted list of learning goals and an introduction. Color diagrams and photographs illustrate critical concepts throughout the chapters. A glossary of key words used in the text can be found at the end of each chapter. There is also an online exercise for students and a test bank of multiple-choice questions which will be useful to students and instructors.

Chapter 1 sets the Earth in the context of the universe and the solar system. Chapters 2–5 present the material science of geology: the nature of minerals and the igneous, sedimentary, and metamorphic rocks that are comprised of these minerals. Chapter 6 covers the unifying theory of plate tectonics, which was formulated in the latter half of the twentieth century. The real-world consequences of plate tectonics: volcanoes, earthquakes, and mountain building form the subjects of Chaps. 7–9. The nature of geologic time and the evolution of living organisms make up the bulk of Chap. 10. Surface processes including weathering, soil development, the hydrologic cycle, rivers and groundwater, deserts, and glaciers are covered in Chaps. 11–14. In Chap. 14, the particularly important topic of climate change in a geological context is discussed. Oceanography, including the topography of the ocean floor, seawater composition, and the ocean currents, which are so important in controlling our climate, compose Chap. 15. The book concludes with a chapter on Earth Resources including metallic ores and energy resources.

Overall, the text constitutes a complete introduction to physical geology. The 16 chapters correlate with the normal length (in weeks) of a semester in a US college or university; hence an instructor could cover approximately one chapter per week, and present material in comprehensible portions over 2 or 3 class periods in that week. Given the impact that society is having on our Earth, and the societal needs for the many resources that the Earth provides, this book is a timely and important resource for today's college students.

Athens, GA, USA
December 2025

Michael Roden

Acknowledgements For a huge undertaking such as this, there are countless people who helped 'see' book to completion. Countless colleagues and friends, particularly Steve Clark, Dan Bulger, and Jay Austin, willingly provided help with discussions, photographs and helped as 'field assistants' to various outcrops and field trips as well as providing me with some of the photographs used in this text. Their help is greatly appreciated, and their names, and those of countless others, are listed in the captions of relevant photographs. I thank my father, late mother, and siblings for instilling in me from a young age the importance of taking education seriously. Without the financial sacrifice of our parents to send us to boarding schools, this undertaking would most likely never have seen the light of day. I am thankful to all my teachers and classmates and the elementary school and two high schools I attended. The different teaching approaches my former professors utilized at the five universities I graduated from helped my understanding of the nuts and bolts of my discipline. These various approaches helped play a role in my decision to put together this textbook, incorporating some of their strategies. Richard Smosna is gratefully acknowledged for selecting me to be the Head Teaching Assistant at WVU! To my friends in "The Basement" at UGA Geology/Geography Building, I deeply appreciate your friendship. Noel Heim (Tufts University) got me started with Adobe Illustrator, and I am extremely appreciative to him for his help. My students and colleagues at Reinhardt University, University of Georgia, Georgia State University, Auburn University, University System of North Carolina, Weatherford College, and Texas Christian University helped in making me a better professor. The year I spent in the oil and gas industry mostly in South Texas and Louisiana, coupled with the ones as a field/mining geologist in Southern Africa, helped make me an all-round geologist. University of Zimbabwe, University of the Witwatersrand, and University of KwaZulu-Natal provided the solid footings for my career. The support and help from Josphine and our three boys saw me to the finishing line, thank you! Finally, I am thankful to the editorial work of Arnette Buettner and her colleagues, particularly Sridevi Purushothaman and Coral Zhou, for their work in seeing this work to 'the printers'. Madanagopal D Deenadayalan and his team worked tirelessly to get this work published! Thank y'all! I thank Dr. Reuben H. Simoyi for his help in my studies in the USA.

Competing Interests The author has no competing interests to declare that are relevant to the content of this manuscript.

What is Physical Geology?

This textbook deals with the subdiscipline of science that encompasses virtually all aspects of the Earth, its origin, evolution, structure, composition, chemical and physical properties, and the materials that it is composed of, called Physical Geology. Physical Geology, a freshman or first-year university/college course, also deals with processes that led to, and continue to be responsible for, the formation of our planet's features. Since studies of the physical and biological sciences are related, Physical Geology is a discipline that overlaps with and incorporates findings from other disciplines such as astronomy, biology, geography, chemistry, and physics. Consequently, Physical Geology also includes subdisciplines such as geomorphology, paleontology, mineralogy, geochemistry, and geophysics. Mineralogy, petrology, structural geology, and sedimentology, among others, are specialist fields of Physical Geology. These topics, and a lot more, are all discussed in this textbook.

Contents

1 Origin of the Universe and Earth's Interior

Abstract

Discussed in this chapter is the history of astronomy, the origin of the Universe, including our solar system, with particular emphasis on our planet, Earth. The Big Bang theory is summarized. Our solar system is composed of planets, other planetary objects smaller than planets, and a star. These form an integral part of the Milky Way galaxy. Properties of planets and planetary objects that comprise our solar system are also discussed and summarized, with particular emphasis on the physical properties, internal structure, and composition of Earth. Key basic data on solar system planets is summarized in tabular format. The chapter ends with a discussion of the scientific method and how it applies to geology.

Learning Goals

Studying this chapter will enable you to:

- List the names of some early Astronomers and their contributions to modern astronomy
- Explain the evolution of the solar system as well as that of planet Earth
- Describe the Big Bang and Nebular theories as well as the Condensation theory
- List and describe the 8 planets, some dwarf planets, the asteroid belt, and the Kuiper belt in our solar system and their relative positions in space
- Specify the orbital and physical properties in which the Jovian and terrestrial planets differ
- Distinguish and describe the unique features of planet Earth in comparison to other planets
- Describe some of the common moons of other planets in the solar system

Supplementary Information The online version contains supplementary material available at https://doi.org/10.1007/978-3-032-20231-4_1.

1.1 Early Ideas in Astronomy and the Development of Modern Astronomy

Have you ever wondered how our planet came into being? How was the Sun formed? What was the origin of the vast quantities of water on our planet as well as the air that we breathe? In Geology, and all other sciences, the aim is to address such questions using natural laws of physics and other natural sciences. **Astronomy** (words in bold appear in the Glossary at end of each Chapter), which is the study of the **Universe**, is involved with probing the origins of the Universe, including our **solar system**, and particularly our planet **Earth**. In this chapter we attempt to address some of these questions. Our solar system can be regarded to be composed of planets, other planetary objects smaller than planets, and a star, which form an integral part of our **galaxy**, the **Milky Way galaxy** (Fig. 1.1). The study of the history and the structure of the Universe is called **cosmology**.

Human beings are thought to have been tracking the motions of celestial objects over 5000 years ago. Some ancient civilizations recorded the motion in space of some planets as well as the "motion" of the Sun, the Moon, and some **comets**. The **geocentric hypothesis** of the Universe, held by ancient astronomers, was developed to explain the arrangement of these planetary bodies. In this hypothesis, the Earth was thought of as being the stationary center of the Universe with the Sun, the Moon, and other planets orbiting around a motionless Earth in perfect circular orbits. This view was supported by Claudius Ptolemy, as well as by most people in Europe during the Middle Ages from c.476–1400 A.D. Another view, the **heliocentric hypothesis**, however, proposed that it is the stationary Sun which

J. B. Chaumba, *Introductory Physical Geology*, Springer Textbooks in Earth Sciences, Geography and Environment,
https://doi.org/10.1007/978-3-032-20231-4_1

Fig. 1.1 An artist's concept depicting the shape of the Milky Way galaxy. We live around a star, the Sun, which is located about two-thirds of the way out from the center where the Galactic Longitudes are centered. (*Credit* https://www.jpl.nasa.gov/images/pia10748-our-milky-way-gets-a-makeover-artist-concept/)

lies at the center of the solar system, with other planets orbiting around it. René Descartes, a seventeenth-century French philosopher, was one of the earliest proponents of the heliocentric model, and so was another French scientist, Pierre Simon de Laplace, at the end of the eighteenth century.

During the fifteenth century, discoveries by people like Nicolas Copernicus and Galileo Galilei lent support to the heliocentric hypothesis. Further, Isaac Newton's explanation of gravity provided the breakthrough needed in understanding the orbiting of planets as being caused by the attractive force exerted on one object by another. Terminology such as **energy** and **matter** was developed by scientists, which later made it possible for astronomers to observe and make measurements in space, enabling **stars** to be interpreted.

Stars, which are accumulations of incandescent gases held together by gravitational attraction in vast systems of millions or billions of stars, emit enormous quantities of light and heat as a result of being heated. **Galaxies**, which can be considered to be building blocks of the Universe, are huge groups of stars that are held together by gravitational forces. An example of a galaxy is the **Milky Way galaxy** (Fig. 1.1), which contains the Sun plus in excess of 300 billion other stars. It is estimated that the visible Universe contains well over a billion galaxies! The Milky Way galaxy, an example of a "spiral in space," and billions of others, display a flattened spiral pattern to them that is composed of curving arms when viewed from a great distance (Fig. 1.1). These curving arms slowly spiral around a center which glows and resembles a disk as shown in Fig. 1.1 where the Galactic Longitudes are centered. The solar system, which is composed of the Sun and the eight planets (Sect. 1.3), currently constitutes just a small fraction of one of the two spiral arms occurring near the outer edge of the Milky Way galaxy. The solar system is estimated to revolve around the center of the Milky Way galaxy only once in 250 million years! The Andromeda galaxy (Fig. 1.2), which lies approximately 2.5 million **light-years** away, is the Milky Way galaxy's largest galactic neighbor. It is estimated that hundreds of billions of other galaxies lie beyond the Milky Way galaxy!

1.2 How was the Universe Formed?

You might already be wondering where and how these galaxies originated, when were these galaxies formed, and were they all formed at the same time? Let us address these questions next. Observations of the shape, location and measurement of **wavelengths** of light emitted by distant galaxies through the use of telescopes by astronomers such as Edwin Hubble approximately a century ago had a huge impact on our understanding of the Universe. These discoveries led to the phenomenon of the **red shift** displayed by distant galaxies relative to light of nearby

Fig. 1.2 Andromeda galaxy (**M31**). At approximately 2.5 million light-years away, the Andromeda galaxy, or M31, is our Milky Way's largest galactic neighbor. The entire galaxy spans 260,000 light-years across—a distance so large, it took 11 different Galaxy Evolution Explorer image segments stitched together to produce this view of the galaxy next door. (*Credit* https://www.jpl.nasa.gov/images/pia15416-andromeda/)

Fig. 1.3 Expansion of the Universe. Due to the expansion of the Universe, the distances between galaxies increase. Image credit: Eugenio Bianchi, Carlo Rovelli & Rocky Kolb. (*Credit* http://phys.org/news/2015-02-fast-universe.html)

stars. These distant stars were interpreted by Edwin Hubble to be moving away from the Earth at great velocities—a phenomenon called the **Doppler effect**. Further, it was concluded by Edwin Hubble that light from all distant galaxies exhibits a red shift (i.e., all distant galaxies are moving away from the Earth!) irrespective of their direction observed from the Earth. This led Edwin Hubble to yet another milestone conclusion that the Universe must still be expanding today! This helped pave the way for the theory of the expanding Universe (Fig. 1.3).

1.2.1 The Big Bang and the Big Bang Theory

The observable universe is estimated to contain 100 billion galaxies based on the Ultra Deep Field, an image taken by the Hubble telescope. Since it is now thought that the universe is composed of billions of expanding galaxies, this expansion must have commenced at some specific time in the past. The origin of the Universe, therefore, should be marked by the time when this fiery expansion occurred. Cosmologists are convinced that this expansion began approximately 14 billion years ago when everything in the Universe was confined to a point of extremely high temperature and density called the primeval fireball. An immense explosion then occurred, marking the **Big Bang**, causing all matter in the Universe to be scattered and strewn outward at very high velocities causing expansion at an extremely high rate as the density and temperature decreased tremendously while the volume increased. The debris thrown around, which was dominantly hydrogen and helium, later cooled and condensed to form galaxies and stars. In one such galaxy, the Milky Way galaxy, the solar system was formed. The **Big Bang theory** states that all energy and matter must have been contracted into an infinitely small point which then caused an explosion. This explosion, thus, marked the beginning of the Universe 14 billion years ago.

It is thought that significant changes occurred to the Universe from the instant after the Big Bang. The Universe must have been extremely dense and hot to the extent that it was composed only of energy soon after the explosion. Hydrogen atoms are thought to have started forming within a few seconds after the explosion, and 3 min after its formation, the temperature of the Universe is interpreted to have dropped to below a billion degrees. Further, at this ~3 min age of the Universe, its diameter is thought of having enlarged to ~53 million km (~35 million mi) allowing fusion of hydrogen atoms to form atoms of helium. The formation of new nuclei within the first 5 min of the Big Bang is thought to have occurred before any stars were formed, and it is termed the Big Bang nucleosynthesis, where only light atoms (<5 atomic number) were formed very rapidly. Later, the Universe would be cool enough to permit chemical bonding to form molecules such as H_2. Further slowing down of atoms and molecules and their accumulation into cloud and dust called **nebulae** was made possible by further expansion and more cooling of the Universe. It is estimated that hydrogen and helium gases comprised the earliest nebulae of the Universe.

1.2.2 Formation of Stars

Cosmologists believe that 200 million years after the birth of the Universe, huge voids of empty space separated vast and slowly spiraling dark nebulae. With the force of gravity acting on the nebulae, more massive regions started to suck in surrounding gas such that these massive regions grew in both mass and density. More gases that were sucked in by the more massive regions ultimately were compacted into smaller, initially spiraling gas which was set in rotational motion around an axis. More gas continued to be attracted to the more massive regions; the gas was crammed into progressively smaller regions, resulting in rotational motion becoming faster. This increased rotational motion caused the nebulae to evolve into a disk shape (Fig. 1.2).

1.2.3 The Nebular Theory

If one imagines a large cloud of interstellar gas and cloud, called nebulae, a light-year across, which then collides with another interstellar cloud or even the explosion of a nearby star due to some external influence, and the nebulae start to contract under the influence of its own gravity. As contraction occurs, the nebulae become hotter and denser to eventually form a star, the Sun, at its center. As suggested by René Descartes, while the Sun was forming in the hot core of the cloud, the planets and their moons formed in the cloud's cooler outer regions. Thus, planets can be thought of as the by-products of the star formation process!

Laplace proved mathematically that conservation of momentum demands that a contracting and rotating cloud must spin faster as it decreases in size. The rotational speed, in turn, causes the shape of the nebulae to change as it collapses, and centrifugal forces caused by the rotation oppose the contraction in directions that are perpendicular to the rotational axis leading to rapid collapsing of the nebulae along the axis. The remaining nebulae eventually flatten into a primitive solar system that is shaped like a pancake. The **solar nebulae** are the spiraling debris that would later form our solar system (Fig. 1.4). The planets would later be formed out of the spinning solar nebulae, which all lie in nearly the same plane, and they all rotate around the Sun in the same nearly circular orbits, and they are all relatively isolated in space. These properties were inherited by the planets from the rotating disk in which they originated, and the **nebular theory** is the idea that the planets were formed from such a disk.

1.2.4 The Condensation Theory of Planet Formation

Modern computer calculations predict that a disk of warm gas would not form lumps of matter that would later evolve into planets. Most astronomers now favor the condensation theory which also rests on the nebular theory. The condensation theory of planet formation invokes the presence of interstellar dust in the solar nebular whereby the space between the stars is strewn with microscopic dust particles, an accumulation of ejected matter of many dead stars. Such dust particles are thought to have formed in cool atmospheres of old stars and then grew by accumulation of more atoms and molecules from interstellar gas within the Milky Way galaxy. Dust grains help to cool warm matter by efficiently radiating its heat away as infrared radiation and as the cloud cools, its molecules move more slowly, reducing the internal pressure and thus allowing the nebular to collapse more easily due to gravity. Further, by acting as **condensation nuclei**, the dust grains greatly speed up formation of planets by collecting more atoms.

Once the solar nebulae had formed and began to cool, dust grains produced condensation nuclei around which matter accumulated (Fig. 1.4), helping the process of accumulation of small lumps of matter to occur at a rapid rate as well as their sticking to other lumps. The lumps grew larger in size, which increased their surface area in the process, which in turn helped accelerate the rate at which new material was swept up. This **accretion** process led to the formation of sequentially larger and larger objects of pebble size, then baseball size, basketball size, cobbles, and then larger and larger bodies. It is estimated that in approximately 100,000 years, objects that were a few kilometers across (the size of small moons) were formed from accretion.

Further, at that time, the gravitational force of the accreted objects was strong enough to attract other nearby objects to form **planetesimals** which are the building blocks of the solar system. Planetesimals have strong gravitational pull such that any material that would not have collided with them would be swept up by them, enlarging planetesimals into fewer but larger **protoplanets**, and ultimately these would in turn form the planets we have today (Fig. 1.4). It is also important to note that, as protoplanets grew, their strong gravitational fields led to the occurrence

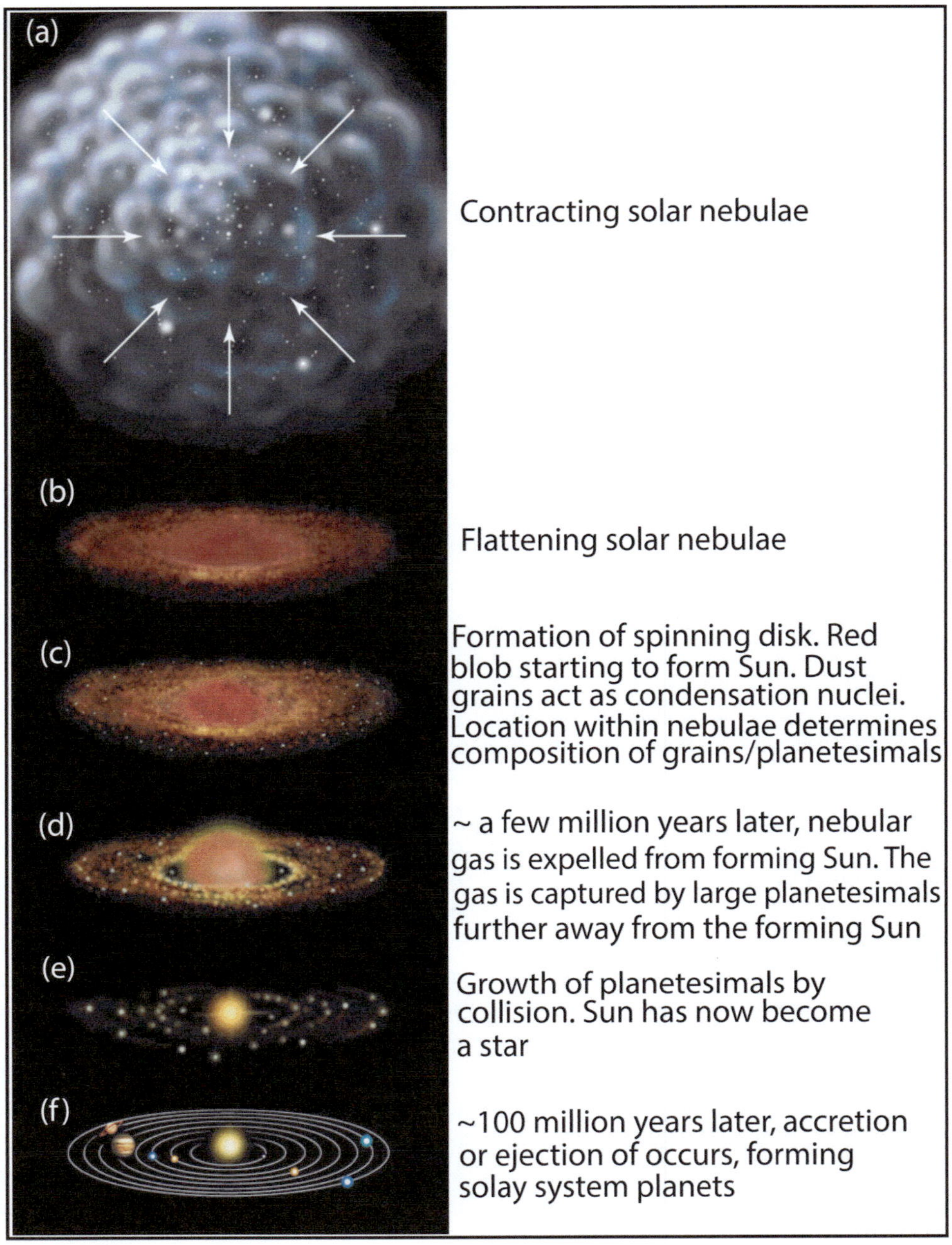

Fig. 1.4 Formation and evolution of the solar system began 4.55 billion years ago with the gravitational collapse of a small part of the solar nebulae. Most of the collapsing nebulae collected in the center, to form the Sun, whereas the rest flattened into a protoplanetary disk from which the planets, moons, asteroids, and other small solar system bodies were formed. (*Credit* https://astroclock2010.wordpress.com/cosmic-timeline-17/)

of several high-speed collisions between planetesimals and protoplanets.

Fragmentation occurred as a result of such collisions due to small objects breaking into smaller chunks that would be then swept up by the larger protoplanets. A relatively small number of these 10- to 100-km fragments were able to escape capture by protoplanets or moons, however, to become asteroids and comets. After approximately 100 million years, the primitive solar system evolved into a glowing **proto-Sun** at the center, the eight planets, and dozens of **proto-moons** that we have today in our solar system. After approximately 1 billion more years, all interplanetary debris in the solar system had cleared. These first billion years of the formation of the solar system were characterized by intense collisions of meteorites and the few impact craters still well-preserved on the planets such as Mercury and our Moon as well as some poorly preserved ones on planet Earth attest to this.

1.3 Differentiation of the Solar System

So far, we have discussed how the solar system formed. We now evaluate whether there are any differences in the planets in the solar system and what likely caused them to address questions such as why we observe planets at differing distances from the Sun, and whether they all are composed of the same materials. To do this, let us consider the temperature profile of the solar nebular. Contraction of the solar system due to gravity caused heating and its flattening into a disk (Fig. 1.4). Density and temperature, therefore, were greatest near the center but much lower in the outer regions. Estimates suggest that the temperature of gas near the core of the contracting nebular was several thousand Kelvins whereas it was only ~100 K at the position where we now find Saturn (a distance of 10 *astronomical units* [A.U.]—1 A.U. is the approximate mean distance between Earth and the Sun which is ~150 million km [~93 million mi]). The extremely high temperatures in the warmer cloud regions caused dust grains to break apart forming molecules which then split into atoms. Thus, the extent of the destruction of gas was dependent on temperature, which therefore also depended on location within the solar nebula. It is estimated that a significant proportion of the original dust in the solar system disappeared at this stage, but the grains in the outer regions likely remained unchanged.

The dusty nebula then radiated away its heat as its temperature decreased everywhere except in the core where the Sun continued to form. Cooling of the gas occurred as new dust grains started to condense out. Thus, even though there were huge amounts of interstellar gas early on, it was partly destroyed, and only to form again later. A critical change, however, had already occurred in that initially the nebula gas was uniformly distributed with dust grains of all compositions but when the gas was reformed later, the grain distribution was much different. A plot of the temperature in various parts of the solar system just before accretion began is shown in Fig. 1.5. As the solar nebula contracted, the regions closest to the proto-Sun recorded the highest temperatures as the only materials to condense out were those that could survive the temperatures there. In the innermost regions where Mercury's present orbit is located, for example, only metallic particles could condense. At approximately 1 A.U. a little further out, many of the heavier elements, e.g., Si, Fe, Mg, and Al, combined with oxygen to form rocks and silicate and/or metallic minerals, accounting for why the innermost planets are made of the most refractory materials. Water ice is interpreted to have only been able to condense beyond ~3.5 A.U. (Fig. 1.5). In the middle and outer regions, beyond 5 A.U. from the center, temperatures were low enough for several abundant gases such as water vapor, methane, and ammonia to condense. Consequently, carbon, nitrogen, and oxygen are the most common materials after hydrogen and helium.

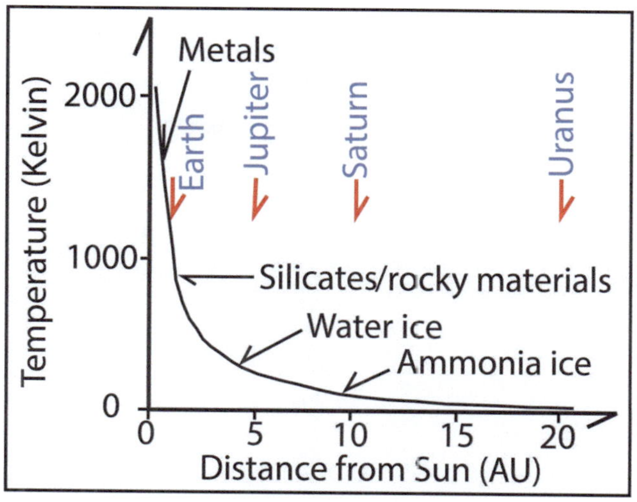

Fig. 1.5 Graph showing temperature variation with distance from the proto-Sun in the early solar nebula. Only metals could condense out of the gaseous state to form particles in the hot central regions. At greater distances from the center, lower temperatures enabled rocky and icy particles to condense. Shown in figure are estimates of temperatures as a function of distance in the collapsing solar nebula and explain the properties of the planets as a function of distance from the proto-Sun. Labels shown indicate the minimum radii at which particles of various types were able to condense out of the solar nebula

At the same location at these great distances further from the center, anywhere gas icy grains could form, they eventually predominated over rocky and metallic substances that also condensed out of the solar nebula and such ancestral particles later led to the formation of the cores of the **Jovian planets**. Consequently, the composition of the Jovian planets is more representative of the Universe as a whole than that of the rocky and metallic inner planets. The inner solar system is rather underrepresented in lighter particles because of the conditions under which they formed, and the outer solar system is not deficient in heavy elements.

1.4 Brief Overview of the Solar System

The Sun and everything else that orbits it make up the solar system. Objects that orbit the Sun include the eight planets which are, from the Sun outwards, Mercury, Venus, Earth, Mars, Jupiter, Saturn, Uranus, Neptune and their moons, as well as the asteroid belt, the Kuiper belt, and comets (Fig. 1.6). The planets' counterclockwise orbits around the Sun are not evenly spaced but they become further and further apart with increasing distance from the Sun. Planets that are located within the Asteroid belt are called the inner planets or **terrestrial planets**, whereas those located between the Asteroid and Kuiper belts are termed the outer

Fig. 1.6 Illustration of solar system showing part of the Sun, planets, some dwarf planets, asteroid belt, Kuiper belt objects, and comets at their relative distances (see Table 1.1) from the Sun. (*Credit* NASA)

planets or **jovian planets**. Let us now first describe the Sun followed by descriptions of each of these planets in the order in which they are encountered as one moves outward from the Sun, noting any similarities and differences between them. It is important to note that the paths of solar system planets are all almost elliptical, and they all move on low-eccentricity orbits around the Sun with the exception of Mercury.

Figure 1.6 compares the Sun with the planets as well as the planets with one another. Inner planets have physical and chemical properties that resemble those of Earth and are commonly termed terrestrial planets from the Latin word *terra* (meaning "Earth" or "land"). These planets are rocky, small, with relatively low masses; they are located near the Sun. They have solid surfaces, and they are made of rocky materials. The largest of the inner planets is Earth. Jovian planets (from *Jove*, name for the Roman god Jupiter) are named after Jupiter, which is the largest of the outer planets. They are much larger than the terrestrial planets and orbit further away from the Sun, and they have no solid surfaces, as light hydrogen and helium gases comprise their outer layers. All solar system planets are now known to have been formed 4.55 billion years ago.

1.4.1 The Sun

The star closest to us is the Sun which is a glowing ball of gas that is powered by nuclear fusion and held together by gravity. The Sun, our parent star, has a radius of ~ 696,000 km. It is the only source of heat and light for Earth's life, as well as providing energy that powers our weather and climate (Fig. 1.6). The surface of the Sun, called the **photosphere**, is the part that emits the radiation seen. The luminosity of the Sun is caused by contin-

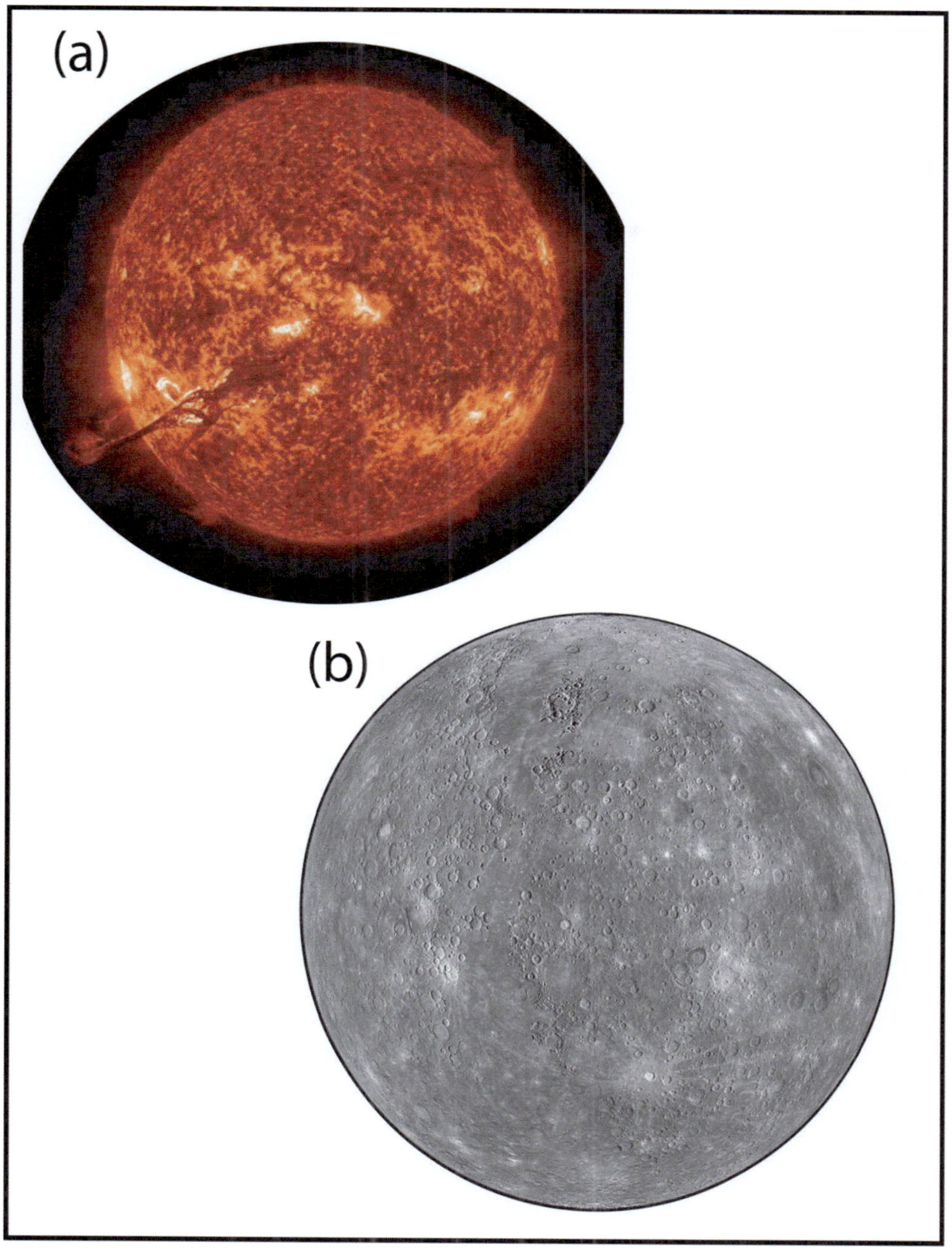

Fig. 1.7 **a**. A long, magnetic filament burst out from the Sun producing one of the best shows seen on Aug. 31, 2012, by the Solar Dynamics Observatory (SDO), which was sparked by a medium-sized (C-class) flare (*Credit* http://www.jpl.nasa.gov/spaceimages/details.php?id=PIA18167). **b**. A composite image of Mercury taken by *MESSENGER* showing the highly cratered surface of the planet (*Credit* http://www.nasa.gov/sites/default/files/thumbnails/image/messenger-3-first_and_last_0.jpg)

uous emission from the photosphere but with characteristic explosive and unpredictable surface activity (Fig. 1.7a).

1.4.2 Mercury

Our knowledge of other planets in the solar system draws from what we know about our home in space—Earth. Mercury, the closest planet to the Sun (Table 1.1, Fig. 1.6), is the smallest one. The surface of Mercury is highly cratered (Fig. 1.7b) (after **crater**—Greek for "bowl"), with an ancient surface that is full of pulverized dust and boulders. Most craters were formed very early in the planet's history, primarily due to meteorite impacts. No tectonic activity occurs on Mercury, which also lacks an atmosphere and experiences extremely high temperatures. Mercury has a mean density of ~ 5400 kg m^{-3}, which is slightly less than Earth's value of 5500 kg m^{-3}, suggesting that Mercury's interior must contain high density materials such as iron. If we factor in Mercury's smaller size compared to Earth's,

Table 1.1 Selected basic data on solar system planets

	Distance from Sun (106 km)	Diameter (km)	Density (kg/m^3)	Gravity (m/s^2)	Rotation Period (hours)	Length of Day (hours)	Orbital Period (days)	Orbital Velocity (km/s)	Mean Temperature (°C)	Number Of Moons
Mercury	57.9	4879	5427	3.7	1407.6	4222.6	88	47.4	167	0
Venus	108.2	12,104	5243	8.9	−5832.5	2802	224.7	35	464	0
Earth	149.6	12,756	5514	9.8	23.9	24	365.2	29.8	15	1
Mars	227.9	6792	3933	3.7	24.6	24.7	687	24.1	−65	2
Jupiter	778.6	142,984	1326	23.1	9.9	9.9	4331	13.1	−110	67
Saturn	1433.5	120,536	687	9	10.7	10.7	10,747	9.7	−140	62
Uranus	2872.5	51,118	1271	8.7	−17.2	17.2	30,589	6.8	−195	27
Neptune	4495.1	49,528	1638	11	16.1	16.1	59,800	5.4	−200	14

(*Data source* NASA)

we are led to conclude that Mercury's interior must be compressed less by the weight of overlying material and that its core should contain a much higher proportion of its mass than does Earth!

Due to Mercury's smaller size in comparison with that of Earth (Fig. 1.6), Mercury's gravitational field is also weaker than Earth's (Mercury's surface gravity is 0.4 times that of Earth's), hence the need by Apollo astronaut to wear those bulky space suits! No appreciable atmospheres have been observed by astronomers on Mercury due to the planet's weak gravitational field. This is attributed to Mercury's low escape velocities (i.e., needed for molecules or atoms to escape from a planet's gravitational force) of 4.2 km s^{-1} compared to Earth's of 11.2 km s^{-1}. The primary atmosphere of Mercury is thought to have escaped very early in the planet's evolution. However, NASA's MESSENGER probe made measurements of a tenuous atmosphere composed of hydrogen and helium which occur in Mercury's atmosphere, as well as atoms of Na, K, and Mg. Mercury's daytime temperature of 700 K falls to 100 K at night, giving the planet a diurnal temperature range of 600 K which is the highest of any solar system planet mainly due to its proximity to the Sun and the lack of an atmosphere's moderating influence.

1.4.3 Venus

Often regarded as Earth's sister planet because of their comparable sizes (Fig. 1.6), Venus (Fig. 1.8a) is similar to Earth not only in size, but also in mass, chemical composition, and density, as well as having an orbit that is close to Earth's (Table 1.1). Considered to have been indistinguishable at the time they were formed, Venus and Earth are now very different: no life is known to exist at Venus, whereas Earth teems with vibrant life; Venus has a dense and hot CO_2 atmosphere which rains sulfuric acid, yet Earth has an atmosphere of nitrogen, oxygen, carbon dioxide, and numerous other gases. The evolutionary paths of the planets must have radically diverged at some point due to each planet's composition and environment, for example.

Often referred to as "the morning star" or "the evening star," Venus is the second planet from the Sun (Table 1.1) and, after the Sun and the Moon, is the third brightest object in the solar system. The radius of Venus measures 6052 km, or 0.95 Earth radii. The average density of Venus is ~ 5200 kg m^{-3}, and it has a mass of 4.9 × 10^{24} kg, or 0.82 that of Earth. With bulk properties interpreted to be similar to Earth (Table 1.1), Venus' chemical composition is also inferred to be similar to Earth's, implying that their evolution and internal structures are comparable. However, Venus' spin is retrograde (opposite to that of the other solar system planets), and its axial tilt (the angle between its equatorial plane and its orbital plane) is 177.4° in comparison to Earth's 23.5°. The retrograde spin of the planet is thought to have been caused by an impact with a large body that reduced Venus' spin close to zero.

Unlike Earth, Venus' atmosphere is very dense, and it has thick and opaque clouds to visible radiation. Venus' surface shows extensive features such as shield volcanoes, lava domes consistent with volcanic activity, and some of this volcanism may still be active today. Some features on Venus' surface such as Cleopatra were formed by meteorite impacts. Venus has no magnetic field of its own, probably due to its very slow rotation and lack of current volcanic activity, and no evidence of tectonic activity has yet been observed on Venus.

There is no plate tectonics known to occur on Venus unlike on Earth. The surface of Venus, however, is now thought to possess evidence of internal activity that pushes up from below resulting in deformations which resemble impact craters composed of a raised ring, like a crown, surrounding a sunken middle, with concentric fractures radi-

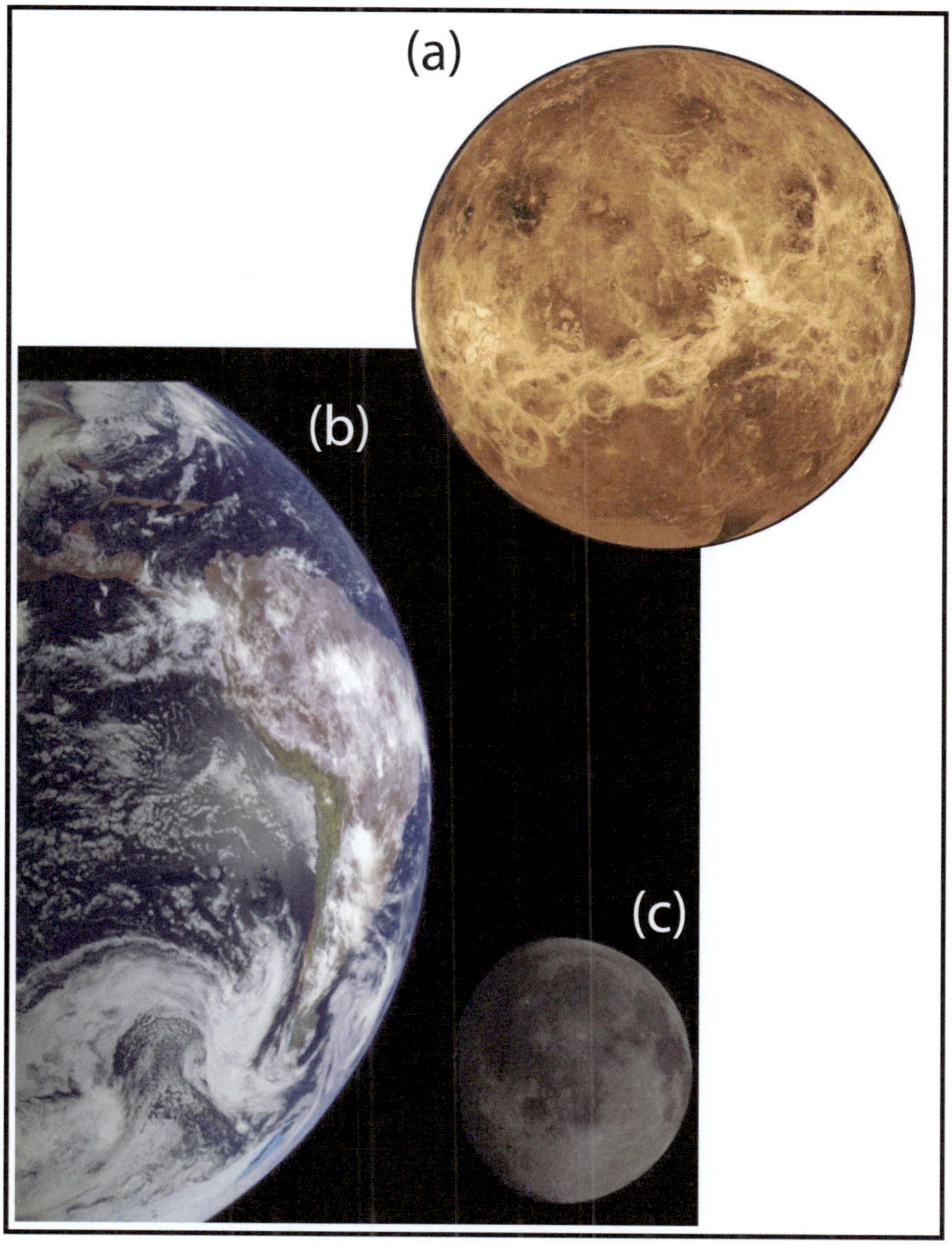

Fig. 1.8 **a** A Venus composite image of the complete radar image collection obtained by the Magellan mission (*Credit* http://nssdc.gsfc.nasa.gov/photo_gallery/photogallery-venus.html). **b** and **c**. View of Earth **b** from space with Moon **c** inserted, showing South America and parts of Central and North America as well as parts of the Pacific and Atlantic Oceans (*Credit* NASA)

ating outwards coronae. Coronae, which are hundreds of kilometers across, are thought to be of volcanic origin and are inferred to be caused by plumes of hot molten material welling up from the interior of the planet. The coronae push the planet's surface upward into a dome that then collapses inward when the plume cools resulting in molten material leaking out of the sides of the collapsed dome to form the ring. Molten material consequently leaks out of the sides of the collapsed dome leading to its leakage and formation of the ring. Although Venus doesn't have tectonic plates, tectonic activity is thought to exist in the form of interactions between mantle plumes and the lithosphere. It is thought by researchers that these interactions may be occurring under the coronae.

1.4.4 Earth

The third planet from the Sun—Earth—is our home in space. Earth (Figs. 1.6 and 1.8b) is by far the most studied of all solar system planets. Studying Earth is critical as knowledge gained from an understanding of this most important planet is crucial in our quest to compare and contrast other planets and their moons with Earth. Earth has an average density of ~5500 kg m^{-3}, an equatorial radius of 6378 km, a mass of 5.97 × 10^{24} kg, and the Moon which orbits it (Table 1.1). Almost all of Earth's mass is found from the surface to the interior of the planet, with the atmosphere and the **magnetosphere** comprising a mass of less than 0.1% of Earth's mass.

So far, Earth is the only planet known to contain liquid water mostly in the oceans covering ~71% of the surface of the planet, as well as minor concentrations, yet very important, water on the continents (**hydrosphere**). In addition to containing water, Earth is unique in that it is composed of an **atmosphere** lying just above its surface that we can breathe! The magnetosphere, occurring at higher altitudes above the surface of the Earth, is a zone of charged particles that are prevented from escaping by the planet's magnetic field. From the surface to its interior, Earth can be subdivided into the crust (which can be differentiated into the continental and oceanic), the mantle, the outer core, and the inner core. These subdivisions are described in greater detail in Sect. 1.5. The hydrosphere and atmosphere are described in Chaps. 14 and 15, respectively. Since the 1960s but only discovered in 2025, a little asteroid named 2025 PN7, measuring approximately 18 to 36 m wide, is orbiting Earth and will continue to do so until 2083. This asteroid is acting as a rare type of celestial companion, or a "quasi-Moon" for Earth, traveling in near sync with Earth until 2083, when it will drift away from Earth.

1.4.5 Moon

Earth's only satellite is the Moon. Like Mercury, the Moon (Figs. 1.6 and 1.8c) has a heavily cratered surface. The Moon is composed of light-colored areas which were thought by early astronomers to resemble continents, and dark areas which were similarly thought to resemble oceans. The dark areas were later shown to be very extensive flat areas that were formed from lava flows, but we still refer to them as *maria* (Latin for "seas," singular: mare). The light-colored areas, which are now known to be composed of land elevated some several km above the maria, were termed *terrae* (Latin for "land") by early astronomers, and they are now called the highlands. Like the surface of Mercury, craters are common on the Moon. Craters, which range in scale from microscopic to several hundreds of kilometers in diameter, have been described from both maria and highlands, although they are much more prevalent on the latter.

The highlands are made up of rocks rich in aluminum and have lower densities (~2900 kg m^{-3}), whereas the maria are made up of rocks that are rich in iron and are denser (~3300 kg m^{-3}). Thus, the maria are made up of material that resembles the lunar's mantle (similar to terrestrial basalt), whereas the highlands are made up of material that resembles the lunar crust. Lunar highland rocks yield ages ranging from 4.4 to 4.0 billion years, whereas ages of maria rocks are 3.9–3.2 billion years old. Based on the impact theory, which postulates collision of the Earth by a large object the size of Mars, this collision is thought to have occurred when Earth was still young and molten. The collision is envisaged to have been a glancing blow, and matter that was dislodged from Earth is thought to have reassembled to form the Moon which then assumed a stable orbit. Earth is assumed to have formed an iron core by the time of the collision, enabling the Moon to end up with a composition comparable to that of the Earth's mantle.

1.4.6 Mars

Mars has properties that are broadly comparable to Earth's and for this reason it is regarded by some scientists as the planet with the second most hospitable conditions for life in the solar system after Earth. Early in its history, Mars (Fig. 1.9a), the fourth planet from the Sun (Fig. 1.6, Table 1.1) is thought to have had blue skies and running water, although today Mars appears as dry and dead as other solar system planets with the exception of Earth. Mars has a mass of 6.4 × 10^{23} kg, a radius of 3400 km, or 0.53 Earth radii, a density of ~3900 kg m^{-3} (Table 1.1), and two tiny moons (Phobos {fear} and Deimos {panic}) that orbit around it. These heavily cratered and irregularly shaped Martian moons have a composition very different from Mars. The Martian moons are thought to be asteroids that were captured by the planet when they were slowed. It is also thought that the Martian moons may represent remnants of a single object that was broken up during capture. Mars density implies a very high density for its 2500 km-diameter core (likely composed of iron sulfide), assuming that Mars's surface rocks are similar to those of other terrestrial planets. Rotation of Mars on its axis is once every 24.6 h to produce a Martian day as long as Earth's. Further, Mars' equator is inclined to the plane of orbit at an angle of 24° which is comparable to Earth's 23.5°.

Deep canyons, extensive sand dunes, and volcanoes, for example, characterize the surface of Mars (Fig. 1.9b). The Martian surface has been explored in great detail by several studies from numerous specific sites by orbiters and also from spacecraft landing sites. The northern hemisphere of Mars, for example, is composed of less cratered 3 billion year-old volcanic plains at low elevation thought to be formed from huge volumes of lava, whereas the southern hemisphere is largely composed of highlands (Fig. 1.9b). These 4 billion-year-old southern hemisphere highlands

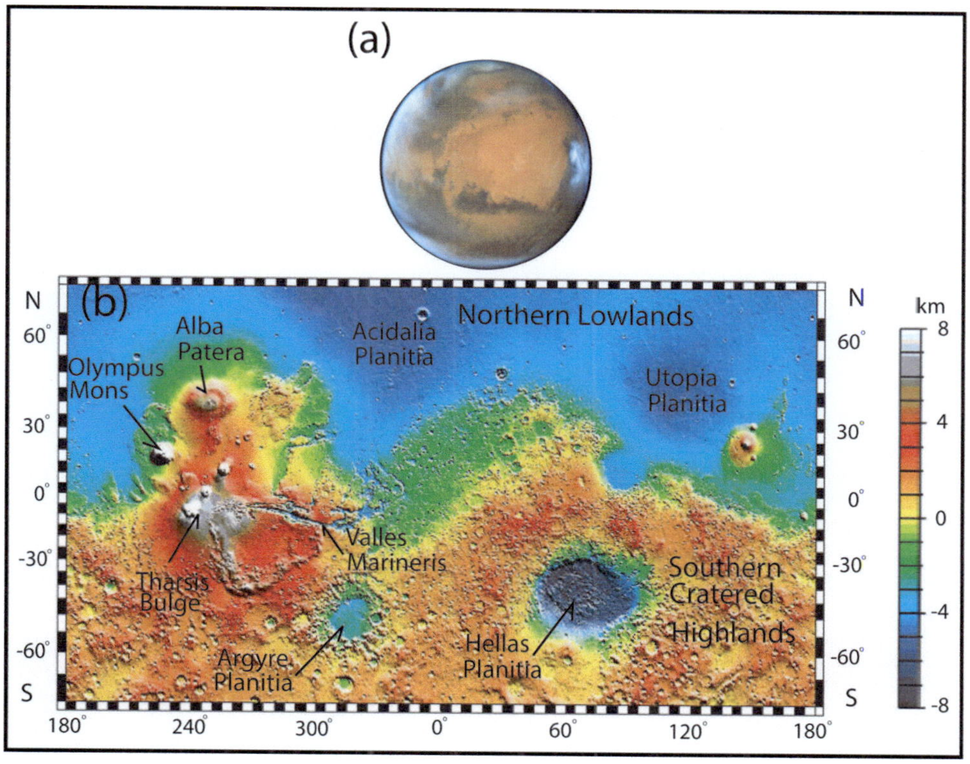

Fig. 1.9 **a**. Image of Mars captured by the Hubble Space Telescope on May 12, 2016, when the planet was 50 million miles from Earth just a few days before Mars' opposition on May 22, when the Sun and Mars will be on exact opposite sides of Earth (*Credit* NASA). **b**. Map showing the topography of Mars. Notice that the northern hemisphere is mostly lower elevation (~5 km below the datum and colored blue), whereas the planet's southern hemisphere is mostly higher elevation (~2 km above the datum and colored red). This "dichotomy" in topography is a big to planetary scientists (*Credit* NASA)

occur at elevations approximately 5 km higher than the volcanic plains to the north but with numerous impact craters such as the large Hellas Planitia and Argyre Planitia. Tharsis bulge, lying on the Martian equator, is rising some 10 km above the Martian surface with less impact craters than those in the Martian highlands suggesting its relatively young age (2–3 billion years old). Valles Marineris (Mariner Valley), considered to be the Grand Canyon of Mars, is a prominent feature seen to the right side of Tharsis bulge (Fig. 1.9b) but it is considered not have been formed by incision by water as in the case of the Grand Canyon, but by splitting and cracking of the surface due to bulging outwards of the Tharsis bulge. No evidence of current tectonic activity on Mars has yet been observed.

Three very large and high shield volcanoes, such as Olympus Mons, occur in the Tharsis bulge area, and they are thought to have been formed from hot spots. Such high Martian volcanoes have been interpreted to be a consequence of the low surface gravity of the planet. In fact, Mars is thought to contain the largest known volcanoes in the solar system, but no evidence for recent volcanic activity has been observed. No water has been found on Mars, but evidence exists that liquid water may have been present on the planet in the form of the occurrence of both runoff channels and outflow channels in the southern highlands and in equatorial regions, respectively. From Martian Landers, we now know that the surface of Mars is composed of huge amounts of iron oxide, and its sky has a pale pink color due to the airborne dust from the rusting surface. The Martian atmosphere is quite thin and is composed of primarily carbon dioxide (95.3%), as well as lesser amounts of nitrogen (2.7%), argon (1.6%), oxygen (0.13%), carbon monoxide (0.07%), and water vapor (0.03%).

1.4.7 Asteroids, Comets, and Meteoroids

Although the solar system has 8 planets, it is estimated that hundreds of thousands of other minor celestial bodies such as **asteroids** (Fig. 1.10a), comets, Kuiper belt objects, and meteoroids also orbit around the Sun (Fig. 1.6). Many of these minor celestial bodies are rocky and of negligible masses that were considered to be debris that was left over from the formation of the solar system, but they are thought to hold unique clues about the solar system origin. Asteroids are small and rocky bodies that are too small to be classified as planets, but they are sometimes also referred to as planetoids or minor planets. In addition to differences

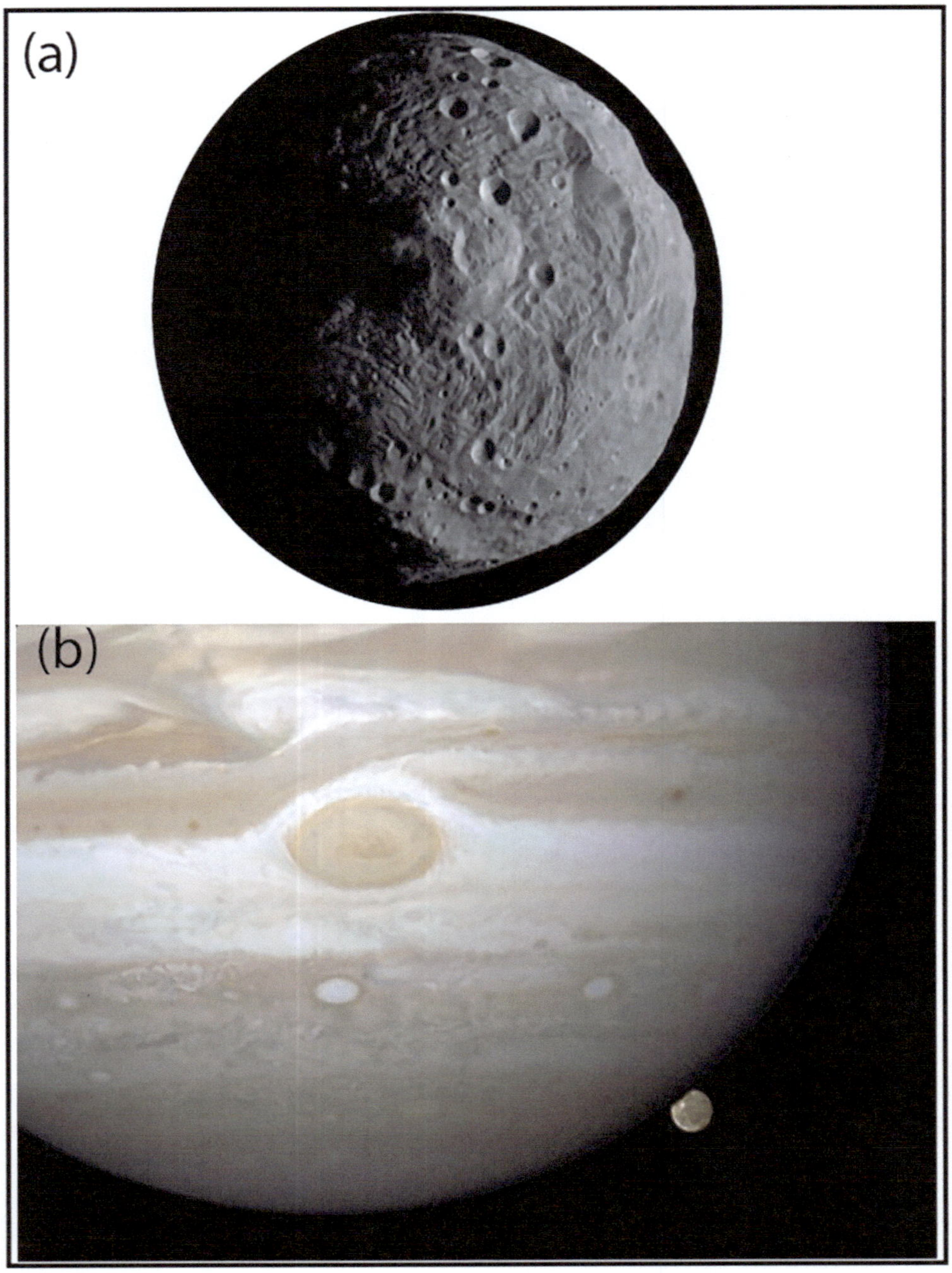

Fig. 1.10 **a** An image of Vesta that was captured by NASA's *Dawn* spacecraft on July 16, 2011. The image was taken from a distance of about 9,500 mi (15,000 km) away. One pixel in the image corresponds to 1.4 km (~0.88 mi) (*Credit* NASA). **b** NASA's Hubble Space Telescope caught Jupiter's Moon Ganymede shown just before it disappears behind this giant planet. Jupiter is so vast that only part of its Southern Hemisphere is seen in this image (*Credit* NASA)

in size, asteroids also have more eccentric orbits around the Sun and are located between Mars and Jupiter (Fig. 1.6). Ceres, which measures 940 km across, is the largest known asteroid, with other more known ones being Pallas and Vesta. Over 500,000 asteroids have now been identified, with the majority of them occurring in a region called the **asteroid belt** that lies approximately halfway between the orbits of Mars and Jupiter, i.e., between 2.1 A.U. and 3.3 A.U. from the Sun (Fig. 1.6). With the exception of a single asteroid, asteroids are known to orbit the Sun in the sense as the planets (prograde orbit).

The occurrence of asteroids within the Asteroid belt has been interpreted to suggest their origin either as primal rocks that failed to aggregate into a planet, or as remnants of a long broken up planet. The strong gravitational field of Jupiter is thought to continuously interfere with asteroids which prevents them from coalescing into a planet. The small sizes of the asteroids inhibit their study by telescopes from Earth (except Ceres); hence little is known about the majority of them. Available evidence suggests that asteroids differ greatly from each other, as well as from the solar system planets and most of their moons. Asteroids are classified into carbonaceous (C-type), which are the least reflective ones and therefore contain large fractions of carbon in their make-up, and silicate (or rocky) material, which are the more reflective ones which contain little carbon in their make-up. Other types of asteroids such as M-type asteroids contain large proportions of iron and nickel in their makeup. C-type asteroids are thought to be constituted of very primitive material that represents the earliest solar system stages and have thus not been subjected to reheating or chemical evolution since their formation 4.55 billion years ago.

Few asteroids possess eccentricities whose paths may intersect with Earth's orbit. These asteroids, termed **Earth-crossing asteroids**, are of interest to us due to their possibilities of collisions with Earth. These stray asteroids have highly elliptical orbits which lie outside the main asteroid belt orbits probably due to the gravitational force of especially Jupiter, and also Mars which deflect them into the orbits of terrestrial planets. *Apollo asteroids* are Earth-crossing asteroids. **Trojan asteroids**, consisting of approximately several hundred asteroids, orbit the Sun at the distance of Jupiter due to this planet's strong gravitational field.

Faint and fuzzy patches of light in the sky that are still several astronomical units away from the Sun are called comets. A **comet** (from the Greek word *kome*—meaning "hair"), a few kilometers in diameter, travels in a highly elliptical orbit that carries it far beyond the known planets with the Sun at one focus to brighten and form an extended tail as it nears the Sun. A comet's brightness and tail diminish as it departs from the Sun until it becomes a faint point light that recedes into the distance. A comet emits no visit light of its own, similar to planets but rather it shines by reflected, or re-emitted, sunlight.

Meteors, or shooting stars, which are often seen on clear nights, are sudden streaks of light caused by friction between incoming interplanetary matter (or cosmic debris) and air molecules in the Earth's atmosphere. A meteor is a brief flash, whereas a comet's tail is characterized by a steady and broad swath of light. Further, a comet's tail is visible for weeks or even months in deep space, whereas a meteor is an event which passes swiftly in the atmosphere of Earth. The incoming interplanetary matter may be a comet, an asteroid, or a **meteoroid** (with sizes of less than 100 m across). Air molecules are excited and heated by the friction, and as the air molecules return to their ground states, they emit light. Both asteroids and meteoroids are chunks of rocky interplanetary debris. The distinction between a meteoroid and an asteroid is only based on size with meteoroids, at less than 100 m in diameter, being smaller than asteroids. Meteoroids are even smaller than comets. Meteoroids often pass through Earth's atmosphere to produce meteors due to the preponderance of meteoroids than comets and asteroids. **Meteorites** are pieces of interplanetary debris that survive the glowing passage through Earth's atmosphere to land on its surface.

1.4.8 Jupiter

The first planet to be encountered outside the orbit of asteroids (Fig. 1.6), Jupiter (Fig. 1.10b), is the largest one in the solar system, and it is also the closest to the Sun of all the Jovian planets. Numerous moons of varying sizes and physical properties that orbit Jupiter are termed the Galilean moons, and they total 67 moons (Table 1.1). The four largest of the Galilean moons are Ganymede, Callisto, Io, and Europa (Table 1.2). On Hubble Space Telescope images of Jupiter, clouds of different thicknesses, chemical composition, and heights (Fig. 1.10b) are visible. Further, on such images, clearly visible are alternating light and dark bands which cross Jupiter parallel to its equator as well as the large oval (the **Great Red Spot**) near the left of center in Fig. 1.10b. Ganymede, shown to the bottom right in Fig. 1.10b, completes an orbit around Jupiter every seven days. Ganymede's orbit is tilted nearly edge-on to Earth such that it routinely can be seen passing in front of, and disappearing behind Jupiter (Fig. 1.10b). Ganymede, which is bigger than Mercury, is composed of rock and ice, and it is the largest moon in our solar system.

Jupiter's mass is 1.9×10^{27} kg, or 318 Earth masses, and a radius of 71,500 km, or 11.2 Earth radii (Table 1.1). With a rotation speed of 9 h and 55 min, Jupiter has the fastest

Table 1.2 Factsheet of the four largest Galilean satellites

	Mass (10^{20} kg)	Radius (km)	Distance from Jupiter (km)	Mean density (kg/m^3)	Density (kg/m^3)	Orbital period (days)
Ganymede	1481.9	2631.2	1,070,000	1940	1900	7.15
Callisto	1075.9	2410.3	1,880,000	1830	1900	16.7
Io	893.2	1821.5	422,00	3530	3500	1.77
Europa	480	1560.8	671,000	3010	3000	3.55

rotation rate in the solar system leading to more flattening and the development of a bigger equatorial bulge. The Giant Red Spot on Jupiter is thought to be a hurricane that is twice the size of Earth and has been present for hundreds of years. This Giant Red Spot is also thought to be the largest of several persistent storms in Jupiter's atmosphere. Molecular hydrogen (H_2—86.1%) and helium (He—13.8%) are the most abundant gases in Jupiter's atmosphere, with minor amounts of methane, ammonia, and water vapor also being present. Hydrogen and helium, in similar proportions, are also thought to make up the interior of the planet as well. The planet's powerful gravity is thought to be responsible for hydrogen and helium being the dominant gases that they are, and it is thought that Jupiter's atmosphere represents its original atmosphere at its formation 4.55 billion years ago.

The banded appearance of Jupiter and other jovian planets, defined by a series of bright **zones** and dark **belts** (Fig. 1.10b) which cross these planets, is thought to be formed by convective motion in the atmospheres of the planets. The light-colored zones are interpreted to lie above rising convective currents in the atmosphere of Jupiter and other planets, whereas the dark belts represent sinking convective currents. The upwelling, warm, and convective gas currents, therefore, represent high-pressure regions, whereas the downwelling, cooler, and convective gas currents represent low-pressure regions. Jupiter has a faint ring that lies approximately 50,000 km above its top cloud layer. This ring is thought to be made up of fragments chipped off, by meteorite impacts, from Jupiter's two small moons of Metis and Adrastea that are located close to this ring.

The internal structure of Jupiter is comprised of three main zones. Below Jupiter's cloud cover, both density and temperature are thought to increase with depth. The outer radius is represented by the top of the cloud layers. From a depth of 100 km down to a depth of 20,000 km, molecular hydrogen predominates, and temperatures here are 300 K. Below the molecular hydrogen zone and surrounding the core is a layer of metallic hydrogen extending from 20,000 km to 60,000 km in depth. Below a depth of 20,000 km, hydrogen is thought to behave as liquid metal with average temperatures of 18,000 K. The core of Jupiter right at the center is composed of icy and rocky material at a depth of 60,0000 km to 70,000 km and experiences extremely high average temperatures in the region of 25,000 K.

Solar storms occur when charged particles from the Sun collide with Earth's upper atmosphere as the cause of auroras on Earth. This collision energizes the gases in the atmosphere, which are responsible for imparting auroras with their distinctive colorful glows referred to as the Northern or Southern Lights. In May 2025, the successor to NASA's Hubble Space Telescope, James Webb Space Telescope released the captured images (Fig. 1.11) of massive auroras encountered at the north pole of Jupiter. Auroras are believed to be caused by charged particles which crash into the planet's atmosphere. These auroras of Jupiter are inferred to be hundreds of times brighter than Earth's Aurora Borealis.

1.4.9 Saturn

Saturn's famous and intricate rings make this planet probably the most beautiful one in the Universe. Strikingly similar to Jupiter in composition, structure, and size, Saturn (Fig. 1.12a) is the second of the Jovian planets encountered further away from the Sun (Fig. 1.6). Saturn orbits the Sun at almost double the distance of Jupiter with an orbital period of 29.4 Earth years and a mass of 5.7×10^{26} kg. The equatorial radius of Saturn is 60,000 km, or

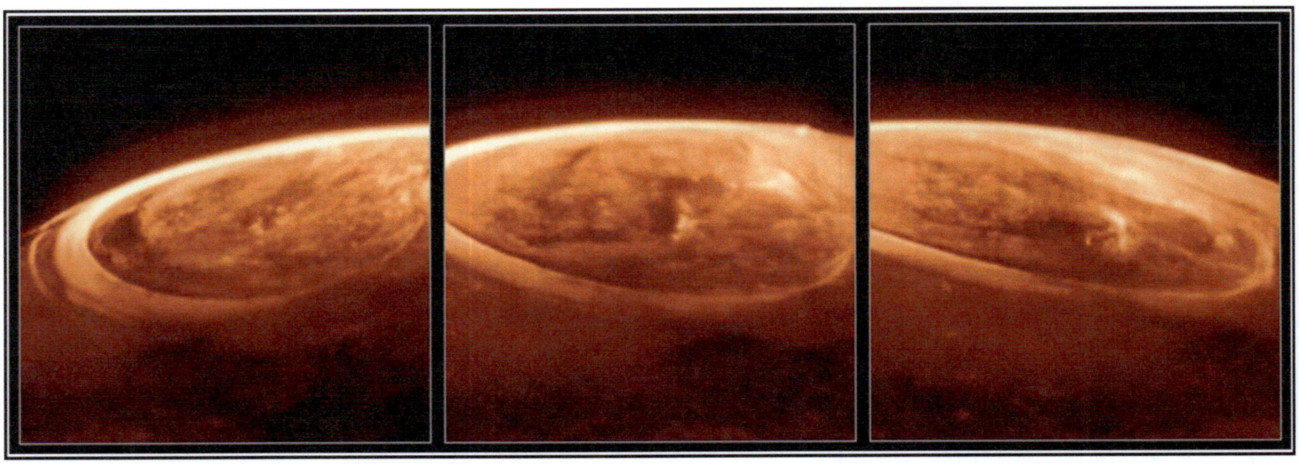

Fig. 1.11 Images of auroras on Jupiter's north pole. (*Credit* NASA)

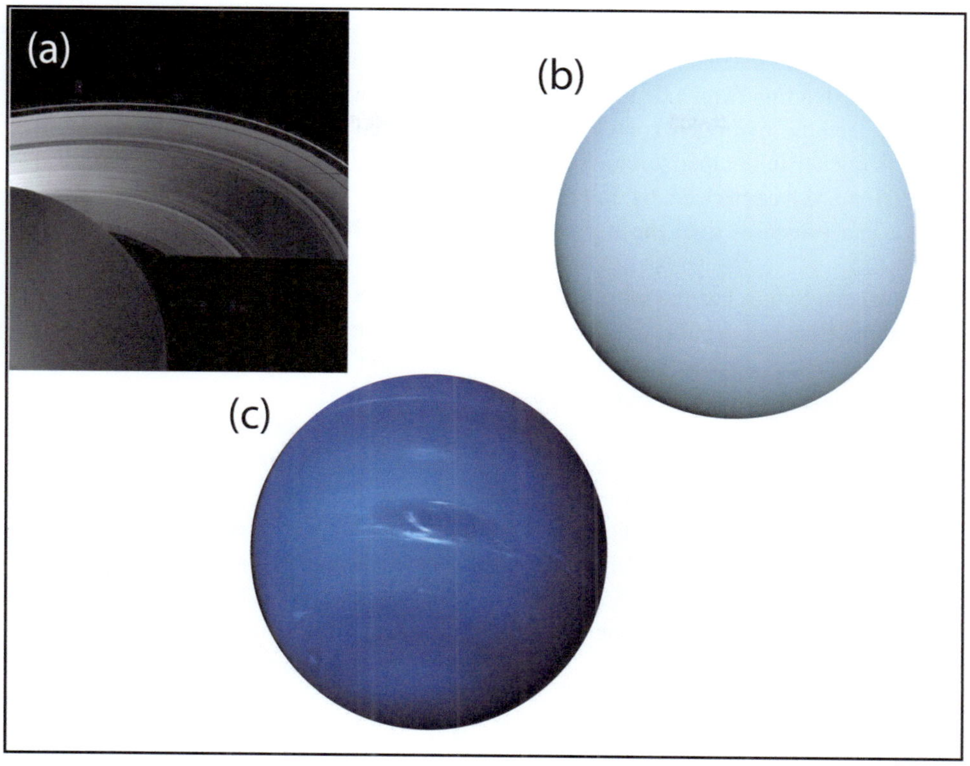

Fig. 1.12 a. A shadow casts on Saturn's A ring alongside the larger shadow cast by its moon Tethys. The image was taken by the *Cassini* spacecraft on June 19, 2009, at a distance of approximately 1.8 million km (1.1 million mi) from Saturn and at a Sun-Saturn-spacecraft. The night side of Saturn here is dimly lit by ringshine (*credit* NASA). b. This is an image of the planet Uranus taken by the spacecraft *Voyager 2* in 1986. (NASA)

9.5 Earth radii, and its density is 700 kg m^{-3} which is less than that of water of 1000 kg m^{-3}. Saturn has the lowest density of all the planets. Such a low density of Saturn is an indication that the planet is composed primarily of molecular hydrogen (H_2—92.4%) and helium (He—7.4%). Other gases such as ammonia (NH_4—0.2%) also occur, and in Saturn's cold upper atmosphere, both liquid and solid forms of ammonia are thought to occur. This low density also helps make the planet the most flattened one in the solar system. Relative to Jupiter, Saturn's lower mass leads to a lower pressure in its interior due to lower pressure compared to that of Jupiter. Thus, just like Jupiter, it is thought that the most abundant elements of hydrogen and helium never escaped from Saturn's atmosphere due to its low temperature and large mass. Saturn, however, has an apparent helium deficit of helium in its atmosphere. Helium is thought to dissolve in liquid nitrogen at the high temperatures and pressures experienced in the interior of Jupiter. However, helium is thought not to dissolve in the interior of Saturn due to the lower internal temperatures but instead it forms droplets. It is thought that helium precipitation is the cause of helium depletion in Saturn's outer layers as the helium condensed out of the surrounding hydrogen.

The beautiful ring system of Saturn lies in the planet's equatorial plane, and they are estimated to be in excess of 200,000 km in diameter but only a few tens of meters in thickness. The ring system's appearance on Saturn changes as it orbits the Sun due to the variation in the angles at which the rings are illuminated. Saturn's clouds are yellowish to tan in color and form bands that are parallel to the planet's equator.

c. Picture of Neptune produced from many images taken by *Voyager 2* in 1989 at a range of 7 million km (~4.4 million mi) from the planet. Also shown in the picture are the Great Dark Spot and its companion bright smudge. Visible to the west limb are the fast-moving bright feature—the "Scooter" and the Little Dark Spot (credit: NASA).

The interior of Saturn, composed dominantly of hydrogen, is broadly comparable to that of Jupiter, although the relative positions of the basic internal parts are different. Below Saturn's cloud cover, both pressure and temperature are thought to increase with depth. The outer radius is represented by the top of the cloud layers. Molecular hydrogen occurs in a zone that extends from a depth of 250 km to 30,000 km, and the temperature here is estimated to be 250 K. Below this molecular hydrogen zone occurs the metallic hydrogen zone which extends from a depth of

30,000 km to 45,000 km deep with a temperature of 8000 K. The core of Saturn is composed of icy and rocky material at a temperature of 12,000 K extending from a depth of 45,000 to 60,000 km. Thus, the metallic hydrogen zone of Saturn is thinner, and its core is thicker in comparison to Jupiter's. However, Saturn's interior is characterized by lower core temperatures, densities, and pressures than those prevalent on Jupiter.

Of all solar system planets, Saturn has the most complex and most extensive system of natural satellites with its total 62 moons (Table 1.1) which are deduced to be covered in snow and ice. Eight of Saturn's largest moons are over 300 km in diameter. Many of the moons of Saturn are thought to be made of mostly of water and ice. The largest of Saturn's moons, Titan, is thought to have an atmosphere that is denser and thicker than Earth's atmosphere, and it is also the most substantial atmosphere of any solar system moons! Titan's atmosphere is composed of molecular hydrogen (H_2), hydrocarbons such as ethane (C_2H_6), propane (C_3H_6), and carbon monoxide (CO).

1.4.10 Uranus

Discovered in 1781 by British astronomer William Hershel, Uranus was the new (and 7th) planet in the solar system to be discovered in over 2000 years. Uranus appears in telescopes as a tiny pale greenish disk with very few surface details (Fig. 1.12b). Uranus's mass is 14.5 Earth masses. An average density of 1300 kg m^{-3} for Uranus implies that a large rocky core constitutes a greater fraction of the planet's mass than do the cores of both Jupiter and Saturn. However, Uranus's core is thought to be comparable to the cores of both Jupiter and Saturn in terms of mass, size, and composition. The rotation period of Uranus, which produces a day of 17.2 h long, is very short. One feature that makes Uranus unique from the rest of the solar system planets is that its axis of rotation lies within a plane of the ecliptic, that is, 98° from the perpendicular. Thus, like that of Venus, Uranus's rotation is retrograde; Uranus lies tipped over on its side relative to the other planets. It is thought that Uranus's unique spin axis was caused by a catastrophic event due to a collision with another huge body early in the planet's history.

The outer atmosphere of Uranus is thought to be similar to the atmospheres of both Jupiter and Saturn. Uranus's outer atmosphere is composed of H_2 (84%), He (14%), CH_4 (2%), and very minor amounts of ammonia. Uranus consists of five large moons plus 22 additional smaller ones (Table 1.1). Uranus has a ring system with a total of 9 rings that was discovered by astronomers in 1977 during a *stellar occultation* when the rings passed in front of a bright star whose light was then momentarily dimmed by those rings. *Voyager 1* confirmed the existence of these rings in 1979. Uranus's rings, unlike the bright and wide rings of Saturn, are dark, widely spaced, and narrow. The dark coloration is interpreted to be due to radiation darkening.

1.4.11 Neptune

Neptune is the 8th planet, and it was discovered, thanks to mathematical solutions, mainly due to its gravity's perturbations on Uranus's orbit and mass. Neptune has an orbital period of 163.7 Earth years and appears as a bluish disk (Fig. 1.12c). Given its great distance from Earth (Table 1.1), surface features of Neptune are difficult to discern with the exception of a few surface markings. Such surface markings are interpreted by astronomers to suggest multi-colored bands of clouds with light blue hues dominating (Fig. 1.12c). Neptune's mass is 17.1 Earth masses. Neptune has an average density of 1600 kg m^{-3}, implying that a large rocky core must constitute a greater fraction of the planet's mass than do the cores of both Jupiter and Saturn. However, Neptune's core is thought to be comparable to the cores of both Jupiter and Saturn in terms of mass, size, and composition. The axis of rotation for Neptune is inclined at 29.6° to a line that is perpendicular to its orbital plane, which is comparable to Saturn's tilt of 27°.

Neptune's interior rotates once every 16.1 h, whereas its atmosphere has an average rotation period of 17.3 h, which is indistinguishable from that of Uranus. The outer atmosphere of Neptune is also thought to be similar to the atmospheres of both Jupiter and Saturn. Neptune's outer atmosphere is composed of H_2 (84%), He (14%), CH_4 (3%), and very minor amounts of ammonia; these concentrations are very comparable to those of Uranus. Neptune has one large moon in Triton, with a longest diameter of 2710 km, and two others with diameters of at least 300 km, whereas the rest of its moons (11) have diameters that are less than 300 km. Like Uranus, Neptune also has a ring system that was only detected by *Voyager 2* in 1989. Neptune's ring system is composed of 5 dark rings, with the dark coloration also thought to be due to radiation darkening. Several storm systems have also been interpreted to occur on Neptune, with the largest such storm, the **Great Dark Spot**, clearly visible on images of the planet (Fig. 1.12c).

1.4.12 Kuiper Belt Objects and Dwarf Planets

The Kuiper belt objects, sometimes termed trans-Neptunian objects, orbit just beyond Neptune's orbit, ~30–50 AU from the Sun (Fig. 1.6). These dark inhabitants of the outer solar system only reflect minute amounts of sunlight and are therefore not easy to observe. Consequently, astronomers only have limited information on Kuiper belt objects. Pluto, considered to be the largest member of this group and also a **dwarf planet**, is the most well-known Kuiper belt object. Pluto's discovery in 1930 was prompted mainly due to perturbations on both Uranus's and Neptune's motions; such perturbations, however, turned out not to be the case after Pluto's discovery! Pluto's orbit around the Sun is elongated, and it is also inclined at 17.2° to the plane of the ecliptic. At perihelion, Pluto lies inside the orbit of Neptune, whereas at aphelion it lies well outside the orbit of Neptune. Pluto orbits near the inner margin of the Kuiper belt objects. Eris, Makemake, and Haumea are other bodies that orbit the Sun and are massive enough to have spherical shapes due to their gravity, but they did not clear their neighborhoods of other bodies, and they are therefore classified as dwarf planets. Icy dwarf planets that orbit beyond Neptune are also referred to as **plutoids**.

1.5 Earth's Internal Structure

Earth evolved in a broadly similar manner to the other terrestrial planets as summarized in Fig. 1.4 during the early history of the solar system. With further cooling, Earth eventually differentiated to form the structure it has today. With accretion of material that formed our planet, temperature of young Earth is thought to have reached the melting points of both iron and nickel. Since both these elements are denser than silicates, they were concentrated at the planet's center to form the core. It is also envisaged that silicates, being less dense than iron and nickel, flowed outwards to produce the mantle and the crust. The mantle is, therefore, composed of iron and nickel in addition to silicates. These processes led to the formation of a differentiated Earth. However, Earth is unique from other terrestrial planets in several ways. Some of these unique ways are that Earth is the only planet known to have liquid water as well as the only planet where life exists. From measurements of the average density of Earth, the planet had to contain much denser rocks and materials (such as metals) than the granites and gabbros found at its surface. Studies of seismic waves that travel at different velocities through the interior of the Earth helped Earth scientists to understand the internal structure of the planet. Since seismic waves travel at different velocities through materials of different compositions, Earth scientists were able to delineate layers made of rocks of different densities. This led to the subdivision of the Earth, from its surface down to the center, into the crust, mantle, and core (Fig. 1.13). The mantle can be further subdivided into the upper and lower mantle, whereas the core can be further subdivided into the outer core and the inner core (Fig. 1.13).

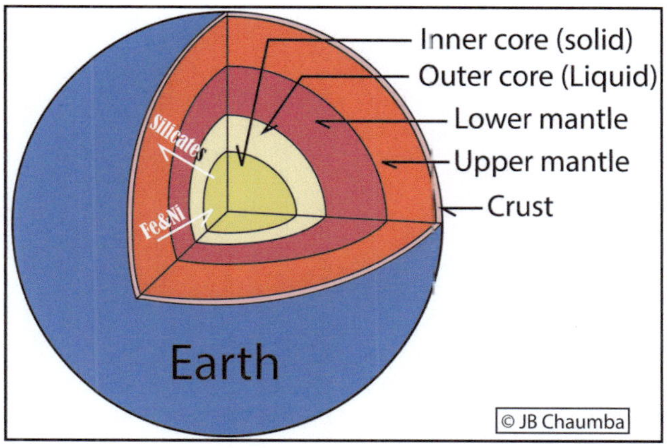

Fig. 1.13 Layered structure of Earth cutaway showing, the crust, upper mantle, lower mantle, outer core, and inner core

The thickness of the outermost layer of Earth, the crust, varies from place to place depending on location. The Croatian seismologist, Andrija Mohorovičić, is credited with discovering the base of the crust based on acceleration in seismic wave velocities at the boundary of the crust and the mantle beneath it (Figs. 1.14 and 1.15). The base of the crust is now termed the Mohorovičić Discontinuity, or simply the **Moho**, in his honor. The crust is thinner beneath oceans (7–10 km [~4–6 mi]) than beneath continents (Fig. 1.15) where it varies from as low as 25 km (~16 mi) where extension has occurred to as high as 70 km (~43 mi) in mountainous regions where collision is occurring but averages 35–40 km (~22–25 mi). Oceanic crust, which occurs beneath oceans, is made up of **mafic** or dark colored rock called **basalt**, which is underlain by coarser-grained **gabbro** but similar in composition. Continental crust, which makes up the continents, is composed of **felsic** or light-colored rock similar to **granite** in composition and **andesite** which is **intermediate** in composition. Thus, oceanic crust is denser than continental crust.

The mantle extends from the base of the crust to a depth of 2885 km (~1800 mi) (Figs. 1.14 and 1.15), and it is Earth's largest layer by volume. An abrupt increase in seismic wave velocity marks the top of the mantle. An **ultramafic** rock called **peridotite** is thought to compose the mantle. The upper mantle extends from the base of the crust to a depth of 660 km (~410 mi) (Fig. 1.14). Since seismic wave velocities are known to abruptly increase further at depth of 410 km (~255 mi) and 660 km (~410 mi) (Fig. 1.14), this zone marks the transitional layer between the upper mantle and the lower mantle beneath it, and it is therefore called the **transition zone** (Fig. 1.14).

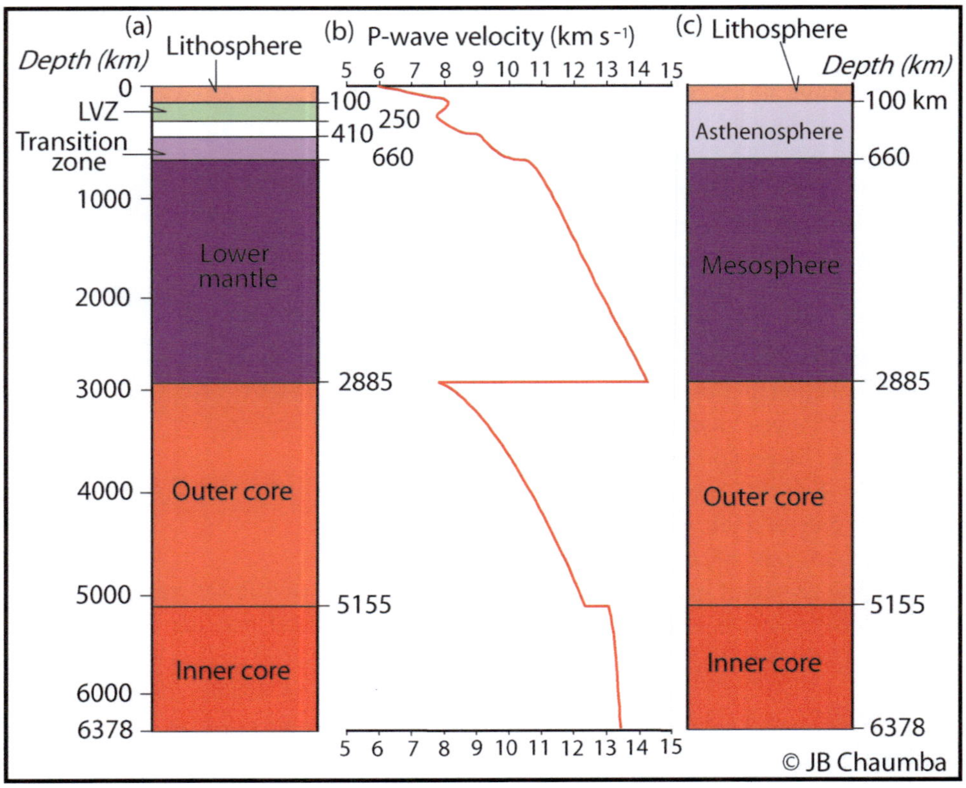

Fig. 1.14 Diagram **a** shows subdivision of the Earth into its major compositional layers. Shown in **b** are changes in seismic wave velocity (P-wave) within Earth. Subdivision of the Earth into its mechanical layers is shown in **c**. LVZ is the low velocity zone

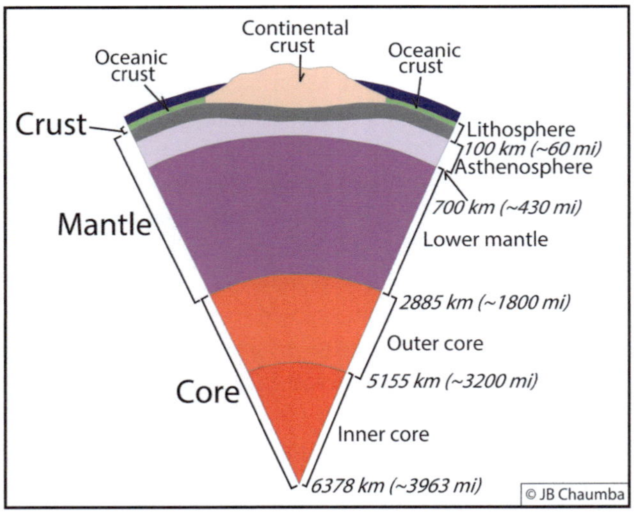

Fig. 1.15 Diagram showing Earth's three major layers of crust, mantle, and core to the left, as well as the subdivisions of these major layers into the continental crust, oceanic crust, lithosphere, asthenosphere, lower mantle, outer core, and inner core to the right

A **low velocity zone** (LVZ) occurs in the upper mantle at depths of 100–250 km (60- ~ 155 mi) deep (Fig. 1.13). The LVZ has anomalously low P-wave velocities due to partial melting which occurs at these depths. The top of the LVZ marks the contact between the strong and rigid **lithosphere** and the weak and soft **asthenosphere** (plastic, flows slowly) below (Fig. 1.14). We will return to the lithosphere and asthenosphere in greater detail in Chaps. 3 and 6. The sudden jump in seismic wave velocity in the transition zone occurs due to sudden increases in rigidity & incompressibility of rocks. Minerals that make up the rock peridotite undergo transformations into new minerals at these depths. The base of the transition zone at a depth of 660 km (~410 mi) also marks the base of the asthenosphere, and it is in contact with the underlying mesosphere or lower mantle (Fig. 1.14). An **ultra-low velocity zone** (ULVZ) occurs at the lowermost mantle, and it is thought to be related to the formation of deep mantle plumes within the lower mantle. We shall discuss more about mantle plumes in Chaps. 3 and 6.

From the base of the mantle at 2885 km (~1800 mi) deep, that is, from the ULVZ, to a depth of 5155 km (~3200 mi), seismic wave velocities increase gradually to where another abrupt increase in seismic wave velocities occurs. This layer is called the **outer core**, and it is composed of alloys of the metal iron which exist in a liquid state due to the very high temperatures experienced here. The liquid state of rocks in the outer core enables the iron alloys to flow, and it is thought that this flowage generates Earth's magnetic field. Below the outer core and at Earth's center occurs the solid **inner core** which is thought to be composed of iron alloys. Although temperatures in the inner core are extremely high, hotter than those experienced in the outer core, the greater depth of the inner core ensures that greater pressures are experienced here which are responsible for the solid state of the iron alloys encountered here.

Another way to view the layers of the Earth we have just discussed is in Figs. 1.13, 1.14 and 1.15 is to consider the strength of these layers. This allows us to distinguish further between the lithosphere and the asthenosphere. Earth's outer layer which extends from the surface down to a depth of 100–150 km (60– ~ 93 mi) is rigid, meaning that it is composed of rocks that can break and bend. We call this outermost layer, which consists of the entire crust (both oceanic and continental) plus the uppermost part of the mantle, the lithosphere (Figs. 1.1 and 1.15). The uppermost part of the mantle included in the lithosphere is called the lithospheric mantle. Lying beneath the lithosphere is a layer composed of rocks that are weak, soft, and plastic, and can flow without breaking, and we call this layer the asthenosphere. The asthenosphere lies wholly within the mantle at depths of greater than 100–150 km (60–~93 mi) (Figs. 1.14 and 1.15). Two types of lithosphere are distinguished: oceanic lithosphere occurs beneath oceanic crust, whereas continental lithosphere occurs beneath continental crust. Oceanic lithosphere extends to depths of 100 km (60 mi), whereas continental lithosphere extends deeper to a depth of 150 km (~93 mi) (Figs. 1.14 and 1.15). The top of the transition zone occurring at a depth of 410 km (~255 mi) or even the base of the transition zone occurring at a depth of 660 km (~410 mi) of (Fig. 1.15) have been assigned by various researchers to mark the bottom of the asthenosphere.

1.6 The Scientific Method and Geology

To unravel past geological events, one needs to conduct a scientific investigation, as this is the method of discovering scientific principles. To carry out a scientific investigation, one begins with some bodies of data from experiments, or from some sets of observations from natural phenomena, or even from some calculations on how the natural world works. The **scientific method** is a sequence of steps taken to evaluate scientific problems that leads to verifiable results using systematic steps (Fig. 1.16). The first step typically taken involves a question about natural phenomena (Fig. 1.16a), which is then followed by formulating a **hypothesis** to explain the observations, data, or calculations (Fig. 1.16b). An explanation of a general concept is an example of a simple hypothesis. However, one can have a hypothesis called a model which describes how a natural system functions. A hypothesis can also be an exact mathematical formula which relates to different kinds of numerical data. Any hypothesis should be susceptible to being tested by other investigators and be verified or nullified by an examination of relevant evidence in order to falsify it.

A hypothesis is used to make predictions about the natural world. This is then followed by conducting experiments to test the hypothesis and determine whether the results of the experiments do not agree, or support, the hypothesis. If the results do not agree with the hypothesis, the hypothesis can be modified to suit the new (and old!) data, otherwise the hypothesis is discarded. If new data consistently supports a hypothesis, that hypothesis can be elevated to a **theory** (Fig. 1.16c) such as the plate tectonics theory or the atomic theory. A theory is an explanation for a set of data that is generally accepted. Theories are also continually tested as technology advances, and a theory that

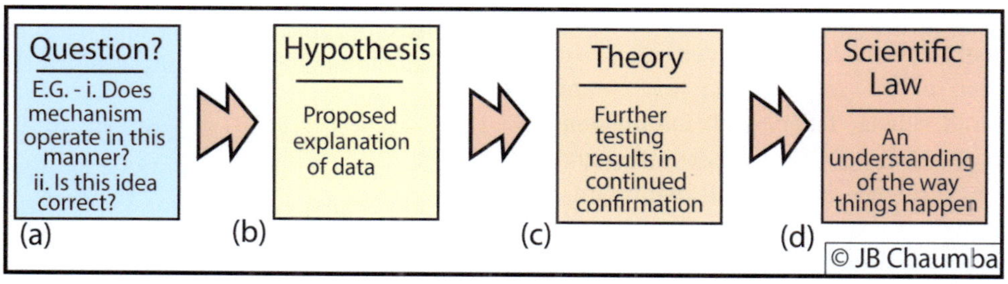

Fig. 1.16 Typical steps that are followed in the scientific method. One begins with a question **a**, followed by gathering of data which is evaluated leading to the formulation of a hypothesis **b**. The hypothesis is then repeated tested leading to: i. its rejection, ii. its revision, or iii. its acceptance. A hypothesis that is consistently supported by new data can rise to the level of a theory **c**. If new data continually supports the theory over a long period of time, it can be elevated to a law or principle **d**

continues to withstand scrutiny can also be raised to a **scientific law** (Fig. 1.16d). The law of gravity is an example of a scientific law.

Glossary

Accretion gradual growth of bodies such as planets from the accumulation of other smaller bodies

Andesite an igneous rock (or magma) that if medium-grained and composed of intermediate amounts of magnesium, iron, feldspar, and silica

Asteroid small and rocky bodies that orbit around the Sun

Asteroid belt zone of the solar system where most asteroids are found that occurs between the orbits of Mars and Jupiter

Asthenosphere a soft layer with the Earth that occurs below the lithosphere that is partially molten and rocks deform by flowing

Astronomy science that deals with the study of the Universe lying above the atmosphere of Earth

Basalt an igneous rock (or magma) that if fine-grained and rich in magnesium and iron but poor in feldspar and silica

Belts downwelling and low-pressure dark regions of in occurring the jovian planets' atmospheres

Big Bang a violent event that marked the beginning of the Universe during which all matter originated

Big Bang theory an explanation that describes how the universe originated starting with a small point and then rapidly expanding over the next 14 billion years to the cosmos that exists today

Condensation nuclei dust grains occurring in the interstellar medium and acts as seeds on which other materials clustered during formation of the solar system

Comet a small object made up of ice and dust with an elliptical orbit around the Sun. Some of the comet's material vaporizes to form a gaseous head and a long tail when it is close to the Sun

Condensation theory the model that is currently favored for the formation of the solar system that combines features of the old nebular theory with information on interstellar dust grains that acted as condensation nuclei

Cosmology branch of science dealing with the origin and development of the universe

Crater bowl shaped depression occurring on surfaces planets of moons formed by collisions with meteorites or other interplanetary debris

Crust the topmost and outermost layer of Earth's interior

Doppler effect change in the frequency or observed wavelength of a wave that is induced by motion

Dwarf planets massive body orbiting the Sun with gravity that causes its shape to be spherical, yet it is not large enough to clear its neighborhood of other objects

Earth third planet from the Sun which forms part of the inner planets; it is our home in space

Energy property of an object that can be transferred to another object or it can be converted into different forms

Earth-crossing asteroids also called Apollo asteroids—are asteroids with an orbit that crosses that of Earth

Felsic igneous rocks or magmas that are composed of relatively low amounts of magnesium and iron but rich in silica and feldspar

Fragmentation breaking up of a large object into numerous smaller pieces due to high velocity impacts between objects

Full moon phase attained by the moon whereby it appears as a complete disk in the sky

Galaxy a collection of a large number of stars that are bound together by gravity

Geocentric hypothesis an early explanation of the solar system that states that Earth lies at the Universe's center and that all other planets and the Sun orbit around it

Granite an intrusive and coarse-grained igneous rock rich in silica and feldspar

Great Dark Spot a system of storm that is visible on Neptune's atmosphere near its equator

Great Red Spot a high-pressure, large, and long-lived system of storm that is visible on Jupiter's atmosphere

Heliocentric hypothesis an explanation of the solar system that states that the Sun lies at its center, with Earth and other planets orbiting around it

Hypothesis an explanation, or conceptual model, for observations, measurements, or data

Inner core a layer which occurs at the center of the planet Earth and below the inner core and it is composed of solid iron alloys

Intermediate igneous rocks or magmas that are composed of moderate amounts of magnesium, iron, silica, and feldspar

Jovian planets the planets of the solar system which consist of Jupiter, Saturn, Uranus, and Neptune which all resemble Jupiter in both physical and chemical composition

Light year distance that light travels in one year at a speed of 300,000 km/s (~186,400 mi/s). One light year is approximately 9.5 trillion km (~6 trillion mi). nebular theory

Lithosphere the topmost layer with the Earth that occurs above the asthenosphere that is rigid and strong and rocks deform by breaking and folding

Low velocity zone a layer within the asthenosphere but near its upper part and below the lithosphere where seismic waves travel at lower velocities due to partial melting

Mafic igneous rocks or magmas that are composed of high amounts of magnesium and iron but composed of low amounts of silica and feldspar

Mantle a layer within the Earth's interior occurring below the crust and above the core

Matter anything that possesses mass

Meteorites meteoroids that survive glowing passage through Earth's atmosphere to land on its surface

Meteoroid Fragment of interplanetary debris before it enters the atmosphere of Earth

Milky Way galaxy a spiral galaxy that contains the Sun and the solar system and the disk of its galaxy is visible as faint bands of light called the Milky Way during the night sky

Moho a discontinuity named for Andrija Mohorovičić which is marked by increased seismic wave velocities and it marks the boundary between the Earth's crust and mantle

Outer core Earth's layer which occurs between the mantle and the inner core and composed of liquid iron alloys

Peridotite igneous rock (or magma) that is coarse-grained and very enriched in magnesium and iron but very low in silica and feldspar

Planetesimal an object in the solar system that had grown to reach the size of small moons with a strong enough gravitational field to influence its neighbor

Photosphere visible surface of the Sun that occurs just above the uppermost layer of the interior of the Sun

Plutoids dwarf planets with orbits beyond Neptune's

Proto-moon lump of material that was formed in the early stages of the formation of the solar system considered to be the predecessor to moons of today

Proto-planet lump of material that was formed in the early stages of the formation of the solar system considered to be the predecessor to planets of today

Proto-Sun central accumulation of material particles in the early stages of the formation of solar system considered to be predecessor to the Sun.

Redshift change in the frequency or observed wavelength of light emitted from a source moving away from an observer induced by motion. Relative recessional motion causes the wave to have longer wavelength (redder) to be observed than it would if it were stationary

Scientific law a statement that is based on continued experimental observations that describes some aspect of natural phenomena

Scientific method steps followed to discover scientific principles whereby a hypothesis is formulated, predictions are made, and tested

Solar nebula spiraling gas that surrounds the proto-Sun during the period of formation of the solar system

Solar system the Sun and all the planets, their moons, and other objects such as asteroids, meteorites, and comets that orbit around it

Star glowing ball of gas that is powered by nuclear fusion in its core and held together by its gravitational force

Terrestrial planets the four planets of Mercury, Venus, Earth, and Mars which resemble Earth in both physical and chemical properties

Transition zone portion of the mantle characterized by numerous abrupt increases in seismic wave velocities

Sun our parent star that is located at the center of the solar system

Theory an explanation for an observation or a set of data that is generally accepted and has bas been tested by the scientific method

Universe all space, time, energy, and matter together

Ultra-low velocity zone a patch at the lowest part of the lower mantle at the lower mantle-outer core boundary where seismic waves travel at very low velocities due to relatively high degrees of partial melting

Ultramafic igneous rocks or magmas that are composed of very high amounts of magnesium and iron but composed of very low amounts of silica and feldspar

Wavelength distance between any two successive crests or troughs at any given point in time

Zones upwelling, high-pressure bright regions in the atmospheres of jovian planets

Minerals

2

Abstract

This chapter begins with providing the complete definition of a mineral, a summary of the atomic structure, atomic bonding, and formation of minerals. Mineral descriptions are accompanied by several photographs of minerals, followed by discussing the physical properties that are utilized in mineral identification. The Mohs hardness scale is explained. Finally, the common silicate and non-silicate rock-forming minerals are described, again with the aid of numerous photographs. Tables of the Mohs scale of hardness of minerals, and of some common non-silicate minerals, are provided.

Learning Goals

Studying this chapter will enable you to:

- Draw and label the atomic structure
- List, compare, and contrast the particles making up an atom
- List and differentiate among ionic, covalent, and metallic bonds
- Describe, define, and distinguish between an atom, an element, and a mineral
- List and describe the common physical properties that are used to identify minerals
- Draw and label a silicon tetrahedron
- Describe, list, and differentiate between silicate minerals and non-silicate minerals
- List the common rock-forming minerals on Earth and mineral groups

Supplementary Information The online version contains supplementary material available at https://doi.org/10.1007/978-3-032-20231-4_2.

2.1 Introduction

Having discussed the origin of our planet in Chap. 1, we now must be familiar with the notion that Earth is one of the four inner planets in our solar system referred to as terrestrial (or rocky) planets. The rocks which make up these inner planets are made up of **minerals**. Throughout history, mankind has always valued minerals for a number of reasons. A number of minerals glitter, whereas other minerals have been prized as gems, jewelry, among other reasons. Over 4000 minerals have been described, with approximately 40 more continuing to be discovered each year. The importance of minerals can be gleaned from the observation that ancient human civilizations such as the Stone Age, the Bronze Age, and the Iron Age, for example, have been defined by human beings' use of mineral materials. Further, minerals make up rocks, and since rocks are the building blocks of the solid part of the Earth (and others like it, such as the terrestrial planets we discussed in Chap. 1), minerals may also be considered to be the building blocks from which our planet is made. Human beings make use of minerals in an infinite number of ways, such as money, building materials, jewelry, gemstones, weapons, metals, ornaments, and specialized uses in computers and dentistry. Consequently, for us to understand the rocks that constitute our planet, we have to understand what minerals are and how they are formed.

2.2 Definition of a Mineral

In Geology, a mineral has a special definition. For any substance to be defined as a mineral, it must satisfy the following parameters:

i. form naturally
ii. solid substance

J. B. Chaumba, *Introductory Physical Geology*, Springer Textbooks in Earth Sciences, Geography and Environment, https://doi.org/10.1007/978-3-032-20231-4_2

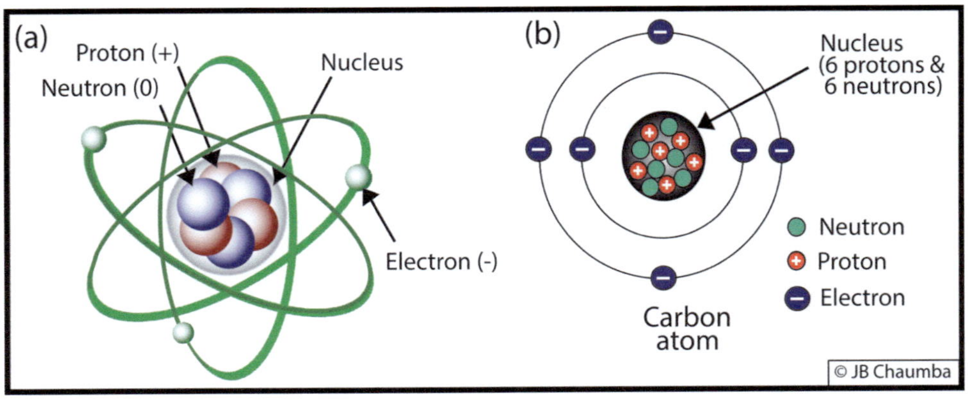

Fig. 2.1 **a** Diagram showing a simplified structure of an atom with a nucleus at the center in which protons and neutrons are located, with electrons orbiting around it. **b** An example of a simplified atomic structure of the element carbon that is composed of 6 protons and 6 neutrons in the nucleus, as well as 6 electrons (in the electron cloud)

iii. possess a crystalline structure
iv. atoms are arranged in an orderly and repetitive manner
v. possess a specific chemical composition with some variation
vi. generally inorganic substance

Thus, in Geology, a mineral can be defined as a naturally occurring inorganic solid that possesses an orderly crystalline structure with a specific chemical composition. The chemical composition, however, allows for minor variations within well defined and specific limits where certain elements can substitute for others. Since minerals have to grow by natural processes, this excludes minerals that can be synthesized in laboratories, called synthetic minerals. Biogenic minerals, those crystalline materials formed by marine organisms such as calcite that make up coral reefs and shells, however, are also considered to be minerals. Further, since only solid substances can be minerals, fluids cannot, although mercury, for example, which occurs in liquid state, is an exception. Glass, which sometimes forms from very rapid cooling of magma (see Chap. 3), such that it lacks a specific regular arrangement of atoms, is an example of a **mineraloid**. Mineraloids are mineral-like substances which, although they may have relatively constant compositions, lack specific crystal structures.

Having defined what a mineral is, the next question one may ask is what a mineral is made up of. Planet Earth is composed of elements that were formed during supernova explosions as well as fusion reactions. A mineral contains one or more **elements**. Elements, like all matter, are made up of **atoms**, which are the building blocks of matter. An atom is the smallest particle of an element that contains all the properties of an element. A property of an element is a characteristic of a substance that makes distinguishing it from other substances possible, such as its ability to form bonds with other atoms (chemical property) or its hardness (physical property). Let us now briefly review the atomic structure.

2.3 Atomic Structure

An atom is the smallest ingredient of an element, and it cannot be further broken down into any simpler form by chemical reactions, heating, or cooling. An atom is composed of a central region called a **nucleus** which is surrounded by a shell in which negatively charged **electrons** "orbit" the nucleus (Fig. 2.1a and b). Within the nucleus of an atom, **protons**—which are positively charged particles, and **neutrons**, are located (Fig. 2.1a and b). Atoms of carbon, for example, have six protons (Fig. 2.1). A neutron is a particle with no charge, and it is also located in the nucleus of an atom (Fig. 2.1) with the exception of hydrogen, as it doesn't need it.

Protons and electrons have charges that are equal in magnitude but of opposite polarity, as shown in Table 2.1 since an atom has an equal number of protons and electrons, it has no net charge. Thus, most substances tend to be electrically neutral. Each shell is characterized by electrons of a specific number (e.g., Fig. 2.1b). The electrons orbit

Table 2.1 Table illustrating the properties of sub-atomic particles

Particle	Location	Charge	Relative charge	Mass	Relative mass (amu[a])
Proton	Nucleus	+1	+1	1.672×10^{-24} g	1
Neutron	Nucleus	Neutral	0	1.674×10^{-24} g	1
Electron	Electron cloud	−1	−1	9.05×10^{-28} g	1/1840 (~0)

[a]amu—atomic mass unit

the nucleus in regions of different levels of energy termed principal shells. The outermost shell of any atom contains electrons that can bond chemically with other atoms. These electrons in the outermost shell of any atom are referred to as **valence electrons**.

The **atomic number** of any element represents the number of protons in the nucleus. The **atomic mass** for a given element is approximately the number of protons plus neutrons (nucleons) in the nucleus. An **isotope** is one of two or more forms of atoms of an element that differ in their number of neutrons. Thus, isotopes have the same atomic number (same number of protons and electrons) but possess different mass numbers due to fewer or more neutrons.

2.4 Elements and the Periodic Table

All atoms of one element are different from the atoms of other elements, and all atoms of one given element are identical. Approximately 90 elements occur naturally although approximately 25 other elements have been synthesized in laboratories. Elements have been arranged on the Periodic Table in such a way that those in columns (called groups) contain similar properties as shown in Fig. 2.2. On the Periodic Table, elements are assigned one-, two-, or three-letter symbols (Fig. 2.2).

2.5 Atomic Bonding and Formation of Minerals

As stated earlier, naturally occurring elements form the building blocks of minerals on Earth. However, **chemical compounds** are formed due to atoms of some elements combining with atoms of other elements. Thus, most minerals can be considered to be chemical compounds that are composed of at least two elements. Not all minerals are chemical compounds, however. Some minerals, like native copper, native gold, native sulfur, and diamond (Fig. 2.3), for example, are only formed from single elements. As already been stated, for most elements, however, most atoms **bond** by gaining, losing, or sharing electrons to form chemical compounds.

At the temperatures and pressures experienced on our planet, most atoms tend to bond and form molecules, metallic substances, or ionic compounds. Atoms become chemically stable when their outermost energy levels are filled up with electrons. Except for hydrogen and helium atoms, which possess the lowest energy levels that require only two electrons in their single energy levels, chemical stability for all other atoms is achieved when eight electrons occupy their outermost energy level. Thus, the **octet rule** requires that the chemical stability of atoms is considered to be only achieved with eight electrons in the outer-

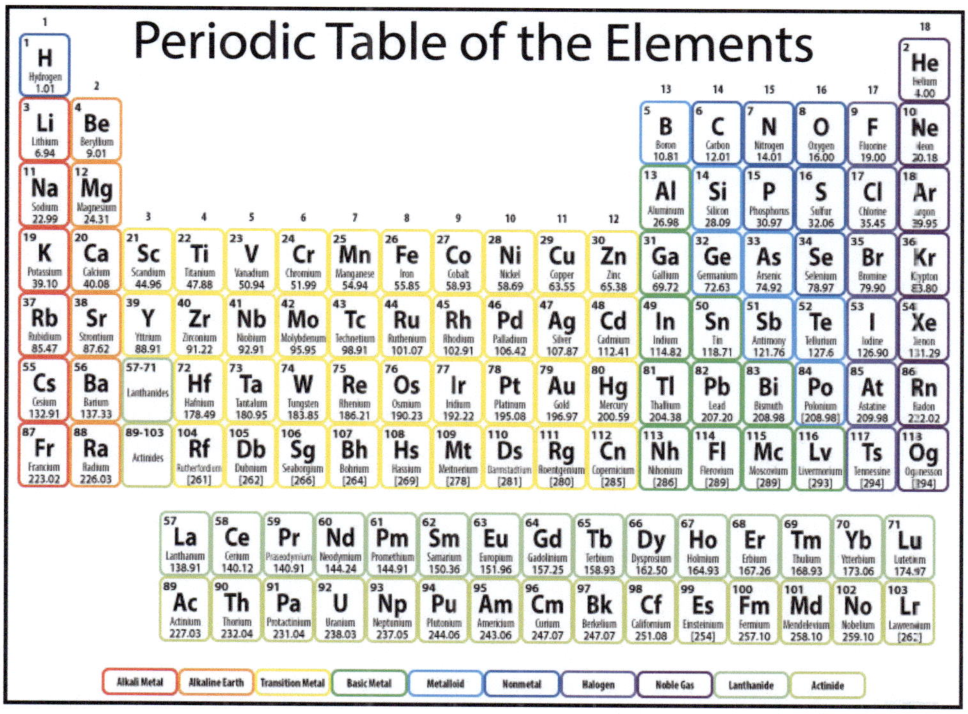

Fig. 2.2 Periodic table of the elements. (*Credit* https://sciencenotes.org/printable-periodic-table/)

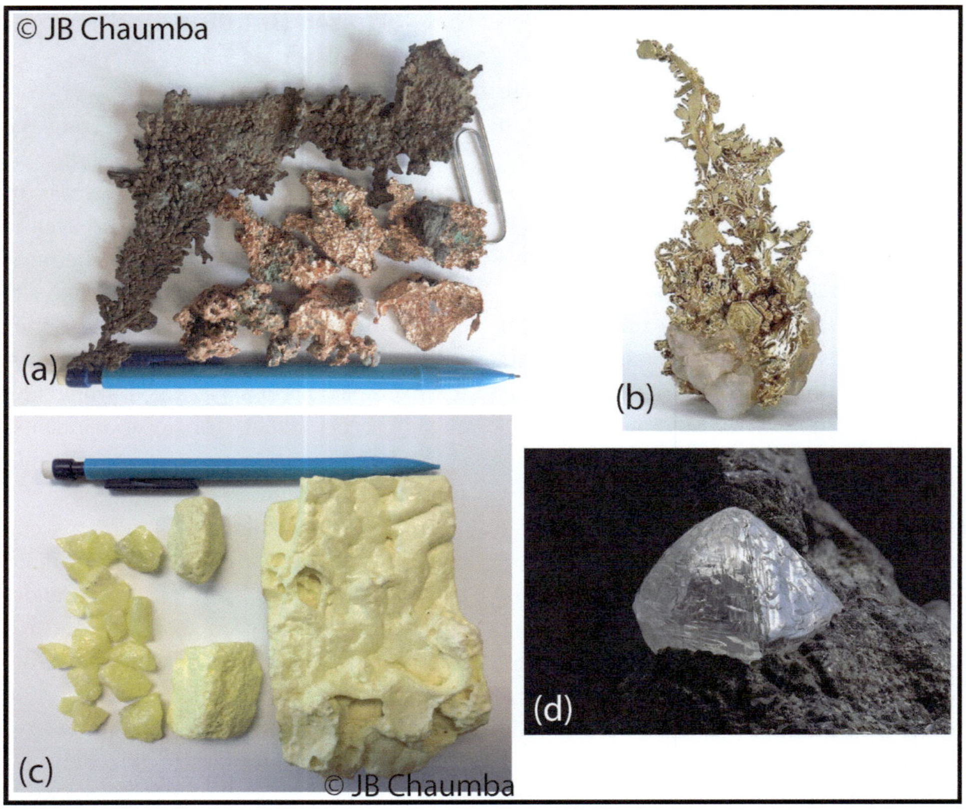

Fig. 2.3 Some examples of the commonly occurring native elements. **a** native copper from Michigan (Photograph by JB Chaumba), **b** specimen of pure gold in a quartz vein, **c** native sulfur crystals. The three coarsest crystals are from the Frasch process, Gulf Coast, Texas, USA, and **d** A gem-quality octahedral diamond crystal in a kimberlite from Diamant De Beers Mine located in Kimberley, South Africa (*Credit* Chinellato Matteo)

most energy level. The number of electrons in the outermost energy level determines whether an atom gains, loses, or shares electrons. A tendency for an atom to lose its electrons is observed when atoms have only one or two electrons in their outermost energy levels (e.g., sodium) in order to eliminate their outermost energy levels that are filled in part to leave the outermost shells with eight electrons. A tendency to gain electrons is observed in atoms that have six or seven electrons in their outermost energy levels (e.g., chlorine) so that they fill their outermost energy level, thereby fulfilling the octet rule.

Sharing of electrons in order to fulfill the octet rule occurs when atoms tend to have three to five electrons in their outermost energy levels (e.g., carbon). Some atoms, which are chemically stable, have atoms with two (e.g., helium) or eight (e.g., noble gases) electrons in their outermost energy levels that are already filled and are thus called inert, as they do not bond with others. A wide variety of chemical bonds are possible due to variation in electron configurations of different elements. Let us now review the three most common chemical bonds that atoms in our planet tend to form.

2.5.1 Ionic Bonding—Electron Transfer

When an atom gains or loses an electron, it no longer remains neutral, but it becomes a positively or negatively charged particle called an **ion**. For a sodium atom (Na), for example, when it loses an electron, it becomes a positively charged sodium ion since it now has one more proton compared to its number of electrons. We can denote this sodium ion with a positive (+) superscript, e.g., Na^+. Similarly, when a chlorine atom gains an electron, it now has more electrons than protons resulting in a negatively charged chlorine ion. The chlorine ion is then denoted with a negative (−) superscript, Cl^-. In general, the number of electrons gained or lost is written before the charge, also as a superscript. As an example, magnesium, which tends to lose 2 electrons during electron sharing, is denoted as Mg^{2+}, and similarly oxygen has a tendency to gain two electrons and is therefore denoted as O^{2-}.

A tendency to lose or gain electrons occurs when two atoms with strong tendencies to gain or lose electrons come in contact with each other so that they satisfy the octet rule and thereby achieve chemical stability. In such cases, the

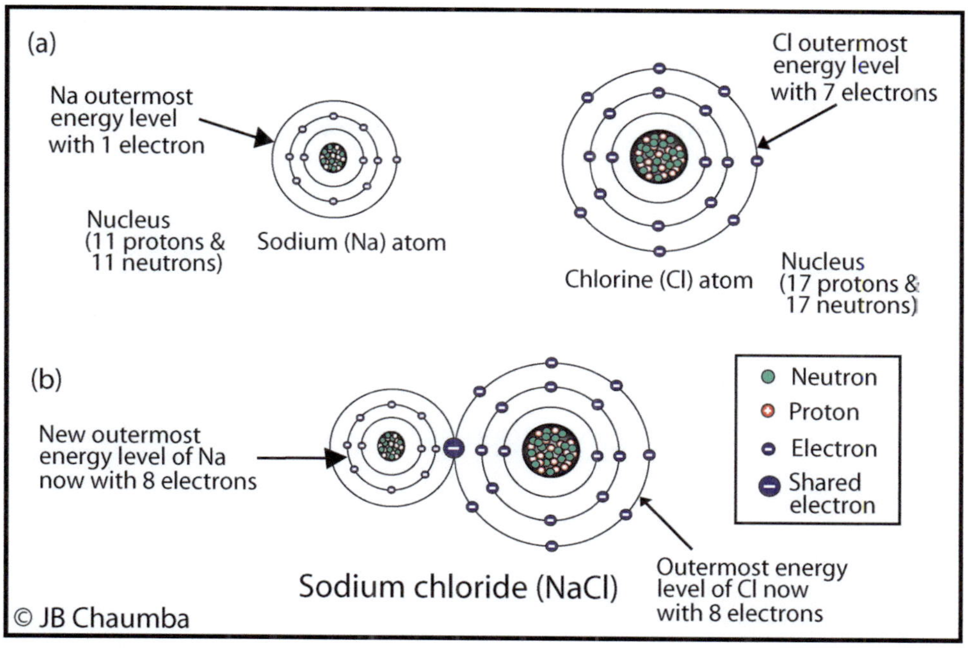

Fig. 2.4 Simplified schematic diagram illustrating the ionic bonding that is formed between sodium (Na) and chlorine (Cl) atoms. A sodium atom, which initially has one electron in its outermost electron shell, donates this electron to a chlorine atom, which initially has seven electrons in its outermost energy level, to form a compound with the outermost energy levels of both atoms that have eight electrons. The ions of Na and Cl bond to form a neutral ionic compound—sodium chloride (NaCl)

atom that loses one or more electrons, or the donor atom, becomes a positively charged ion (or **cation**), whereas the atom that receives one or more electrons, or the receiving atom, becomes a negatively charged ion (an **anion**). Attraction between these oppositely charged ions forms an **ionic bond**. For sodium and chlorine, for example, the ionic bond that is formed when attraction between these ions results in the electrically neutral sodium chloride compound (NaCl), illustrated in Fig. 2.4.

Sodium, with a strong tendency to lose the electron in its third and outermost energy level (Fig. 2.4.a), loses it to form a sodium cation, Na^+, eliminating the third energy level (Fig. 2.4b). The second energy level of the Na atom, which is now the outermost energy level, now has eight electrons (Fig. 2.4b). Chlorine, which has a strong tendency to gain an electron to add to its outermost energy level with seven electrons (Fig. 2.4a), gains one electron to become a chlorine anion and also fills in outermost energy level with eight electrons to satisfy the octet rule (Fig. 2.4b). Sodium chloride (NaCl), which is a compound formed from the ionic bonding between sodium and chlorine (Fig. 2.4), is electrically neutral. Component compounds such as sodium and chlorine discussed above tend to be characterized by chemical and physical properties that are often very different from those of the ionic compounds that are formed. Sodium chloride, or halite (the common table salt), has very different chemical and physical properties from those of either sodium or chlorine.

2.5.2 Covalent Bonding—Electron Sharing

Sharing of electrons typically occurs when atoms have their outermost energy levels that are half full, with three, four, or five electrons. In order to achieve chemical stability, these atoms also have a tendency to share electrons with comparable atoms with electrons in their outermost energy levels that are also half-filled. The outermost energy levels of the atoms can then be filled up with electrons through electron sharing to form atoms, whereby the outermost energy levels overlap to produce very strong bonds called **covalent bonds**. Indeed, covalent bonds tend to be among the strongest of any atomic bonds that can be formed. It is possible to have at least two atoms of the same element form covalent bonds with each other, as in the case of hydrogen or diamond.

During covalent bonding, electron configuration tends to change so that the two electrons in two hydrogen atoms, for example, occupy the space between the two atoms (Fig. 2.5). The force holding the hydrogen atoms is caused by the attraction between the oppositely charged particles, although no ions are formed in hydrogen atoms (Fig. 2.5a). The protons with a positive charge in the nuclei and the negatively charged electrons surrounding the nuclei of the hydrogen atoms form the opposite charges that cause the attractive force (Fig. 2.5b). The covalent bonding between the two atoms of hydrogen forms a hydrogen molecule (H_2) (Fig. 2.5b).

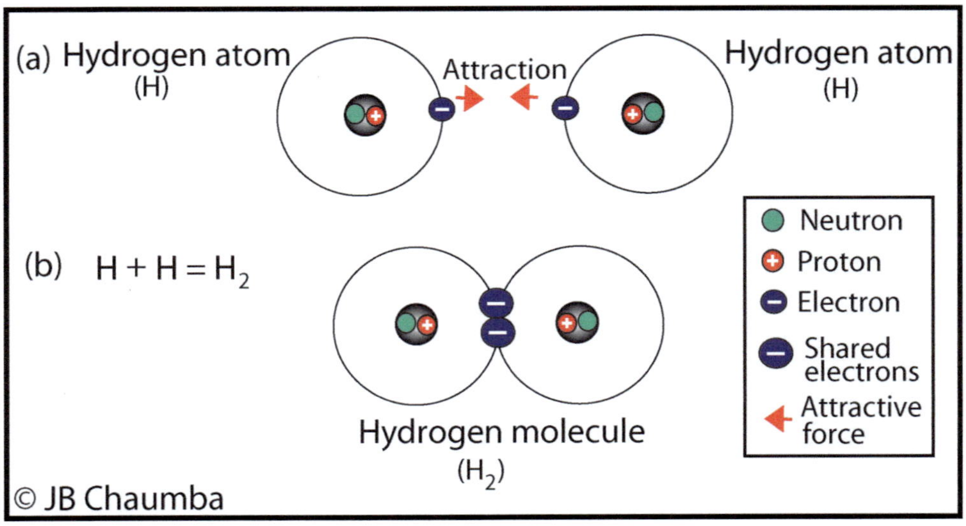

Fig. 2.5 Hydrogen atoms only need two electrons in their outermost energy level to reach the structure of the noble gas helium (**a**). The covalent bond that is formed between these two hydrogen atoms holds them together because the pair of electrons is attracted to the two hydrogen nuclei (**b**)

2.5.3 Metallic Bonding—"cloud" of Electrons Free to Move

When atoms of elements that donate electrons pack together, which they have a tendency to do, one such atom typically gets surrounded by eight or sometimes twelve other atoms. A "cloud" of free electrons that exists between the positively charged nuclei is then free to move throughout the whole metallic structure (Fig. 2.6). The force of attraction that occurs between the "cloud" of negatively charged electrons and a cluster of positively charged nuclei occurs to form what is termed **metallic bonding**. Metallic bonds are responsible for the unique properties of metals such as being excellent conductors of electricity due to the freely moving "cloud" of electrons. Further, metals can be hammered or pressed permanently into thin sheets without cracking or breaking (malleable), or they can be drawn into thin wires (ductile).

Other types of bonds such as intermolecular bonding also occur. For example, the strong covalent bonds that tend to hold water molecules together also attach themselves weakly to other molecules by **intermolecular bonds**. In water molecules, each hydrogen atom forms a single covalent bond, whereas each oxygen atom forms two covalent

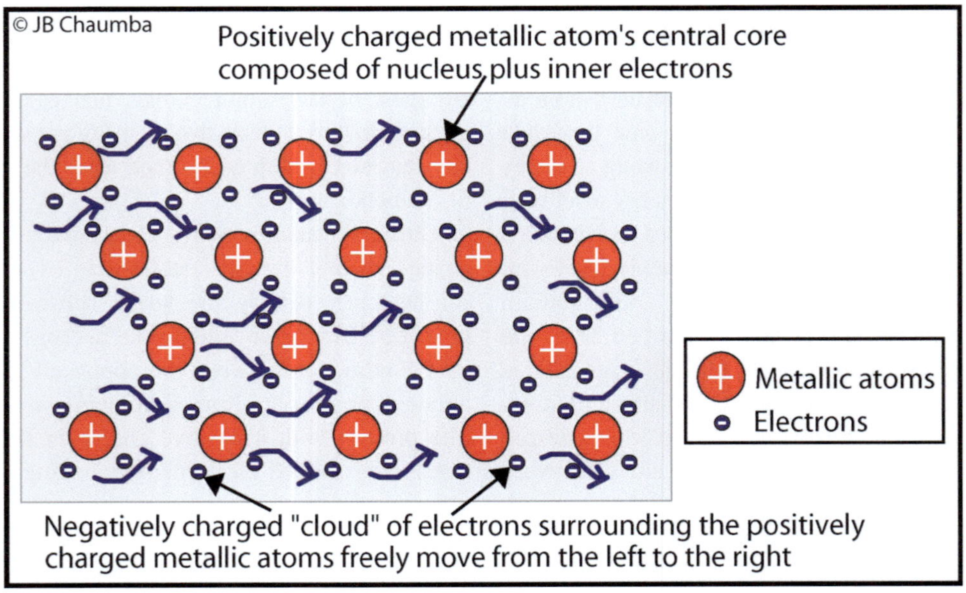

Fig. 2.6 Metallic atoms surrounded by a "cloud" of negatively charged electrons can freely move throughout the structure

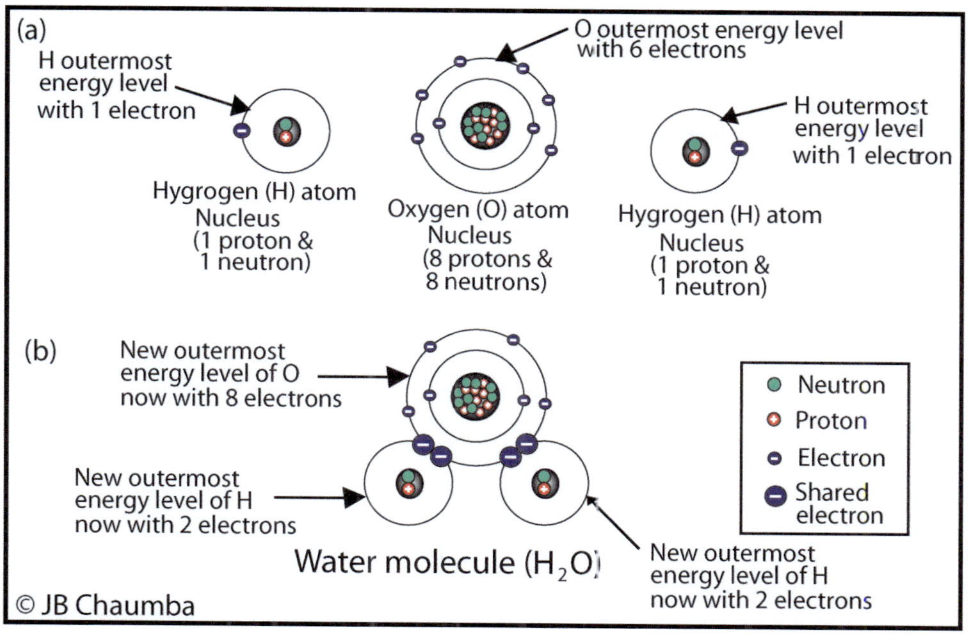

Fig. 2.7 In a water molecule (H_2O), each hydrogen atom forms a single covalent bond, whereas each oxygen atom forms two covalent bonds. The two hydrogen atoms and one oxygen atom **a**, bond to form a water molecule by sharing electrons **b**

bonds (Fig. 2.7). Thus, two pairs of electrons are shared in a molecule of water (H_2O) (Fig. 2.7). This bonding results from an uneven distribution of electrons as the oxygen atom's nuclei produces a positive charge which is stronger than that produced by the hydrogen atom's nuclei. Thus, the shared electrons spend more time near the oxygen atom's nuclei to which they are more attracted to, resulting in a weak negative charge developing on the water molecule's oxygen side due to the presence of the negatively charged electrons. The electrons, therefore, spend less time near the hydrogen atom's nuclei producing a weak positive charge on the hydrogen side of the molecule. The resultant weakly charged regions have a tendency to attract oppositely charged regions of molecules close by to form, within such molecules, weak **hydrogen bonds**. These weak intermolecular attractions, called **van der Waals bonds**, are responsible for substances that may be strongly bonded to either dissolve in water or to sufficiently bond layers of atoms or atoms together in some minerals like graphite. Thus, the manner in which atoms in minerals are bonded is important in determining the chemical and physical properties of minerals.

2.6 Properties Useful in Mineral Identification

Crystals are formed due to the bonding of ions and atoms of minerals, resulting in geometric shapes that are regular. The external expression of a crystal is caused by the orderly arrangement of atoms in the interior of the crystal to form a three-dimensional unit. The external symmetry that is expressed by crystal faces permits us to infer geometric patterns of atoms in mineral crystal structures. In order to identify minerals, the quickest and easiest way is by the use of diagnostic physical properties. A few simple chemical tests, however, are also helpful. Below, some of the basic properties commonly applied to identify minerals are reviewed.

2.6.1 Hardness

Hardness is defined as a mineral's resistance to scratching, and it is determined by comparing the hardness of the mineral to that of a known substance. A numerical mineral hardness rating scale was developed by the German mineralogist Friedrich Mohs, and it ranks from the softest mineral (talc), which is rated 1 to the hardest one (diamond) with a rating of 10 (Table 2.2). Comparing the hardness of minerals to some other common materials (Table 2.2) allows a simplified scale to be established. A mineral's hardness is an indication of the strength of its bonds, as scratching a mineral removes atoms from its surface, thereby breaking the bonds holding these atoms.

One must be careful not to confuse a scratch on a mineral with a powder mark from one's fingernail, a penny, a nail, or glass. It is important to note that a scratch cannot be removed by rubbing. It is important, too, to note that the Mohs scale is not a linear scale. Corundum, with a Mohs

Table 2.2 Mohs scale of hardness of minerals and additional observations

Mohs hardness	Mineral	Reference materials	Notes
1	Talc		Scratched with ease by fingernail; greasy feel
2	Gypsum	Fingernail (2.2)	Scratched by fingernail
3	Calcite	Copper penny (~3.2)	Barely scratched by copper coin; scratched very easily with a knife
4	Fluorite	Pocket knife (~5.1); glass plate (~5.5)	Easily scratched with knife
5	Apatite	Steel file (~6.5)	Can't be scratched with knife
6	Orthoclase	Streak plate (~7.0)	Easily scratches glass
7	Quartz		Barely scratches glass; can't be scratched with knife
8	Topaz	Masonry drill bit (8.5)	Very easily scratches glass
9	Corundum		Cuts glass
10	Diamond		Used to cut glass

hardness value of 9, for example, is twice as hard as topaz with a Mohs hardness value of 8, whereas diamond, with a Mohs hardness value of 10, is almost four times as hard as corundum (Fig. 2.8).

Any mineral can scratch itself as well as scratch any other mineral with a Mohs hardness value lower than itself. Diamond, for example, can scratch itself, and it can also scratch any other mineral. The Mohs hardness scale is compared with the absolute hardness that measured with a sclerometer is shown in Fig. 2.8. Corundum, the mineral with the second highest hardness on the Mohs scale, easily scratches glass (Fig. 2.9).

2.6.2 Cleavage

A mineral is said to have **cleavage** if it breaks along a flat and smooth surface. Cleavage surfaces appear bright and tend to reflect light all at once in a single flash. Some minerals tend to have atomic bonds that are weaker in their crystal structures. When stresses are applied to such minerals, they tend to easily break along the weak bonds, and cleavage is the tendency of these minerals to break, or cleave, along such planes of weak atomic bonds. Most mineral specimens found in laboratories are pieces broken off from larger samples. The number of cleavage planes

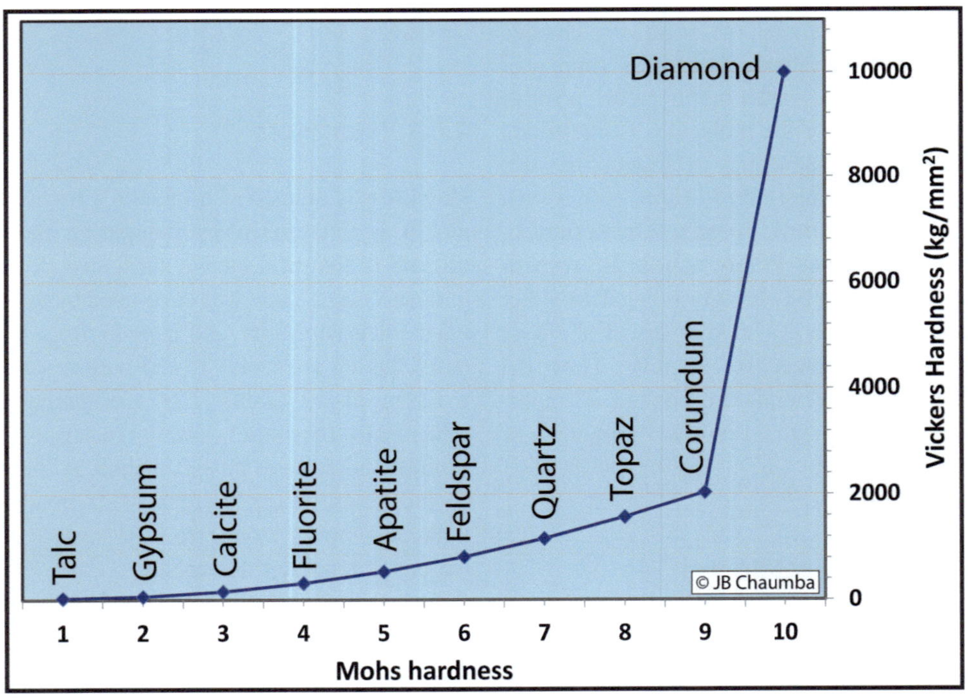

Fig. 2.8 Plot comparing the Mohs hardness scale of minerals versus the Vickers hardness of minerals. Mohs hardness refers to mineral's resistance to being scratched, whereas Vickers hardness refers to resistance to indentation under pressure

Fig. 2.9 Corundum crystals and a scratch made on glass (Mohs hardness of 5.5) with one of the corundum crystals

and the angle at which the cleavage planes intersect in minerals with two or more cleavage sets is important in mineral identification. Micas, with only one cleavage plane (Fig. 2.10), have the simplest cleavage type, and they have only one cleavage direction from the weak atomic bonds in that direction.

Some minerals like feldspars have two cleavage sets that intersect at right angles (Fig. 2.10), whereas some minerals like hornblende also have two cleavage sets, but they intersect at 120°. Still other minerals like halite have three cleavage planes that intersect at 90° (Fig. 2.10) or not at right angles, such as calcite (Fig. 2.10). The mineral fluorite has cleavage surfaces that intersect at right angles (Fig. 2.10). Cleavage surfaces may be confused with crystal faces. Crystal faces typically form regular polygonal shapes such as cubes, prisms, or octahedrons. Crystal faces are surfaces along which minerals grow, as shown by corundum crystals (Fig. 2.9).

2.6.3 Luster

The appearance of a mineral in reflected light is called **luster**, another useful property in identifying minerals. Regardless of their color, a number of minerals look like metals, and they reflect light, as do gold, chrome, or steel (and they therefore have a **metallic** luster). Minerals such as galena and native copper (Fig. 2.3a), for example, tend to develop a dark coating when they are exposed to the atmosphere resulting in surfaces not as shiny (i.e., **submetallic** luster) as metallic minerals. These minerals resemble dull metals like rusted iron. The majority of minerals have a glassy, waxy, pearly, oily, or greasy appearance, which is often not easy to separate, and these minerals are said to have a **non-metallic** luster. Halite and galena, two examples of non-metallic and metallic minerals, respectively, are shown in Fig. 2.11.

2.6.4 Streak

The color of a mineral when finely powdered is called **streak**. Streak is often useful in mineral identification as it is usually constant although the color of the mineral (such as fluorite, for example) may vary. One can determine the streak of a mineral by rubbing it on a piece of unglazed porcelain (called a streak plate) resulting in the hard and rough surface of the streak plate producing fine powder of the mineral. Both black and white ceramic tiles are commonly used at the same time to test for streak in order to obtain good streak colors. For minerals with a metallic luster (e.g., Fig. 2.11b), the colored streaks produced are usually useful properties in identifying minerals, such as in the case of graphite (Fig. 2.12). In contradistinction, white streaks, called light-colored streaks that are produced by non-metallic minerals, are generally unimportant. Minerals that are harder than the streak plate will scratch the porcelain and thus leave no streak.

2.6.5 Color

Color is typically the first characteristic that is observed during mineral identification. Unfortunately, color is the most unreliable diagnostic property because a single mineral can exist in many different colors, and different minerals can have the same color. A good example is quartz, which can be smoky, milky, clear, yellow, rose, brown, or violet (called amethyst) (e.g., Fig. 2.13a). Fluorite is another mineral which occurs in several different colors (Fig. 2.13b). Color variations are often caused by impurities in the mineral's crystalline structure, and minerals containing iron, for example, are often dark, whereas those that contain high concentrations of aluminum are light in color. Further, chemical impurities also tend to give color to minerals.

2.6.6 Density/Specific Gravity

The weight or mass of a material per unit volume is called **density**. Units of density are expressed as g/cm^3. The standard for comparison, pure water at 4 °C, has a density of 1 g/cm^3. Minerals range in density from 2.0 g/cm^3, i.e., twice as dense as water, to 15 g/cm^3 or higher although most minerals have densities that fall in the 2.5–2.8 g/cm^3 range. For example, quartz, one of the common minerals on the continents, has a density of 2.65 g/cm^3. Gold (Fig. 2.3b), which has the highest density of 19 g/cm^3, tends to also be diagnosed based on this property, among others.

Fig. 2.10 Images of some minerals with cleavages showing the number of cleavage planes or directions, if more than unity, together with the angles at which those cleavage planes intersect. **a** Muscovite and **b** biotite both have a perfect basal cleavage plane typical of micas. **c** Galena has three good cleavage directions intersecting at 90°, **d** plagioclase feldspar has two good cleavage directions intersecting at right angles, **e** augite, a pyroxene, has two good cleavages also intersecting at right angles, **f** Hornblende, an amphibole, as two good cleavage planes which intersect angles of 60 and 120°, **g** calcite has 3 perfect cleavage planes which do not intersect at right angles whereas fluorite (h) had four good cleavage planes which also don't intersect at right angles

2.6.7 Crystal Habit/shape

A **crystal** is a regular geometrical shape that is developed when atoms and ions form bonds to produce a mineral. **Crystal habit** or shape is a typical or common shape of certain minerals, and it can be a useful characteristic of these minerals. A common crystal shape can be useful in identifying some minerals. The mineral pyrite (e.g., Fig. 2.14a), for example, has a tendency to occur in numerous characteristic crystal shapes and habits such as cubes, pyritohedrons, and octahedrons (or any combinations of these forms) and penetration twins. For other minerals, at

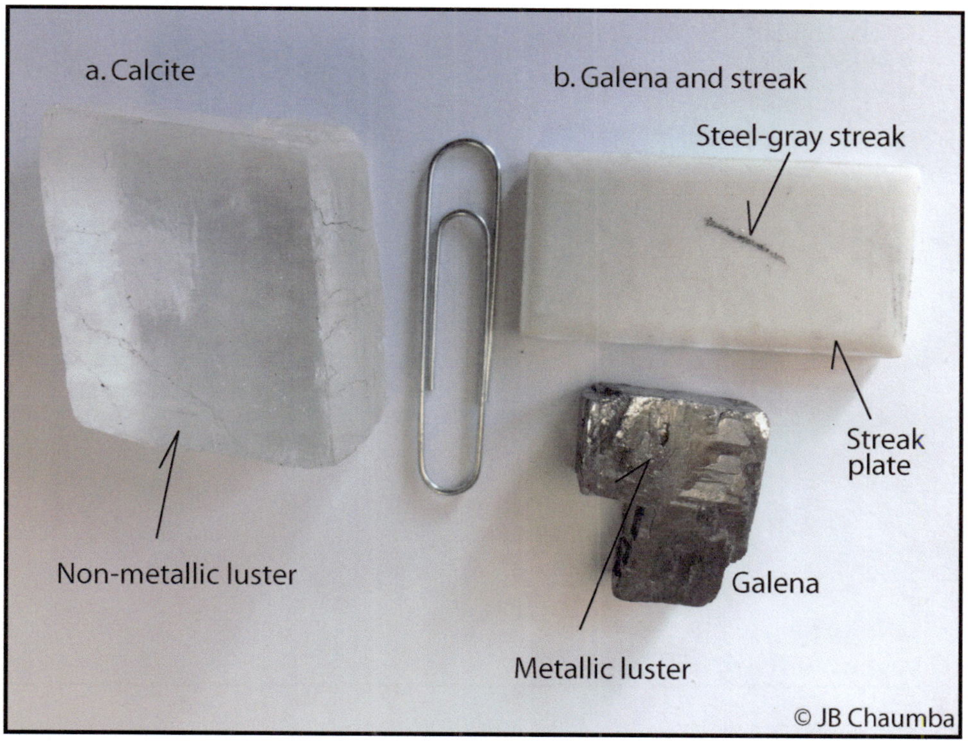

Fig. 2.11 Image of the mineral halite which has a non-metallic luster **a** and galena **b** which has a metallic luster. The steel-gray streak of galena can also be seen in **b**

Fig. 2.12 Graphite specimens and their characteristic silver-gray to black streak

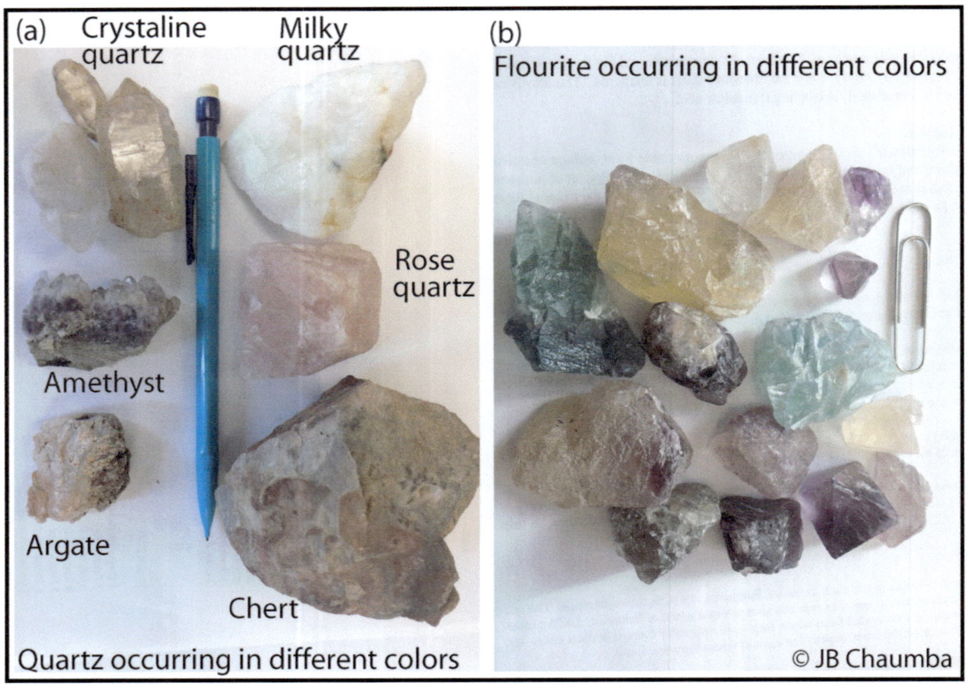

Fig. 2.13 Images of specimens of quartz **a** and fluorite **b** occurring in several different colors. Using color alone to identify either of these minerals can be easily misleading

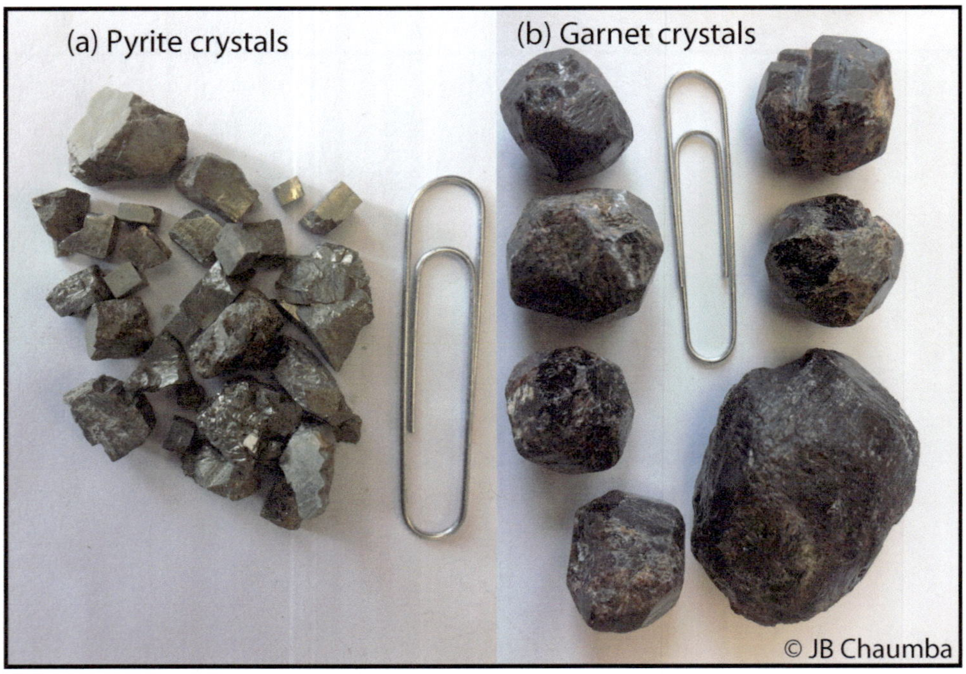

Fig. 2.14 A. Crystal of pyrite tend to occur in several different shapes **a** such as cubes, pyritohedrons, and octahedrons, whereas crystals of garnet tend to occur in dodecahedral shapes **b**. Both pyrite and garnet lack cleavage and exhibit irregular fracture

least two different shapes can be the characteristic shape of an individual or aggregates of crystals. Fluorite and halite crystals occur in cubic shapes, whereas garnet crystals tend to form dodecahedron shapes and lack cleavage (Fig. 2.14b). Further, some minerals are elongated in one direction, some have a tendency to grow in sheets, and still others tend to have equal dimensions in three dimensions, for example.

2.6.8 Fracture

A **fracture** is formed when a mineral breaks along an irregular or bumpy surface. Thus, a fracture is any mineral break that is not cleavage. Since a fracture surface is rough, it usually does not reflect light as well as a cleavage surface. Glass, for example, breaks along a smooth but curved surface to produce a conchoidal fracture. Obsidian and quartz typically produce conchoidal fractures upon breaking (Fig. 2.15).

Fig. 2.15 Conchoidal fracture exhibited by obsidian **a** and rose quartz **b**

2.6.9 Miscellaneous Properties

Some minerals which have a high proportion of iron, such as magnetite, tend to attract or be attracted by a magnet, and can be easily identified using a magnet (Fig. 2.16). Lodestone, a variety of magnetite that is highly magnetic, can easily attract tiny pieces of objects made of iron such as paper clips. Other minerals that are made up of carbonate ions (CO_3^{2-}) readily fizz, or **effervesce**, i.e., to produce carbon dioxide gas when a drop of dilute hydrochloric acid is applied to them, especially on a freshly broken surface. The carbon dioxide gas produced is observed as fizzing or bubbling on the surface (Fig. 2.17).

Some minerals such as calcite exhibit an optical property called **double refraction** whereby a ray of ordinary that enters them is split into two which then travel in different directions. Still other minerals can be identified by how they taste. Halite, for example, can quickly be identified through its salty taste when one licks a freshly broken surface of the mineral. A mineral-like talc has a soapy feel, whereas graphite has a greasy feel. Another mineral-like sulfur can easily be diagnosed because of its rotten egg smell; the same smell that is felt on the streak of many sulfur-bearing minerals.

2.7 Mineral Groups

We have already noted that in excess of 4000 minerals have already been described. Studying them all would be a daunting task! However, this job falls to mineralogists and for our purposes we only concern ourselves with a few dozen that are commonly encountered. We refer to these few abundant minerals as **rock-forming minerals**. In addition, we also have some economic minerals that are very useful that are often studied with the rock-forming minerals. Out of the ~90 naturally occurring elements, the lion's share of the rock-forming minerals is composed of only eight elements: oxygen, silicon, aluminum, iron, calcium, sodium, potassium, and magnesium. The abundances of these eight rock-forming minerals in the Earth's continental crust are oxygen = 46.6%, silicon = 27.7%, aluminum = 8.1%, iron = 5.0%, calcium = 3.6%, sodium = 2.8%, potassium = 2.6%, and magnesium = 2.1%, with other minerals constituting the remaining 1.5% (Fig. 2.18). Oxygen and silicon, being the two most abundant of the eight major elements (Fig. 2.18), tend to easily bond to form what is called the **silicon-oxygen tetrahedron**.

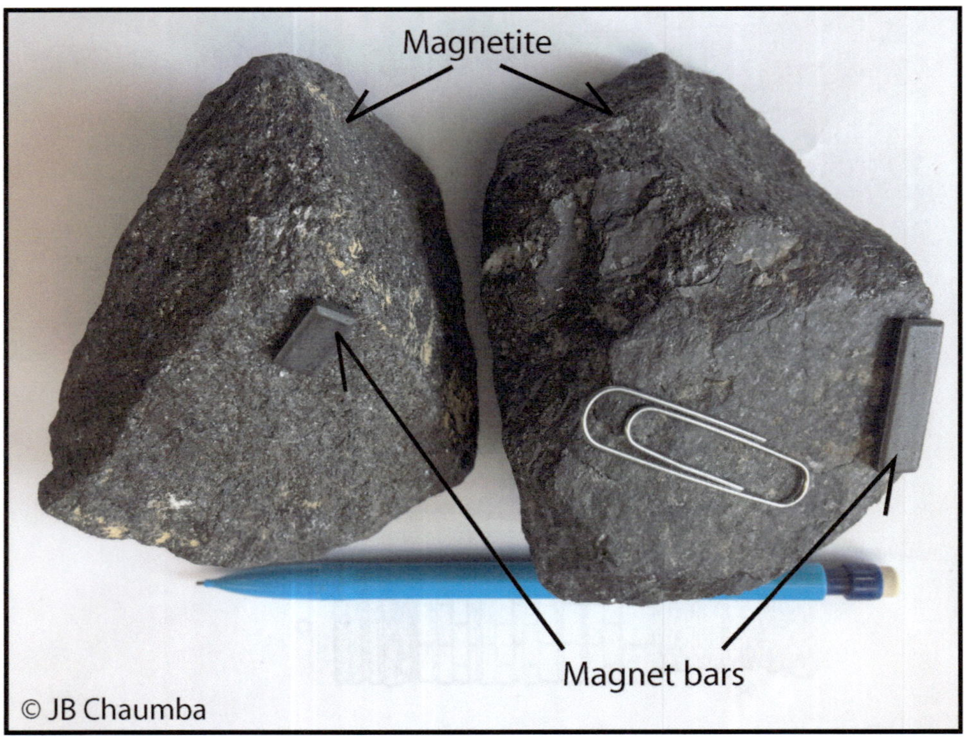

Fig. 2.16 The mineral magnetite is magnetic, and it tends to attract magnetic substances

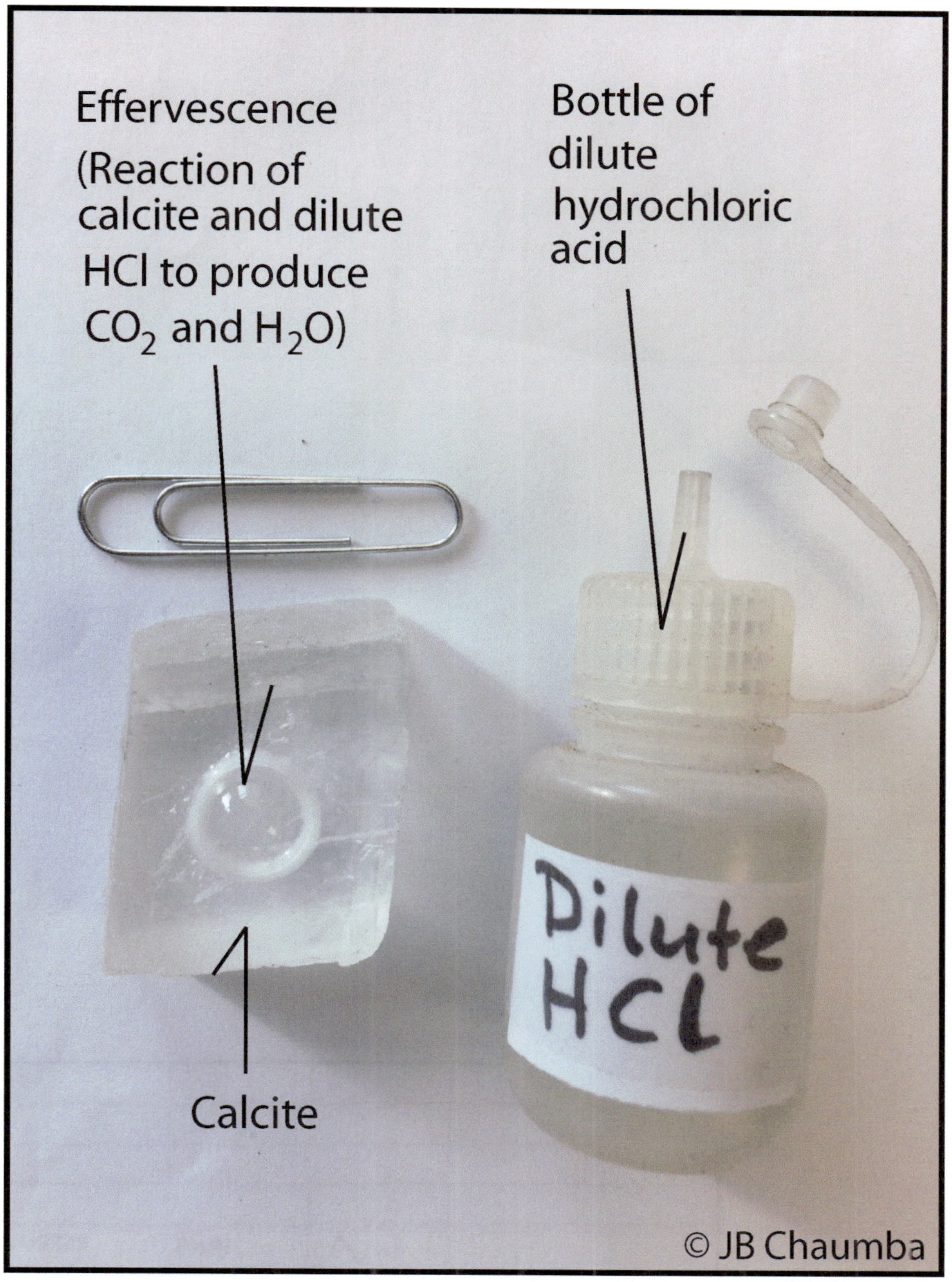

Fig. 2.17 Calcite tends to react with dilute hydrochloric acid to produce CO_2 gas (termed effervescence or bubbling)

2.7.1 Silicate Minerals

The silicon tetrahedron, a combination of four oxygen atoms and one silicon atom (Fig. 2.19) is the basic building block of **silicate minerals**. In excess of 800 silicate minerals have been described, and the significance of these silicate minerals can be borne in mind from their ubiquity: they account for over 90% of the crust of our planet. Examples of silicate minerals include olivine, quartz, and feldspar. In the silicon tetrahedron, planes that are drawn through the oxygen atoms form a tetrahedron (Fig. 2.19a). An oxygen ion has a charge of –2, whereas a silicon ion has a charge of + 4 (Fig. 2.19a). Since there are 4 oxygen ions and one silicon ion in the silicon tetrahedron, this results in a net charge of –4 for the silica tetrahedron (Fig. 2.19).

The silicon-oxygen tetrahedra are arranged and linked together in a variety of ways in different silicate minerals. Shown in Fig. 2.20 are the other major silicate groups and an example of one of the common minerals in each group. Single silicon-oxygen tetrahedra units can be arranged and linked together to form very complex frame-

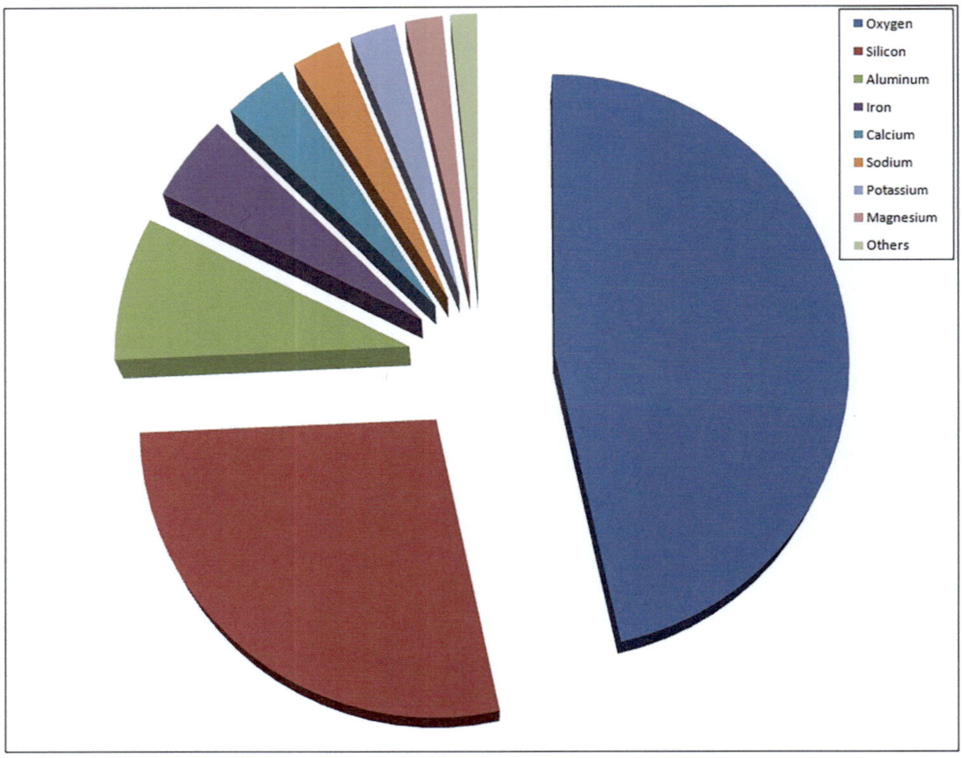

Fig. 2.18 Pie chart showing the 8 major elements that constitute the continental crust of the Earth

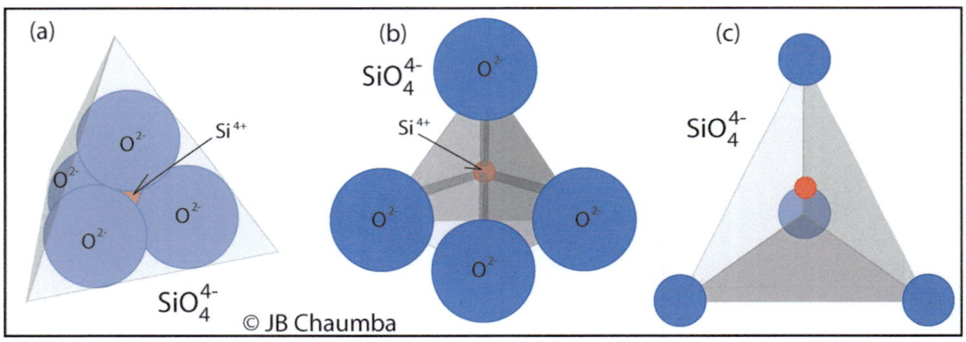

Fig. 2.19 Three representations of the silicon-oxygen tetrahedron (SiO_4^{4-}), the fundamental building block of silicates. **a** ball model, **b** ball-and-stick model, **c** expanded ball model. The relatively smaller silicon atom occupies the center of the tetrahedron, whereas the much larger oxygens occupy the corners

works (Fig. 2.20). The simplest silicate structure that can be formed is when isolated silicon-oxygen tetrahedra are bonded to magnesium and/or iron ions to form the mineral olivine (Fig. 2.20). In olivine, the silicate structure is joined by elements such as magnesium and iron (Fig. 2.20). In other silicates, other elements such as calcium, sodium, and potassium can also join the silicate structures. By sharing oxygen atoms at the corners of the silicon-oxygen tetrahedra, several different ways in which the silicon-oxygen tetrahedra are joined into chains, sheets, and three-dimensional frameworks are possible, and some of them are also shown in Fig. 2.20.

Single chains are formed when two oxygen atoms are shared to link the silicon-oxygen tetrahedra whereas in double chains two or three oxygen atoms are shared to join the silicon-oxygen tetrahedra (Fig. 2.20). In sheet silicates, three oxygen atoms are shared by the silicon-oxygen tetrahedra to form two-dimensional sheets (Fig. 2.20). The space between sheets tends to be occupied by water molecules or other ions.

Some framework silicates such as feldspar and quartz are very important in the continental crust, and all four oxygen atoms are shared with oxygen atoms that form neighboring silicon-oxygen tetrahedra. Being the most abundant mineral

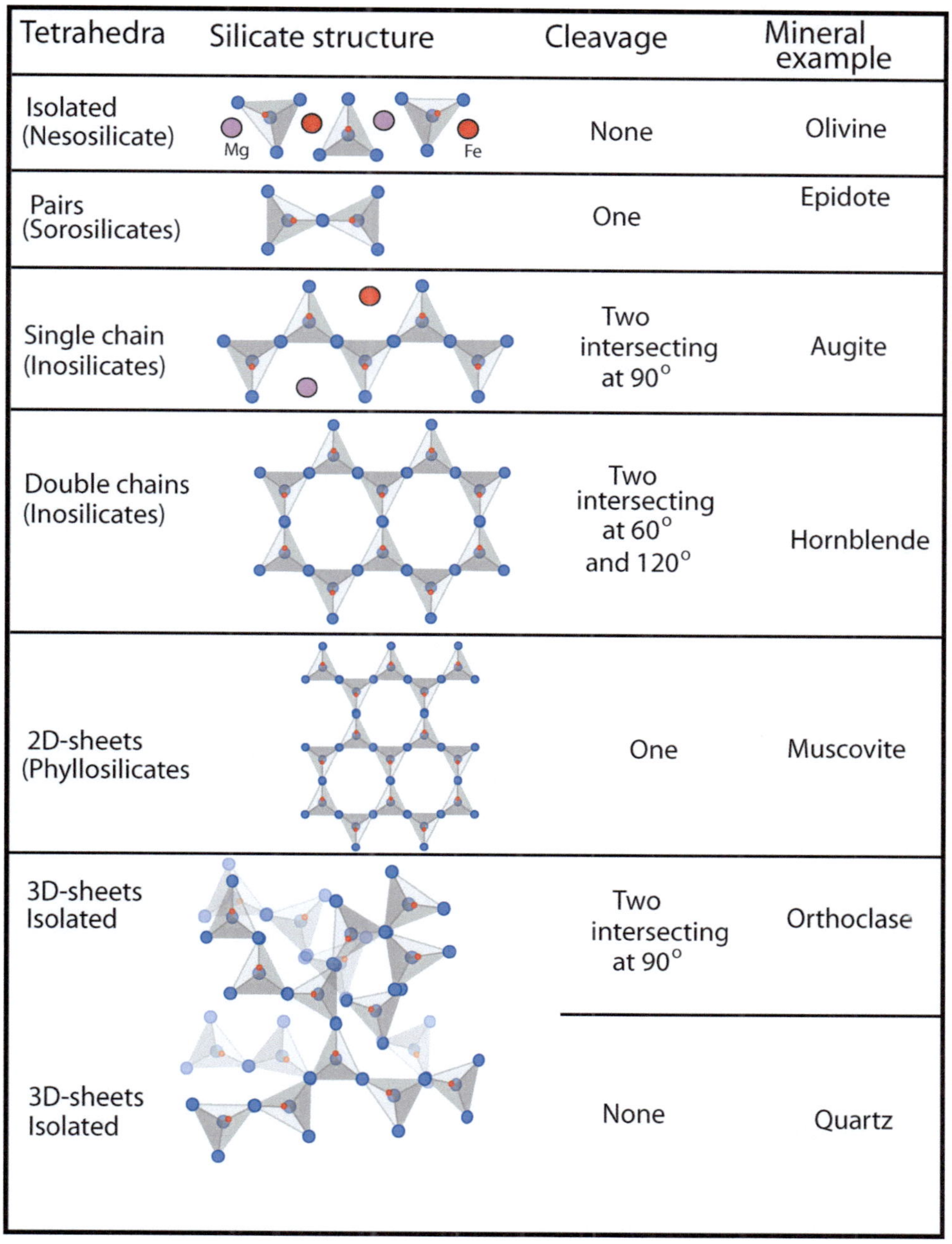

Fig. 2.20 Diagram showing the silicate structure, cleavage, and an example of some of the common silicate minerals. The silicate structure increases in complexity from the top to the bottom of the chart. Note that for 3D-sheets, the schematic structure shown is not representative of any mineral

in the crust of the Earth, plagioclase is a very important rock-forming mineral. Quartz, one of the minerals that is more resistant to weathering, helps form the sandy beaches on most coastlines and on most riverbeds, among other places.

Also included in Fig. 2.20 is information on the presence of cleavage planes, number of cleavages, and the angles at which the cleavage planes intersect where more than one cleavage plane exists. Since each mineral group is characterized by a particular silicate structure, the internal structure of the mineral is therefore reflected in the cleavage of that mineral. The cleavage in silicate minerals forms when the mineral tends to break between different tetrahedra rather than across it due to the strong silicon-oxygen bonds. Quartz, for example, has three-dimensional silicon-oxygen tetrahedral structures and thus has strong silicon-oxygen bonds in all three directions.

Table 2.3 Some common non-silicate minerals

Mineral group	Anion/element	Chemical formula	Mineral name
Native elements	Single element	Cu	Copper
Carbonate	CO_3^{2-}	$CaCO_3$	Calcite
Halide	Cl^-, F^-, Br^-	NaCl	Sodium chloride
Sulfates	SO_4^{2-}	$CaSO_4$	Anhydrite
Oxides	O^{2-}	Fe_2O_3	Hematite
Sulfides	S^{2-}	FeS_2	Pyrite

A sheet silicate such as muscovite, for example, has a sheet structure due to the sheeted bonds of the silicon-oxygen tetrahedra and thus tends to form a basal cleavage between the sheets but not across them. The perfect cleavage occurring in micas results from weak bonds occurring between plates called van der Waals bonds. Some important sheet silicates such as clays incorporate water in their structures, and the water lubricates the sheets, thereby making clays easy to work with in pottery. The final process in pottery is when heat is applied, and this heat drives the water off to form durable and rigid structures such as pots and plates.

Some silicates are light colored, and examples of common light-colored silicates are quartz, muscovite, and feldspars—the most common minerals in the crust of the Earth. Due to its resistance to weathering, quartz is the most common mineral on the Earth's surface. Pure quartz lacks other ions that can absorb light, and it therefore tends to be transparent. However, quartz tends to trap other ions when it crystallizes, hence its wide range of colors. Examples of common dark colored silicates are biotite, augite (a pyroxene), hornblende (an amphibole), and olivine.

2.7.2 Common Non-Silicate Minerals

Non-silicate minerals, constituting only ~ 8% of the Earth's continental crust, include native elements (e.g., Fig. 2.3), **carbonates**, **oxides**, **sulfides**, **sulfates**, and **halides** (Table 2.3). A number of negatively charged ions or complex ion groups comprise the different non-silicates (Table 2.3). Native elements are those elements which occur not naturally combined with others such as silver, platinum, diamonds, graphite, and copper (e.g., Table 2.3). Metallic bonds are responsible for bonding the metal atoms in most native elements.

In carbonates (Table 2.3 and Fig. 2.21), the anionic group is CO_3^{2-}, and calcite and dolomite are formed when carbon and/or magnesium bond with this CO_3^{2-} anionic group (Table 2.3). Calcite ($CaCO_3$) is a common carbonate mineral. In halides (Table 2.3 and Fig. 2.22a), the anion can be chloride (Cl^-), fluoride (F^-), or bromide (Br^-) (i.e., halogens) and when sodium and calcium bond with these halides, table salt (NaCl), fluorite (CaF_2), and caesium bromide (CsBr), respectively, are formed.

The anionic group SO_4^{2-} is bonded by metal cations in sulfates. The sulfate anhydrite ($CaSO_4$) is formed when Ca is bonded with SO_4^{2-}. Gypsum ($CaSO_4 \cdot 2H_2O$) (Fig. 2.22b) is a soft hydrated calcium sulfate. Most sulfates tend to form when they precipitate from solutions at or near the surface of our planet. When metal cations are bonded to O^{2-} anions, oxides (Table 2.3 and Fig. 2.23) are formed. The oxide hematite (Fe_2O_3) is formed when Fe^{2+} is bonded to O^{2-} (Table 2.3). Finally, in sulfides (S^{2-}), the sulfide anion is bonded by metal cations. The most common sulfide (Table 2.3 and Fig. 2.24), pyrite (FeS_2), for example, is formed when Fe^{2+} is bonded to S^{2-} (Table 2.3).

Fig. 2.21 Examples of carbonates: **a** calcite, **b** crystals of calcite from Blue Ball Quarry, Blue Ball, Pennsylvania, USA, and **c** siderite ($FeCO_3$) from Roxbury, Connecticut, USA

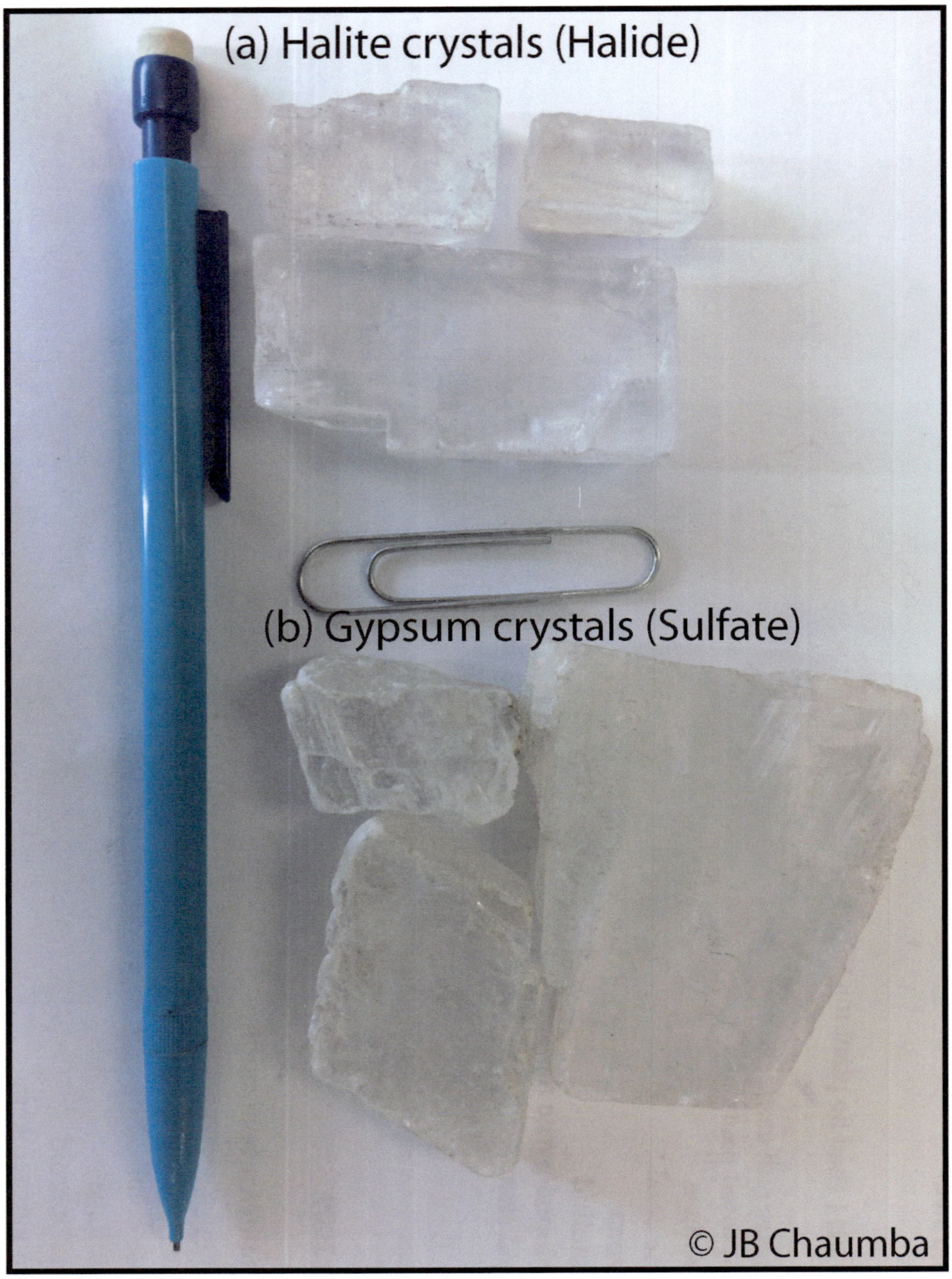

Fig. 2.22 **a** shows halite, an example of a halide whereas **b** shows gypsum, an example of a sulfate

Fig. 2.23 Examples of oxides: **a** botryoidal habit hematite or "kidney" ore, **b** fossiliferous hematite from Rochester, New York, USA, **c** corundum crystals, and **d** banded iron formation

Fig. 2.24 Some examples of sulfides: **a** bornite, **b**, pyrite from Gilpin, Colorado, USA, and **c** chalcopyrite from Rouyn District, Quebec, Canada

Glossary

Anion a negatively charge ion

Atom the smallest particle of an element that retains its chemical properties

Atomic mass the total number of both protons and neutron located within the nucleus of an atom

Atomic number the number of protons located in the nucleus of an atom

Bond to combine and form chemical compounds

Carbonates group of minerals whose anionic group is CO_3^{2-}

Cation a positively charge ion

Chemical compound a substance that is produced from combination of two or more elements in specific proportions. Compounds are different from the properties of their constituent elements

Cleavage tendency of minerals to split along planes of weakness due to weak bonds along those surfaces

Color an object's property of producing different sensations on the eye due to the way it reflects or emits light.

Covalent bond bonding produced by sharing of electrons in atoms with about half-filled electrons in their outer energy levels

Crystal a regular shape that is developed due to bonding of atoms and ions to form a mineral

Crystal habit shape that is common to, or characteristic of, certain crystals or crystal aggregates

Density a measure of a mineral's weight per given volume, typically expressed in g/cm^3

Electrons negatively charged particles with negligible masses that orbit around the nucleus of an atom

Element a form that cannot be broken down into a simpler form

Energy level specific regions occurring around the nucleus of an atom where electrons reside

Fracture minerals with strong bonds in all directions tend to break randomly to form curved or jagged surfaces

Halides group of minerals whose anionic group is Cl^-, F^-, Br^-, for example.

Hardness the resistance of a mineral to scratching

Hydrogen bond a bond formed due to attraction of nearby regions of oppositely charged regions such as that between the negative weakly charged region of the oxygen atom's side and the weakly charged positive region of the hydrogen atoms' nuclei in water

Intermolecular bond a bond formed as a result of an uneven distribution of atoms

Ion a charged particle of an atom that has either gained or lost electron(s)

Ionic bond a bond formed due to the attraction between oppositely charged ions

Isotope atoms of the same element possessing the same number of protons but different number of neutrons in their nuclei

Luster appearance of light that is reflected from a mineral surface

Metallic bond a bond formed due to the attraction of a negatively charged electron "cloud" and clusters of positively charged nuclei

Mineral an inorganic and naturally occurring substance with a specific chemical composition

Mineraloid a substance that possesses a relatively constant composition that resembles a mineral but it lacks an orderly arrangement of atoms to give it a specific crystal structure

Native element a mass of a single metal occurring naturally

Non-silicate group of minerals that lack silica in their structures and comprise ~ 4% of the crust of the Earth

Nucleus small, heavy, and central part of an atom containing most of the atom's mass and positive charge

Neutron a particle with a mass of 1.674×10^{-24} g but with no charge that is located in the nucleus of an atom

Octet rule atoms have a tendency to combine so that each atom has eight electrons in its outermost shell

Oxides group of minerals whose anionic group is O^{2-}

Proton a particle with a mass of 1.672×10^{-24} g and a positive charge that is located in the nucleus of an atom

Rock-forming minerals one of a few dozen minerals that compose the Earth's crust and mantle

Silicates group of minerals whose anionic group is SiO_4^{4-} and comprise ~ 96% of the Earth's crust

Silicon-oxygen tetrahedron four-faced and pyramidal structure formed from one silicon cation at the center surrounded by four oxygen anions that is the building block of silicate minerals

Streak the color of a mineral in powdered form which is constant for any mineral

Sulfates group of minerals whose anionic group is SO_4^{2-}

Sulphides group of minerals whose anionic group is S^{2-}

Tetrahedron see **Silicon-oxygen tetrahedron**

Valence electrons electrons that occupy the highest energy level of an atom and that are involved in forming chemical compounds

Van der Waal bond a weak intermolecular attraction formed from a slightly negative side caused by electrons that temporarily group on the same side of an atom's nucleus and attract nuclei from neighboring atoms, whereas the electron-poor side with a positive charge can briefly attract electrons of neighboring atoms

3 Origin of Magma and Igneous Rocks

Abstract

The origin of magmas is described with the use of average geothermal gradients for the upper mantle. Classification of igneous rocks and magmas is accompanied by the use of tables and photographs of the several types of common igneous rocks. Bowen's reaction series and magmatic differentiation are also described and illustrated by means of diagrams. Various types of melting are discussed using illustrations, followed by descriptions of plutonic, intermediate, and extrusive igneous rocks. A discussion of the economic importance of igneous rocks concludes the chapter.

Learning Goals

Studying this chapter will enable you to:

- Define geothermal gradient
- Describe the various ways through which magmas can be generated
- Describe how melting occurs and why magma moves toward the Earth's surface
- Differentiate between plutonic and volcanic rocks, and provide examples of each
- Describe vesicular igneous rocks and how they were formed
- Classify igneous rocks based on texture, composition, and environment of formation
- Explain how grain size of igneous rocks is influenced by rate of cooling
- Differentiate magma from lava and list their characteristics
- Describe how magma viscosity affects the rate of magma flow toward the Earth's surface.

3.1 Introduction

Some of the minerals that we discussed in Chap. 2 can combine in differing proportions to make up rocks. A **rock** can be defined as a coherent solid that occurs naturally, and it is composed of mineral aggregates or, rarely, glass. Some rocks, however, are made up of non-mineral matter. Examples of rocks which are not made up of minerals are coal, obsidian, and pumice. Coal is organic, whereas obsidian is volcanic glass. Pumice is a very light rock material that contains tiny gas bubbles formed by the hardening of lava with high contents of **volatiles** (water and gases) discharged by a volcano.

Studying rocks is of importance since they contain lots of clues about processes that occurred during the history of planet Earth and other similar planets as well as clues on the environments under which the rocks were formed. In addition, knowledge of the different types of rocks, their composition, as well as where they occur on our planet, is of importance in locating mineral, rock, and energy resources on Earth and similar planets. If heat is continually applied to any rock, that rock will eventually melt to form **magma**. But how do magmas form and where do they come from? Let us briefly discuss this next.

3.2 Generation of Magmas

3.2.1 Where Do Magmas Come from?

We discussed in Chap. 1 that the Earth's outer core is the only part of our planet that is molten. However, being silicate melts, magmas are not thought to originate from the outer core as they are composed mostly of Fe with some

Supplementary Information The online version contains supplementary material available at https://doi.org/10.1007/978-3-032-20231-4_3.

J. B. Chaumba, *Introductory Physical Geology*, Springer Textbooks in Earth Sciences, Geography and Environment,
https://doi.org/10.1007/978-3-032-20231-4_3

Ni, and thus it lacks the correct chemical composition to be the source of magmas. Similarly, within the ocean basins, magmas are also unlikely to be generated by melting of the oceanic crust as most magmas erupted in the ocean basins are basaltic in composition. It would require nearly total melting to generate mafic magmas by melting of the mafic crust, which is very unlikely.

Both mafic and felsic magmas are erupted on, and intruded into, the continental crust. Mafic magmas are unlikely to originate from the continental crust, given its felsic average composition. More felsic magmas, such as andesitic and rhyolitic magmas, however, may originate from melting of the continental crust. Mafic magmas, therefore, are thought to be generated from the mantle. With the exception of the magmas generated within the continental crust, magmas are generally thought to originate from melting of mantle peridotite in the mantle.

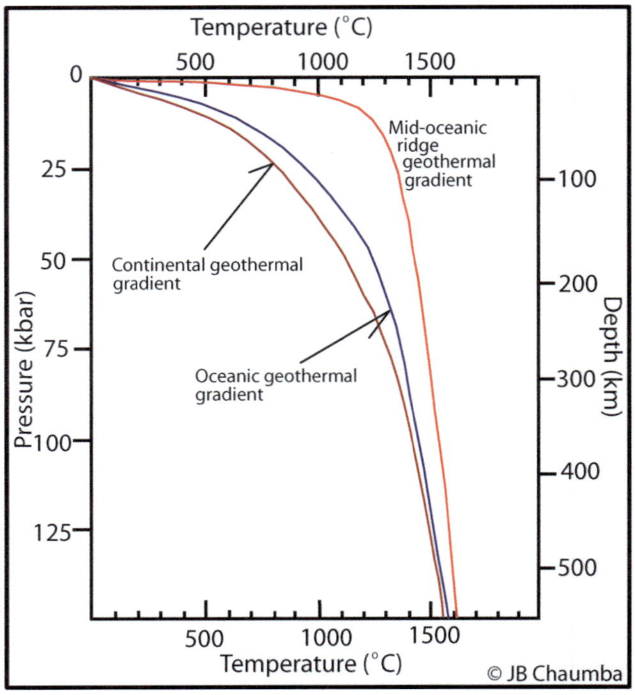

Fig. 3.1 Plot showing representative average geothermal gradients for the upper mantle from the surface to a depth of ~500 km for continental lithosphere, for oceanic lithosphere, and for mid-ocean ridge tectonic settings

3.2.2 How Magmas Are Generated

Since we do not have magmas being erupted everywhere on our planet, special cases must be required to cause melting to occur. The variation of temperature with depth or pressure within the Earth, called the normal **geothermal gradient**, varies depending on whether one is on the continents or on oceanic crust, as shown in Fig. 3.1. At shallow levels, the normal geothermal gradient is high at mid-ocean ridges, and slightly higher in the oceanic crust than in the continental crust (Fig. 3.1). Based on seismic evidence, the mantle is inferred to be solid under normal conditions.

To generate a melt, either the geothermal gradient must be raised so that it is above the peridotite solidus, or the temperature of the peridotite solidus must be reduced. In either scenario, we have to get the temperature, which is expressed by the geothermal gradient, into the partial melt region (Fig. 3.2b). Thus, a small increase in temperature, or a slight decrease in the peridotite solidus, is likely to occur and may result in limited degrees of melting, called partial melting (Fig. 3.2b). Once a partial melt has formed, the liquid portion can be easily separated from the remaining solids since liquids generally have a lower density than solids, and they are also more mobile.

Let us now examine the conditions under which partial melting can occur.

3.2.3 Raising the Geothermal Gradient

Radioactive heat—Elements like U, Th, K, and Rb have radioactive isotopes. During radioactive decay, sub-atomic particles are released by the decaying isotope and move outward until they collide with other atomic particles. Upon collision, the kinetic energy of the moving particles is converted to heat. If this heat cannot be conducted away, then the temperature will rise. Most of the heat within the Earth is generated by radioactive decay, and this is the general reason why temperature increases with depth in the Earth. Most radioactive isotopes, however, are concentrated in the crust. Although there are some areas in the continental crust where high concentrations of radioactive elements have locally raised the temperature, at least high enough to cause metamorphism, this is a rare occurrence. It is even more unlikely that areas of high radioactive element concentrations develop within the mantle. Thus, concentrations of radioactive elements are not likely to cause melting.

3.2.4 Decompression Melting

Decompression melting occurs as a result of convection. Convection can be defined as a type of heat transfer whereby the heat moves together with the material. If the temperature gradient is high enough that material at depth expands so that its density is lower than the material above it, this can induce convection. This creates an unstable situation, and the lower density and hotter material rises, and it is then replaced by descending cooler material which generates convection currents or cells.

In an area where convection currents rise, hotter material at depth rises, carrying with it its heat. As convection currents rise to lower pressure (decompression) (Fig. 3.2b),

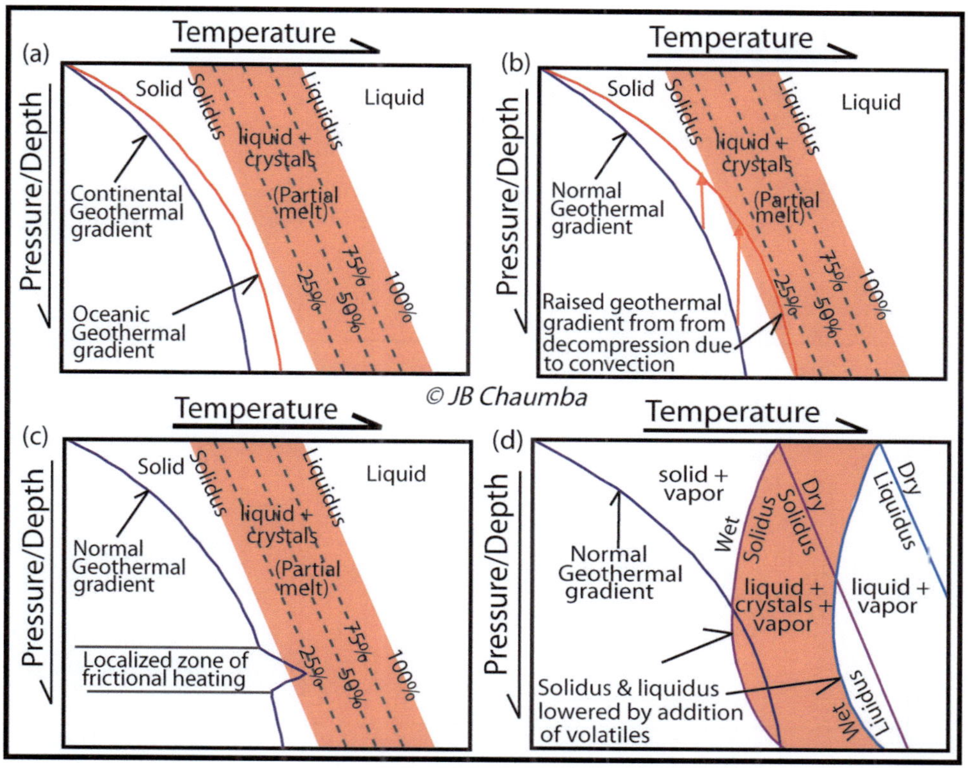

Fig. 3.2 **a** Plot showing the variation of temperature with pressure (or depth) of both continental and oceanic geothermal gradients. Note that both continental and oceanic geothermal gradients do not intersect the peridotite solidus, indicating that under normal circumstances no melting occurs. **b** decompression melting caused by convection currents, whereby hotter material at depth may rise carrying its heat with it. **c** localized rise in temperature within the zone where sliding or shearing causes a localized spike on the geothermal gradient may cause local temperatures to rise above the solidus to initiate melting, **d** Flux melting occurs when the temperature of the liquidus is lowered, without having to raise the geothermal gradient, to such an extent that solidus temperature becomes lower than the geothermal gradient resulting in partial melting

they tend to cool, but they still have higher temperatures than their surroundings. Thus, decompression causes the local geothermal gradient to be raised (Fig. 3.2b). If this new geothermal gradient reaches temperatures that are greater than the peridotite solidus, partial melting, and thus the generation of magma, occurs.

Volatile components such as water and carbon dioxide, because of their high mobility, may be easily added to the mantle, particularly at **subduction zones**, where an oceanic lithosphere is pushed beneath continental lithosphere or sometimes beneath another oceanic lithosphere. This addition of volatiles can occur because the oceanic lithosphere is in contact with seawater. In addition, water may be in the oceanic crust both due to weathering which can produce hydrous minerals such as clay minerals, which contain water, and carbonate minerals, which contain carbon dioxide, and may also be in the pore spaces in the subducted oceanic lithosphere. Further, oceanic sediments may be deposited on the basaltic oceanic lithosphere which is produced at oceanic ridges. As the oceanic lithosphere descends into the mantle at subduction zones, its temperature increases with depth, leading to metamorphism of the subducted oceanic lithosphere, including the sediments. Addition of fluids such as water and carbon dioxide, either to the descending lithosphere or to the mantle overlying the subducted lithosphere, tends to lower the solidus and liquidus temperatures enough, causing partial melting to occur (Fig. 3.3c, d).

It is thought that in the continental crust the normal geothermal gradient will be high enough to cause melting despite the occurrence of hydrous and carbonate minerals occurring in many rocks of the continental lithosphere. In addition, continental rocks are at lower temperatures than those in the mantle, and they have much higher viscosities, convective decompression is unlikely to occur in the continental lithosphere rocks. Evidence exists that rocks of the continental lithosphere sometimes melt through a process called **crustal anatexis**.

Crustal anatexis may occur when basaltic magmas, which are generated in the mantle by decompression melting, frictional heat, or flux melting, rise into the continental lithosphere carrying heat with them. These basaltic magmas have a higher density than the continental crust, and they may not reach the surface but instead intrude and cool slowly

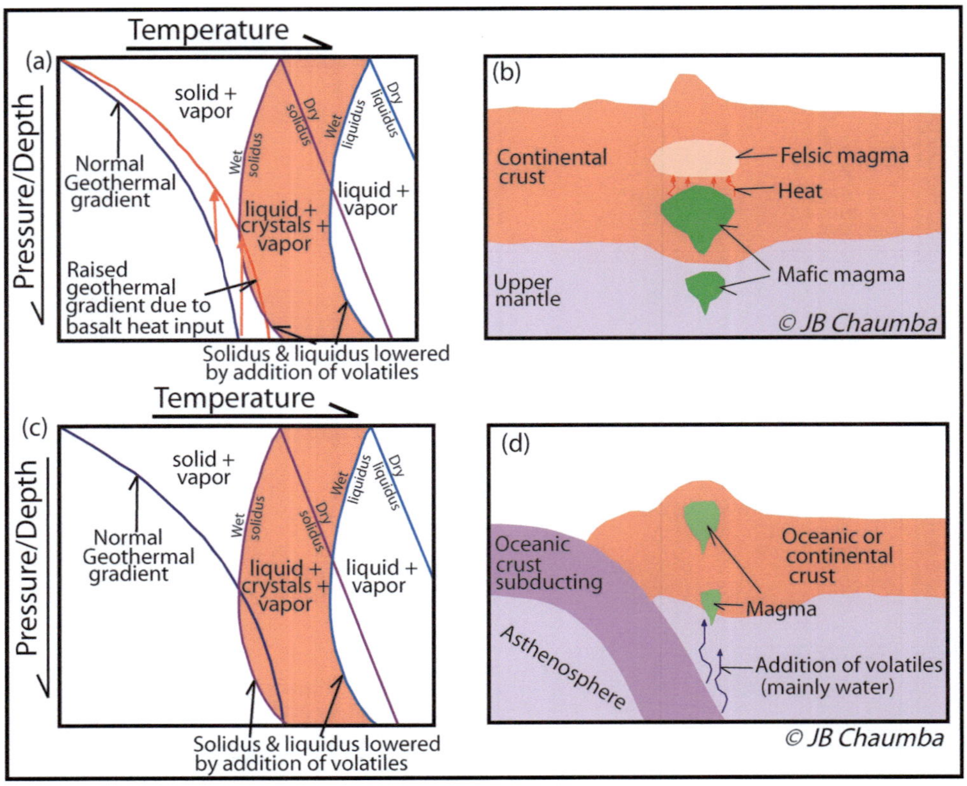

Fig. 3.3 **a** Decompression melting caused by convection currents whereby hotter material at depth may rise carrying its heat with it. **b** diagram illustrating anatexis melting caused by magma rising into the continental lithosphere carrying heat with them. **c** The temperature of the liquidus is lowered, without having to raise the geothermal gradient resulting in the solidus temperature being lower than the geothermal gradient which results in partial melting. **d** As the oceanic lithosphere subducts into the mantle at subduction zones, its temperature increases with increasing depth. Additional fluids such as water and carbon dioxide, either to the subducted lithosphere or the mantle overlying the subducted lithosphere tend to lower the solidus and liquidus temperatures enough, initiating partial melting

at depth. Upon cooling, these basaltic magmas release heat into the crust, raising the geothermal gradient, i.e., increase the local temperature (Fig. 3.3a, b). Repeated intrusions of the mantle-derived basaltic magmas into the same area of the continental crust tend to cause further increases in temperature, and this eventually causes the geothermal gradient to become higher than the wet solidus of the crustal material, leading to partial melting of the continental crust (Fig. 3.3a, b).

3.2.5 Frictional Heat

In areas where rocks slide past one another, such as at the base of the lithosphere or at subduction zones, heat could be generated by friction. If this heat cannot be conducted away fast enough, then it may cause a localized rise in temperature within the zone where the sliding or shearing is taking place. This could cause a localized spike on the geothermal gradient that could cause local temperatures to rise above the solidus.

3.2.6 Lowering the Temperature of Solidus

Addition of volatiles such as water and carbon dioxide to the mantle may cause the solidus and liquidus temperatures to be lowered (Fig. 3.2d). When this happens, the temperature of the liquidus may be lowered, without having to raise the geothermal gradient, to such an extent that the solidus temperature may be lower than the geothermal gradient which may result in partial melting (Fig. 3.2d). This type of melting is known as **flux melting**.

Once magmas have been generated by partial melting, they can easily ascend since fluids are more mobile and they generally have lower densities than solids in which they are formed, and those overlying the newly formed magmas. The magmas rise and cool in the process due to reduction in temperature of the Earth with decreasing depth. Further, the magmas that can be trapped within the crust cool much more slowly, due to the loss of heat to surrounding rocks, until it eventually solidifies. The magma that reaches the Earth's surface, however, quickly loses its heat due to coming in contact with either water or air which are characterized by much lower temperatures than that of the magma.

3.3 Plutonic and Extrusive Igneous Rocks

Now that we have discussed how magmas are generated, the next logical step is to discuss **igneous rocks**, which are rocks that are formed from cooling of the generated magmas we have just discussed. An igneous rock is any rock that is formed when magma, or molten material, is formed within the Earth's upper mantle and crust and then later solidifies at shallower crustal levels. Some of the magma cools within the Earth's interior, whereas some of it may make its way to the surface of the Earth where it cools down and solidifies.

During cooling of the magma, ions that were once mobile start to crystallize and form minerals, and this process continues until a rock is formed—an igneous rock—by transforming the magma into rock. Magma that reaches the Earth's surface is called **lava**. Thus, igneous rocks can be classified based on where the molten material cooled and solidified. If magma cools down and solidifies within the Earth's interior, **plutonic** (after the Roman god of the underworld) or **intrusive** igneous rocks are formed (Fig. 3.4). If the magma reaches the Earth's surface, typically via fractures or pipes to form lava, **volcanic** or **extrusive** igneous rocks are formed when lava cools down and solidifies here (Fig. 3.4).

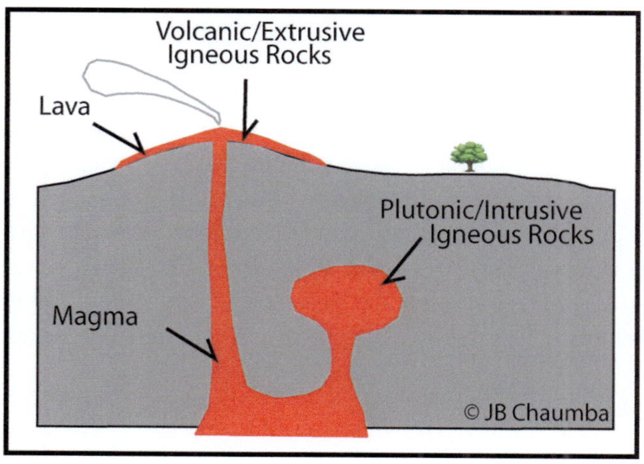

Fig. 3.4 Diagram illustrating the environments of formation of volcanic (or extrusive) and plutonic (or intrusive) igneous rocks. Also shown in the diagram is magma within the Earth's crust and lava when the magma reaches the Earth's surface

3.4 Influence of Magma Type on the Type of Igneous Rock

Silicate minerals, which contain the silicon-oxygen tetrahedron which we discussed in Chap. 2, dominantly make up igneous rocks. Silica is composed of silicon and oxygen, and the silica content, expressed as weight percent SiO_2, is often used to refer to how much silica a magma, or igneous rock, is composed of. Igneous rocks (or magma types) can be subdivided into ultramafic, mafic, intermediate, and felsic (or silicic) with increasing weight percent (wt. %) SiO_2 content, as shown in Table 3.1.

Ultramafic rocks are rocks derived from the words ultra (meaning extreme) and mafic (ma—*ma*gnesium; fic—*f*erric which is the Latin word for iron). Although ultramafic rocks contain very low SiO_2 contents, they contain extremely high magnesium oxide contents and are also characterized by extremely high proportions of dark colored minerals which make them appear very dark in color. **Mafic** rocks (sometimes referred to as rocks of basaltic composition) are rocks which contain lower magnesium oxide contents, but higher SiO_2 contents (Table 3.1) compared to ultramafic rocks. Igneous rocks that are dominantly composed of light-colored minerals such as quartz and potassium feldspar are called **felsic** (derived from fel—*fel*dspar and sic—*si*lica or quartz). Felsic rocks, or rocks of granitic composition, (Table 3.1) may also contain very low amounts of mafic minerals such as biotite and amphibole, in addition to other felsic minerals like muscovite. Minerals such as magnesium and iron have a tendency to make mafic rocks denser than felsic rocks. **Intermediate** rocks, or rocks of andesitic composition, are characterized by both magnesium and SiO_2 contents which fall between those of mafic and felsic rocks (Table 3.1).

We have already discussed that once a magma is generated in the mantle or lower crust from partial melting, it rises due to its density being lower than that of the surrounding rocks. The rate at which the magma ascends depends on its **viscosity**, which is the resistance of a magma to flow. The higher the viscosity of a magma, the more the resistance it has to flowing, and magmas with low viscosities tend to flow much more easily. With increasing temperature, viscosity decreases, and as temperature decreases, viscosity increases. Very hot magmas (ultramafic) flow much more easily (are "runny") like water due to their low viscosities whereas relatively cool magmas (felsic) are "thick" like honey. Consequently, felsic magmas have high viscosities, and they tend to flow much more slowly (Table 3.1). Felsic magmas may have eruption temperatures as low as 650 °C (~1200°F) whereas

Table 3.1 Classification of magma types and igneous rocks

Type of magma	Ultramafic	Mafic	Intermediate	Felsic
Weight % SiO_2	<45	45–52	52–65	>65
Magma viscosity	Very low	Low	Medium	High
Igneous rock	Peridotite/komatiite	Gabbro/basalt	Diorite/andesite	Granite/rhyolite
Rock type	Plutonic/volcanic	Plutonic/volcanic	Plutonic/volcanic	Plutonic/volcanic
Viscosity	Very low	low	Medium	High

basaltic or mafic magmas erupt at high temperatures of at least 1050 °C (~1920°F). Intermediate magmas are characterized by medium viscosities and eruption temperatures (Table 3.1).

3.5 Influence of Cooling Rate/texture on the Type of Igneous Rock

The grain size of igneous rocks is strongly dependent on the rate of cooling of the magma (Table 3.2). In the case of plutonic igneous rocks, which cool deep within the crust, the grain size tends to be coarse because they lose heat to the surroundings much more slowly. Temperatures of surrounding rocks deep within the crust are generally a few hundred degrees (Fig. 3.1), and the magma thus tends to cool down over long time periods, giving ions enough time to nucleate and form large crystals which ensure that coarse-grained igneous rocks are formed. The opposite is true, however, in the case of extrusive igneous rocks, which form from rapid crystallization of lava at the Earth's surface or on the ocean floor. At these settings, the lava quickly loses its heat to the atmosphere or to seawater, resulting in very fine-grained rocks being formed as ions lack time required to nucleate and form large crystals. In some cases, the magma may cool much more rapidly such that there is no time even for ions to form any crystalline substance but the magma quenches to form glass. Volcanic eruptions that are violent often form volcanic ashes which are tiny glass shards, a type of volcanic glass.

Other igneous rocks termed **hypabyssal** are formed when magma solidifies in minor intrusions under conditions that are intermediate between those of plutonic and extrusive igneous such as dikes and sills (Table 3.2 and Fig. 3.4). Hypabyssal rocks, therefore, possess textures that intermediate between those of coarse-grained plutonic igneous rocks and those of fine-grained extrusive igneous rocks (Table 3.2 and Fig. 3.4). If igneous rocks are formed from magma of the same composition but at different cooling rates, they have coarse- and fine-grained equivalent rocks. For example, gabbro is formed from slow cooling of mafic magma at depth, whereas basalt is formed from rapid cooling at the Earth's surface of magma of the same composition (Table 3.2 and Fig. 3.4). For felsic magma of the same composition, rhyolite forms when rapid cooling occurs at the Earth's surface whereas a granite is formed when slow cooling occurs at depth (Table 3.2 and Fig. 3.4). If the magma is of intermediate composition, an andesite is formed when cooling occurs quickly at the surface of the Earth, but a granite is formed from slow cooling at depth (Table 3.2 and Fig. 3.4). A diorite, which is typically medium-grained and of intermediate composition, is formed when magmas of any composition cool at shallow depths (Table 3.2 and Fig. 3.4). An andesite is also the volcanic equivalence of a diorite (Table 3.2 and Figs. 3.4 and 3.5).

Examples of gabbros, basalts, **vesicular** basalt, and pumice are shown in Fig. 3.6a–d, respectively. A vesicular basalt is formed from magmas that contain a high proportion of gases which get trapped in rocks as the magma cools down, or crystallizes, resulting in basalts with numerous openings such as those shown in Fig. 3.6c. As magmas migrate toward the Earth's surface, they experience drops in pressure that allow water and gases in the lava to form bubbles. These bubbles are then frozen in the lava as vesicles if the bubbles do not get large enough to pop. Vesicles that are filled with secondary minerals such as calcite, quartz, or zeolite, which form after the rock was formed and long after the flow cooled, are called **amygdaloids**. Such secondary minerals are commonly white:

Table 3.2 Classification of igneous rocks and textures

Igneous rock class	Cooling rate	Texture	Ultramafic	Mafic/basaltic	Intermediate/andesitic	Felsic/GRANITIC
Plutonic	Slow	Coarse-grained	Peridotite	Gabbro	Diorite	Granite
Hypabyssal	Intermediate	Medium-grained			Diorite	
Extrusive	Fast	Fine-grained	Komatiite	Basalt	Andesite	Rhyolite
Extrusive	Very fast	Very fine-grained	Komatiite	Pillow basalt		Obsidian

Fig. 3.5 Examples of fine-grained igneous rocks (rhyolite, andesite, and basalt), medium-grained rock (diorite), and coarse-grained rocks (granite and gabbro). Granite is the coarse-grained equivalent of the fine-grained rocks of rhyolite and andesite, whereas gabbro is the coarse-grained equivalent of the fine-grained rock basalt

quartz, calcite, or zeolite. Another rock type with vesicles is **pumice** (Fig. 3.6d), and in pumice the vesicles form when gas bubbles are trapped in the rock during rapid cooling of a magma that is rich in volatiles. The magma cools very quickly such that atoms in the magma are not able to nucleate into a crystalline structure. Pumice is therefore an example of a mineraloid, i.e., an amorphous substance.

Glassy texture is formed when lava is erupted either into the air or it flows underwater causing rapid cooling such that its ions lack time to form any crystals. Two types of volcanic glass are pumice and obsidian. Pumice is formed when volatiles escape quickly from felsic magmas. The high viscosity of felsic magmas prevents the volatiles from escaping easily leading to their trapping as numerous gas bubbles in the solidifying magma (Fig. 3.6c, d). Magmas with gas bubbles, or frothy magmas, are often associated with violent eruptions due to the expansion of the frothy magma. Upon cooling, the frothy magma forms pumice. **Obsidian** is formed from very rapid cooling of magma with high silica contents but containing less volatiles than magmas which solidify to produce pumice. Obsidian displays fracture due to the lack of any crystal structure (and therefore cleavage planes), a consequence of the absence of any regular arrangement of atoms caused by very rapid cooling. Since obsidian displays conchoidal fracture (see also Chap. 2), it was widely used as knife blades and arrowheads by ancient civilizations worldwide.

Certain igneous rocks have textures whereby very coarse-grained crystals occur in a groundmass of fine- to coarse-grained visible to invisible crystals to form what is termed a **porphyritic or fragmental** texture (Figs. 3.7a and 3.8). Porphyritic textures are formed when magmas cool very slowly at depth to form the very coarse-grained crystals followed by relatively more rapid cooling at or nearer the Earth's surface to form the fine- to coarse-grained or invis-

Fig. 3.6 Images of gabbro **a**, basalt **b**, vesicular basalt **c**, and pumice **d**. Vesicular basalt and pumice are formed from magmas that contain volatiles

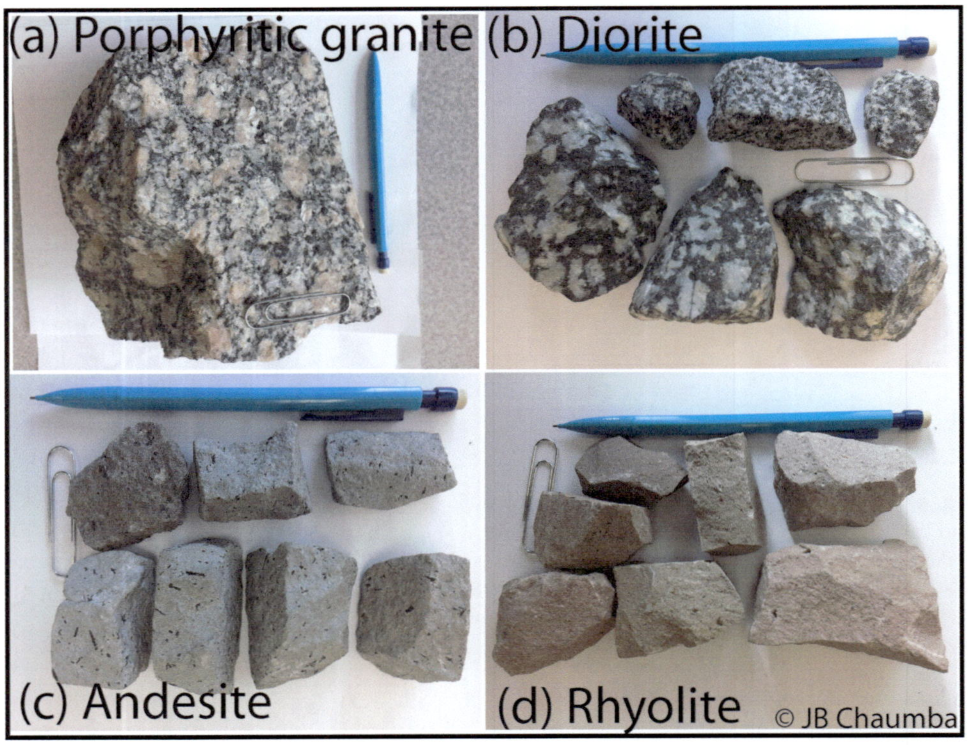

Fig. 3.7 Images of porphyritic granite **a**, diorite **b**, andesite **c**, and rhyolite **d**. Vesicular basalt and pumice are formed from magmas that contain volatiles

Fig. 3.8 Pyroclastic rocks range in grain size from volcanic ash from Mount St. Helens May 18, 1980 eruption (Photo credit: USGS) **a**, photograph of welded tuff **b**, through lapilli **c** to volcanic bombs (credit: Kristi Diller), and **d** photograph of volcanic bomb

ible crystals when the mixture of the early formed crystals and the remaining magma rises to or nearer the Earth's surface. We have already discussed that diorite (Fig. 3.7b), a typically medium-grained rock of intermediate composition, is formed when magmas of any composition cool at shallow depths. Examples of rock samples that form from rapid cooling at the Earth's surface of magmas of intermediate composition such as andesites are shown in Fig. 3.7c, whereas samples of rhyolite (Fig. 3.7d) are produced when felsic magmas cool rapidly at the Earth's surface.

Rocks that display a pyroclastic texture are formed from the consolidation of rock fragments when violent or explosive explosions occur, ejecting pulverized fragments of glass, rock, and lava through the vent. When viscous magma rich in volatiles erupts violently, it produces ash and dust particles when the ash is blown into very fine-grained fragments of glass (Fig. 3.8a). The very fine-grained rock fragments that are ejected from these pyroclastic eruptions are called volcanic ash (Fig. 3.8a). **Tuff** or welded tuff (Fig. 3.8b) is formed when hot ash falls in the glass shards fuse. Tuff is composed of glass fragments that fused together when still hot, and once a molten blob, tuff displays a pyroclastic texture. Lapilli (Fig. 3.8c), sometimes called "little stones," are coarse-grained ejecta the size of marbles. Volcanic bombs (Fig. 3.8d), which are masses of molten rock coarser than 64 mm, have a tendency to assume a streamlined shape due to friction as they are hurtled through the atmosphere. Very coarse-grained angular blocks of rocks may also be ejected from the vent of walls of a volcano, and these are called blocks. Volcanic blocks are very coarse-grained particles produced from hardened lava, and volcanic bombs are ejected incandescent lava. Both volcanic bombs and blocks tend to fall near the vent of a volcano due to their large size, although they are sometimes ejected much away from the vent.

3.6 Bowen's Reaction Series

Neil L. Bowen, a geochemist working at the Carnegie Institute for Science in Washington, D.C., showed almost a hundred years ago that the common range of igneous rocks may be formed from the same magma of basaltic composition over a temperature range of a few hundred degrees. This concept is called Bowen's reaction series. When magma is allowed to cool slowly in a laboratory setting to mimic processes that may occur in a magma chamber within the Earth's crust, certain minerals were cbserved by Bowen to always crystallize at relatively high temperatures. As temperature drops, other minerals were observed to crystallize in sequences (Fig. 3.9). Thus, what Bowen observed is that specific minerals form at specific temperatures during cooling of a magma.

At very high temperatures olivine **crystallizes** first, followed by the crystallization of pyroxene and calcium-rich plagioclase feldspar when temperature cools further (Fig. 3.9). At intermediate temperatures, amphibole and biotite were observed to crystallize, together with plagioclase feldspar with approximately equal proportions of calcium and sodium (Fig. 3.9). At relatively lower temperatures, after most of the magma has crystallized, muscovite

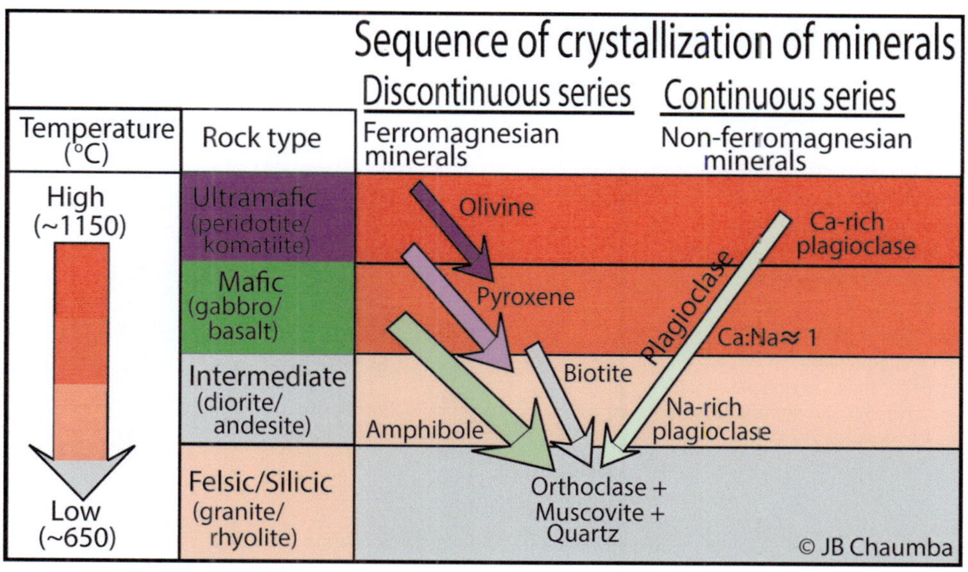

Fig. 3.9 Bowen's reaction series. Shown in the diagram is the range of temperatures (in degrees Celsius) where crystallization occurs in a body of magma to form the sequence of minerals shown. The different rock types that are formed from the crystallizing minerals can range from ultramafic, mafic, intermediate, and felsic rocks. The discontinuous series is composed of ferromagnesian minerals, whereas the continuous series is composed of plagioclase feldspar

and sodium-rich plagioclase feldspar, and potassium feldspar were observed to crystallize (Fig. 3.9). During the final stages of the crystallization of the magma at low temperatures, quartz crystallized from the last remaining magma. Thus, as temperature decreases, the mafic index of rocks that are formed ranges from ultramafic through mafic and then intermediate to felsic rocks (Fig. 3.9). The general progression of the minerals that form can be separated into two branches; one composed of mafic minerals which only crystallize at certain specific temperatures called the discontinuous series, and the other composed of plagioclase feldspar called the continuous series (Fig. 3.9).

Bowen's reaction series may help explain why different rock compositions can be emitted from a single volcano, suggesting that a single magma type can evolve, or change, to form different varieties of igneous rocks through crystallization. Since olivine is usually the first mineral to crystallize from a cooling magma, whereas quartz is the last mineral to crystallize, Bowen's reaction series also helps explain why these two minerals rarely are found together in the same rock. By the time quartz will start to crystallize from a magma, olivine would have long ceased to crystallize. Further, the observed sets of minerals which tend to occur in the same igneous rock are supported by their comparable temperature ranges in Bowen's reaction series. Minerals such as quartz, muscovite, potassium feldspar, and sodium-rich plagioclase feldspar, which are located at the low temperature part of Bowen's reaction series (Fig. 3.9), for example, are all tend to be common minerals in a granite. Thus, minerals which form in the same temperature range in Bowen's reaction series have a tendency to occur together, lending support to Bowen's conclusions.

As magma crystallizes, certain elements are removed from the magma to form the minerals being formed. Crystallization leads to the remaining magma being changed in composition by depleting the magma of those elements which form the early formed minerals. During a process called fractional crystallization, the early formed minerals separate from the magma leading to different minerals and, thus, different types of igneous rocks being formed. In a scenario called crystal settling, the early formed minerals, which tend to be heavier than the magma as they are richer in ferromagnesian minerals, sink to settle at the bottom of the magma chamber (Fig. 3.10). The remaining magma's composition then changes, and different minerals will then be formed as crystallization continues. In some cases, the remaining magma may be forced into any cracks that are formed due to intrusion of the magma into the surrounding rocks where it will also crystallize to form rocks that are different in composition from the early formed rocks (Fig. 3.10). Magmatic **differentiation** refers to the process whereby secondary magmas are formed from magma of starting composition termed the parent magma.

Figure 3.11a and b show a porphyritic granite with megacrystic (very coarse-grained) sodium-rich plagioclase leucocratic feldspar crystals, coarse-grained pinkish alkali feldspar crystals, and coarse-grained gray quartz crystals from the Lilesville granite, Anson County, North Carolina. Also present are dark amphibole and biotite crystals. Relatively finer grained of these minerals are also present in the granite, hence its porphyritic texture. The granite forms part of numerous plutons that were emplaced in the Appalachian orogen during the formation of the supercontinent Pangea.

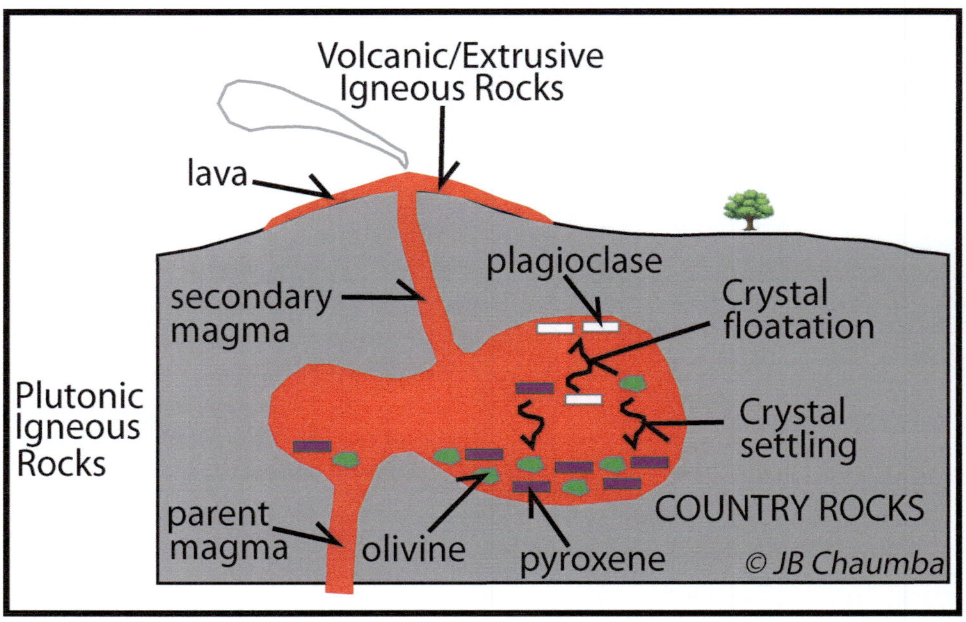

Fig. 3.10 Diagram illustrating differentiation of magma, crystal settling, and crystal floatation. Early formed ferromagnesian minerals such as olivine and pyroxene, which are denser than the magma, crystallize and then sink to and settle at the bottom of the magma chamber, whereas non-ferromagnesian minerals such as plagioclase, which are less dense than the magma, float and accumulate at the top of the magma chamber

Fig. 3.11 Images of an outcrop of the Lilesville porphyritic granite (**a**) and a broken piece of the granite (**b**)

Shown in Fig. 3.12 is an outcrop of basalts what were generated from huge outpourings of lava than extended across much of the ancient continental mass Gondwana ~183 million years ago. These basalts marked the start of the breakup of Gondwana, and some of the remnants of these comprise the Lebombo mountains as well as the Lesotho highlands in Southern Africa.

3.7 Economic Importance of Igneous Rocks

The economic importance chiefly lies in metallic minerals resources. Metals such as the platinum-group elements, nickel, copper, gold, silver, and other metals tend to occur in association with igneous rocks. Large, layered intrusions, which are composed of large ultramafic mafic rocks, are generally thought to have been formed due to magmatic differentiation. These layered intrusions tend to act as host to most of these metals, as well as chromite, vanadiferous magnetite, asbestos, magnesite, and many other commodities. Most of the ore deposits in layered intrusions were either formed by crystal setting or were concentrated by fluids of magmatic origin.

Intrusions composed of granitic rocks contain deposits of tin, tungsten, gold, lithium, uranium, beryllium, and other metals. These minerals that are in granitic intrusions are generally thought to have been formed as a product of the cooling and crystallization of granitic magmas, leaving the residual melt to be concentrated in heavy metals such as gold and silver, as well as other elements such as rare Earth elements. Further, volatiles such as water also get concentrated in the residual melt during fractional crystallization of the granitic magma, eventually crystallizing in an environment that promotes migration of ions leading to the formation of unusually coarse-grained crystals (up to a few meters in length). A **pegmatite**, which is a very coarse-grained granitic rock consisting of the minerals feldspar, quartz, and muscovite, and it is formed from late-stage crystallization of granitic magmas. Pegmatites are principal sources of the semiprecious gems of beryl, topaz, and tourmaline, as well as being sources of rare Earth elements, lithium, gold, and silver. All the three minerals of feldspar, quartz, and muscovite in a pegmatite are economically important as feldspar is useful in the manufacturing of ceramics, and muscovite glitters in addition to being useful in electrical insulation. Quartz is useful in the manufacturing of ceramics, glass, and for foundry molds in the casting metal.

Fig. 3.12 Scenic view showing a hiker with 183 million-year-old rocks of the Drakensberg Group basalts comprising the cliff in the background (credit: R.E. Diamond)

Glossary

Amygdaloid a vesicle filled with a mineral in a volcanic rock

Crustal anataxis mantle-derived basaltic magmas which are intruded into the continental crust causing the geothermal gradient to become higher than the wet solidus of the crustal material, resulting in partial melting of the continental crust

Crystallize the process of formation and growth of a crystalline solid from a melt

Decompression melting partial melting which occurs as a consequence of an ascending body of mantle material caused by a reduction in confining pressure

Differentiation any process in which a melt cools down and solidifies into at least two different rock types. Differentiation refers to the gradual separation of an originally homogeneous planet such as Earth, upon cooling, into different layers such as the crust, the mantle, and the core.

Extrusive igneous rock a rock that is formed from lava when it flows away from a vent and rapidly cools down at the Earth's surface to form a fine-grained rock

Felsic igneous rocks or melts that are dominantly composed of the elements feldspar and quartz (silica)

Fux melting when which occurs volatiles, such as water and/or carbon dioxide, are added to solid rock at high temperatures, causing it to melt at relatively lower temperatures than it would melt when it is free of volatiles

Fragmental *see porphyritic*

Geothermal gradient denotes the rate at which temperature increases from the surface with depth

Glassy texture denotes the texture of an igneous rock such as obsidian which lacks any crystals due to very rapid cooling

Igneous rock any rock that is formed from the cooling and solidification of molten materials such as magma or lava

Intermediate rock the composition of color index of an igneous rock which lies between felsic and mafic, sometimes referred to as andesitic rocks

Intrusive igneous rock a rock that is formed from magma when it trapped in a chamber deep within the Earth's crust and slowly cools down to form a coarse-grained rock

Lapilli pyroclastic material composed of particles ranging from 2 to 64 mm in diameters consisting of lots of fragments of clumps of ash, pumice, or frozen clots of lava

Lava magma that flows out from a vent onto the surface of the Earth

Mafic igneous rocks or melts that are dominantly composed of the elements magnesium and iron but poor in silica

Magma any molten material below the surface of the Earth

Obsidian an igneous rock of felsic composition and composed of volcanic glass

Pegmatite a very coarse-grained intrusive igneous rock composed of crystals usually larger than 2.5 cm (1 inch) in size and formed from late-stages of crystallization of granitic magmas

Plutonic igneous rock *see intrusive igneous rock*

Porphyritic texture an igneous rock texture which is composed of very coarse-grained crystals, or phenocrysts, embedded in a relatively finer-grained groundmass

Pumice a vesicular and glassy extrusive igneous rock which contains numerous pore spaces

Radioactive heat heat produced by unstable atomic nuclei due to spontaneous disintegration of the atomic nuclei when alpha particles, beta particles, gamma rays, and other particles are released, transforming nuclei into lighter and new element in the process

Rock a naturally occurring solid composed of aggregates of minerals or, rarely one mineral, or sometimes a mass of glass

Subduction zone a boundary that marks the zone where collision occurs between two tectonic plates, where an oceanic plate is pushed into the mantle beneath either another oceanic plate or a continental plate

Ultramafic an igneous rock or magma that is enriched in magnesium and iron but consisting of very low silica composition

Vesicle openings or holes in extrusive igneous rocks that is formed when magmas rich in volatiles preserve some of the bubbles

Viscosity the resistance of a melt to flow

Volatiles dissolved components of magma such as water vapor and carbon dioxide that form at magmatic temperature which expand upon reaching relatively low pressures at and close to the surface

Volcanic igneous rock *see extrusive igneous rock*

4 Sedimentary Rocks

Abstract

The chapter begins with describing sediments, the ingredients of most sedimentary rocks, and their lithification with increasing depth of burial to form sedimentary rocks. The four types of sedimentary rocks (detrital, chemical, biochemical, and organic) are discussed in detail using tables, diagrams, and photographs. Descriptions of various types of sedimentary structures follow, as well as the different types of depositional environments of sedimentary rocks accompanied by diagrams and photographs, are provided. A summary diagram of all the depositional sedimentary environments, each with a representative photograph, is included. Marine transgressions and regressions accompanied by illustrations are discussed with the aid of illustrations. The chapter ends with a discussion of the economic importance of sedimentary rocks.

Learning Goals

Studying this chapter will enable you to:

- Define sedimentary rocks
- Recognize the common sedimentary rocks and describe how they were likely formed
- List and define the processes that form sedimentary rocks from sediments
- Differentiate the categories of sedimentary rocks
- Describe and interpret the environment of deposition of common sedimentary rocks
- Explain what makes sedimentary rocks different from other rock types
- Interpret past changes in sea level from examining layering, compositional, and textural changes in sedimentary rock sequences
- Recognize the depositional environments
- Relate sedimentary rocks to their depositional environments.

4.1 Introduction

The daily influences of the atmosphere are responsible for the slow disintegration and decomposing of minerals and rocks that are exposed at or near the surface of the Earth—a process termed **weathering**. We shall discuss weathering in more detail in Chap. 10. Weathering produces loose materials that have a tendency to move downslope due to gravity. The products of weathering, which are commonly particles and dissolved substances termed **sediments** (derived from the Latin word *sedimentum* which means "settling"), are also transported by agents of erosion such as running water, waves, wind, or glaciers, depending on the region where the weathering occurs. The sediments, which are the ingredients of the bulk of sedimentary rocks, are eventually deposited in a depositional basin, with most sediments being deposited in the oceans, and also in inland seas, lakes, river floodplains, desert basin, and sand dunes, among others, to eventually form **sedimentary rocks**. Sedimentary rocks and sediments that form them, thus, form layers (or beds) which accumulate as a "cover" that is underlain by a "basement" of igneous and metamorphic rocks. Rock fragments being transported to depositional basins tend to be broken down into smaller fragments due to collision with each other, as well as collision with the bottom and sides of any river channel or glacial valleys where the transportation occurs. It is estimated that sedimentary rocks, constituting a small fraction (about 5% by volume) of the outermost 10 miles (16 km), are also very important in that they comprise 75% of the rock exposures on the Earth's continents.

Supplementary Information The online version contains supplementary material available at https://doi.org/10.1007/978-3-032-20231-4_4.

J. B. Chaumba, *Introductory Physical Geology*, Springer Textbooks in Earth Sciences, Geography and Environment,
https://doi.org/10.1007/978-3-032-20231-4_4

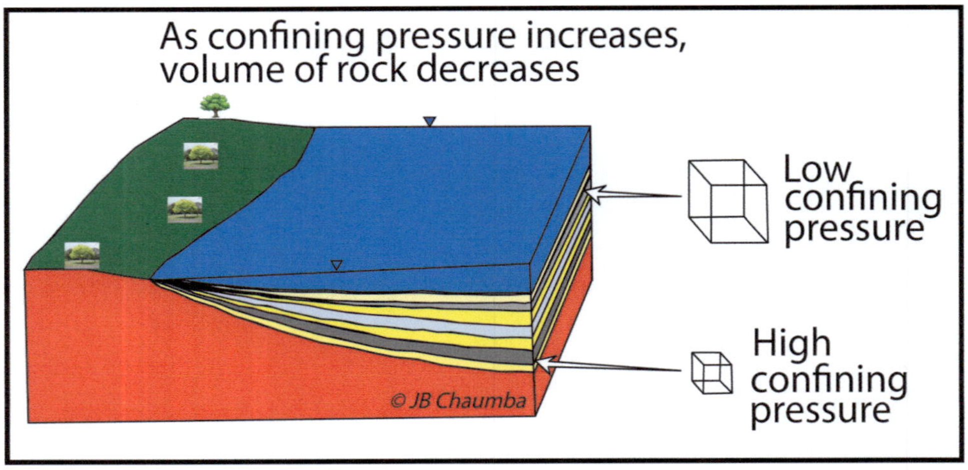

Fig. 4.1 As earlier deposited sediments are progressively buried by those deposited later, the weight of later-deposited sediments (which are at low confining pressures) squeezes both air and water out of the sediments at deeper levels which are at high confining pressures leading to reduction in the volume of the rock

Some sedimentary rocks may also be formed when fragments of shells and shells are cemented together. Further, some sedimentary rocks are formed either from chemical precipitation of minerals directly from water or from accumulation and subsequent alteration of organic matter that is derived from once living organisms. After deposition, the sediments are then lithified (Fig. 4.1), a process which transforms the earlier deposited sediments into sedimentary rock. The process of **lithification** occurs due to both compaction and cementation. **Compaction** is caused by the weight of the overlying sediments as more sediments are deposited above earlier formed ones, which expels both water and air deposited with the sediments leading to reduction in open spaces between the accumulated sediments. Compaction, thus, permits the deposited broken down rock fragments to pack more tightly due to increase in confining pressure (inward-pressing force in all directions) (Fig. 4.1). **Cementation** occurs as percolating groundwater leads to infilling of the pores within the sediments by mineral matter such as calcite, silica, or iron oxides. Cementation, therefore, fills any remaining spaces between the deposited broken down fragments by forming a "cement" which acts as the glue holding the particles together. Both compaction and cementation act to transform the deposited sediments into solid rock.

There are several reasons why sedimentary rocks are important to study. Energy resources such as coal, natural gas, and petroleum are hosted in sedimentary rocks. Natural gas and petroleum often occur in pore spaces within sedimentary rocks. Some fertilizers, which are useful in agriculture, and elements such as aluminum, manganese, and iron, as well as some non-metallic mineral resources such as calcite, gypsum, and halite, also occur in sedimentary rocks.

Studies which involve reconstruction of the history of our planet mostly rely on studies of sedimentary rocks. Fossils, which tend to predominantly occur in association with sedimentary rocks, are essential in studies of evolution on our planet. Studying sedimentary rocks permits Earth Scientists to make interpretations of what the environments at the Earth's surface were like in the past. Furthermore, studying sedimentary rocks enables Earth Scientists to decipher the depositional environments in which sediments were deposited, and to make interpretations on the method of transport of the sediments as well as how far the sediments were transported from their source regions. Let us first review the ways in which sedimentary rocks are classified, before we briefly discuss some common structures displayed by sedimentary rocks. This will be followed by a brief discussion of the depositional sedimentary basins in which different sedimentary rocks were formed. We finish the chapter with a brief discussion of transgression and regression and how they can be inferred in the rock record.

4.2 Classification of Sedimentary Rocks

In this textbook, sedimentary rocks are classified into four different categories based on how they were formed. Sedimentary rocks that are formed from broken fragments (termed *detritus*) of pre-existing rocks are called detrital or clastic sedimentary rocks. Sedimentary rocks that are formed from minerals that precipitate directly (or settle out) from any body of water are called chemical sedimentary rocks. Some sedimentary rocks composed of shells are classified as biochemical sedimentary rocks. Finally, some sedimentary rocks which are composed of plants and other organisms are called organic sedimentary rocks. Let us now discuss these classes of sedimentary rocks.

4.2.1 Detrital Sedimentary Rocks

The mode of formation of various sedimentary rocks is useful in their classification. The size of particles (Table 3.1), however, is the main determinant in classifying detrital

Table 4.1 Table showing classification of detrital sediments and rocks

Classification of detrital sediments and rocks				
Size of particles (mm)*	Particle name	Mud/Sand/Gravel	Sedimentary rock	
< 0.002	Clay	Mud	Shale	**Mudstone**
0.002–0.05	Silt		Siltstone	
0.05–2	Sand	Sand	Sandstone	
2–4	Granule	Gravel	Conglomerate or Breccia	**Conglomerate** (if particles are rounded) or **Breccia** (if particles are angular)
4–64	Pebble			
64–256	Cobble			
> 256	Boulder			

* 1 mm = 0.039 in

sedimentary rocks. Particle sizes range from some very fine-grained clays and fine-grained silts which form the finest grained sedimentary rocks called shales and siltstones (which are also referred to as mudstones (Table 4.1, Fig. 4.2a, b)), through medium to coarse-grained sands which form sandstones, to very coarse-grained granules, pebbles, cobbles, and boulders which form conglomerates and breccias (Table 4.1).

Mudstones (Table 4.1, Fig. 4.2a, b) are the most widespread sedimentary rocks on Earth, constituting over 50% of all sedimentary rocks. Shale (Fig. 4.2a) finds uses in brickmaking and ceramics (for example, in the manufacturing of pottery and tiles). Shale can also be mixed with sand, calcium carbonate, and gypsum in the manufacturing of Portland cement which is essential in construction.

Sandstone (Table 4.1, Fig. 4.2c), the medium to coarse-grained detrital sedimentary rock, constitutes about a quarter of all sedimentary rocks and the quartz in sandstone is typically surrounded by cements of silica or carbonates. There are three different types of sandstone. The first one, quartz arenite (Table 4.1, Fig. 4.2d), which is composed of at least 90% grains of quartz, contains very little *matrix*. Matrix is the finer-grained material that surrounds the coarser-grained grains. The second type of sandstone is arkose (or feldspathic sandstone), which is usually pinkish to reddish sandstone and containing > 35% feldspar (Fig. 4.2d).

Sandstones tend to display a wide range in color which depends on the composition of the rock. The color of sandstone may also often provide some clues to the environment of their formation. Some argillaceous (clay-bearing) sandstones are often colored gray to blue. However, sandstone is often light tan in color (Fig. 4.2c) due to the presence of minerals that are light colored. Further, other elements may create various shades of red (Fig. 4.3a), caused by iron oxide (rust) in other sandstones, whereas other sandstones may possess a purple hue color that is caused by the presence of the mineral manganese.

Graywacke (or lithic sandstone) (Fig. 4.3b), the third type of sandstone, is composed of a mixture of grains of quartz and feldspar and it is often dark gray or green in color. Graywackes contain numerous dark fragments of rock, predominantly of volcanic origin, and a matrix of fine-grained clay and mica. Graywackes typically contain feldspar, and they are poorly sorted with angular grains which point to their formation near their source areas. Sandstones are important economically as reservoirs of energy resources such as oil and natural gas, and act as aquifers in which groundwater is stored and pumped from. Further, sandstones are also useful as a building stone.

Very coarse-grained detrital sedimentary rocks such as conglomerates and breccias are differentiated based on the roundness or angularity of their particles. Rounded grains characterize conglomerates (Table 4.1, Fig. 4.2e) whereas angular grains characterize breccias (Table 4.1, Fig. 4.2f). Conglomerates, with rounded (or spherical) particle shapes, suggest particle transport by forceful currents for very long distances from their source regions whereas breccias, with angular grain shapes, suggest short transport distances from their source regions. Both conglomerates and breccias contain fine-grained matrix that is often cemented by silica, calcium carbonate, or iron oxide. *Sphericity* is a measure of the closeness of particles in sediments/rocks to a sphere, an indication that the sharp edges and corners have been smoothed by transportation in a fluid or by the action of waves, for example. *Angularity* is a measure of the degree to which particles in sediments/rock have angular or sharp edges and corners.

Although, as stated earlier, the size of the particles is the chief determinant in classifying detrital sedimentary rocks, in some cases the composition of the rock is also employed to classify these rocks. An example is quartz sandstone which is predominantly composed of quartz particles. Further, a silty sandstone, for example, is a rock dominated by sand sized particles although it also contains appreciable proportions of silt sized particles too.

It is worth to note that all detrital rocks, irrespective of whether they are composed of mineral grains of fragments of shale, are composed of solid particles that are compacted

Fig. 4.2 Detrital sedimentary rocks. Images of very fine- and fine-grained mudstones (shale (**a**) and siltstone (**b**)), which show an increase in grain size of sediments from shale to siltstone. Sandstone (**c**) is dominated by quartz particles whereas arkose (**d**) contains some appreciate feldspars. Both sandstone and arkose are medium grained, whereas conglomerate (**e**) and breccia (**f**) are both coarse-grained detrital sedimentary rocks. Conglomerate contains coarse-grained rounded particles whereas breccia is composed of coarse-grained angular particles. Grain size, thus, shows an increase from mudstones through sandstones and quartz arenites to conglomerates and breccias

Fig. 4.3 Some types of sandstones. **a** Image of red sandstone from Zion Canyon, Utah. **b** Graywacke from near Grafton, New York

Fig. 4.4 Some images of chemical sedimentary rocks. **a** rock gypsum, **b** rock salt, **c** chert, and **d** dolostone

or cemented together and therefore they possess a clastic texture. A rock such as sandstone is composed of sand-sized particles, that is, the particles are of the same grain size. This is referred to as **sorting**. Rocks that are well-sorted, therefore, are composed of uniform particle sizes whereas poorly sorted sediments or rocks such as graywacke contain particles of different particle sizes which can range from fine- through medium- to coarse-grained. Sorting of sediments is made possible by the transporting medium, such as water of wind. Fine grained particles such as clay and silt settle in still waters, where they accumulate in settings such as in lakes, swamps, and lagoons. Fast moving water and glaciers are capable of transporting very coarse-grained particles such as gravel, whereas slow moving rivers and wind can transport coarse- to fine-grained particles.

4.2.2 Chemical Sedimentary Rocks

Chemical sedimentary rocks are formed during settling out (chemical precipitation) of ions dissolved from solution in water solutions such as oceans, seas, and lakes due to physical processes. Chemical sedimentary rocks possess a crystalline texture formed both from the precipitation stage of the rocks and also from recrystallization when new minerals form later when new minerals form from the earlier formed ones. Chemical sedimentary rocks are composed of both coarse-grained crystals as well as very fine-grained minerals that cannot be discerned with the naked eye. The chemical composition of chemical sedimentary rocks is useful in classifying chemical sedimentary rocks into evaporites, travertine, and chert.

Evaporites form from saltwater that has exceeded its capacity to carry dissolved ions which then form particles which either settle out or grow at the surface of the water body. The salt water must be supersaturated in dissolved ions, and this can occur where excessive evaporation causes removal of water from a body of water at a faster rate than that at which new water enters it, such as in lagoons or in desert lakes.

Rock gypsum (Fig. 4.4a) is a common evaporite sedimentary rock. Since evaporites form due to the evaporation of water from water bodies, evaporites refer to all deposits of salt with the proportion of evaporation being important: rock gypsum is formed when 80% of the water evaporates whereas rock salt (formed from the mineral halite) (Fig. 4.4b) is formed when 90% evaporation of water occurs.

Travertine (Fig. 4.5), a rock that is composed of crystalline calcium carbonate ($CaCO_3$) and often referred to as chemical limestone, is formed when chemical precipitation from groundwater seeps out at the ground surface. The seeping can occur either in springs of cold or hot water, or even on the walls of caves caused by degassing of groundwater. Evaporation of water can also lead to increased concentration of calcium carbonate which leads to precipitation of travertine. Thus, it is important to note that travertine, which is a type of chemical limestone, may be formed in several different ways. One way of forming travertine

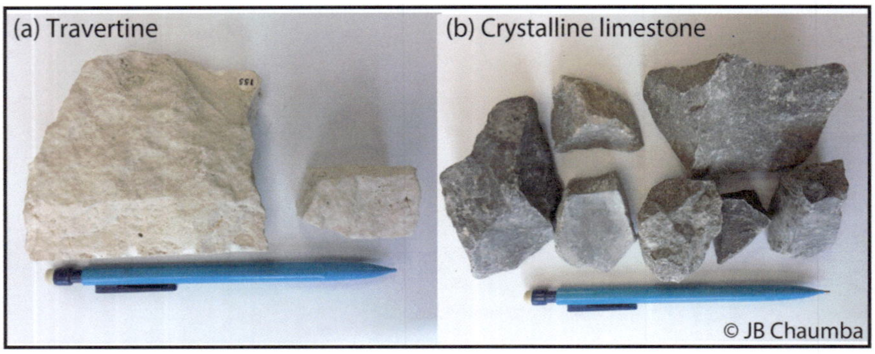

Fig. 4.5 Chalk is a relatively soft, fine-grained sedimentary rock largely composed of calcite. **a** travertine and **b** crystalline limestone

occurs in caves (as speleothems—cave formations). As has just been described above, travertine may also be produced along carbonate terraces such as those that occur near hot springs.

Chert (Fig. 4.4c) occurs in several different colors such as red, brown, gray, green, black, and white. Impurities in the chert cause the different colors. When silica-rich sediment buries a forest, such as that in the Petrified Wood National Forest, a type of chert called petrified wood is formed as the wood cellulose is replaced by cryptocrystalline quartz. The shape of the wood cells, as well as the growth rings in the wood, is retained. Another type of chert, agate, has a tendency to precipitate in concentric rings inside hollows in rocks to form a striped appearance; variations in the composition of the impurities are thought to be responsible for the striped appearance in agate. Cryptocrystalline quartz can gradually replace crystals of calcite in limestone after formation of the latter, to form what is term replacement chert.

Dolostone, another rock composed of calcium carbonate, is composed of the mineral dolomite ($CaMg[CO_3]_2$) (Fig. 4.4d). A chemical reaction which occurs between groundwater which contains magnesium and calcite is thought to be the source of the magnesium that is found in dolostone.

4.2.3 Biochemical Sedimentary Rocks

Indirect chemical precipitation in water bodies can occur when a number of organisms can extract dissolved ions from these bodies of water to form shells. The shells are, thus, formed when several plants and animals in water extract mineral matter that is dissolved in these water bodies. After the organisms die, the shells accumulate at the bottom of these water bodies where they can be later lithified to form biochemical sedimentary rocks (Fig. 4.6). Corals and algae produce shells of $CaCO_3$ which crystallize either as aragonite or calcite (only the crystal structure differentiates them apart as they have the same chemical composition). Limestone forms from biochemical sediments that are secreted by marine organisms. It is estimated that almost 90% of limestone forms from the secretion of biochemical sediments by marine organisms, whereas limestone that forms by direct precipitation from seawater constitute the remainder of a little over 10%.

Aragonite and calcite are biochemical limestones and since the material that forms limestone accumulates in different ways, several textures of limestone exist. Some limestones are formed from corals growing in place, whereas other limestones are formed from shells of **plankton** (or floaters) precipitating from water. Still other limestones are formed from transported and fragmented debris of shells, and carbonate mud can also form other types of limestone. For example, shells can be broken apart by waves to form clasts of carbonate which can be later cemented to form limestones. Limestones that are composed of shell fragments of fossil shells that are visible are termed *fossiliferous limestones* (Fig. 4.6a) and they are distinguished from limestones that are composed of very fine-grained carbonate mud termed *micrites.*

A biochemical limestone that is made up mostly of ooids (or ooliths), of sand-sized particles that have concentric rings of $CaCO_3$, is called oolitic limestone (Fig. 4.6b). The concentric rings are formed around grains of sand or shell fragments that are thought to have been rolled around on the shallow sea floor, accumulating layer after layer of limestone. Chalk (Fig. 4.6c) is a type of porous and soft limestone composed of microscopic shells of plankton. The White Chalk Cliffs in southeastern England are excellent examples of chalk. Coquina (Fig. 6d) is an easily identifiable biochemical limestone since it is coarse-grained and is composed entirely of shells and fragments of shells that are loosely cemented.

Biochemical chert is formed from planktons that secrete silica and accumulate on the bottom of the ocean followed by dissolution of the shells to produce a gel rich in silica. When the silica gel solidifies, it forms chert. Some cryptocrystalline quartz, for example, can be formed from deposition of plankton that secretes shells of silica.

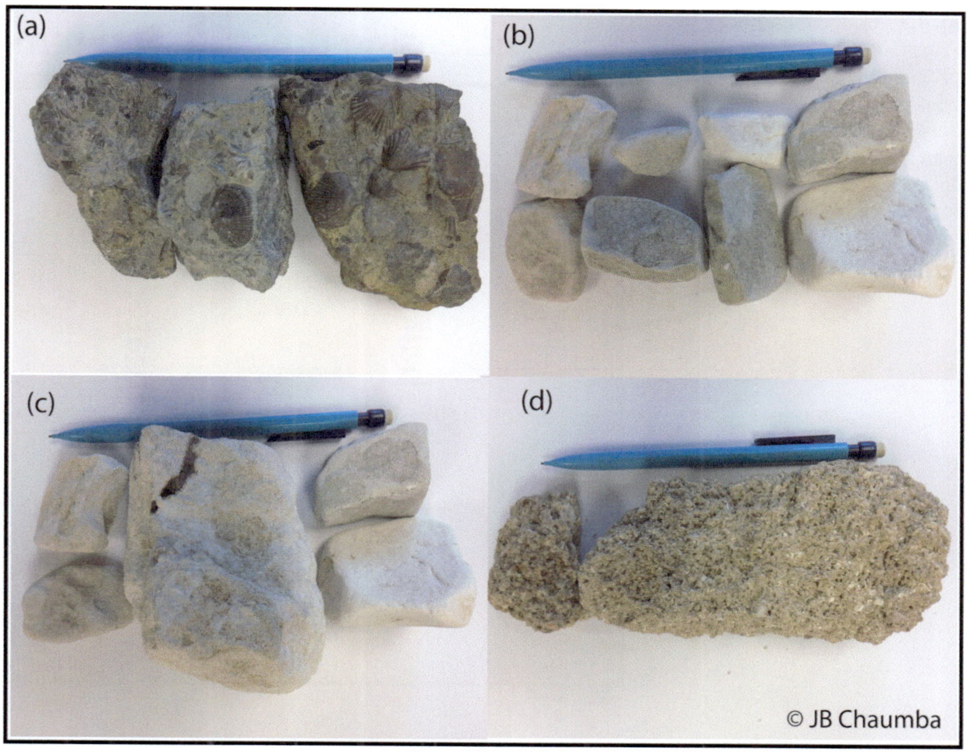

Fig. 4.6 Images some biochemical sedimentary rocks of fossiliferous limestone (**a**), oolitic limestone (**b**), chalk (**c**), and coquina (**d**)

4.2.4 Organic Sedimentary Rocks

Sediments that are rich in organic debris can be lithified to form sedimentary rocks rich in organic matter called organic sedimentary rocks. Examples of organic sedimentary rocks are coal and oil shale. Coal is thought to form from burial, over long time periods, of large quantities of plants. This burial is inferred to occur in an environment such as a swamp that is deficient in oxygen which prevents decaying (or oxidation) of the buried plants. As more material is buried above the earlier buried plants, increased pressure and temperature conditions increase the carbon content of the coal by driving off volatiles such as water, hydrogen, ammonia, carbon dioxide, and sulfur, as well as other impurities like silica (Fig. 4.7). The combustible content of coal is related to the carbon content, which ranges from 40 to 90% carbon.

Oil shales contain organic material which ranges in concentration from 15–40% kerogen as well as clay. Fats and proteins derived from plankton and algae are the source of the kerogen found in oil shales, whereas the organic material, which is responsible for the black color of oil shale, transforms to form the kerogen.

4.3 Sedimentary Structures

Since sedimentary rocks form at the surface of the Earth, they are invaluable in studying Earth history. Sedimentary rocks record the environmental conditions such as the temperature of the Earth experienced by the sediments, and the composition of the atmosphere at the time the rocks were formed. As each **stratum** (or layer) of the sedimentary rock is formed, it records the nature of the prevailing conditions at the Earth's surface. Consequently, strata (or several beds), can be studied to reveal any changes in the nature of the environment through time, rendering sedimentary rocks of paramount importance in investigating past climatic environments, remains of past life, or traces made by once living organisms (or **fossils**).

A **sedimentary structure** refers to the feature(s) that were formed at the surface of strata during deposition, as well as to the manner in which the grains are arranged within the strata. Strata can vary from very thin beds to those that can reach tens of meters in thickness. Strata are separated by flat surfaces termed bedding planes which mark either the end or beginning of each cycle of sedimentation. Bedding, or stratification, refers to the arrangement of sediments to form a sequence of beds. Sedimentary structures contain pieces of information that are used to infer the environment in which sediments were deposited. A number of sedimentary structures are formed when clastic sediments are deposited due to the activity of rivers, waves, and wind. These sedimentary structures, or bedforms, are formed at the interface between the transporting media and the sediments. A few examples of sedimentary structures are discussed next.

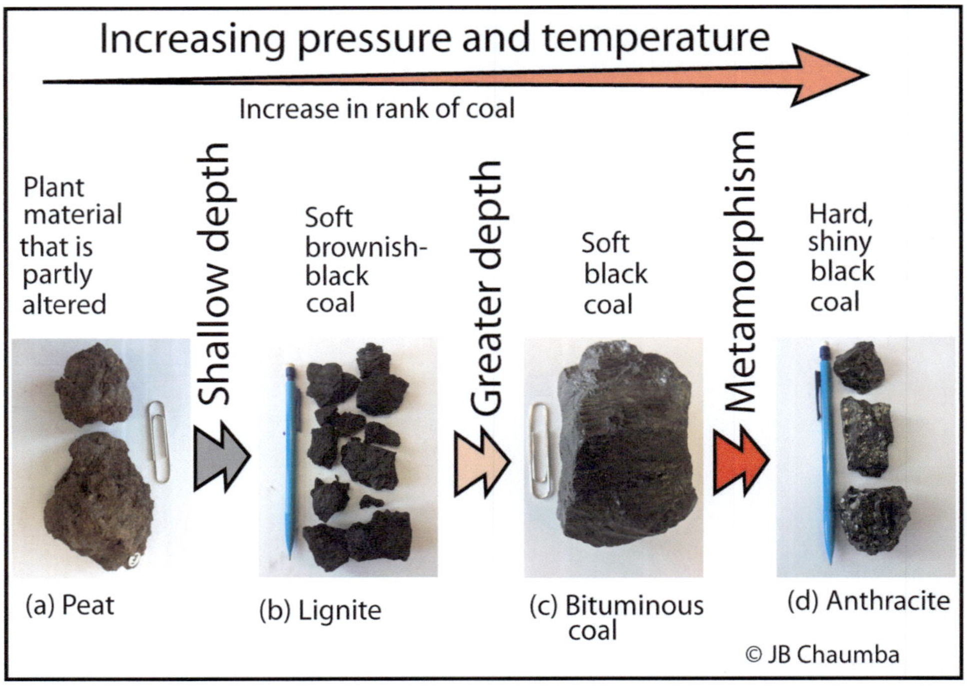

Fig. 4.7 Cartoon illustrating the successive stages inferred to operate in the formation of coal. Temperature and pressure are thought to increase, accompanied by an increase in the carbon content (and hence the rank or grade of coal) from peat (**a**), lignite (**b**), bituminous coal (**c**), and anthracite (**d**)

4.3.1 Bed Surface Markings

During, or soon after, deposition when the sediment is still soft, some features such as **mud cracks** (or desiccation cracks) and **scour marks** develop on the surfaces of beds. When a layer of mud or clay dries up and shrinks after deposition, cracks that are roughly hexagonal in shape form plates that tend to curl upwards at their edges. Mud cracks are the openings between these cracks (Fig. 4.8a). Mud cracks indicate drying after deposition. Conditions that lead to the formation of mud cracks are common along muddy margins of lakes and in tidal zones, for example. Mud cracks are formed on the Earth's surface, indicating exposure to air of the mud or clay at the time that it was formed.

Further, as the current flows over the surface of sediments, small troughs, termed scour marks, may be eroded (Fig. 4.8b). Scour marks form parallel to the flow of current. Fossils, already mentioned above, such as footprints or imprints of shells, may also form on the bedding surfaces.

4.3.2 Ripple Marks

Elongated but relatively small ridges that form in sand on the surface of bedding planes and they are aligned perpendicular to the direction of flow of current are features called **ripple marks**. Symmetric ripple marks (Fig. 4.8c), which form due to back-and-forth (oscillation) motion of water beneath waves, and asymmetric ripple marks (Fig. 4.8d), which form from unidirectional motion from either air or water currents, are the two types of ripple marks. Ripple marks are useful in determining the direction in which the current that formed those ripples moved. Bedding planes of ancient rocks typically contain ripple marks, so do modern beaches. A dune, although much larger, resembles a ripple mark. Dunes can form on riverbeds (tens of centimeters to several meters high) whereas much larger dunes (tens of meters to over 100 m) form in desert regions due to wind deposition. The surface where the ripple marks occur also indicates the direction in which the topmost layer was deposited (called younging), and it is therefore labeled as the top of the sedimentary sequence (Fig. 4.8b–d).

4.3.3 Cross Beds

Cross beds are internal laminations in which one set of laminations is inclined relative to other set (Fig. 4.9a). Cross bedding forms when currents of water or air move in one direction (unidirectional) whereby erosion occurs on the windward (upcurrent) side and deposition of the eroded material occurs on the leeward (downcurrent) side. Classical examples of cross bedding are displayed by sandstones exposed in the Navajo National Park in the U.S. state of Utah (Fig. 4.9b). Gravity is important in the migration of ripples or dunes. Particles that are transported to the crest of the cross beds tend to slip down the leeward (or slip face) side, resulting in ripple or dune migration in the downcurrent direction. It is the slip face, with its internal curved surfaces, which is responsible for the cross bedding.

Fig. 4.8 **a** Scorching temperatures which sometimes spend a couple of summer months above 40 °C (104°F) form these mud (or desiccation) cracks on Rock Creek, Johnson County, Texas, USA. **b** An illustration of a sole mark, an example of a scour mark, showing the sole mark which resembles the sole of a shoe as it appears on the bottom of the younger sandstone bed which overlies the older shale bed in this example. Diagrams illustrating symmetrical ripples (**c**) and asymmetrical ripples (**d**). From the ripples shown in (**c**) and (**d**), one can determine which way strata are younging

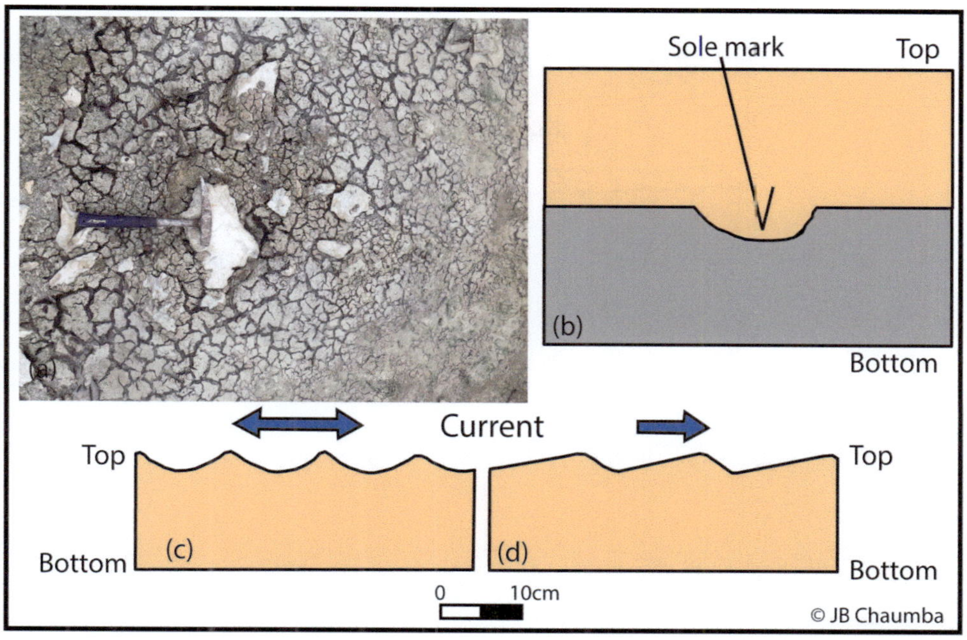

Cross bedding can be formed due to wind or water depositing clasts in rivers, deltas, and dunes. Wind may pile sand into dunes, and water currents also produce dunes. From cross bedding, or cross stratification, it is possible to infer a depositional environment that is dominated by either water or wind currents. The direction in which the cross beds dip is an indicator of the flow direction of the current which deposited the clasts (Figs. 4.9a). Mud cracks, symmetric ripples, graded beds, and certain types of cross beds tend to occur at the surface of beds and are often useful in determining which way the beds are younging. Deformation of strata can result in overturning of the sedimentary rock sequence, such that the younger beds are now below the earlier deposited one (Fig. 4.9c). Where strata have been deformed and overturned (e.g. Fig. 4.9c), studying surface markings can help unravel the sequence in strata which were originally deposited.

4.3.4 Turbidity Currents and Graded Beds

Some sediments may be deposited in unstable areas such as those laid down on submarine slopes. In the event of a triggering mechanism, such as a hurricane or an earthquake, the sediments can then easily flow downslope at great speed due to gravity. As the sediments migrate downslope, they tend to mix with seawater to form a material that is denser than seawater termed a **turbidity current** (*turbid* = cloudy and filled with sediments).

As the turbidity current flows from the steeper continental slope to the bottom of the ocean which has a more gentle slope, its velocity is greatly reduced, resulting in the deposition of the particles that it will be transporting. These sediments are not all deposited simultaneously, however. The very coarse-grained particles settle out of the turbidity current first since they are the heaviest, and they are then followed by progressively finer-grained sediments which are deposited above the coarser-grained particles which had accumulated earlier (Fig. 4.9d, e). The finest grained particles (clay) settle out last and are deposited on top of the rest of the sediments since clay particles can be carried in suspension the longest due to their very fine grain size. The result of this process is the formation of **graded beds** (Fig. 4.9d, e). Thus, graded beds are formed when flowing water sorts particles by size creating graded beds whereby the coarsest grains are at the bottom, followed by successively finer grained sediments that accumulate toward the top (Fig. 4.9d, e). Graded beds, or **turbidites**, therefore, are deposits that are formed as a result of turbidity currents. Graded bedding can also be formed from faster settling of coarser and heavier grains in a mixture of poorly sorted sediments.

4.4 Depositional Sedimentary Environments

A depositional sedimentary environment is the condition in which sediments accumulated. For example, sediments can accumulate in different settings which range from terrestrial through coastal to marine environments. The composition of the sediments, their grain sizes, sedimentary structures, sorting, and the type of fossils in the sediments are useful in identifying the type of depositional sedimentary basin in which sediments accumulated. The transporting media of

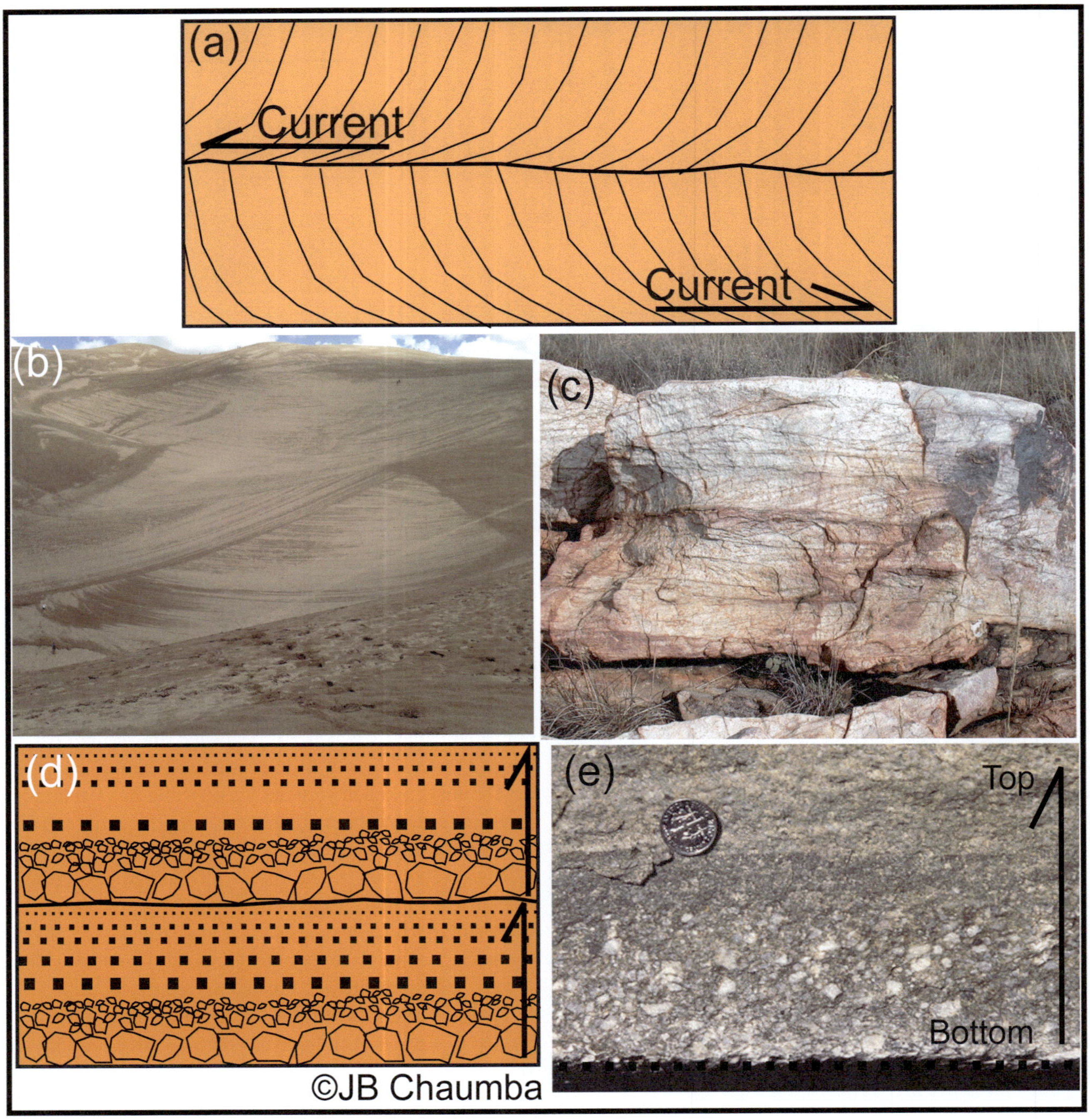

Fig. 4.9 **a** Diagram illustrating cross bedding in undeformed strata. **b** Image of cross bedding in the Great Sand Dunes National Park, Colorado, USA (credit: T. Gesualdo). **c** A cross-sectional view of overturned cross beds in the Witwatersrand Quartzite, Vredefort impact structure, South Africa (credit: D.L. Reid). **d** Diagram illustrating graded bedding which forms from running water sorting particles by their sizes. Graded bedding is common in sediments deposited by turbidity currents. Arrows indicate direction in which the sediments are younging. **e** Image of graded bedding (credit: J.W. Merck, Jr.). Arrows indicate direction in which the sediments are younging

the sediments (waves, swift flowing or still water, ice, wind) is often helpful in determining the prevailing climate during the time the sediments accumulated. Studying the sedimentary strata (its succession) may provide clues on whether the sediments were deposited in shallow water or deeper water, along a beach or in a lake environment. Sediments tend to be deposited in three different environments, namely continental (or terrestrial), transitional, and marine (ocean) (Table 4.2). Let us now briefly examine these three depositional environments and some examples of each, and how they can be distinguished from each other.

Table 4.2 Table summarizing some depositional environments discussed in the text

Depositional environment	Example	Examples/common features
Continental	1. Fluvial	Meandering and braided stream
	2. Desert	Sand dune, playa lake, alluvial fan
	3. Mountain stream	Alluvial fan
	4. Lacustrine	Bottomset, foreset, topset
	5. Glacial	Marves, moraines
	6. Cave	Travertine
Transitional	1. Estuarine	Sand, silt, and clay
	2. Barrier island	Sand dune
	3. Lagoon	Silt and clay
	4. Tidal flats	Mollusks, salt-tolerant plants
	5. Delta	Distributaries
Marine	1. Shallow marine	Continental shelf
	2. Deep marine	Pelagic clay
	3. Carbonate	Limestone, dolostone, micrite
	4. Evaporites*	Rock salt and rock gypsum

**Note* that evaporites can form in a variety of depositional environments such as marine depositional environments and also in arid regions in lagoons, tidal flats, playa lakes, and saline lakes

4.4.1 Continental Depositional Environments

Continental (or terrestrial) depositional environments form on landmasses (Table 4.2). Sediments that are deposited on the continents, either on dry land, or under and adjacent to freshwater are predominantly detrital in nature. Any iron-bearing sediments that accumulate on the continents can react with dissolved oxygen in freshwater resulting in the formation of iron-oxide minerals to form reddish colored sediments termed red beds. Let us now discuss some examples of terrestrial depositional environments. Continental depositional environments form sedimentary rocks that have highly variable compositions, sedimentary structures, and textures.

4.4.1.1 Fluvial (River) Depositional Environments

Fast flowing rivers can transport and deposit sand and gravel. Sand, silt, and clay that accumulate along the banks, bars, and floodplains of rivers (Fig. 4.10a, b) are some classic examples of fluvial deposits. Thus, fluvial deposits occur along the course of streams, and discharge (the quantity of water passing a certain point on the stream in a given time), water velocity, nature of the sediments being transported, and the shape of the river channel all determine the particle size of the sediments that are deposited. Sands, silts, and clays have a tendency to grade into one another in fluvial deposits. The type and amount of sediments that are deposited in fluvial environments depend on the discharge, water velocity, and the shape of the stream channel (Fig. 4.10a, b).

4.4.1.2 Desert (Aeolian) Depositional Environments

Wind is the principal medium of transporting sediments in desert environments. Strong winds, which characterize desert environments due to their little plant cover, are capable of eroding, transporting, and depositing sediments. Sand dunes (Fig. 4.10c), alluvial fans, and playa lake (Fig. 4.10d) deposits are all deposited in desert environments. However, since air has a density that is much lower compared to the density of water, it chiefly transports particles of sand, silt, and clay. Coarser sand particles can be transported via **saltation** whereby the particles are carried in suspension for only short distances and then they bounce off the surface. Some coarse-grained sand particles are transported by rolling along the surface of the desert. The inability of the wind to transport very coarse-grained particles results in wind deposits being well-sorted. Very fine-grained clay tends to be transported mostly in suspension.

4.4.1.3 Mountain Stream Depositional Environments

Sediments that are transported by rivers accumulate quickly when rapidly flowing streams emerge from mountainous areas onto a flat plain (Fig. 4.10e) due to the drastic reduction in water velocity. Abrupt deposition of sediments which occurs on flat plains can lead to the formation of an **alluvial fan** (Fig. 4.10e). The sediments deposited by moun-

(a)
(b)
(c)
(d)
(e)
(f)
(g)
(h)

◀**Fig. 4.10** Some examples of continental depositional environments. **a** Bend of meandering displayed by the Orange River, Namibia/South Africa. Note the different generations of old meanders in the foreground and background. **b** Braided channel displayed by tributary to the Nyanyadzi River in Manicaland, Zimbabwe. Photo credit: Kelvin Moyana. **c** Sand dunes, Namib Desert, Namibia (credit: D.L. Reid). **d** Sossus Vlei evaporites along the Tsauchab River valley after water breaches the Namib Sand Sea. **e** Image of small alluvial fans in Kuiseb Canyon, Namib Desert, Namibia. **f** Amistad Lake in southwestern Texas. The Lake straddles the US-Mexico border within canyons of Cretaceous limestone. **g** Image of fluted moraine. A set of low ridges formed parallel to ice flow, meter-scale in width, 100 m-scale in length. The ridges commonly extend down-valley from a boulder, and consist of till (Photo credit: Swissedu.ch). **h** A view of Chinhoyi Caves, Mashonaland West, Zimbabwe (credit: Explore Tourism Zimbabwe)

tain streams transport are poorly stratified and poorly sorted alluvial fan deposits. Alluvial fans can be recognized by their wedge-shaped cross-section as well as their lobed form (Fig. 4.10e).

4.4.1.4 Lacustrine (Lake) Depositional Environments

These are deposits that accumulate from standing water bodies, and they are often finely layered (laminations) and well-sorted. Lakes (Fig. 4.10f), which are landlocked bodies of water, tend to be dominated by silt and clay, although a variety of other types of sediments may occur depending on climate, water depth, and the lithology of the area surrounding the lake. Lacustrine deposits often contain organic matter and thus they are often dark in color. Since swamps tend to be characterized by lush vegetation, lacustrine deposits that accumulate here often consist of organic rich mud and peat. The presence of fossils of freshwater organisms, mud cracks, as well as tracks made by vertebrates can be indicators of lacustrine depositional environments. Further, lacustrine environments contain diatoms which are siliceous algae.

4.4.1.5 Glacial Depositional Environments

The glacial environment may include a variety of other environments, such as stream, lake, and even shallow marine. Glaciers are capable of transporting and depositing large volumes of very poorly sorted sediments (Fig. 4.10g). All sediments deposited in glacial environments, or **drift**, are composed of particles which range from very coarse-grained boulders to gravel and sand to very fine-grained ones such as clay (Fig. 4.10g). Non-stratified and poorly sorted deposit (such as moraines) that is deposited directly by glaciers is termed **till**. This results in deposits that show a variety in their texture, structure, and composition. Glacial meltwater can erode and transport the drift and deposit them beyond the margins of the glacier to form **outwash** that are similar to those deposited by rivers (well rounded, well sorted, and coarse-grained deposits). Where glacial deposits accumulate in glacial lakes, fine-grained, well-sorted, and graded deposits which contain alternating dark and light laminations, or **varves**, accumulate. A varve, or a dark–light couplet, is thought to result from spring and summer accumulation of silt and clay sediments which form the light layer and the dark layer results from accumulations of clay and organic matter resulting from freezing over of the lake. Thus, varves are interpreted to represent yearly accumulations of sediments. Icebergs often release some dropstones which are common in varves and they often disturb the laminations.

4.4.1.6 Cave Environments

Chemical sedimentary rocks are the dominant rocks in cave environments (Fig. 4.10h). Travertine can precipitate from groundwater and it gets deposited on the ceilings, walls, and floors of caves. Further, droppings, and bones of animals that live in caves such as bats also form cave sediments. We will discuss cave environments in more detail in Chap. 11 that deals with running water and groundwater.

4.4.2 Transitional Depositional Environments

Transitional depositional environments, or the shore, are found between continental and marine depositional environments. Here, a merge occurs between continental and marine depositional processes. The transitional depositional environment changes constantly due to the occurrence of waves, currents, tides, and storms which break down shells and mineral and rock particles. The finer grained particles are transported to the sea, whereas the coarser grained particles are left behind resulting in the deposition of more resistant sand that is well-sorted and well-rounded. Deltas and huge accumulations of sand or gravel (beaches) of the shoreline are encountered. In addition, tidal flats (or mud flats), which are coastal areas that are covered by mud, tend to be flooded alternately as most tidal flats are within the intertidal zone. Consequently, water is drained from tidal flats due to tidal activity.

4.4.2.1 Estuarine Depositional Environments

Estuaries are the mouths of rivers and lagoons that are drowned by the sea. In estuaries (Fig. 4.11g), freshwater from rivers and saltwater from the oceans mix, resulting in the formation of brackish (salty) water. When mixing occurs, lighter freshwater is forced to flow oceanward with the denser saltwater underneath it. Estuaries are thought to form in two ways. Some estuaries are interpreted to have

(a)
(b)
Pamlico Sound
(c)
Atlantic Ocean
(d)
(e)
(f)
(g)

◂**Fig. 4.11** Some examples of transitional depositional environments. **a** Image of Chowan River estuary, Albemarle Sound at Edenton, NC, USA (credit: Inner Banks Inn & Restaurant, Edenton, NC). **b** Image of part of a barrier island (Jekyll Island, Georgia, USA) (credit: Explore St. Simon's Island). **c** The town of Duck, NC (built-up area in center of image) lies on the Outer Banks, NC, a narrow strip of sand surrounded by water on both sides. Pamlico Sound, a lagoon is in the background (credit: Duck, NC Real Estate, Village Realty). **d** Ripples on Great Salt Lake, UT, USA (credit: D.T. Nelson). **e** Tidal flats, Jekyll Island, Georgia, USA (credit: Explore St. Simon's Island). **f** Murky waters spilling out into Gulf of Mexico where deposition of suspended sediment forms the Mississippi River Delta, Gulf of Mexico, USA Image credit: USGS EROS Data Center Satellite Systems Branch as part of the Earth as Art II image series (Photo credit: Jesse Allen—NASA), **g** An estuary is a partially enclosed body of water where streams or rivers enter and mix freshwater with saltwater. This mixture of freshwater and saltwater is called brackish water. Conditions in an estuary are ever-changing and provide unique environments that support a diverse community of wildlife and plants (credit: SAriverauthority.Org)

been formed by a rise in sea level at the end of the Pleistocene Ice Age when glaciers melted and the water accumulated in the oceans. Examples of the estuaries include the Chesapeake Bay and Delaware Bay. Other estuaries are thought to have been formed due to subsidence of coastal areas which allowed ocean water to rise, resulting in ocean water migrating upriver valleys. In areas with weak tidal currents, estuaries tend to trap large quantities of sediment. In areas characterized by swift tides, sand is the common estuarine deposit whereas silt and clay deposits are common in protected estuaries.

The uniqueness of bay and estuary systems can be illustrated using the San Antonio Bay.

Circular tides characterize the San Antonio Bay, causing different levels of the content of salt, or salinity as well as also driving the cycling of nutrients. A partially enclosed body of water where streams or rivers enter and mix freshwater with saltwater is an estuary (Fig. 4.11g). The freshwater and saltwater mixture is referred to as brackish water. In an estuary, conditions are ever-changing, which act to provide unique environments that support diverse communities of both wildlife and plants.

4.4.2.2 Barrier Islands

When sediments are deposited offshore, roughly elongated narrow islands that are somewhat parallel to the coastline called barrier islands (Fig. 4.11b, c) are formed. Barrier islands may exceed 100 km in length, and most are only a few kilometers or less in width. Sandy sediments predominate in this high energy environment, where the sediment is constantly moved by waves and longshore currents. Along most of the U.S. Atlantic Coast and along the Gulf of Mexico coast, most of the barrier islands have been extensively developed and they attract numerous tourists throughout the year.

4.4.2.3 Lagoons

Lagoons are shallow water bodies separating barrier islands from the mainland (Fig. 4.11c). Since lagoons lie behind barrier islands, they are protected from strong ocean waves and currents. As a result, lagoon deposits are composed of fine-grained sediments. Silts and clays that are found in lagoons are often extensively burrowed by worms and mollusks, with shells and borings of the latter usually being preserved when the sediments are lithified. Organic mud, peat, and fine-grained sediments from the continents transported to the lagoons by rivers also form lagoon deposits. Further, marine sediments which are transported and deposited by storm surges also form lagoon deposits.

4.4.2.4 Tidal Flats

On the landward side of lagoons are almost featureless, low-lying plains called tidal flats (Fig. 4.11d, e). They are tidal because they are constantly inundated and drained by tides, usually twice daily depending on the location. Depending on the location, tides rise and fall from several inches to several feet vertically, over just a few hours. Because lagoons are largely protected from the open ocean, they gradually fill with clay and silt brought by nearby streams. Gentle tidal currents distribute the sediment across the lagoon. The layers of sediment build upward until they reach the low tide. The constant rise and fall of the water level makes tidal flats a challenging environment for many organisms. Yet, salt-tolerant plants, mollusks, crustaceans, and worms thrive in many tidal flat environments. Where tidal flats prograde, a sequence of rocks that grade upwards from sand through sand and mud to mud is formed. Such a sequence of rocks produces sets of cross beds (symmetric ripples) (Fig. 4.11d) that dip in opposite directions.

4.4.2.5 Delta Depositional Environments

A delta (Figs. 4.11f, g, 4.12) is a fan-shaped body of sediments that are well sorted. These sediments are deposited by rivers due to sudden reduction in water velocity where the river empties into the ocean (Fig. 4.11f) or a lake (Fig. 4.11g). Deltas that form at the coast tend to **prograde**, or build seaward (or lakeward), if sediments are deposited at a faster rate than that at which they are eroded by coastal currents, tides, and waves. Progradation is the formation of a vertical sequence of rocks that increase in grain size from fine-grained beds at the bottom to coarse-grained ones at the top. Small and simple deltas can also be formed when rivers empty their waters into lakes. Deltas in lakes are subdivided into a threefold classification composed of bottomsets, foresets, and topsets (Fig. 4.12). Bottomsets are fine-grained and they are formed at some distance from the shore and foresets (medium- to –coarse-grained) are later deposited over the bottomsets, which in turn will eventually be overlain by coarse- to very coarse-grained topsets when deposition occurs in channels (Fig. 4.12). Marine fossils (those of lake-dwelling organism) may be present in bottomset beds, with fossils of land plants and animals likely to be present on topset beds.

Marine deltas are more complex due to the influence of waves, tides, and fluvial processes. Marine deltas that are dominated by rivers, such as the Mississippi River delta, tend to be characterized by long distributary channels which extend far into the ocean. Deltas that are dominated by waves such as those of the Nile River in Egypt and the São Francisco River in Brazil, also have distributary channels. However, as the oceanward margin of such wave-dominated deltas progrades, they tend to be simultaneously modified by the action of waves. Some deltas, such as the Ganges River delta in Bangladesh, are dominated by tides. At tide-dominated deltas, the characteristic features here are the tidal sand bodies which form parallel to the direction in which the tide flows.

4.4.3 Marine Depositional Environments

Marine depositional environments, which vary depending on the depth of sediment deposition, encompass the continental shelf, slope and rise, as well as the deep ocean floor (Fig. 4.13). Most of the detrital sediments that are derived from the continents eventually get deposited in the marine environments. Some sediment types, however, originate from the marine environment. Examples of marine-derived sediments include limestones and evaporites which are found in the shallow marine depositional environments (continental shelf). Marine depositional environments can be subdivided into shallow marine, continental slope, and deep marine environments.

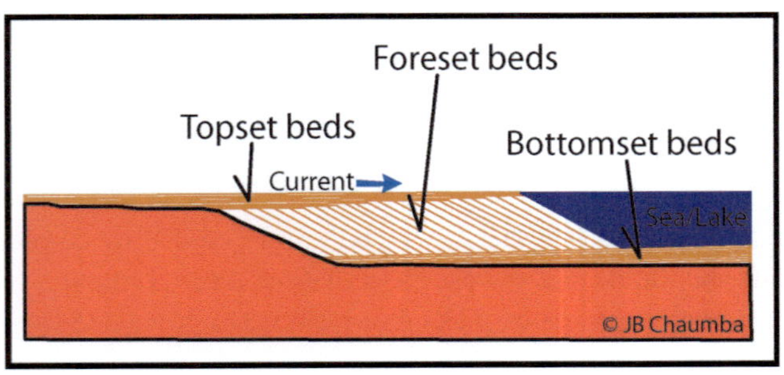

Fig. 4.12 Diagram showing topsets, cross bedding (foresets), and bottomsets in a delta. The cross beds are deposited above the bottomsets, whereas topsets are deposited above the foresets

4.4.3.1 Shallow Marine Depositional Environments

The continental shelf (Fig. 4.13) is part of the shallow marine environment. The continental shelf depositional environment, which can be defined as the submerged edges of the continents and continues to the coastal slope (Fig. 4.13), composes a shallow ocean environment which occurs near the shore. The part of continental shelf that is stirred up by tides and waves is the high-energy inner part, and it is composed of sand that forms huge dunes. These sand dunes have cross beds and bedding planes that display ripple marks formed by wave activity. The outer, low-energy part of the continental shelf is composed dominantly of mud which contains marine fossils. The inner-outer transition part of the continental shelf is characterized by interfingering layers of sand and mud.

Sediments that are eroded from the continents and carried to the sea in streams eventually are transported across the continental shelf, and some are deposited on them. The nature of the source rock on adjacent landmasses, the elevation of source areas, the distance from shore, and the presence of carbonate secreting organisms, for example, influence the kind of sediment deposited on the continental shelf. Because larger grains are heavier than small ones, they tend to drop out sooner and are deposited closer to shore. Also, shallow-water currents and wave action keep finer particles in suspension and carry them farther out into the ocean. For these reasons, sediment deposited in the shallow marine environment tends to be coarser than material accumulating in deeper parts of the ocean. Sand and silt are thus common in the shallow marine environment.

A continental slope is an area of the ocean floor that extends from the oceanward edge of the continental shelf down to the depths of the ocean (Fig. 4.13). From the upper boundary of the continental slope which is sharply defined, the surface of the ocean floor drops to depths ranging from ~1400 (4600 ft) to ~3200 (10,500 ft) meters. At these great depths, the slope of the ocean floor tends to be gentle. The

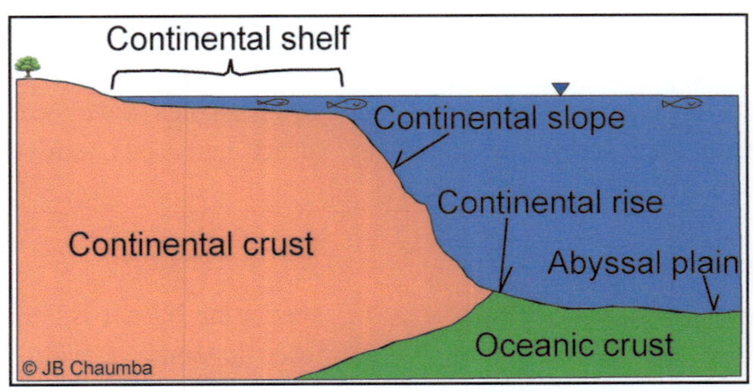

Fig. 4.13 Diagram showing the continental shelf, continental slope, continental rise, and abyssal plain

continental rise is composed of a less-pronounced slope (Fig. 4.13).

The shallow marine environment occurs at depths not exceeding 200 m (700 ft). Continental shelves can range in width from less than 1 km (0.6 mi) wide to hundreds of kilometers into the ocean. Sediments that are derived from the continents are ultimately carried oceanward by tides and waves to be dumped on the continental shelf. Plant and animal life is encountered in the upper part of the shallow marine environments which is penetrated by sunlight to depths of ~100 m (330 ft) below the ocean surface. Thus, sands and muds that are rich in carbonate mix with the diverse forms of plant and animal remains and they are common sediments of shallow marine depositional environments.

4.4.3.2 Deep Marine Depositional Environments

The seafloor beyond the continental rise is almost entirely covered by fine-grained sediments. In the deep marine environment far from the continents, most of the sediment that settles onto the ocean floor is very fine clay, volcanic ash, calcareous or siliceous shells of marine microorganism, and wind-blown dust from the continents. Exceptions, however, can occur and they include sporadic occurrences of coarser sediments that are carried down continental slopes into the deeper parts of the ocean by turbidity currents as well as coarse debris that may be dropped by melting icebergs into the deep-water environment. Clay that covers most of the deeper parts of the ocean floor is termed pelagic clay.

4.4.3.3 Carbonate Depositional Environments

Limestone and dolostone (altered limestone) are the carbonate rocks that are widespread. Micrite, a microcrystalline carbonate mud, is present as limestones composed of sand- and gravel-sized particles, or mixed terrigenous/carbonate sediments (Fig. 4.14a). Terrigenous sediments are shallow marine sediments derived from the continents. The carbonate particles are not transported from elsewhere, but they form in the depositional environment although waves and local currents are responsible for any sedimentary structures and sorting that may occur. A consequence of **marine regression** (or drop in sea level) is progradation of terrigenous sediments whereby a blanket of siliciclastic sediments forms over the exposed carbonates where carbonate topographic lows are encountered, limiting dissolution by rainwater to the carbonate topographic highs (Fig. 4.14b).

Mud cracks are also common in micrites that form along shorelines. It is important here to note that although the majority of limestones form in warm and shallow oceans either on carbonate shelves or on carbonate platforms that rise from the ocean depths, there are some limestones that are formed in lacustrine depositional environments on the continents. An important parameter for the deposition of limestone is that of the almost total absence of detrital sediments. The ocean tends to be warmer where there are fewer continent-derived sediments, and lime muds of biochemical origin accumulate. In shallow and warm ocean environments, coral reefs flourish. Here, sunlight can penetrate down to the ocean floor, providing heat energy to warm the water as well as light energy needed for photosynthesis. Algae and other plants also thrive, forming food for animals that swim and those that live at the bottom of the ocean.

4.4.3.4 Evaporite Depositional Environments

Evaporite deposits often form in closed marine basins where the rate of evaporation exceeds the rate of inflow (Fig. 4.14c, d, e). Figure 4.14c, d, e Three models have been proposed for deposition of marine evaporites in basins of restricted water circulation. In the first model, evaporites are inferred to form from evaporation from a deep-water deep basin setting (Fig. 4.14c). In the second model, extensive thin-shelf deposits are thought to form from evaporation in a shallow-water deep basin associated with shallow, ephemeral seas (Fig. 4.14d). Finally, evaporation from a shallow-water shallow basin can form lead to the formation of evaporites. Evaporite deposits associated with model three (Fig. 4.14e) often display a repeated sequence of minerals which is interpreted to indicate cyclic conditions with the resulting minerals being determined by solubility (Fig. 4.14e). Some of the most common evaporate minerals include calcite, gypsum, anhydrite, and halite. It is important to note here that non-marine evaporites are formed by streams that flow into closed basins (especially in arid regions) giving rise to deposits of evaporites such as borates, nitrates, and sodium carbonates, comparable to the deposits which occur in Utah and in southern California (USA).

Although evaporites, which include rock gypsum and rock salt, form in a variety of depositional environments such as saline and playa lakes, the majority of them are deposited in the ocean. Extensive evaporite deposits also occur in the U.S. Gulf Coast states, and in the U.S. states of Michigan, Ohio, and New York and in the Mediterranean Sea. Examples of areas with favorable conditions for the formation of evaporates include arid regions (e.g. Persian Gulf). Table 4.3 summarizes continental, transitional, and marine depositional environments. Examples of each depositional environment are shown in the form of photographs accompanied by a brief explanation of the locality of the environment of deposition.

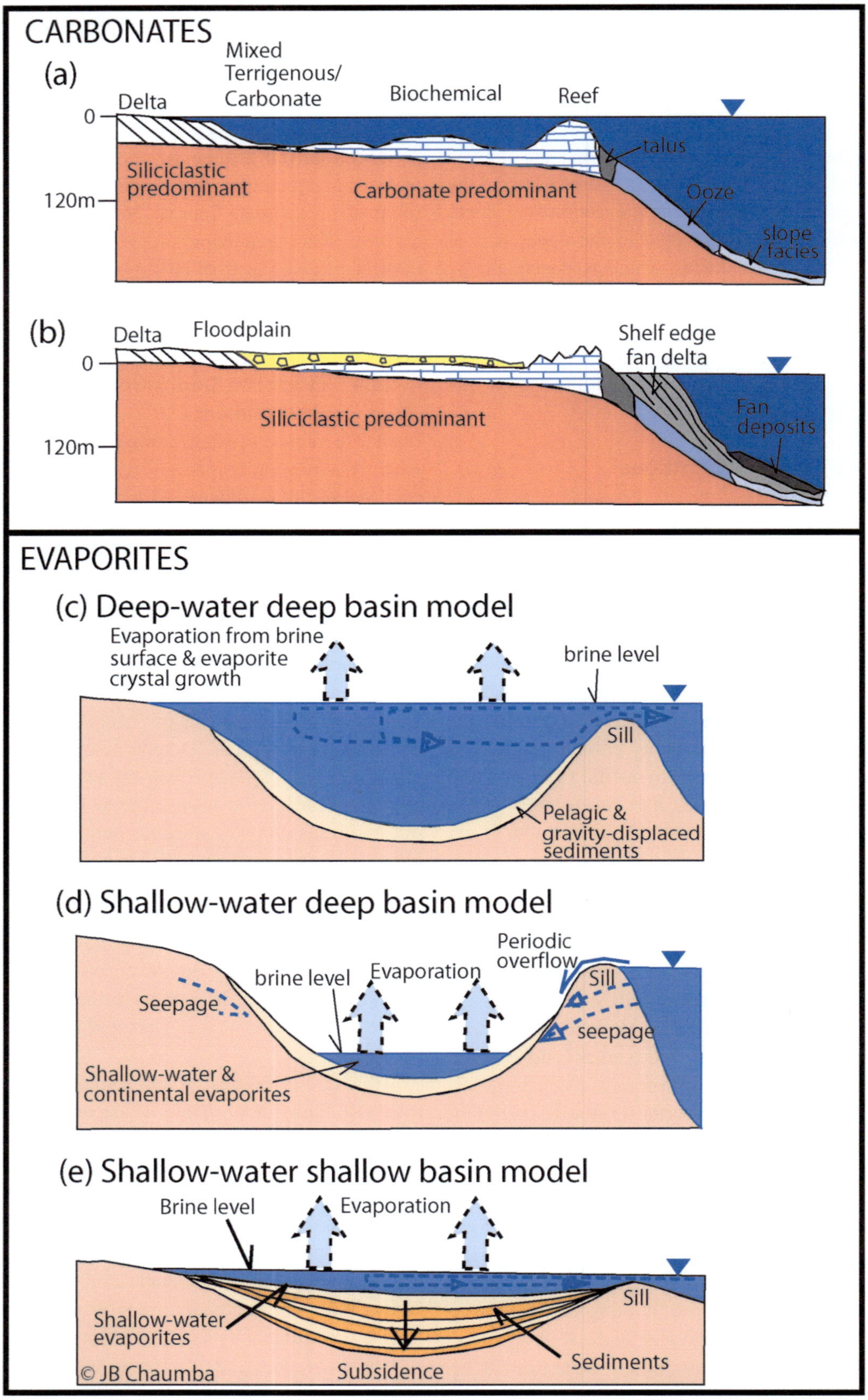

Fig. 4.14 **a** and **b** Marine regression can result in siliciclastic deposition and its influence on carbonate rocks based on an example from the Northeast Australian Platform. **c**, **d**, **e** Three models for the deposition of marine evaporite deposits in basins of restricted water circulation

Table 4.3 Summary and pictures of continental, marine, and transitional depositional environments

Environment	Example	Illustration	Explanation
Continental	1. Fluvial		Stream deposited sediments, Namibia
	2. Desert		Silversands beach, Bettys Bay, So. Africa
	3. Mountain stream		Mountain stream ideposits in Namibia
	4. Lacustrine		Lacustrine deposits in Aussenkehr, Lesotho
	5. Glacial		Franz Josef Glacier, New Zealand, 2006
	6. Cave	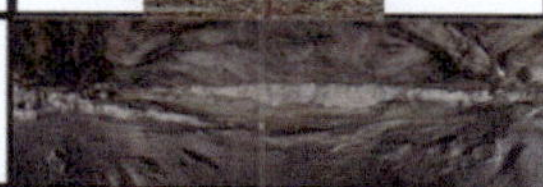	Woodstock Cave, Cape Town, SA
Transitional	1. Estuarine		Qwuloolt estuary, Western Australia
	2. Barrier island		Ocracoke, North Carorilna, USA
	3. Lagoon		Big Lagoon, Western Australia
	4. Tidal flats		Little St. Simons Island, Georgia, USA
	5. Delta		River delta in Puget Sound, WA, USA
Marine	1. Shallow marine		Rocky coast near Cape Town, South Africa
	2. Deep marine		Once on sea floor, then uplifted & eroded, NZ
	3. Carbonate		Thick Campanian chalk Isle of Wight, England
	4. Evaporites		Valley of the Tsauchab River, Namibia

Bulk of photographs credit: D.L. Reid; River delta in Puget Sound, Washington state, USA (credit: Puget Sound Nearshore Ecosystem Restoration Project)

4.5 Marine Transgression and Regression

Facies, or some parts of rock bodies that can possess characteristics from which the depositional environment can be inferred, can provide clues on how these facies shifted throughout time. Any changes in sea level relative to land which controlled the succession of sediments in depositional sedimentary environments can, thus, be investigated. When sea level rises (**marine transgression**), it slowly transgresses the land depositing a nearshore sand facies, an offshore mud facies, and a far-offshore carbonate facies on the ocean floor (Fig. 4.15a, b, c). The boundaries of these facies shift continent ward as shoreline advances inland to produce a sequence of beds whereby coarser-grained beds are overlain by progressively finer-grained one, or an **onlap sequence** (Fig. 4.15a, b, c).

Eventually, sea level can drop (marine regression), and during this stage, the boundaries of the facies migrate in the same direction as the shoreline to create a sequence of beds whereby finer-grained beds are overlain by progressively coarser-grained ones, or an **offlap sequence** (Fig. 4.15d). During a drop in sea level, the recently deposited beds can be eroded away, resulting in a loss of part of the sequence (Fig. 4.15d). Any ancient sea level advances and retreats can thus be recognized through investigating the onlap and offlap sequences (Fig. 4.15).

4.6 Economic Importance of Sedimentary Rocks

The economic importance of sedimentary rocks is varied. These rocks are of great importance in the construction industry. Because sedimentary rocks are relatively soft, they can easily be cut into blocks that are used for various purposes in construction. Quartz, which is also found in sedimentary rocks, is used in the glass manufacturing industry and rock gypsum is useful in making plaster. Limestone is of great importance in making cement. Energy resources are critical in society such as coal, natural gas,

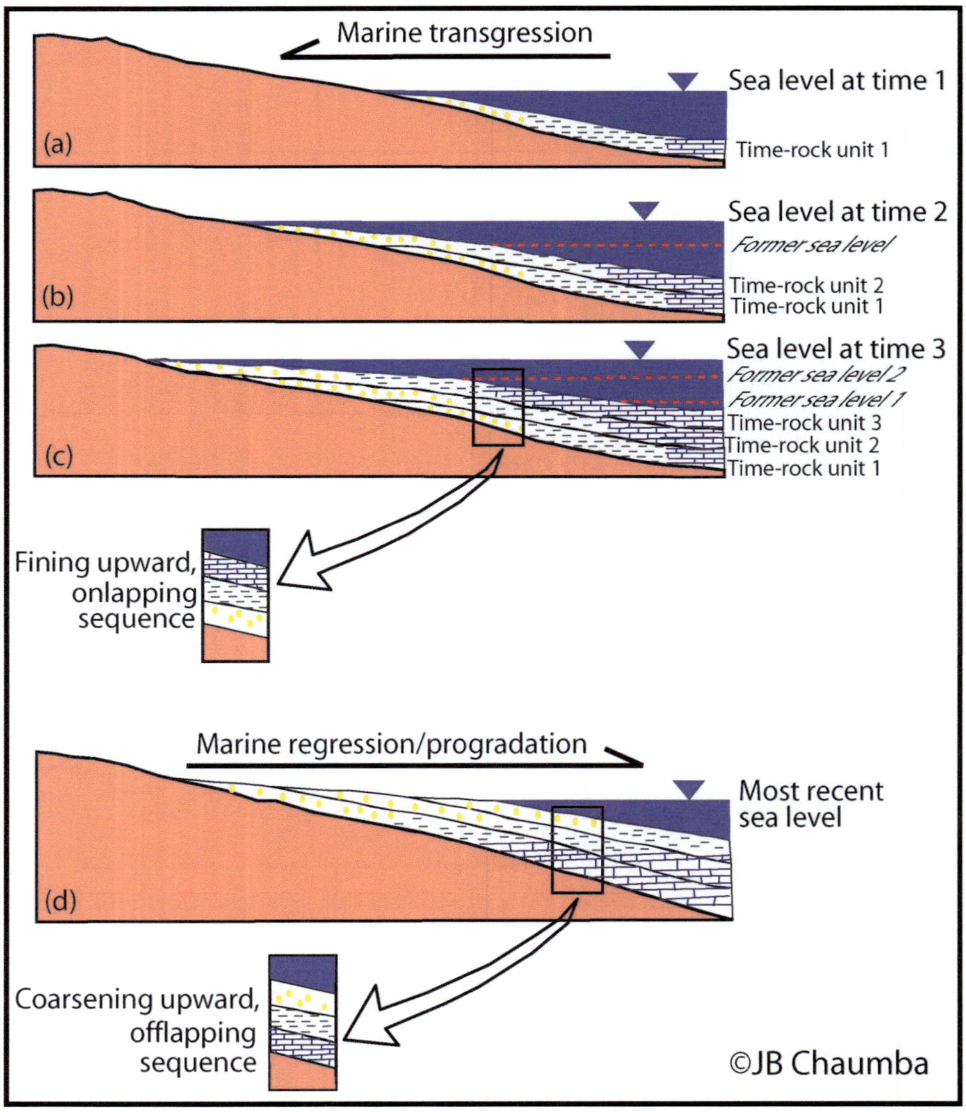

Fig. 4.15 **a** Marine transgression. During an advance in sea level, finer-grained offshore beds overlie coarser-grained nearshore facies. Nearshore beds are progressively displaced away from any marine reference point. **b** Marine regression. During a retreat in sea level, coarser-grained nearshore beds overlie finer-grained offshore facies. Sandy nearshore beds are progressively displaced towards any marine reference point

oil, uranium, among others, formed in sedimentary rocks. Black shales are often of economic importance as sources of petroleum products and metals.

Glossary

Alluvial fan a fan-shaped accumulation of alluvium that is deposited when the flow velocity of a river decreases

Cementation a type of lithification whereby new minerals precipitating from groundwater help to partially or completely fill the pore spaces between the clasts causing the clasts to be attached (cemented) together

Compaction a type of lithification whereby the weight of overlying sediments squeezes water and air out reducing the volume of the rocks or sediments due to tighter packing of clasts

Cross beds laminations in layers of sedimentary rocks that are inclined to the main bedding and indicate the slip faces of dunes or ripples

Drift any accumulation of sorted clastic sediments that have been deposited by meltwaters associated with glaciers

Facies a sequence of sedimentary rocks and structures that indicate a particular environment of deposition

Fossil any remnant of a once living organism, or a trace made by a once living organism that was preserved in rocks of sediments

Graded beds layers of sediments that are formed when turbidity currents deposit their load to form beds that range in grain size from very coarse-grains at the bottom to very very-grained ones at the top

Lithification a process that involves the transformation of sediments into rock and through compaction and cementation

Marine transgression a phase during which sea level rises relative to the land causing the shoreline to move toward higher ground, resulting in flooding coastal areas. Marine transgression can also be caused either by the land sinking or the ocean basins being filled with water (or decreasing in capacity)

Marine regression a phase occurring when areas of a submerged ocean floor become exposed above the ocean level. The opposite event, which is the opposite of marine transgression which occurs when flooding from the ocean covers previously exposed land areas

Mud cracks sedimentary structures that are formed as muddy sediments dry and contract

Offlap sequence A sequence of inclined strata that was deposited during a marine regression whereby each stratum is succeeded laterally by progressively younger units, to indicate the direction in which the sea retreated

Onlap sequence progressive spreading of successive beds of strata over an increasingly wider area, usually due to rising sea level or gradual subsidence of a land area and results in an overlapping stratigraphic relationship between the sedimentary succession and the underlying rocks forming the basement.

Outwash meltwater that escapes from the terminal zone of glaciers results in braided streams which show marked variations in discharge from season to season. Outwash can also refer to stratified gravels and sands that are deposited at or near the margins of glaciers.

Plankton very small animal and plants that float in the ocean or in lakes

Prograde growth of the delta of a river further out into the sea or lake over time

Ripple marks subparallel furrows and wavy ridges that form on the surfaces sand or mud due to the action of water or wind, and are sometimes fossilized

Saltation movement of sediments whereby particles bounce off the surface and repeatedly knocking back the particles into either water of air

Scour marks sedimentary features that are formed due to erosion of a sediment surface by current that flows over it resulting in the soft but cohesive mud or sand surface being sculptured and reshaped

Sediment loose particles formed from the weathering and erosion of mineral and rock by precipitation from water

Sedimentary rock any rock that forms at or near the Earth's surface from products of weathering of preexisting rock due to transportation, deposition, and lithification.

Sedimentary structure a unique form or shape that is produced by sediments at the time of their deposition

Sorting a measure of the proportion of size of clasts in sediments or rock that is composed of uniform size. Well-sorted sediments are composed of class of the same size, whereas poorly sorted sediments/rocks are composed of clasts of different sizes

Stratum till poorly sorted sediments that are deposited directly by glaciers

Turbidites graded beds that are deposited by turbidity currents that form from the bottom of a submarine slope

Turbidity current a submarine current of sediment-laden water that races down submarine slopes

Varves very thin layers of sediments deposited by glaciers composed of silt layers derived from flooding in the spring alternating with clay layers derived from the lake's freezing over during still standing water during the spring

Weathering process by which solid rock is broken up and chemically altered to form sediment

5 Metamorphic Rocks and the Rock Cycle

Abstract

This chapter begins by defining metamorphism and the roles played by increasing heat and pressure, as well as hydrothermal fluids, in the formation of metamorphic tocks. Diagrams are also utilized to explain the causes of metamorphism. With the aid of both diagrams and photographs, the origin of diverse types of metamorphic rocks is described. The classification of metamorphic rocks into foliated and nonfoliated rocks is accompanied using tables, diagrams, and photographs. The intensity of metamorphism of low-, medium-, and high-grade is described. The types of metamorphism under which different metamorphic rocks are formed, such as regional, contact, subduction, hydrothermal, dynamic, and pyrometamorphism, are described. Finally, the rock cycle, metamorphic facies, and the economic importance of metamorphic rocks, are described.

Learning Goals

Studying this chapter will enable you to:

- List the common metamorphic rocks
- Explain how different metamorphic rocks form
- Contrast foliated and nonfoliated metamorphic rocks provide examples
- Explain how metamorphic rocks are classified
- List and describe the agents of metamorphism
- Distinguish the different metamorphic settings
- Explain the rock cycle
- Sketch and label the various pathways through the rock cycle.

Supplementary Information The online version contains supplementary material available at https://doi.org/10.1007/978-3-032-20231-4_5.

5.1 Introduction

When any igneous or sedimentary rock that we have discussed so far in the previous two chapters is subjected to conditions that are different from those they experienced during their formation, they are transformed, that is, they are metamorphosed, to form new rocks. The new rocks are stable under the new conditions of increased pressure and temperature than those experienced when the rock was originally formed. We refer to rocks formed under these increased pressure and temperature conditions as **metamorphic** (from the Greek words *meta*—change and *morphe*—form or shape) **rocks**. These conditions where metamorphic rocks are formed typically lie between those forming igneous and sedimentary rocks. **Metamorphism** (to change from), thus, can therefore be described as the process in which heat, pressure, and the chemical reactions that occur within the Earth's crust change the form/shape and/or the mineral composition of the *protolith* (**parent rock**) with no melting of the rock involved. In addition to pre-existing igneous and sedimentary rocks, even any previously formed metamorphic rock can also be re-metamorphosed to form a new metamorphic rock! Any metamorphic rock, therefore, has a pre-existing rock from which it was formed.

5.2 Definition of Metamorphism

As we discussed in Chap. 3, igneous rocks form at high temperatures (and sometimes pressures) and minerals that form igneous rocks tend to be stable at those high temperatures. Similarly, minerals that form at low temperatures and pressures that are experienced at the Earth's surface are

J. B. Chaumba, *Introductory Physical Geology*, Springer Textbooks in Earth Sciences, Geography and Environment,
https://doi.org/10.1007/978-3-032-20231-4_5

also stable at these relatively low temperatures and pressures. Once igneous and sedimentary rocks are subjected to different environments than those from which they were stable, such as being pushed to greater depth with the Earth's crust, they experience temperature and pressure conditions that are totally different from those under which they were formed. The minerals forming up these rocks, consequently, become unstable in the conditions that the rocks have now been exposed to resulting in their changing to new ones that are now stable under the new conditions. Plutonic igneous rocks, for example, become unstable under conditions prevalent at the Earth's surface such that they begin to weather. Minerals making up sedimentary rocks, which form at relatively low temperatures and pressures at the surface of the Earth, become unstable at high temperatures and pressures resulting in these minerals reacting to form new ones that are stable at these high temperature and pressure conditions.

New textures occur when rocks are metamorphosed. These new temperature and pressure conditions occur to rocks without the involvement of any melting to the rock being changed. Metamorphism, therefore, is the process that transforms earlier existing rocks into new minerals (and thus, rocks) and textures in their solid states (i.e., without melting to form magma). The nature in which existing minerals are transformed into new ones differs as some relatively unstable minerals such as micas easily transform into new minerals that are stable at the new conditions. More stable minerals such as quartz can recrystallize into forms that are stable at high temperature and pressure conditions. Further, ions and atoms also migrate within the metamorphic rocks to form varieties that tolerate high temperature and pressure conditions. Let us bear in that some unstable minerals that are produced during metamorphism tend to react with each other to form new stable minerals which were not present in the parent rock. The energy provided for breaking the bonds between some atoms and ions is provided by the increased heat and pressure. However, not all the bonds can be broken during metamorphism; if that occurs, melting would be initiated which forms magma which later cools down to form igneous rocks. The chemical composition of the new metamorphic rock is broadly comparable to that of the parent rock, with the exception of volatiles such as water and carbon dioxide, for example, which can escape during metamorphism.

Let us make a note here that most metamorphic reactions occur in the presence of hot water fluids termed hydrothermal fluids, which act to accelerate these reactions. Some of these fluids are produced from heating of the rocks which undergo metamorphism (protoliths) which causes the water that was originally bonded to some of these fluids to be released. Some of the water may either have penetrated into the parent rock from groundwater or incorporated together with the sediments at the time of their deposition. Some of the hydrothermal fluids, however, can also be absorbed by some minerals during metamorphic reactions.

5.3 Causes of Metamorphism

A number of factors such as increasing heat, or increasing pressure, or increasing differential stress, as well as the involvement of hydrothermal fluids can cause metamorphism. Each of these factors, or a combination of some of them, can cause metamorphism independently or simultaneously, although each factor tends to vary greatly from location to location depending on the setting at which metamorphism occurs.

5.3.1 The Role of Heat

Weathering, transportation, and deposition tend to result in several different products being lumped together in a depositional sedimentary environment without much happening to them at relatively low temperatures. The increasing depth of burial due to geothermal gradient (see Chap. 3) results in heating of these sediments leading to significant changes to the rocks. Other sources of heat in the Earth's crust are radioactivity, conduction of heat from the mantle to the crust, and convection. Consequently, the single critical agent that drives metamorphism is increasing heat as it is responsible for triggering chemical reactions in rocks resulting in recrystallization of any existing minerals which form new minerals that are stable at increased temperature (and pressure) conditions. This thermal energy is caused either by the increased temperature of the rocks due to intrusion of magma or by an increase in both temperature and pressure.

Recall from Chap. 3 that rocks that form at the surface of the Earth tend to experience a gradual increase in both temperature and pressure with increasing depth at a rate roughly equating to 25 °C every kilometer. Consequently, at a depth of ~8 km, minerals such as clays begin to recrystallize into white micas and chlorite that are stable at corresponding temperatures of ~200 °C whereas numerous other minerals such as quartz and feldspar would still be stable at these conditions. Within the interior of stable lithospheric plates on Earth, rocks occurring at great depths tend to continually remain at the same high temperatures and pressures provided they remain at the same locations and similar chemical compositions. Increasing heat from low to higher temperatures is the main cause of metamorphism, as some rocks at high temperatures and pressures tend to be stable as long as conditions are not changed. Thus, a change in the temperature (and pressure) conditions of rocks in what causes metamorphism.

5.3.2 The Role of Pressure

As the thickness of overlying rocks increases, pressure has a tendency to also increase with depth of burial. As **confining** (or lithostatic) **pressure** (or stress) increases, the pressure of buried rocks also increases, causing a body of rock with a certain volume at shallow depth to become progressively smaller, but denser, with increasing depth of burial (Fig. 5.1a). Thus, the overall shape of the body of rock being buried to deeper levels maintains its shape but it is more compressed due to increasing lithostatic pressure (Fig. 5.1a). Confining pressure is comparable to pressure experienced in fluids as an inward-pressing force is exerted equally in all directions. Confining pressure occurs whereby a more compact rock with greater density is produced due to closing of the pore spaces between the mineral grains. Recrystallization of minerals to form more compact crystalline mineral varieties also occurs as depth of burial of the rocks increases. Consequently, new minerals can also be formed due to increasing lithostatic pressures.

At settings such as plate boundaries, however, pressure experienced by rocks due to plate movements tends to be greater in a certain direction than others, to produce **directed pressure** (stress). Directed pressures result in the shape of the rock being changed as rocks tend to be shortened in the direction in which the maximum pressure is applied and simultaneously causing the rock to be lengthened is the direction perpendicular to that of maximum pressure (Fig. 5.1b, c). A rock that has been affected by directed stresses tends to have any existing elongated minerals and features, for example layering or banding, to be folded (bend). Directed pressure, however, does not lead to the formation of new miners. When new minerals are formed due to increased lithostatic pressures during metamorphism accompanied by directed stresses, they are aligned in directions perpendicular to that of the maximum stresses. **Foliation**, the preferred orientation of these minerals, is common in layered metamorphosed rocks that are composed of platy (e.g., micas) and elongated (e.g., amphiboles) minerals.

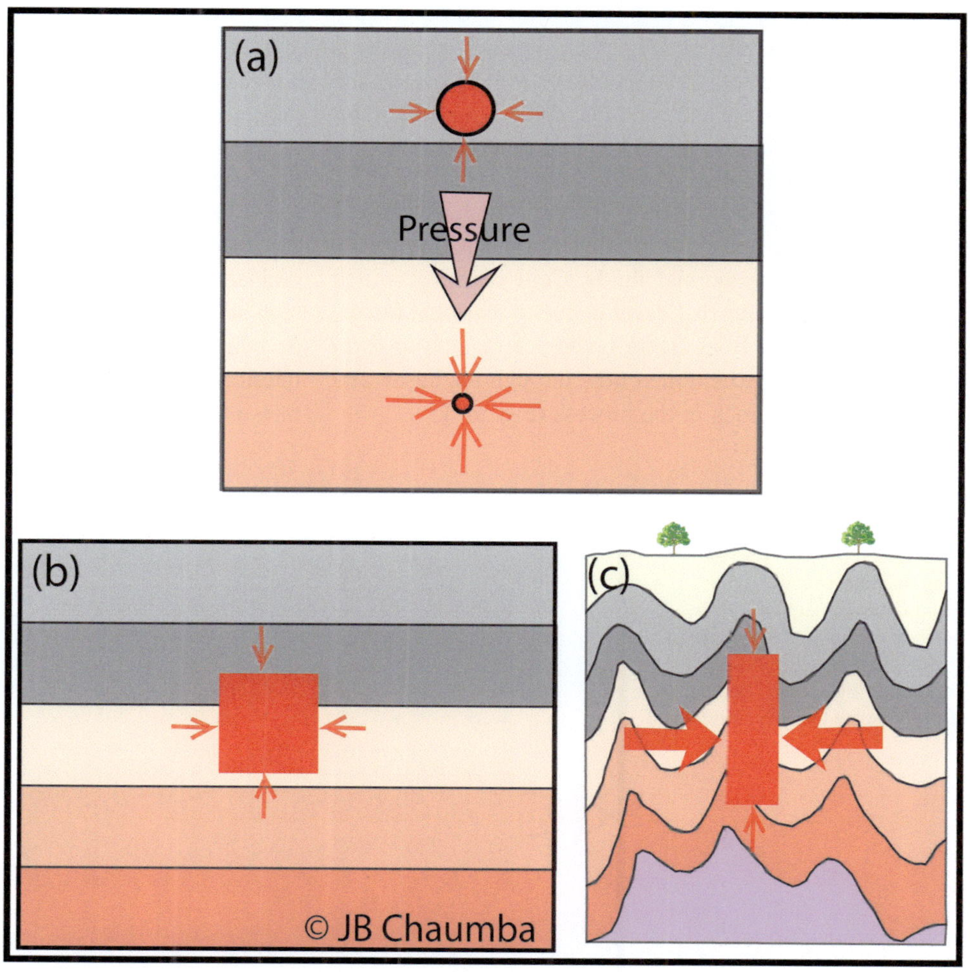

Fig. 5.1 **a** Diagram illustrating the variation of lithostatic pressure with depth. At shallow depth, rocks are subjected to low lithostatic pressures which act on a particular volume of rock equally in all directions (**a**). At greater depth, the volume of rock decreases due to increased inward pressing forces in all directions but maintains its shape (**a**). At plate boundaries, directed stresses that are applied to a particular volume of rock (**b**) result in distorted shapes where it thins in the direction of maximum stress and gets elongated perpendicular to this direction (**c**). The red arrows and their sizes denote the direction and relative size of stresses

5.3.3 The Role of Hydrothermal Fluids

Fluids that are present in the Earth's crust can enable metamorphism to occur in rocks that were initially formed in the absence of any fluids when a change in environment occurs. The fluids, which tend to be composed mainly of water vapor, carbon dioxide, and other volatiles, promote migration of ions and act as catalysts in chemical reactions and recrystallization of minerals. The hotter the fluids, the faster the chemical reactions tend to occur resulting in chemical composition of the original rock being altered. Some ions can be picked up from certain locations and are transported for great distances to a new environment where they can react with completely different rocks from their source regions which permit faster precipitation from the solution of these ions. This can result in a change of the chemical composition of rocks resulting in **metasomatism**, which occurs when some ions are exchanged between any two adjacent but different rocks due to hot fluids permeating through them. Further, the shape and arrangement of mineral grains is also changed.

An example of metamorphism due to hydrothermal fluids can be observed in rocks that have been hydrothermally altered at spreading centers (Fig. 5.2). Basalt is an igneous rock which is stable at high temperatures as it is formed due to melting of the mantle and contains minerals with no water in their crystalline structures. When hydrothermal fluids are introduced, basalt becomes unstable leading to reactions that produce hydrous (water-bearing) minerals such as chlorite and amphibole; an example of metamorphism due to the introduction of hydrothermal fluids. Cold seawater, which descends through faults that develop due to tensional forces, gets heated by heat from rising bodies of magma to form hydrothermal fluids which then flow through the basalt and metamorphoses the rock in the process (Fig. 5.2).

5.4 Metamorphic Textures

Recall that in Chap. 3 we discussed that texture describes the shape, size, and arrangement of mineral constituents in a rock. In describing metamorphic rocks, the constituent minerals, their size, shape, and arrangement all provide information on the extent (or degree) of metamorphism that affected a rock. Thus, metamorphic rocks can typically be described as whether the constituent minerals, which form during metamorphism, possess a preferred orientation (foliated) or not (nonfoliated). Let us now describe these foliated and nonfoliated textures next.

Foliated textures are used to describe either platy minerals which have a preferred orientation or some penetrative structures which are in rocks, or both. Although foliation tends to occur commonly in foliated metamorphic rocks, it can also occur in deformed igneous and sedimentary rocks. Foliation in metamorphic rocks forms in response to directed stresses which produce folded rocks (Fig. 5.3a, b). The protolith shale, for example, can be progressively metamorphosed to form slaty cleavage (which is exhibited by slate), crenulation cleavage (which is exhibited by phyllite), schistocity (which is exhibited by schist), and banding (which is exhibited by gneiss) (Fig. 5.3a), whereas the metamorphism of quartz porphyry or granite progressively forms schist and gneiss (Fig. 5.3b).

Metamorphic rocks that are composed of minerals that do not exhibit a foliated texture are said to possess a nonfoliated texture. Nonfoliated textures form where either the metamorphic rock contains new minerals (and often coarser-grained than those in their protoliths) which originated by recrystallization during metamorphism to form coarse grains or completely new ones (Fig. 5.4a), or the new grains interlock with each other (Fig. 5.4b). An example is the metamorphism of limestone which results in recrystalliza-

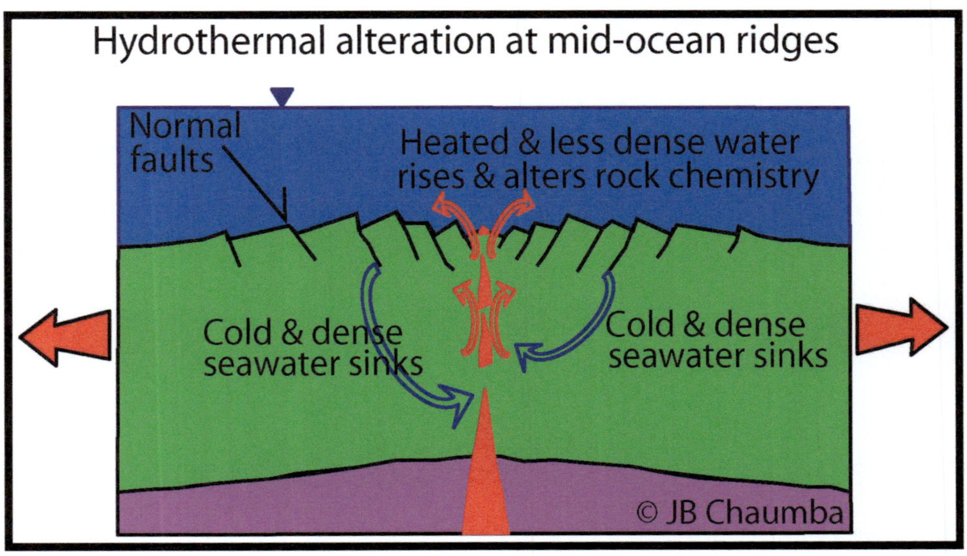

Fig. 5.2 Diagram illustrating sinking cold seawater which is then warmed by heat from the rising magma (which accumulates in magma chambers) forming hydrothermal fluids which metamorphose mid-ocean ridge basalt. Red arrows indicate tensional stresses

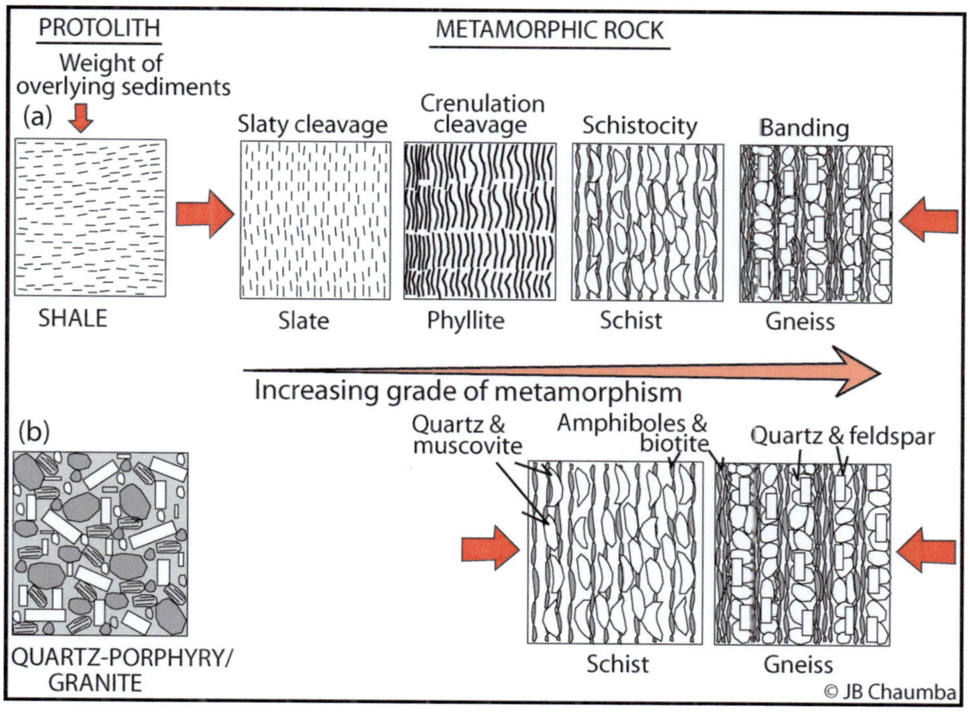

Fig. 5.3 Illustration of textures associated with foliated metamorphic rocks. (**a**) the protolith shale can be progressively metamorphosed to form slaty cleavage (exhibited by slate), crenulation cleavage (exhibited by phyllite), schistosity (exhibited by schist), and banding (exhibited by gneiss) whereas metamorphism of quartz-porphyry or granite forms schist and gneiss (**b**)

tion of calcite crystals and limestone to form marble in which the crystals tend to be coarser grained and display an interlocking texture (Fig. 5.4b). Metamorphism of quart sandstone to form quartzite occurs in a broadly comparable manner (Fig. 5.4b), and both protoliths of limestone and quartz sandstone have chemical compositions that are simple. The new crystals in the metamorphosed rocks tend to show a random orientation and are dominantly equidimensional (Figs. 5.4b).

It is worth noting that, in addition to the foliated and nonfoliated textures we have just discussed, metamorphic rocks can also exhibit textures that are very similar to the protolith at low grades of metamorphism. At higher metamorphic grades, the textures that form have a tendency to produce rocks with very little or no resemblance at all to their protoliths.

5.5 Classification of Metamorphic Rocks

Metamorphic rocks tend to be classified according to whether they exhibit foliation or not, and the minerals that constitute the rocks (Table 5.1).

5.5.1 Foliated Metamorphic Rocks

Regional low-grade metamorphism of mudstones (which are dominated by clay minerals) produces **slate** (Table 5.1; Fig. 5.5a, b), the finest grained of all foliated metamorphic rocks. However, when volcanic ash is metamorphosed, it also produces slate. The variation in the color of slate can be appreciated by a comparison of Figs. 5a and b, red slate and gray slate, respectively. Gray slate forms shale which contains organic matter whereas red slate is composed of iron-oxide. Other slates can occur in shades of green (chlorite-bearing), brown, and purple. Due to its fine-scale foliation termed **slaty cleavage**, slate readily cleaves, or splits, into thin slabs which are durable and possess great tensile strength. This property makes slate suitable for flooring, roofing, and flagging.

Phyllite, which is formed when regional low-medium grade metamorphism of mudstone occurs, is characterized by a preferred orientation of flakes of fine-grained white micas (Table 5.1; Fig. 5.5c, d). The increase in metamorphic conditions which occur when phyllite is formed compared to slate is responsible for the growth of the white micas. Some phyllites show **crenulation cleavage** (Fig. 5.5d), which is a type of fabric that also develops in some phyllites, schists, and gneiss from the application of at least two directions of stresses which results in the formation of foliations that are superimposed. Similar to slate, wide color variations in phyllite occur, but phyllite possesses both a wavy surface and glossy sheen which are lacking in slate. Further, chlorite and muscovite minerals in phyllites tend to be coarser-grained than those which occur in slate.

Medium-grade regional metamorphism of mudstones, slate, phyllite, and even granite porphyry results in the formation of a medium- to coarse-grained rock termed a **schist** (Table 5.1; Fig. 5.5e, f). Schists, thus, form at higher metamorphic conditions than those which produce phyl-

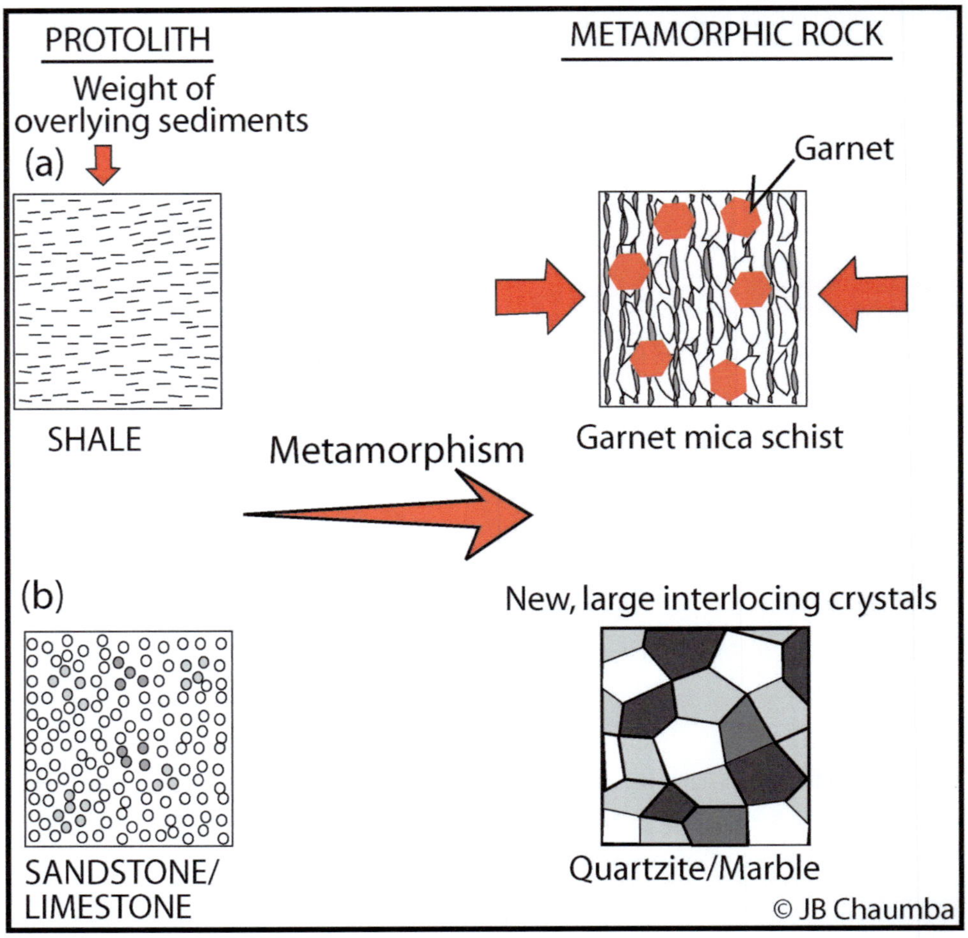

Fig. 5.4 Sketches of textures of nonfoliated metamorphic rocks. **a** New garnet crystals which grew during metamorphism of clay protolith. **b** metamorphism of quartz sandstone to form quartzite with new coarse-grained interlocking quartz crystals

Table 5.1 Table showing the different types of metamorphic rocks and their protoliths

Texture	Grain size	Minerals	Type of metamorphism	Protolith	Name
FOLIATED					
Mineral alignment	Fine	Mica, Qz, Fsp, Amp, Grt	Regional (low-grade)	shale	Slate
	Fine to medium	Mica, Qz, Fsp, Amp, Grt	Regional (low- to medium-grade)	shale	Phyllite
		Mica, Qz, Fsp, Amp, Grt, Pxn	Regional (medium-grade)	Shale/Qz porphyry	Schist
Banding	Medium to coarse	Mica, Qz, Fsp, Amp, Grt, Pxn	Regional (high-grade)	Shale/granite	Gneiss
NONFOLIATED					
Not foliated	Fine	Carbon	Regional/Contact	Bituminous coal	Anthracite coal
	Fine	Wide variety	Contact	Wide variety	Hornfels
	Fine to coarse	Calcite, dolomite	Regional/Contact	Limestone/dolostone	Marble
	Fine to coarse	Quartz	Regional/Contact	Quartz sandstone	Quartzite
	Coarse	Wide variety	Regional/Contact	Conglomerate	Metaconglomerate

Mineral abbreviations: Qz—quartz; Fsp—feldspar; Amp—amphibole; Grt—garnet; Pxn—pyroxene

Fig. 5.5 Foliated metamorphic rocks. **a** Red slate, **b** Gray slate from Bangor, Pennsylvania, USA. **c** red phyllite, **d** green phyllite showing fine-scale crenulation cleavage **e** mica schist from Manhattan, New York, USA. **g** biotite gneiss from Uxbridge, Massachusetts. USA. **h** biotite gneiss from the Piedmont, Southern Appalachians, USA

lites. Schists display a type of foliation characterized by layering in a medium- to coarse-grained rock which displays nearly parallel alignments of muscovite and biotite called **schistosity**. Mica schists are dominated by muscovite and biotite, and other schists can be dominated by other minerals and the dominant mineral is used as a prefix; for example, garnet schist contains appreciable amounts of garnets. Other varieties of schists exist. Amphibolite schists (Fig. 5.5f), for example, are dominated by amphibole and plagioclase feldspars with very little or no quartz.

A banded or layered metamorphic rock composed of granular and elongated minerals is termed a **gneiss** (Table 5.1, Fig. 5.5g, h). The banding is formed by alternating bands of felsic (or light)-colored layers dominated by felsic minerals such as quartz and feldspar and mafic (or dark)-colored layer dominated by the mafic minerals' biotite and amphibole. Muscovite is also common in gneisses. The banding (or layering) can vary in thickness from millimeters to meters, and this gneissic or compositional banding is responsible for imparting the striped appearance which is distinctive in gneisses. Gneisses form from regional metamorphism at high-grade conditions of granite, shale, slate, and schist. Specific names may be given to different gneisses based on the minerals present. A corundum gneiss (Fig. 5.5g), contains appreciable amounts of the mineral corundum, whereas a biotite gneiss (Fig. 5.5h) is dominated by biotite.

5.5.2 Nonfoliated Metamorphic Rocks

As discussed earlier, during metamorphism, some minerals recrystallize from initially smaller ones and grow in grain size (Fig. 5.4b). When these new minerals lack any parallel alignment, it can be interpreted that such rocks likely formed in the absence of directed stresses. Metamorphic rocks that lack foliation and possess equant minerals which grew during metamorphism are termed nonfoliated metamorphic rocks (Table 5.1). Let us now describe some common nonfoliated rocks next.

Metamorphism of calcite-bearing rocks such as limestone and dolostone under medium- to high-grade conditions produces **marble**. Marble (Table 5.1, Figs. 5.4b, 5.6a, b) is composed of nearly uniform, medium- to coarse-grained and interlocking crystals of calcite, a relatively soft mineral with a hardness of 3 on the Mohs hardness scale (see Chap. 2). Marble often forms from protoliths containing numerous varieties of impurities which are responsible for imparting different color variations (such as white, gray, green, pink, and black) in the rock. Further, the uniform texture and softness of marble make it attractive to sculptors who produce sculptures ranging from very small to large, smooth, and detailed ones. Examples of famous sculptures which are typically made of white marble include Michelangelo, in Florence, Italy, and the Lincoln Memorial in Washington, DC. The disadvantage of using marble in sculpturing is that it absorbs skin oils when touched, causing yellow brownish stains. Further, marble is a less-resistant rock and marble monuments that are exposed to the atmosphere are also easily corroded. It must be noted here that metamorphism of impure limestones or dolostones produces marble which displays color banding. Further, some marble can be sheared or flow plastically during deformation that is accompanied by metamorphism to form marble which displays impressive banding.

Medium- to high-grade metamorphism of a hard protolith such as quartz sandstone results in the formation of **quartzite** (Table 5.1, Fig. 5.6c). Recrystallization of the quartz crystals during metamorphism of the protolith quartz sandstone forms new, interlocking, and coarser-grained quartz crystals (Fig. 5.4b) in quartzite where, upon fracturing, the cracks penetrate the new recrystallized crystals as opposed to the unaffected hard and durable quartz crystals in the parent quartz sandstone rock. In quartz sandstone, any cracks that develop tend to curve around any crystals of quartz. Quartzite appears more glassy compared to the whitish-colored, sand sized, and sandpaper-like surface of quartz sandstone. The presence of iron oxides and other impurities are responsible for the color variations in quartzite, which can be reddish (Fig. 5.6c), greenish, whitish, grayish, or even maroon in color.

Amphibolite (Fig. 5.6d) is a coarse-grained metamorphic rock composed of amphibole (typically actino-

Fig. 5.6 Nonfoliated metamorphic rocks. **a** pink marble from Tate, Georgia, USA. **b** white marble from West Rutland, Vermont, USA. **c** quartzite, **d** non-foliated amphibolite

Fig. 5.7 Nonfoliated metamorphic rocks. **a** Hornfels from Edgewater, New Jersey, USA. **b** metaconglomerate

lite and hornblende) and plagioclase. Amphibolites which lack foliation are produced from contact metamorphism of basalt through recrystallization formed from medium-grade contact metamorphism.

Hornfels (derived from a German word "hornfels" for "horn stone") is a nonfoliated fine-grained metamorphic rock formed by contact metamorphism when a hot igneous body comes into contact with shale or any other rock rich in clay. Hornfels (Table 5.1, Fig. 5.7a), thus, represents an equivalent of the original rock that has been metamorphosed due to heat. The source heat can be a body of magma such as a pluton, a dike, or a sill. The minerals in a hornfels are determined by the protolith as well as the grade of metamorphism. Any inequant minerals that are encountered in a hornfels lack any preferred orientation.

Regional metamorphism under high-grade conditions can form a **metaconglomerate** where any original sedimentary structures (for example, bedding) are lacking but the clasts show elongation. A metaconglomerate (Table 5.1, Fig. 5.7b) is a nonfoliated coarse-grained rock which is composed of rounded to subangular clasts occurring in finer-grained minerals of quartz, chlorite, and mica. Some metaconglomerates with high concentrations of quartz may be useful as local building stones.

Anthracite coal (Table 5.1, Fig. 4.7d), which occurs in association with strongly deformed sedimentary rocks that were subjected to medium- to high-grade conditions, is a hard and compact rock or mineral (possessing a submetallic luster). Anthracite is black to steel gray in color, which is hard and brittle, and displays conchoidal fracture. Unlike other varieties of coal, anthracite is clean to the touch.

5.6 Grade of Metamorphism

Metamorphic grade is a measure of the extent or degree of difference between the parent rock and the metamorphic rock, and it is a way of describing the intensity or rank of metamorphism. The grade of metamorphism is dependent on the temperature at which the rocks were metamorphosed. The temperature at which rocks are metamorphosed determines the formation of any new minerals which grow during metamorphism as well as the extent of any recrystallization which can occur. Any diffusion that occurs as well as any chemical bonds that are formed during metamorphism largely depends on temperature. As the grade of metamorphism increases, the rate of recrystallization increases, and coarser-grained metamorphic rocks also form (Table 5.1, Fig. 5.3). The metamorphic grade, thus, is a general indication of the temperature and pressure conditions (or facies) in which metamorphism occurred.

Low-grade metamorphism is characterized by metamorphic rocks which form in the 300–400 °C temperature range, medium-grade metamorphism occurs in the 400–600 °C temperature range, and high-grade metamorphism is characteristic of metamorphic rocks that form at temperatures in excess of 600 °C. New metamorphic minerals are formed as the grade of metamorphism increases. Consequently, the grade of metamorphism can be delineated from the types of minerals constituting different metamorphic rocks.

5.7 Metamorphic Settings

The interaction of temperature, pressure, and hydrothermal fluids in different settings leads to the formation of different types of metamorphisms. Depending on the extent of the area affected, the temperature, pressure, and the involvement of hydrothermal fluids, we can distinguish different settings of metamorphism that we discuss next.

5.7.1 Regional or Dynamothermal Metamorphism

Regional or dynamothermal metamorphism (referred to hereafter as regional metamorphism) is responsible for the formation of metamorphic rocks that are either exposed in some proportions of continents and those in stable inte-

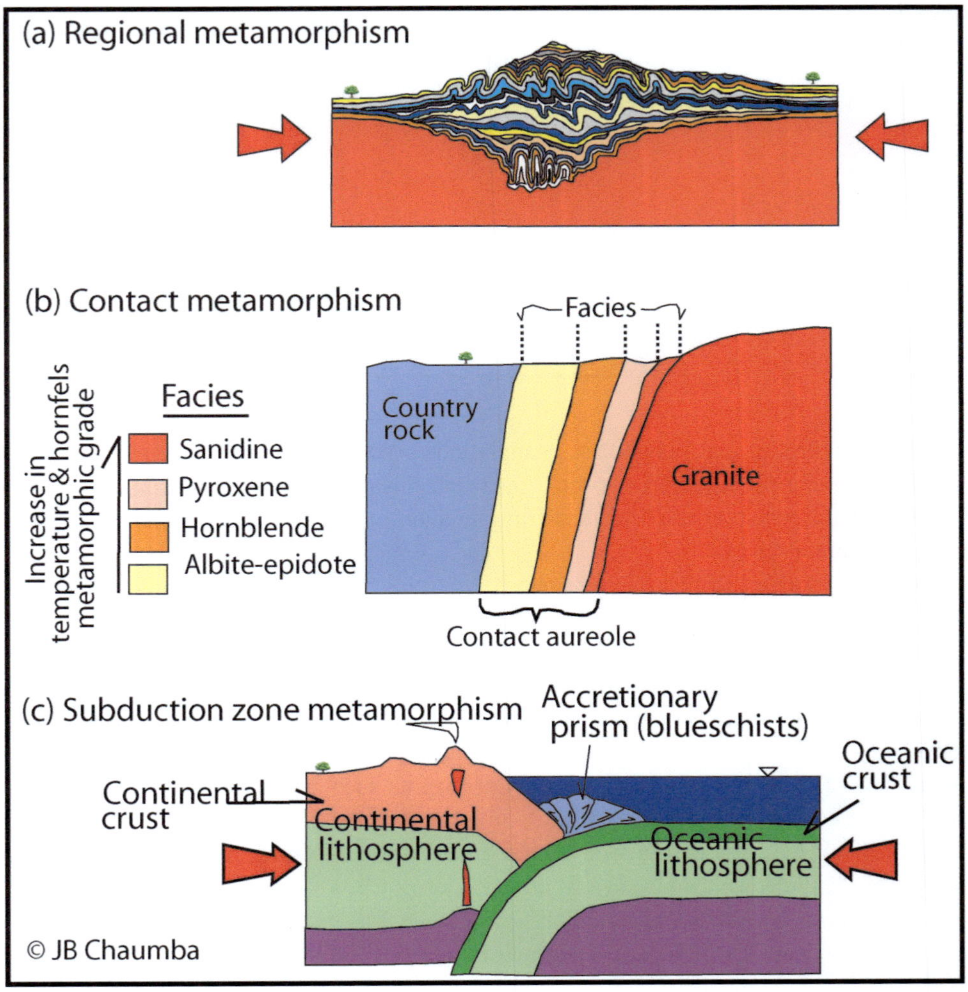

Fig. 5.8 Illustration of tome types of metamorphic settings. **a** compressional stresses squeeze rocks resulting in folds and metamorphism. **b** Contact metamorphism caused by heating of country rocks due to intrusion of hot magma which metamorphoses rocks producing a contact aureole. **c** when a continental plate and an oceanic plate collide, the ocean plate subducts, sediments scrapped off from the overriding plate are dragged down and undergo high-pressure but low temperature metamorphism in the process to form blueschists within the accretionary prism

riors of continents that are covered by a thin veneer of sedimentary rocks. Further, almost all rocks that underlie all mountain belts are metamorphic rocks that were also formed due to regional metamorphism. Regional metamorphism occurs when large volumes of rocks are subjected to high temperatures and immense pressures during large-scale deformation that accompanies mountain building processes (Fig. 5.8a).

5.7.2 Contact or Thermal Metamorphism

Contact (or thermal) Metamorphism takes place when crustal rocks come into contact with magma or lava. The rise in temperature within the surrounding rock (or underlying rock in the case of lava) caused by the intrusion of magma or lava flow is responsible for baking the existing rock adjacent to the intrusion (termed the wall rock) whose chemical composition is then altered (Fig. 5.8b). Further, hydrothermal fluids also continually flow from the country rocks into and out of the intrusion helping alter the chemical composition of the wall rocks (Fig. 5.8b). Rocks in contact with the intrusion are altered the most (highest-grade contact metamorphic rocks), with the rocks further away from the intrusion being altered the least (lowest-grade contact metamorphic rocks) or remain unaffected, depending on the volume of the intrusion and the distance from it (Fig. 5.8b). This decrease in the effects of contact or thermal metamorphism (hereafter termed contact metamorphism) due to the intrusion results in the formation of distinct zones of contact metamorphic rocks around the intrusion termed a *contact (or metamorphic) aureole.* Since neither pressure nor directed stresses play a role in forming contact metamorphic rocks, hornfels (the nonfoliated metamorphic rocks we have already described) are formed here. Hornfels is thus associated with tectonic settings where intrusive igneous rocks occur, such as at both divergent and convergent plate boundaries, as well as in areas where hotspots are encountered. Where clay has been subjected to contact metamorphism, the high-grade rocks are used in the manufacture of porcelain, whereas medium-grade rocks are used in pottery. Further, low-grade rocks from contact metamorphic settings are used in making bricks which are useful in construction.

5.7.3 Subduction Zone Metamorphism

At subduction zones, one oceanic plate is pushed beneath another one (Fig. 5.8c) which results in the oceanic crust being recycled back into the mantle whereas some continental crust is generated by the magma which accumulates above the subducting slab to form an arc. Dehydration of minerals in the subducting slab as it descends generates the arc magmas. Thus, subduction zone settings host some unusual and rare rock types such as eclogites and blueschists which form under high-pressure but relatively low-temperature conditions at accretionary prisms or wedges (Fig. 5.8c). Blueschists display foliation formed from both compressional stresses and shearing that develops in the accretionary prism (Fig. 5.8c). The protoliths of eclogite and blueschist rocks are thought to be basalt that makes up oceanic crust which is then subducted. Although most of the eclogite and blueschists plunge deep into the mantle, small volumes of these rocks may be exhumed along the top of the subducting slab to ultimately reach the surface. Accretionary prisms can reach thicknesses in excess of 20 km, resulting in rocks at the base undergoing high pressure but low-temperature **subduction zone metamorphism** due to the cool temperatures of the subducting oceanic slab beneath (Fig. 5.8c).

5.7.4 Burial Metamorphism

Burial metamorphism, which occurs when rocks are buried to appreciable depths, is experienced by earlier deposited rock layers at the base of sedimentary depositional basins due to increased overburden pressure. As more sedimentary beds are deposited above the earlier deposited ones, an increase in temperature and pressure leads to chemical reactions to occur as a consequence of geothermal gradient, resulting in metamorphism of layers of rocks at the base when the overburden exceeds approximately 10 km depending on the prevailing geothermal gradient. At these depths, recrystallization is triggered in response to both lithostatic pressure and increased temperatures from geothermal gradient. Rocks that undergo burial metamorphism lack foliation due to the absence of directed stresses. Thus, no plate tectonic processes are involved in burial metamorphism.

5.7.5 Hydrothermal Metamorphism

When pre-existing rocks are chemically altered by heated water, **hydrothermal metamorphism** occurs. On continental crustal rocks that have been intruded by magma bodies (Fig. 5.9a), the heated water is derived either from groundwater that would have percolated downwards from the surface (Fig. 5.9a), or from the magma itself (magmatic water).

At a spreading plate boundary such as mid-oceanic ridges, seawater percolates cracks associated with normal faulting at these settings, along which it descends and gets heated by magma intruded at mid-oceanic ridges (Fig. 5.9b). The seawater, now heated to temperatures of at least 350 °C, easily dissolves elements and hydrothermally alters the mafic rocks of the oceanic plate to form hydrous metamorphic minerals such as chlorite, epidote, amphibole, talc, and serpentine, becoming less dense and rising to the surface to complete the hydrothermal circulation currents (Fig. 5.9b). Hydrothermal metamorphism in both cases (Fig. 5.9a, b) involves groundwater/seawater that comes in contact with, or very close to, the intrusion and gets heated causing its density to decrease, and subsequently rising towards the surface/ocean floor. The heated and rising groundwater/seawater is then replaced with cold and denser groundwater/seawater, triggering hydrothermal circulation that is responsible for hydrothermal metamorphism of the country rocks, and formation of metallic mineral deposits (Fig. 5.9a, b).

5.7.6 Dynamic or Fault-Zone Metamorphism

Dynamic or fault-zone metamorphism occurs along faults where rocks grind past each other (Fig. 5.10a). At such faults, directed pressures and frictional heat are generated which trigger dynamic or fault-zone metamorphism generally limited to short distances from the faults (Fig. 5.10a). Dynamic metamorphism often occurs within the topmost 15 km of the crust where fracturing and recrystallization of the original rock in fault zones causes finer-grained minerals, or even powder, to form from coarser grained ones during brittle deformation. Rocks that form in fault zones, such as mylonites (Fig. 5.10a), develop foliation with an orientation that is in conformity to the strike of the fault. At deeper levels within the crust, rocks undergo plastic or ductile deformation and do not fracture. Dynamic deformation does not require a change in neither temperature nor pressure; rather it occurs solely from movement at fault zones, hence the name dynamic or fracture-zone deformation.

5.7.7 Shock or Impact Metamorphism

When meteorites collide with a planet such as Earth, extremely high temperatures and pressures are generated resulting in **shock or impact metamorphism**. The impact which occurs during collision is envisaged to "shock" the

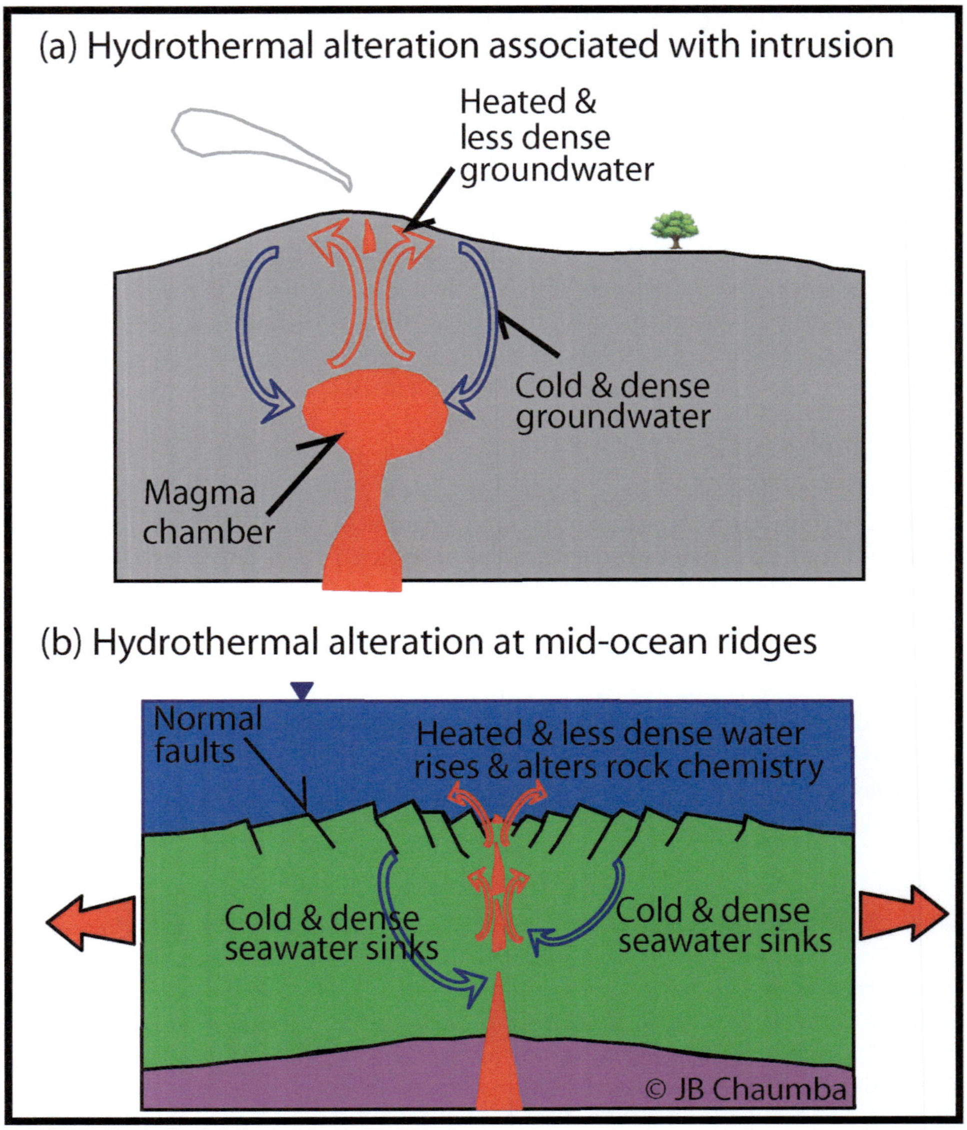

Fig. 5.9 Diagrams illustrating hydrothermal metamorphism associated with magmatic intrusion into continental crust (**a**) and oceanic crust at mid-ocean ridges (**b**). In a continental crustal setting, cold groundwater replaces heated water that comes in contact with the hot body of magma (**a**) whereas cold seawater that comes into contact with hot magma is heated, rises, and is replaced by more cold seawater to trigger convection cells in an oceanic crustal setting (**b**). In both cases, ions and dissolved metals from country rocks are dissolved and concentrated along veins resulting in the formation of metallic mineral deposits

minerals that are struck in a way that results in recrystallization and shattering of those minerals (Fig. 5.10b). A shock wave that is inferred to be caused by very high compressional stresses is generated at the site of impact and it is then propagated deeper into the crust (Fig. 5.10b). At the site of impact, vaporization and melting (Fig. 5.10b) is thought to occur from the heat that is generated by the force of the impact. An asteroid or comet tends to collide with the surface at an angle of 45 degrees, generating the highest shock pressures where the impact occurs. Shock pressures that are generated by the impact decrease with distance from the point of impact from pressures in excess of 50 GPa (gigapascals) to pressures less than 1 GPa further away from the impact point (Fig. 5.10b). Examples of high pressure minerals that are formed from shock metamorphism and that are commonly encountered within and near impact craters are stishovite and coesite.

5.7.8 Pyrometamorphism

Where ultra-high temperature and relatively low pressure conditions are encountered, such as where beds of coal burn underground leading to recrystallization of rocks that they are in contact with, is a type of metamorphism termed **pyrometamorphism**. Whenever there are some very powerful bolts of lightning strikes, the immense heat that is associated with them can also lead to recrystallization of surface sediments and rocks that they strike.

Consequently, in most of the settings where metamorphic rocks are exposed, such as in **shields** (Fig. 5.11), they tend to show evidence of deformation and/or the presence of associated intrusive igneous rocks. Shields are formed from an amalgamation of a number of cratons, such as the Zimbabwean craton (Fig. 5.11). Most of the continental crust is therefore composed of metamorphic rocks

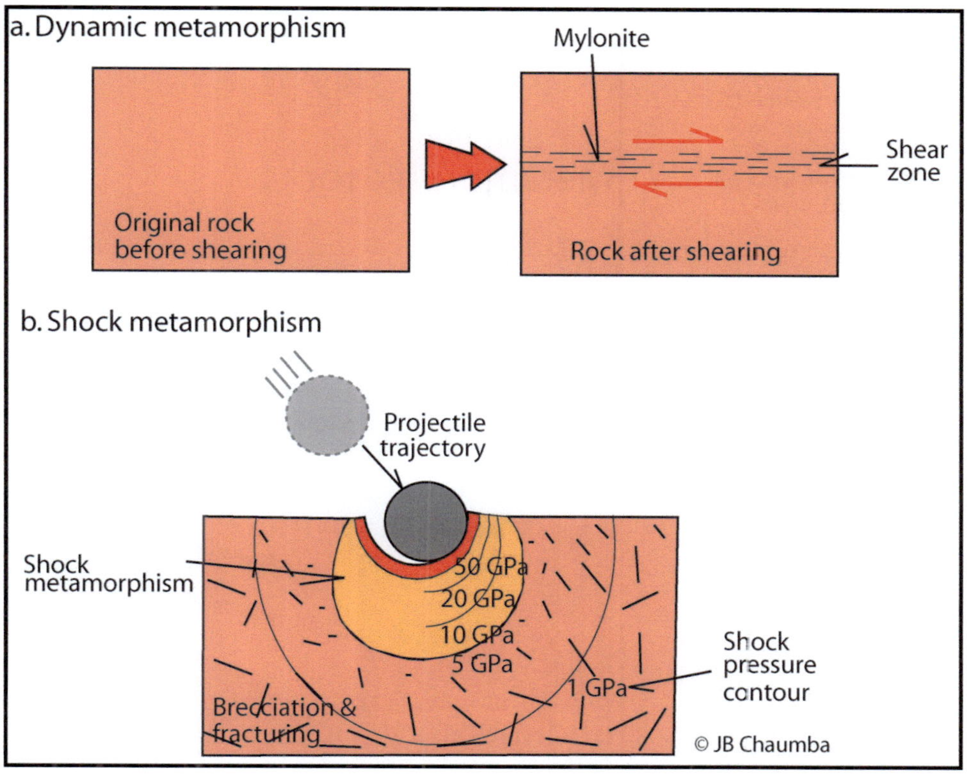

Fig. 5.10 Dynamic metamorphism occurs when a rock is sheared under metamorphic conditions which cause recrystallization of original minerals into finer-grained ones or even powder (**a**). An illustration of shock metamorphism occurring near the point of impact where a zone of vaporized material develops that is surrounded in turn by zones of melted material (i.e., shock-metamorphosed area) and fractured material. Labeled flow lines (in gigapascals—GPa) depict flow of material across the area affected by the impact

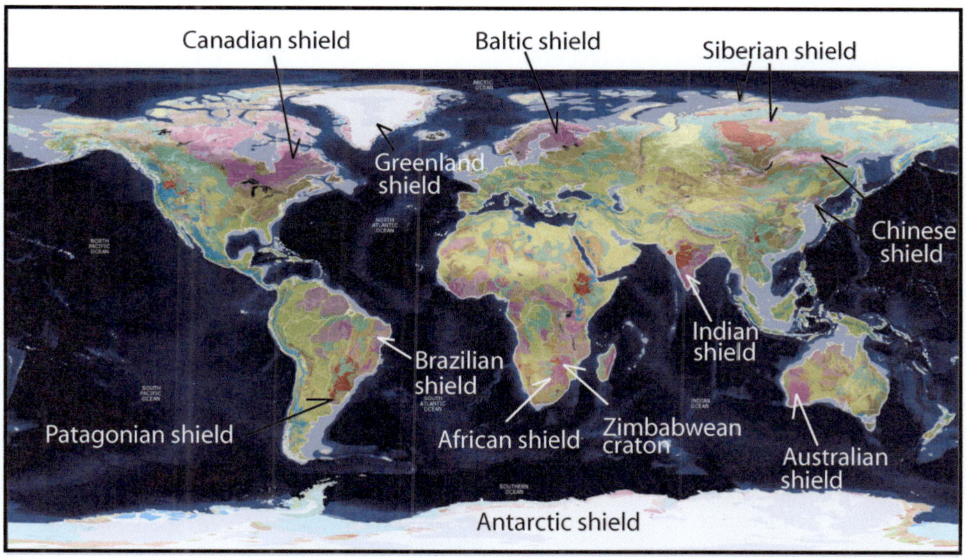

Fig. 5.11 Summarized geologic map of the world showing the distribution of shield areas of Precambrian age which are composed dominantly of metamorphosed rocks (credit: http://portal.onegeology.org/OnegeologyGlobal/). Cratons, such as the Zimbabwean craton on the African shield, are parts of shields

together with associated intrusive igneous rocks (Fig. 5.11). Some metamorphic rocks are also exposed in ancient mountain belts where erosion has removed the overlying sediments resulting in uplifting of once deeply buried metamorphic rocks. Exceptions occur, however, such as when rocks undergo pyrometamorphism as well as burial, dynamic, or shock metamorphism where there is a lack of associated intrusive activity. However, these types of metamorphism result in the formation of relatively small volumes of metamorphic rocks.

5.8 The Rock Cycle

Let us imagine that we start off with the molten material, or magma, and then examine how the **rock cycle** works. Recall that magma forms from partial melting on the Earth's upper mantle or crust (as we discussed in Chap. 3). Magma, being less dense than the surrounding rock where it is formed in the upper mantle and crust, tends to rise towards the surface of the Earth. Some of the magma makes its way to the Earth's surface where it erupts and becomes lava, followed

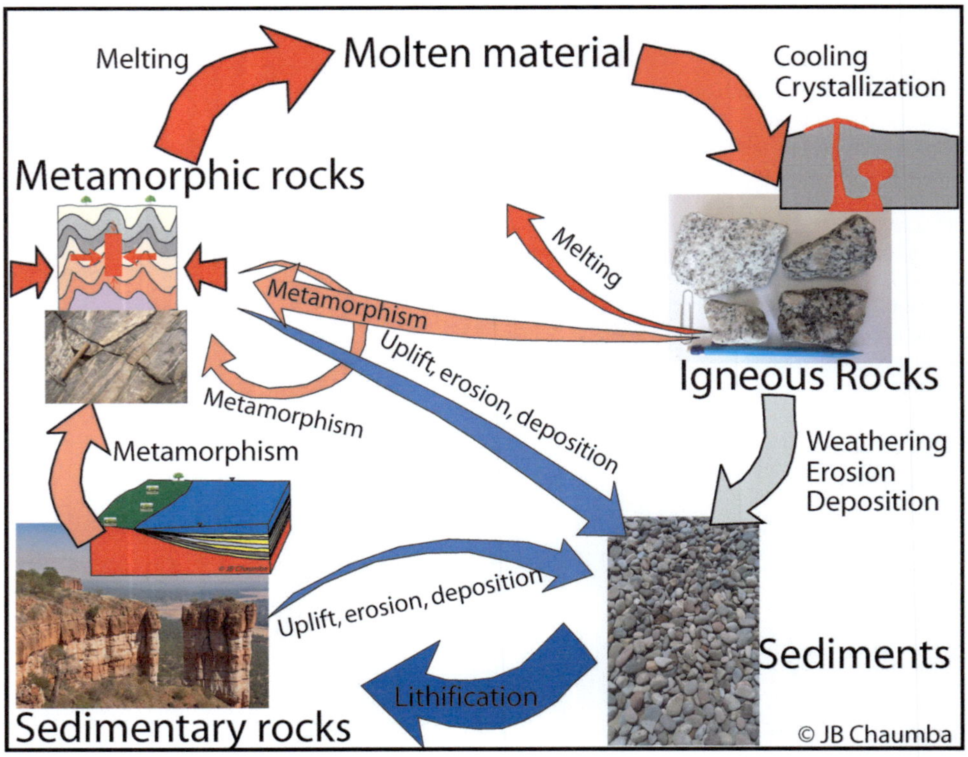

Fig. 5.12 The rock cycle and its several pathways. The wider and bigger arrow represent the "normal" cycle, whereas the narrower and smaller arrows represent "shortcuts" of alternative pathways that can occur in the rock cycle

by rapid cooling and solidification to form extrusive igneous rocks (Fig. 5.12). Some of the molten material, however, solidifies much more slowly within the Earth's crust to form plutonic igneous rocks. Igneous rocks exposed at the Earth's surface are subject to the influence of the atmosphere which results in the igneous rock slowly breaking down–a process termed weathering. Weathered material is then transported either by water, wind, and ice and deposited in sedimentary depositional environments as sediments (Fig. 5.12).

As more sediments are piled above earlier deposited ones, lithification turns them into sedimentary rocks (Fig. 5.12). With continued burial, some of the sedimentary rocks are subjected to high enough temperatures and pressures such that they are changed into metamorphic rocks (Fig. 5.12). If temperature and pressure continue to increase, melting can occur resulting in the formation of magma which rises upwards to eventually cool down and form igneous rocks. What we have just described: magma → igneous rock → sediments → sedimentary rock → metamorphic rock → magma is considered a "normal" cycle of the rock cycle. The cycle never ends, and it can be started anywhere. In nature, however, alternative pathways to the "normal" cycle just described can, and do, occur. For example, a sedimentary rock can be uplifted after it has been formed, followed by erosion and deposition to form sediments which are then lithified to form new sedimentary rocks, bypassing the formation of both metamorphic and igneous rocks (Fig. 5.12). Numerous such alternative or "short-cuts" to the "normal" rock cycle can occur in nature, as shown in Fig. 5.12. The rock cycle, thus, clearly illustrates the relationships which exist between the three rock families of igneous, sedimentary, and metamorphic. Based on the rock cycle, therefore, one can expect to encounter sedimentary rocks in basins. However, it should be borne in mind that igneous rocks (Chap. 3) can also be encountered in sedimentary basins either as sill or dikes (Chap. 4).

5.9 Metamorphic Facies

During metamorphism metamorphic rocks do not, generally, change their chemical composition. Exceptions, however, can occur such as in the formation of skarns where metasomatism takes place. Any changes in mineral assemblages that occur during metamorphism are due to changes in conditions of temperature and pressure during metamorphism. Consequently, the mineral assemblages that occur are indicators of the temperature and pressure that the rocks experienced during metamorphism (Fig. 5.13). **Metamorphic facies** indicate the pressure and temperature environment (Fig. 5.13) that the rocks in question were subjected to during metamorphism. Metamorphic facies, thus, are mineral assemblages that are formed in comparable metamorphic environments. Metamorphic facies can be envisaged as minerals that form at specific metamorphic conditions which are used in the determination of

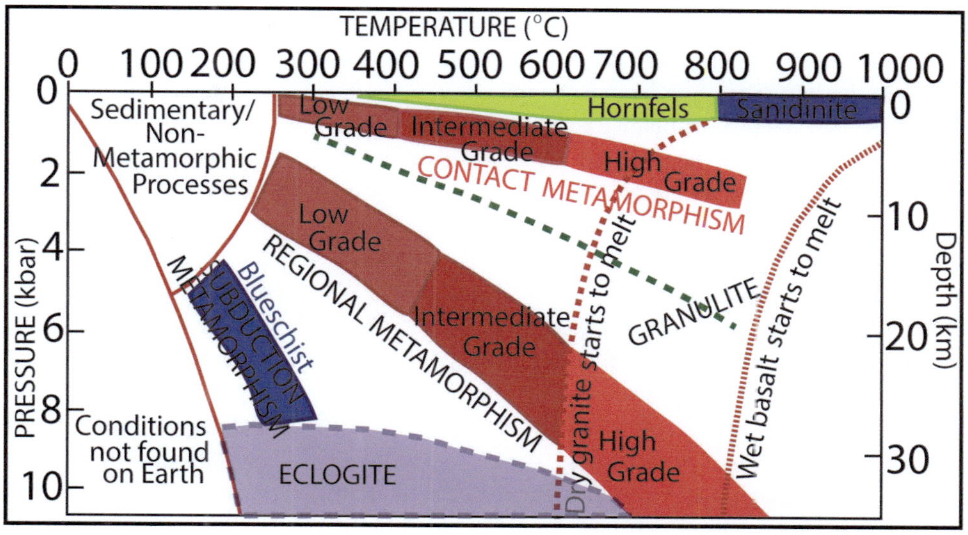

Fig. 5.13 Metamorphic facies

metamorphic grade. Any geothermal gradient present during metamorphism determines the sequence of metamorphic facies observed in any metamorphic terrain. The geothermal gradient that is revealed during metamorphism in any sequence of metamorphic facies in any metamorphic terrain is that it is dependent on the geothermal gradient that was present during metamorphism.

5.10 Economic Importance of Metamorphic Rocks

Hard metamorphic rocks such as marble are used in the construction industry whereas other metamorphic rocks such as slate and marble are useful as building materials. Talc is useful in many ways such as in cosmetics, and as lubricants and paints. Metamorphic minerals such as garnets are used as gemstones and abrasives. Asbestos, which is formed from hydrous metamorphism of ultramafic igneous rocks, is useful in insulation and fireproofing.

Glossary

Accretionary prism Rocks and sediments shaped like a wedge that are scrapped of the subducting slab and attached to the overriding continental plate

Accretionary wedge *see accretionary prism*

Amphibolite a coarse-grained metamorphic rock composed dominantly of the minerals plagioclase and green, brown, or black amphibole

Anthracite a compact and hard variety of coal with a submetallic luster that contains the highest carbon content and the fewest impurities, of all types of coal except for graphite

Banding a metamorphic rock displaying compositional layering defined by alternating bands of mafic colored minerals such as biotite and hornblende separated by felsic minerals such as quartz, feldspar, and muscovite

Burial metamorphism a type of metamorphism which occurs when earlier deposited sedimentary rocks are buried deeper to undergo even higher pressures and temperatures that are exerted by the weight of overlying sedimentary rocks and sediments

Confining pressure the stress or pressure that is imposed on layers of rocks by the weight of overlying material, and it also sometimes referred to as lithostatic pressure/stress or vertical stress or overburden pressure.

Contact metamorphism metamorphism induced in rocks due to introduction of heat, such as when magma comes into contact with crustal rocks

Craton an ancient and stable block of continental crust occurring within stable segments of continents, especially in the interior

Crenulation cleavage a type of fabric that is displayed by some metamorphic rocks such as phyllite and schist whereby two or more stress directions result in the formation of the superimposed foliations

Deformation change induced by folding and/or fracturing that occurs in the form and position of objects such as rocks and sediments

Dynamic metamorphism metamorphism that occurs as a consequence of movement of rocks such as along a shear zone, with no increase in pressure and temperature

Foliation a type of fabric in metamorphic rocks whereby layers of rocks and/or minerals are parallel to each other

Fracture-zone metamorphism *see dynamic metamorphism*

Gneiss a banded metamorphic with alternating bands of mafic minerals separated by felsic minerals

Hornfels a contact metamorphic rock that is formed by metamorphism of any clay-rich rock such as mudstone due to the intrusion of an igneous body

Hydrothermal metamorphism a type of metamorphism **which** occurs when heated and mineral laden waters interact with a surrounding pre-existing rock at low pressures and relatively low temperature, chemically altering the country rocks

Impact metamorphism *see shock metamorphism*

Lithostatic pressure/stress *see confining pressure*

Marble a nonfoliated metamorphic rock that is formed from metamorphism of limestone or other calcite-bearing protoliths

Metaconglomerate a metamorphic rock containing pebbles or larger clasts embedded in a matrix of sand, silt, or clay that is formed from regional metamorphism of conglomerate

Metamorphic rock any rock that has been changed in shape or form due to increased conditions of temperature, pressure, and hydrothermal fluids

Metamorphic facies set of mineral assemblages in metamorphic rocks that were formed under similar conditions of pressure and temperature representing a set of minerals indicating conditions during which metamorphism occurred and from which the pressure and temperature under which the rock formed is inferred

Metamorphism changes occurring to rocks due to increased heat, pressure, and the introduction of hydrothermal fluids resulting in the formation of new minerals and texture but without any melting involved

Metasomatism a type of metamorphism that involves addition and/or removal of components from the original rock assemblage

Parent rock *see protolith*

Phyllite a fine- to medium-grained foliated metamorphic rock that is intermediate between slate and schist in the degree of metamorphism

Protolith original rock that has undergone metamorphism

Pyrometamorphism a type of metamorphism caused by heat that is generated by either fossil fuel fire of lightning strikes

Regional metamorphism a type of metamorphism occurring in a large area that is caused by subjecting rocks to great depth accompanied by compressional stresses associated with mountain building at convergent plate boundaries

Rock cycle a continuous succession of events that leads to the transformation of rocks and sediments from one type to another

Schist a coarse-grained metamorphic rock formed under medium- to high-grade conditions and composed of layers of different minerals that can be split into irregular thin plates

Schistosity a type of foliation displayed by the parallel arrangement of platy mineral grains such as muscovite and biotite in a coarse-grained rock, with quartz and feldspar (and sometimes garnet, kyanite, staurolite, and sillimanite) being the other minerals present

Shield an area of preserved ancient continental crustal rocks usually located in the interior part of a continent

Shock metamorphism a type of impact metamorphism which occurs when extremely high pressures and temperatures are generated due to a meteorite impact leading to metamorphism and deformation of the underlying rocks

Slate a very fine-grained, foliated, metamorphic rock that is formed when a protolith of clay or volcanic ash undergoes low-grade regional metamorphism

slaty cleavage a type pervasive and parallel foliation, characteristic of phyllite (and sometimes slate) that is defined by fine-grained platy minerals such as chlorite in a perpendicular direction to that of maximum stress

Subduction zone metamorphism a type of metamorphism that is characterized by high-ultrahigh pressures but relatively low temperatures in stability zones of subducted oceanic crust leading to the formation of rare rocks such as blueschists and eclogites.

Thermal metamorphism *see contact metamorphism*

Weathering disintegration and decomposition of any mineral and/or rock that is exposed or near the surface of the Earth

6 Plate Tectonics

Abstract

Evidence used by Alfred Wegener in support of his continental drift hypothesis is discussed and summarized. The lack of a mechanism to explain the movement of continents, however, led to skepticism to the acceptance of the continental drift hypothesis. Additional evidence that was later discovered after Wegener's death for continental drift is discussed. Wegener's hypothesis but encompassing a lot more data the mechanism of plate motion, called the theory of plate tectonics, is summarized. Additional evidence uncovered in support of Wegener's hypothesis includes better fit of the continents based on continental edges, seafloor spreading, paleomagnetism, hot spots and mantle plumes, seamounts and guyots, for example. The plate tectonic theory explains how our planet works. Types of plate boundaries, summarized in the form of table, as well as black smokers, white smokers, and rift valleys are examples of other features, form due to plate tectonics. Mantle convection, slab pull, and ridge push, which drive plate motion, are summarized.

Learning Goals

Studying this chapter will enable you to

- list and describe Alfred Wegener's evidence in support of continental drift hypothesis,
- explain the main objections to continental drift hypothesis,
- differentiate the lithosphere from the asthenosphere,
- describe and sketch the three types of plate boundaries,
- summarize the theory of plate tectonics, and list the evidence in support of the theory,
- describe mantle convection and hot spots,
- list and explain the methods that are used to measure both relative and absolute motions of plates,
- summarize the three driving mechanisms of plate motion, and
- locate where earthquakes and igneous activity occur in the context of plate tectonics.

6.1 Introduction

Most people are aware that earthquakes and volcanic eruptions tend to occur in countries such as Japan, Nepal, Italy, Chile, and the US states of Hawaii and Alaska, for example, whereas very few earthquakes occur in countries such as Australia, Zimbabwe, and France, as well as in the US states of Georgia and Wisconsin, for example. Some areas of the continents are relatively flat, such as the US states of Kansas and Missouri, whereas other countries such as Switzerland and some US states of West Virginia and Nevada are entirely located within mountain ranges. Why do we have all these variations and why do they occur? An explanation for these and numerous other related questions which ultimately led to the theory of plate tectonics (from the Greek word *tektonikos* meaning "build") can be traced to the 1960s. The journey that Earth scientists took to reach this theory is a long one. Before we discuss more about the theory of plate tectonics, let us first take a look at the key hypothesis of continental drift that led to its formulation and its elevation to a theory.

6.2 Continental Drift Hypothesis

Alfred Wegener (1880–1930) was born in Berlin, Germany. A polar scientist who studied geophysics and meteorology, Wegener pondered why continents produced a good fit in a jigsaw puzzle manner as well as why paleoclimate indica-

Supplementary Information The online version contains supplementary material available at https://doi.org/10.1007/978-3-032-20231-4_6.

J. B. Chaumba, *Introductory Physical Geology*, Springer Textbooks in Earth Sciences, Geography and Environment,
https://doi.org/10.1007/978-3-032-20231-4_6

tors, truncated geologic features, as well as identical plant and animal fossils were located on opposite sides of the Atlantic Ocean. The interpretation at that time was that the landmasses that are now separated by oceans were once connected by now sunken land bridges. Wegener hypothesized that most, or all, of the continents once formed a single landmass called a **supercontinent** about 300 million years ago. Wegener termed this single continent **Pangea** (Greek for "entire land") (Fig. 6.1a), and he suggested that Pangaea later split into the separate continents which "drifted" away from each other. For comparison purposes, a map of the present distribution of continents is shown in Fig. 6.1b. These ideas were published in his 1915 book titled *The Origin of Continents and Oceans*. Wegener, however, was not the first scientist to hypothesize that all the continents had once been connected, but he was the first one to compile several pieces of evidence to support this hypothesis. One geologist named Edward Suez, for example, has also made a somewhat similar suggestion but for today's southern continents of Africa, South America, Australia as well as India which today is located in the Northern Hemisphere. Let us briefly review some of the evidence gathered by Wegener in support of his continental drift hypothesis.

6.3 Supporting Evidence for Continental Drift

Varieties of evidence that Alfred Wegener used in support of his continental drift hypothesis are described and illustrated next.

6.3.1 Fit of the Continents

It is thought that from the sixteenth and seventeenth centuries, voyages of exploration of the New World, which resulted in the production of rudimentary maps, led some scholars to note the complementary shapes of the coastlines of Europe and Africa with those of North and South America (Fig. 6.1a).

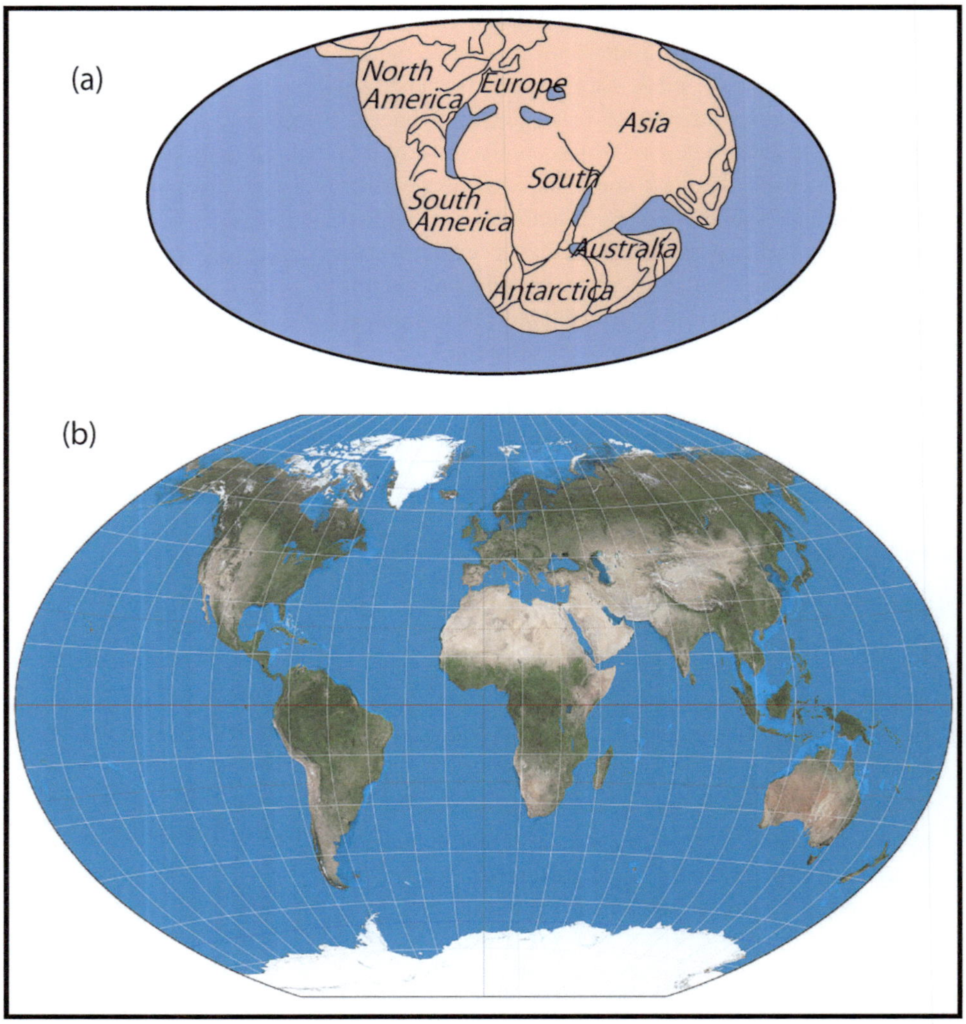

Fig. 6.1 **a** Alfred Wegener's map of the supercontinent Pangaea. **b** Map showing the present position of the continents. *Credit* National Geographic

Wegener also observed the presence of major mountains of comparable age in eastern North America (the northern Appalachians) and in northwestern Europe (Caledonides) (Fig. 6.2). These mountain belts of Devonian age were similarly extensively folded and faulted, metamorphosed, and if North America and Europe are reconstructed into a fit, the Caledonides form a continuation of the Appalachian Mountains (Fig. 6.2). The Etendeka basalts of western southern Africa and the Paraña basalts of eastern South America were both formed approximately 130 million years ago, and they appear to be cut off in two unequal portions at the coast if these two continents are fit together. Further, rocks of the Karoo Supergroup in western South Africa closely resemble those of the Santa Catarina Supergroup in eastern

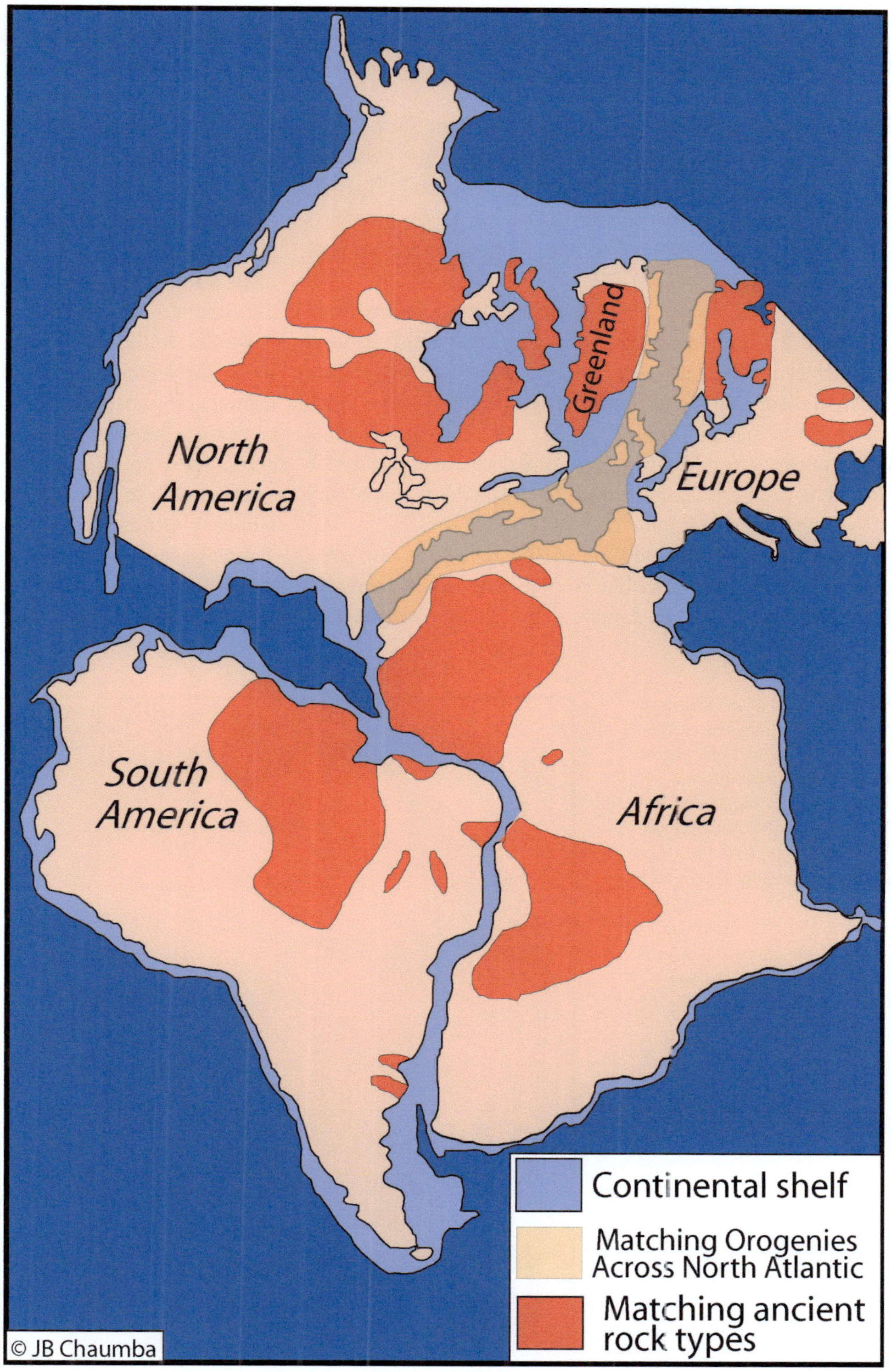

Fig. 6.2 Map showing Africa, South America, and parts of both North America and Europe showing the better fit (although some gaps and overlaps, not shown here, occur) between the continents based on the continental shelf, and the distribution of matching ancient rock types of cratons and rocks of the Appalachian and Caledonide orogenies across the North Atlantic

Brazil. These and other clues indicated that the continents had once been joined together as part of a larger supercontinent. Wegener first presented his ideas in 1912, and they were improved upon with successive editions of his book.

It must be noted that continents and oceans were believed to have fixed positions up until the 1960s; hence, Wegener's suggestion that continents "had moved" was not something that other Earth scientists would readily accept.

6.3.2 Fossil Evidence

Based on this close fit between the continents, especially between those Africa and South America (Figs. 6.1a and 6.2), Wegener reasoned that the similar organisms (Fig. 6.3) on different continents once thrived on a single landmass which later broke into the different continents that are now separated by ocean basins.

Wegener was also aware of some problems that had been described by previous researchers on the subject. Fossils of the extinct reptile *Lystrosaurus* (Fig. 6.3a) are only found in Antarctica, India, and South Africa (Fig. 6.3). The name *Lystrosaurus* means "shovel reptile," and it is thought that *Lystrosaurus* was a herbivore that thrived on land during the early Triassic Period, and it must have lacked the capability to swim across any ocean.

Fossils of *Cynognathus* (Fig. 6.3b), meaning "dog jaw" and an extinct mammal-like reptile, are found only in South Africa and South America (Fig. 6.3). *Cynognathus*, which is also extinct, is interpreted to have roamed during the Triassic period. *Cynognathus* also was incapable of migrating across any ocean as it was a land species.

The extinct *Mesosaurus* (Fig. 6.3c) is known to have been a type of reptile which thrived during the early Permian Period and its remains are found solely in South Africa and eastern South America (Fig. 6.3). *Mesosaurus* is thought to be similar to the modern crocodile. If the continents had never been joined together in the past, there is no possibility that *Mesosaurus* would have the capability to swim across the Atlantic Ocean because it was a coastal animal. *Mesosaurus* is interpreted to have propelled itself through water with its long hind legs and limber tail.

Glossopteris (Fig. 6.3d), a seed-bearing woody tree, is perhaps the most important fossil evidence found. So named in Greek due to its tongue-shaped leaves, *Glossopteris* is an extinct seed fern which reached 30 m in height, and it thrived in the Permian. *Glossopteris* fossils are found in all the Southern Hemisphere continents of Africa, Antarctica, Australia, India, and South America (Fig. 6.3). *Glossopteris* seeds are bulky and are therefore very unlikely to have drifted across oceans. The continents are interpreted to have been once joined for Glossopteris has been so widespread across these southern continents.

A number of fossil groups, therefore, had identical populations separated by thousands of kilometers of ocean. Thus, it seemed very difficult to account for such striking parallel evolution which led paleontologists to suggest that unconvincing land bridges once connected the conti-

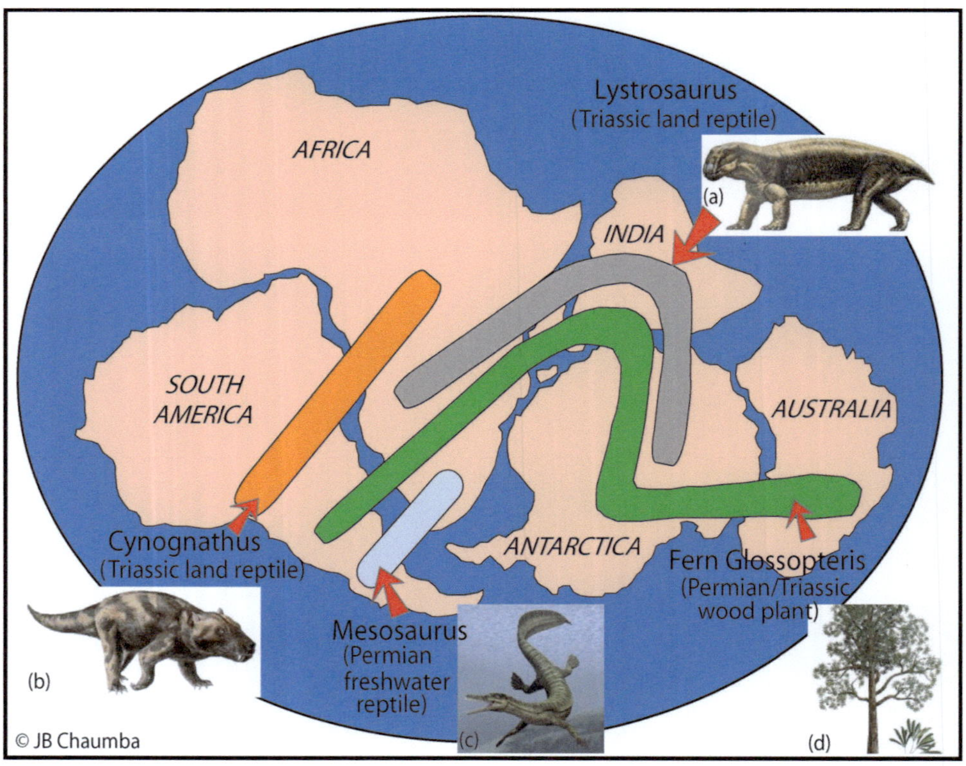

Fig. 6.3 Map of the southern continents (Gondwana) at the time they were part of Pangaea showing the distribution of fossils of *Lystrosaurus* (**a**), *Cynognathus* (**b**), *Mesosaurus* (**c**), and *Glossopteris* (**d**). These continents are now separated by oceans, and the fossils did not have the ability to cross the oceans if the continents had not been joined together at some time in their history

nents. Wegener, however, argued that rising and falling land bridges in the oceans were not possible since the ocean crust is composed of denser rock than which makes up the continents. The denser oceanic crust could not rise up above sea level to form land bridges, and if the land bridge was composed of less dense continental rock, it would not sink into the oceanic crust as it was less dense.

6.3.3 Paleoclimate Evidence

Wegener's meteorology background came in handy when he was studying the distribution of past climates, or paleoclimate indicators, in sedimentary rocks to support his continental drift hypothesis. The occurrence of extensive coal deposits in Upper Paleozoic (Carboniferous and Permian) rocks in northwestern Europe strata implied that the coal deposits were likely to have been formed in hot and wet climates of equatorial regions. Further, the presence of glacial tillites in rocks of the same age in equatorial Africa led Wegener to interpret that the continents must have moved (Africa from the polar region to the equatorial region and Europe from near the equator to northern high latitudes).

From studies of once glaciated rocks, Wegener demonstrated that sections of South Africa, South America, Australia, and India were all once glaciated (Fig. 6.4) at approximately the same time. The glacial striations on these

Fig. 6.4 Map of paleoclimate evidence for continental drift showing the locations of swamps and associated coal deposits, salt deposits, tropical reefs, desert sand, glaciated areas, and glacial striations on the southern continents which formed Gondwana. Photo credit of desert sand: Desert Safari World

once glaciated rocks, however, did not originate in the same location. Rearranging of the continents to their pre-drift positions, however, suggests that the striations were caused by a single glacier which originated at the South Pole (Fig. 6.4). The observation that the fossils matched across continents and that they are in different climatic areas was interpreted to indicate that the past climates (**paleoclimate**s) could be mapped across Pangaea as well. Further, Earth scientists pondered over glacial evidence in South Africa and Australia which occur in areas that are located in tropical and subtropical hot regions. When the continents are reconstructed to their pre-drift positions, it can be observed that the Antarctic Ice Sheet extends across the Gondwana continents (Fig. 6.4). The break-up of **Pangea** resulted in the Gondwana continents moving to different locations which resulted in melting of the glaciers, but the glacial evidence was preserved.

6.4 Skepticism to Continental Drift Hypothesis

How the continents are thought to have "drifted" is what chiefly led Wegener's hypothesis to be rejected by the majority of Earth scientists, particularly those based in the Northern Hemisphere. Wegener thought that the continents were moving through the Earth's crust much like in a manner comparable to icebreakers that plow through ice sheets. Tidal and centrifugal forces were thought by Wegener to be responsible for drifting the continents through the oceanic crust. Reaction to Wegener's hypothesis was, consequently, mostly hostile and harsh. Part of the problem, thus, was that Wegener had no convincing mechanism which accounted for how the continents moved. Opponents of Wegener argued that plowing of continents through oceanic crust would result in distortion of the continents, and that tidal and centrifugal forces were inadequate to cause the continents to move.

Wegener's continental drift hypothesis, however, also had some proponents. Alexander L. Du Toit (1878–1948), a South African geologist who mapped the Karoo Supergroup in South Africa which compared closely with rocks he later mapped in South America. Du Toit published a book titled *A Geological Comparison of South America with South Africa* in 1927, and another one titled *Our Wandering Continents: An Hypothesis of Continental Drifting* in 1937 in support of Wegener's hypothesis. Du Toit's work updated the work of Wegener and he further proposed that, in addition to Pangea, there also existed another landmass made up of Africa, South America, Antarctica, India, and Australia, called **Gondwanaland** in the Paleozoic (e.g., Fig. 6.3). Yet another supporter of Wegener, the Swiss geologist Émile Argand, envisaged continental collisions as the best cause for the folding and faulted rocks occurring in the Swiss Alps where he worked.

After Wegener's death in 1930, a huge body of work which involved mapping and exploring the great system of **oceanic ridges**, which are sites of frequent earthquakes where molten rock rises from below the crust and hardens into new crust, was carried out by Earth scientists. The majority of this additional work was conducted on the Earth's oceanic crust during the 1950s and the 1960s, and it prompted Earth scientists to revisit Wegener's continental drift hypothesis. By the late 1960s, an all-encompassing theory which incorporates continental drift as well as other hypotheses called **plate tectonics** was well received and accepted by almost all Earth scientists. A mechanism to account for how continents moved was found in the seafloor-spreading hypothesis. In short, the plate tectonics theory suggests that the rigid and strong lithosphere (Fig. 6.5) (both the continental crust and the oceanic crust as well as the uppermost part of the mantle form solid **tectonic plates**) "floats" on the weak and soft asthenosphere (Fig. 6.5) (the underlying hot layer is under such tremendous heat and pressure that it behaves as a viscous material that is capable of flowing).

6.5 Evidence for Plate Tectonics Theory

Overwhelming evidence in support of the theory of plate tectonics is now described, in addition to that which Wegener had listed, and Wegener's evidence was also improved upon. Rather than invoking several independent hypotheses to explain large-scale phenomena patterns such as volcanic activity, earthquakes, and folding, for example, which were tailored to specific regions, plate tectonics is an all-encompassing theory explaining how the outer portion of the Earth works.

6.5.1 Better Fit of the Continents

A British geophysicist, Sir Edward Bullard and co-workers, used computer calculations and noted that continents fit better at a depth of 914 m (3000 ft) below sea level rather than using the shorelines (Fig. 6.2). A depth of 914 m was interpreted to represent the true edge of the continents, and it corresponds to approximately half the depth between the shoreline and that of the ocean basins. Matching the continents at this depth, or the continental edge, produced an even better fit of the continents (Fig. 6.2), lending support to Wegener's continental drift hypothesis. After Bullard and his co-workers had published their work, however, it became

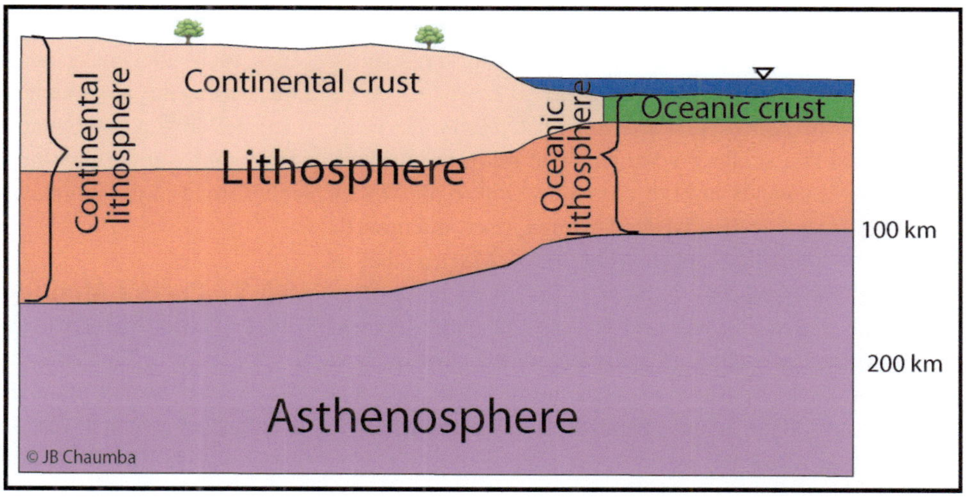

Fig. 6.5 The lithosphere is composed of the continental crust, the oceanic crust, as well as the uppermost part of the mantle. The asthenosphere, which underlies the lithosphere, is the hot layer that is under tremendous heat and pressure that is capable of flowing

known that similar work to what they did had already been published some thirty years earlier by Boris Choubert, a French geologist, in a rather obscure French journal and in French. Choubert's findings had, thus, unfortunately largely remained virtually unknown to the wider scientific community.

6.5.2 Seafloor Spreading

Shortly after the conclusion of the Second World War, vessels that were equipped with sonar traversed the oceans measuring depth profiles of the seafloor beneath them. The survey data was used to create three-dimensional relief maps of the ocean floor, and, by 1953, an American oceanic cartographer named Marie Tharp created the first of several maps that revealed the presence of an underwater mountain range more than 16,000 km (10,000 miles) in length in the Atlantic Ocean—the Mid-Atlantic Ridge (MAR). It was later discovered that the MAR was, in fact, only a small segment of an undersea mountain system termed a mid-ocean ridge (MOR) (Fig. 6.6) that is approximately 64,000 km (40,000 mi) in length.

Harry Hess postulated that molten material from the mantle continuously wells up along the crests of MORs such that it can be transported for distances of ~80,000 km (50,000 miles) from the crests. This mountain chain is called the Mid-Atlantic Ridge. Further, it was also discovered that right at the center of this MOR, a rift (called an **axial rift valley**) measuring ~32 to 48 km (20–30 mi) wide and ~1.6 km (1 mi) deep, extends along the crest of the MOR (Fig. 6.6). Earth scientists also made another discovery that the ridge itself is offset, along **transform faults** (Fig. 6.6) where adjacent plates move in opposite directions (Fig. 6.6), in many places along its length. Extending further away from the offset ridges that are separated by transform faults are **fracture zones** where fractures penetrate from transform faults into a plate which is moving in the same direction

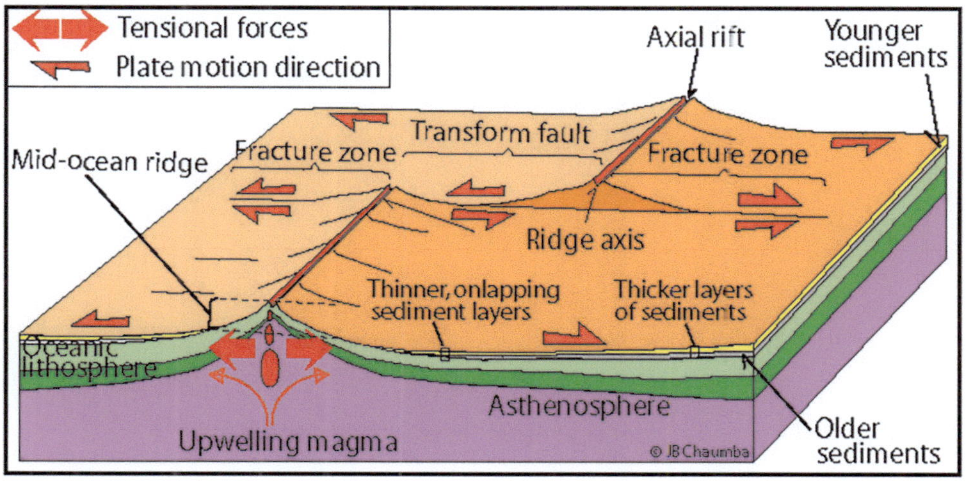

Fig. 6.6 Diagram illustrating transform faults, fracture zones, and tensional forces at a spreading center. The youngest oceanic lithosphere rocks are found at the ridge, and the age increases with increasing distance from the oceanic ridge. The thickness of both the oceanic lithosphere and the overlying sediments also increases with increasing distance from the mid-oceanic ridge

(Fig. 6.6). It is along these crests of ridges and the associated transform faults that the loci of almost all the shallow earthquakes occur in MOR areas.

In 1962, an American Earth scientist, Harry H. Hess, hypothesized that seafloor was created at mid-ocean ridges, from where it spreads in both directions (Fig. 6.6). Molten upper mantle rock, called basaltic magma, rises and upwells beneath the spreading center (Fig. 6.6). Magma that is intruded at the spreading axis cools and hardens to form new oceanic crust that is welded onto the earlier-formed oceanic crust. Spreading is believed to be caused by a number of stresses discussed later in this Chapter, and the upwelling of the mantle beneath the spreading axis is thought to be a passive response to plate separation. The earliest-formed, and thus the oldest, oceanic crust borders the continental crust where the oldest and densest oceanic crust is eventually pushed downward into the mantle, or subducts, at oceanic trenches. It is the oldest, and also the densest oceanic crust at oceanic trenches, that subducts into the mantle at steeply dipping regions which are sites of intermediate and deep earthquakes.

Radiometric dates of samples obtained from the deep ocean floor have shown that basaltic oceanic crust becomes progressively younger as the mid-ocean ridge is approached (Fig. 6.7). Further, it was also shown by Earth scientists that the thickness of sediments overlying the oceanic crust becomes thinner near the ridge, and the youngest layers of sediments are both nearest the ridge. Further away from the ridge, the youngest sediment layers overlie the oldest sediment layers which become progressively older as the distance from the ridge increases. The oldest marine sediments that have so far been recovered by a variety of methods such as deep-sea drilling, dredging, and coring, are not older than 200 million years in age. From dating both oceanic and continental rocks, it was found that rocks making up the ocean floor are considerably younger than rocks that make up the continents. Rocks making up the continents are much older than those constituting the oceanic crust; continental rocks yield ages of over 3 billion years old, supporting the idea that older oceanic crust is reabsorbed in the mantle at subduction zones.

6.5.3 Earth's Magnetic Field and Paleomagnetism

Before we discuss plate tectonic evidence from paleomagnetism, let us first discuss Earth's magnetic field. It is thought that Earth's molten outer core contains liquid iron (see Chap. 1) which circulates and generates its magnetic field. This magnetic field is a dipole (two charges), resembling the magnetic field of a bar magnet. The central locations of the region where the magnetic field lines start and finish are designated Earth's magnetic poles. Earth's magnetic and geographic poles are not exactly aligned since they arise from different mechanisms (Fig. 6.8). Just as the geographic north and south poles indicate imaginary points where Earth's rotation axis intercepts the north and south poles of Earth's surface, the magnetic north and south poles similarly indicate the imaginary points where the magnetic north and south poles intersect the Earth's surface, respectively (Fig. 6.8).

Earth's magnetic poles constantly change location relative to Earth's geographic poles, and the magnetic South Pole currently lies approximately 12° degrees distant from the Geographic North Pole (Fig. 6.8). The magnetic South Pole currently sits in the Arctic Ocean north of the US state of Alaska; hence, the north end on a compass does not precisely point toward geographic north currently but instead points towards Alaska. Consequently, the **magnetic**

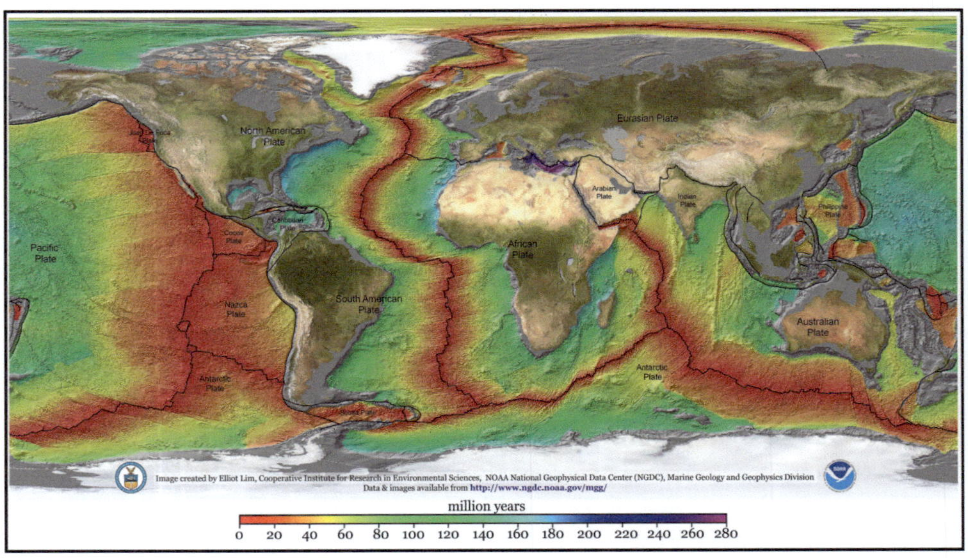

Fig. 6.7 Map showing the age of Earth's oceanic lithosphere. The age becomes progressively younger as the oceanic ridges are approached. *Credit* NOAA

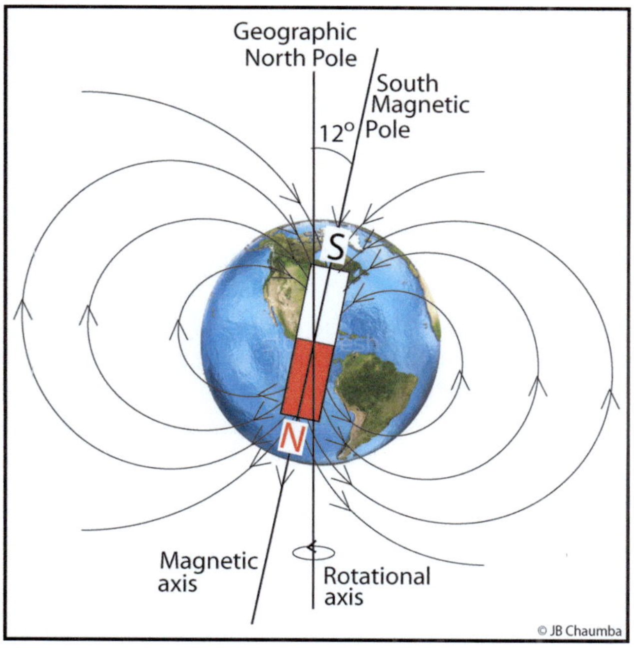

Fig. 6.8 The magnetic field of planet Earth is interpreted as consisting of a giant bar magnet oriented 12° from the Geographic North with its imaginary lines of force as depicted in the diagram. The bar magnet has its N end at the South Pole and its S end at the North Pole, with its lines of force originating from the magnetic north end and terminating at the magnetic south end as shown

declination is a measure of the angle between the line of longitude at any given location and the direction that a compass needle points at that location. The south magnetic pole has been determined from measurements to be wandering across the Arctic Ocean towards Russia at average rates of 55 km/year. Magnetic poles are known to wander within the 70° latitude from the geographic poles.

Researchers who study seismic waves called seismologists showed, in the mid-1960s, that Earth's magnetic field displayed periodic reversals in polarity during periods of thousands of years, giving birth to paleomagnetism. Magnetic surveys conducted by magnetometers, which are highly sensitive instruments that measure Earth's magnetic forces, near the mid-ocean ridge (MOR) showed elongated patterns of normal and reversed polarity of the ocean floor. A timescale for "normal" and "reversed" polarity was established when radiometric dating of rocks displaying normal and reverse polarity was carried out. Results of radiometric dating showed that approximately 171 magnetic reversals occurred in the past 76 million years. Further, the normal and reversed bands parallel the axial rift valley, and they display a symmetrical distribution as mirror images on either side of the axial rift valley (Fig. 6.9). Consequently, Earth's magnetic history is recorded by the spreading ocean crust to form a continuous record of the growth of the oceanic crust. **Seafloor spreading**, thus, is ocean crust that forms along mid-ocean ridges from where it grows and spreads out laterally away from these MORs (Fig. 6.9).

As the molten magma solidifies upon cooling at MORs, it pushes away earlier solidified magma from the flanks of the ridges (Fig. 6.9). This process causes spreading away from the crests, creating successively younger ocean floor at the ridge crest. As material is pushed away from the

Fig. 6.9 Diagram illustrating how striped magnetic anomalies at the mid-ocean ridges are formed. As two plates move away from each other at the spreading ridge, magma that is injected beneath the ridge cools and solidifies, making a recording (or "freezing in") of the direction of the magnetic field at the time it solidified. With passage of time, the magnetic field changes its polarity, and this new polarity is then recorded in the new magma that is intruded and solidifies at the ridge axis. With continued changes in the polarity of the magnetic field, a striped magnetic pattern is formed that is parallel to the ridge axis. Red arrows indicate the direction of plate motion

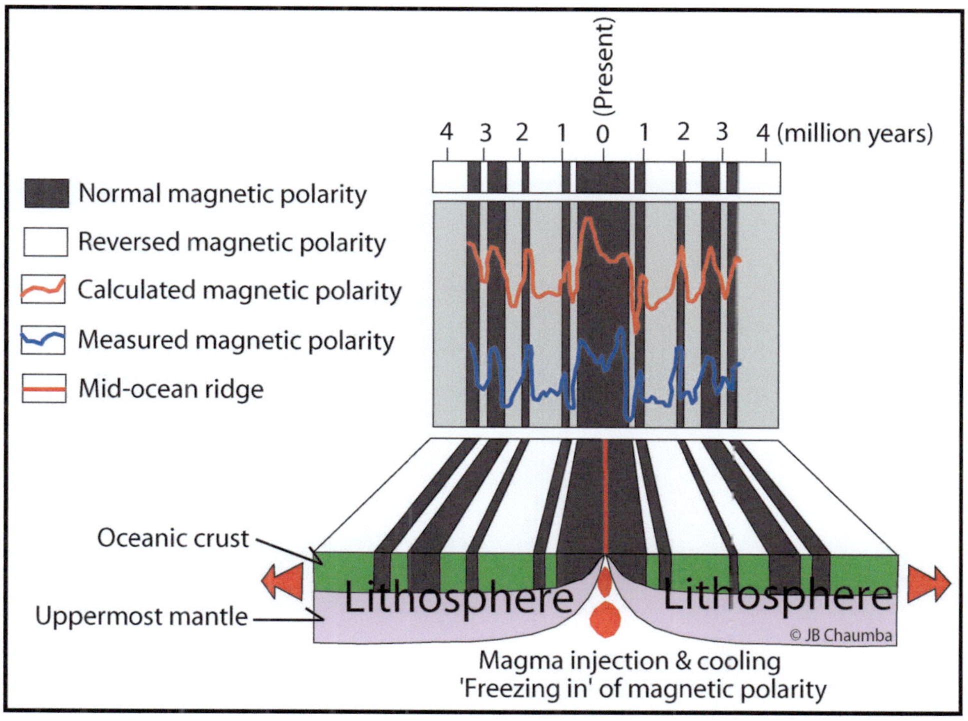

ridge crests, drifting apart of the continents occurs, which can result in continents that were once juxtaposed being separated by large ocean basins. As an example, continents that border the Atlantic Ocean drifted away from the Mid-Atlantic Ridge at rates of 1–2 cm (0.4–0.8 inch) per year, which increased the width of the ocean basin by twice this rate. In the Pacific Ocean, for example, continents that surround this ocean are bordered by deep-seated trench systems leading to the ocean floor being subducted.

A large body of evidence lends support to the seafloor spreading hypothesis. Results from thermal probe studies, for example, have shown that heat flow through sediments at the bottom of the oceans is broadly comparable to that through the continental crust with the exception of heat flow over the mid-ocean ridges. Over some MORs, heat flow measurements were observed to be three to four times the normal values measured away from these ridges. The anomalously high values were interpreted to be caused by molten material that was intruded near the ridge crests. Further, ridge crests have been observed to be characterized by anomalously low seismic wave velocities, which are thought to be caused by upwelling magma, microfracturing, and thermal expansion of the ridge crests. Three different patterns of seafloor spreading are depicted by the Pacific Ocean (Fig. 6.10a), the Arctic Ocean (Fig. 6.10b), and the Atlantic Ocean (Fig. 6.10c) which show the relative ages of the oceanic crust. The youngest regions, which straddle the oceanic ridges, are colored red, whereas the oldest regions, which are located furthest away from the oceanic ridges, are colored blue (Fig. 6.10).

A chron, which is the interval of time between reversals of polarity of Earth's magnetic field, denotes the interval of time that is represented by a magnetostratigraphic polarity unit. A chron represents a certain period of time where the magnetic field of Earth was in predominantly a "normal" position (as today) or in a predominantly "reversed" position (Fig. 6.11a). Chrons are numbered numerically starting from today and then increasing into the past (Fig. 6.11a). A chron is the time equivalent to a chronozone or a polarity zone. When the interval lasted less than 200,000 years, a chron is referred to as a "subchron" (Fig. 6.11a). The time scale of Earth's magnetic anomalies in the last 4 million years can be represented by a series of normal and reversed polarities as shown in Fig. 6.11a.

Rising magma assumes the polarity of Earth's magnetic field before it solidifies (Fig. 6.11b) either into oceanic crust (Fig. 6.9) or on the continent (Fig. 6.11b). At spreading centers, this crust is separated into parallel rock bands by successive pulses of magma. When Earth's magnetic field undergoes a reversal, the change in polarity is recorded in the magma, which contributes to the alternating pattern of magnetic striping on the seafloor. When material is still molten, minerals within the molten material have a random orientation (Fig. 6.11b) as the temperature of the magma is above the Curie temperature of 585 °C (~1085°F). However, as the magma cools below the Curie temperature, some iron-bearing minerals such as magnetite become magnetized and assume the orientation of the Earth's prevailing magnetic field, making a record of the orientation of this magnetic field at the time the rock solidified. Thus, a rock composed of such minerals preserves the orientation of the Earth's dipole millions or hundreds of millions of years ago at the time of its formation and is paleomagnetic, meaning that it possesses a property called **paleomagnetism**.

Magnetic stripes on the ocean floor are detected by towing a magnetometer across the sea surface to pick up changes in magnetic field strength, or "anomalies," caused by the changes in the magnetization of the ocean crust. Normal magnetization areas of the ocean crust add to the overall magnetic field whereas reversed magnetization

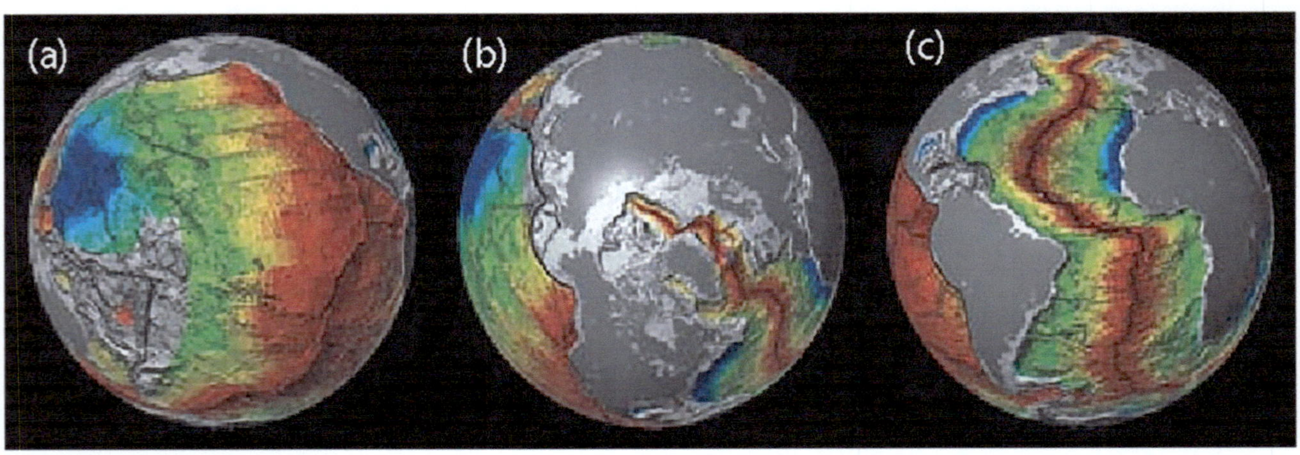

Fig. 6.10 Three images of Earth showing different patterns of seafloor spreading in the Pacific Ocean (**a**), Arctic Ocean (**b**), and Atlantic Ocean (**c**) which results in the relative age of oceanic crust in the three oceans. The color red denotes the youngest regions, whereas the color blue denotes the oldest regions. Credit *U.S. Dept. of Commerce/NOAA*

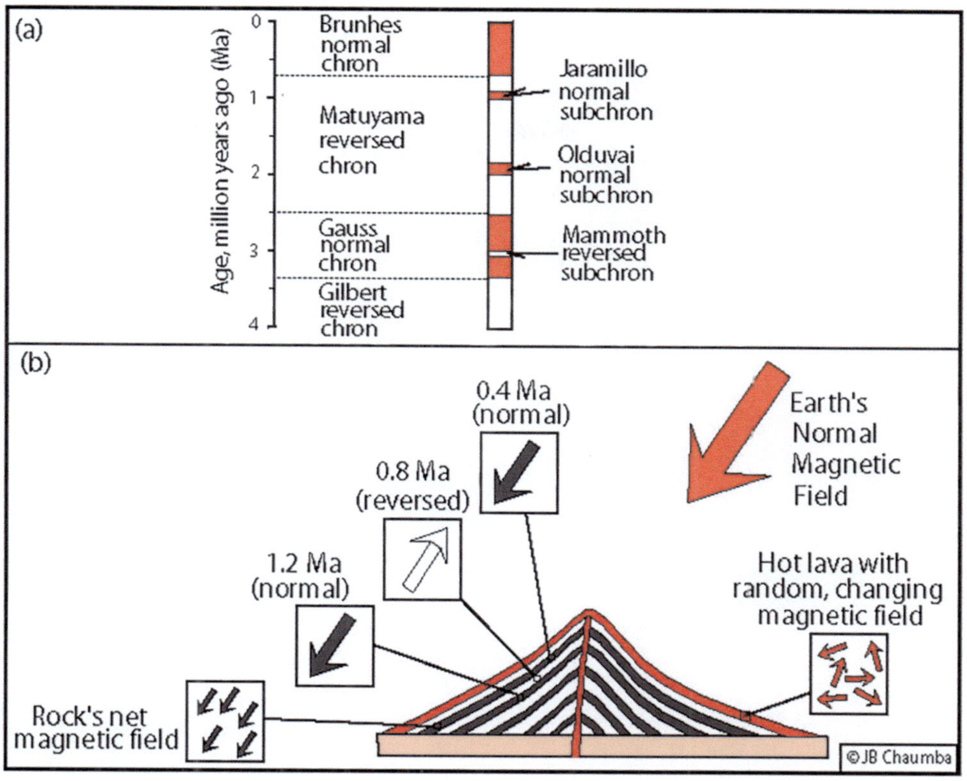

Fig. 6.11 **a** Magnetic reversal chronology of Earth over the past 4 million years. A polarity chron is the time interval between successive magnetic reversals. Chrons are the major divisions which lasted approximately a million years whereas subchrons are the subdivisions of chrons based on reversals that lasted a few hundred thousand to a few million years. **b** The big orange arrow represents Earth's magnetic field which presents the dipole pointing to the south magnetic pole from the north magnetic pole. An erupting volcano can build up successive layers of lava over time, with some layers having normal polarity whereas other layers have reversed polarity

cancels out some of that magnetic field. Consequently, magnetic anomalies, where the magnetic field is stronger or weaker than expected, are produced. Radiometric dating enables magnetic reversals of different ages to be recognized from normal magnetic anomalies.

Paleomagnetism, thus, can be defined as the study of the intensity and orientation of the earth's magnetic field as preserved in the magnetic orientation of certain minerals found in ancient rocks. Paleomagnetic studies of rocks and sediments deposited on the ocean floor have shown that the orientation of the earth's magnetic field has alternated frequently throughout geologic time. Today, the north-seeking end of the compass needle points toward the present North Pole (i.e., the south magnetic pole). Rocks that crystallize today with the current magnetic polarity preserved in those rocks possess "normal" polarity. Studies have shown that this has not always been the case. During some periods in the geologic past, the north-seeking end of the compass needle has been interpreted to have pointed southward, yielding periods of "reversed" polarity resulting in the production of this alternation between normal and reverse polarity, or magnetic "flip-flops."

When magnetic minerals get dislodged from their parent rocks due to weathering and erosion, they are eventually deposited in depositional sedimentary basins where they align themselves parallel to Earth's prevailing magnetic field upon settling in standing water. When the sediments are lithified, magnetization is preserved in the lithified rock. Any changes in the orientation of Earth's magnetic field through time can be traced by studying rock specimens of different ages and then determining the alignment of their magnetic fields.

6.5.4 Apparent Polar Wandering

Earth scientists had long assumed that continents, throughout geologic history, had fixed positions. By obtaining **paleopoles** of rocks of different ages that were closely spaced, Earth scientists observed that the results defined a winding line which they termed an apparent polar-wander path. Paleopoles are Earth's magnetic poles as they were positioned at particular times in the geological past. The perceived movement of paleomagnetic fields of Earth relative to the continents while regarding the continents as fixed in position is called **apparent polar wander**. Apparent polar wander is often displayed on latitude and longitude maps as a path that connects the locations of geomagnetic poles that are inferred from paleomagnetic techniques at different times (Fig. 6.12). Rocks from different continents were found to have different apparent polar-wander paths, which eventually led Earth scientists to the conclusion that it was not the poles which wandered, but rather it was the continents which wandered relative to

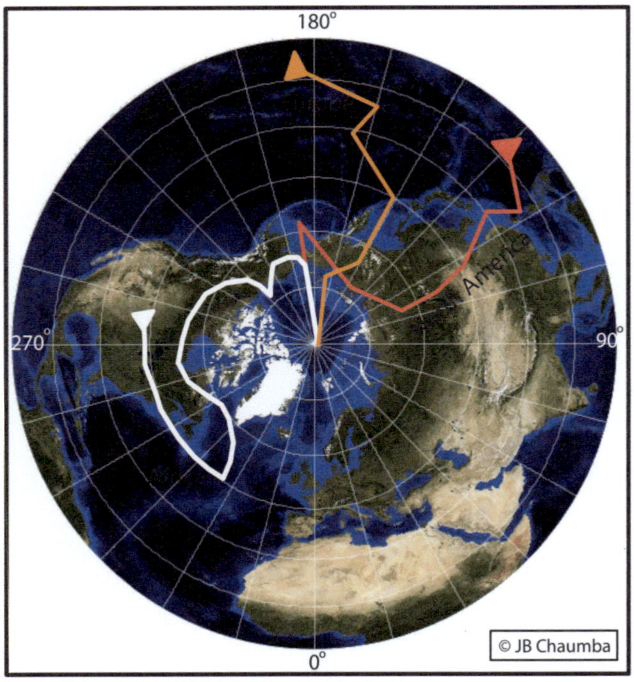

Fig. 6.12 Map of the Northern Hemisphere centered over the Geographic North Pole showing successions of paleopoles defining apparent polar-wander paths for Africa, North America, and Europe (map credit: Intergovernmental Committee on Surveying and Mapping, ICSM)

a fixed pole. Each continent, therefore, had drifted following its unique path even though each continent had also drifted relative to other continents (Fig. 6.12).

Earth scientists are still accumulating more data to determine why magnetic poles move. Current understanding, however, is that the relative movement of the magnetic pole can be due to drifting of the continents, or polar wandering, or a combination of continental drifting and polar wandering. The magnetic poles are now interpreted to rarely move far from Earth's geographic poles (Fig. 6.12).

6.5.5 Hot Spots and Mantle Plumes

J. Tuzo Wilson, a Canadian Earth scientist, first proposed **mantle plumes** in 1963 and observed that active volcanoes (those that are erupting or can erupt in the future) tend to be located where a chain of extinct volcanoes terminates. A mantle plume can be envisaged as an upwelling of abnormally hot and buoyant rock which can rise as it is less dense than surrounding rock. The head of mantle plumes, termed the plume head, can partly melt upon reaching shallow depths (decompression melting, see Chap. 3). Consequently, a mantle plume is often envisaged as the cause of volcanic hot spots, such as Hawaii, Afar, Tristan da Cunha, or Iceland, as well as **large igneous provinces** (LIPs) such as the Karoo, Deccan, Columbia River basalts, and Siberian LIPs. An LIP is a region characterized by extensive volumes of both extrusive and intrusive igneous rocks which accumulated over relatively short time periods. Some such volcanic regions are thought to represent unusually large volumes of volcanism near plate boundaries (for example, Iceland and Afar) or in LIPs. Other LIPs are located far from tectonic plate boundaries. A number of volcanoes and intraplate LIPs are not constrained in their location by boundaries of plates, as is the case with most volcanoes (for example, Hawaii and Tristan da Cunha). Volcanoes and LIPs whose locations are independent of plate boundaries are termed **hot spots**. When melting occurs at the base of the lithosphere, probably due to either heat transfer and/or decompression melting, magma that is generated can form a chain of extinct volcanoes termed a **hot spot track** (Fig. 6.13a) such as the Hawaiian-Emperor seamount chain (Figs. 6.13 and 6.14). A hot spot track is thought to be produced as the lithosphere slowly travels over a mantle plume whose position is fixed. Although hot spots are encountered over continental lithosphere (such as the Cameroun hot spot in Africa and the Yellowstone hot spot in North America), the majority of them occur within oceanic lithosphere.

A mantle plume originates when hot rock is generated at either the core-mantle boundary or as a consequence of upper mantle convection from where it rises through the mantle as a diapir to the base of the lithosphere. The concept that mantle plumes are fixed relative to one another, and anchored at the core-mantle boundary, can account for the time-progressive chains of older volcanoes that extend out from hot spots such as the Hawaiian-Emperor seamount chain (Fig. 6.13).

Hot spots, the majority of them being located far from plate boundaries (Fig. 6.14), have also been attributed to result from intraplate extension locations that are thought to be fed by underlying mantle that is anomalously hot compared to the surrounding mantle since their position on the Earth's surface is independent of tectonic plate boundaries. In addition to hot spots thought to be due to mantle plumes that rise as thermal diapirs from the core-mantle boundary, lithospheric extension is also interpreted to permit passive rising of magma from shallow depths. This hypothesis, thus, asserts that the mantle source beneath them is not anomalously hot and considers the term "hot spot," therefore, to be a misnomer. Some further examples of hot spots are Iceland, Reunion, and Galapagos (Fig. 6.14). It is also inferred by some Earth scientists that the breakup of Pangaea may have been triggered by the large number of hot spots in the Atlantic Ocean (Fig. 6.14).

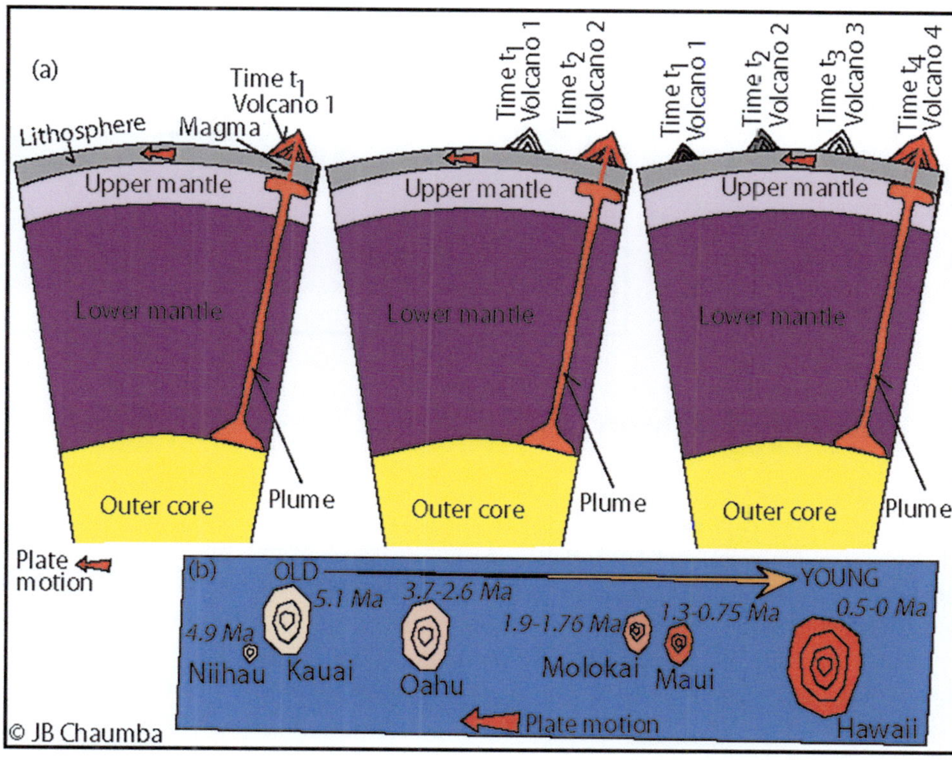

Fig. 6.13 **a** Diagram illustrating a mantle plume originating at the core-mantle boundary which melts due to decompression leading to progressive stages in the formation of a hot spot track with increasing time from t_1 (earliest) to t_4 (latest). **b** Diagram depicting the Hawaiian chain of islands and their ages which increase from right to left, in the direction of plate motion. The youngest rocks are found on Hawaii, whereas the oldest ones occur on Kauai

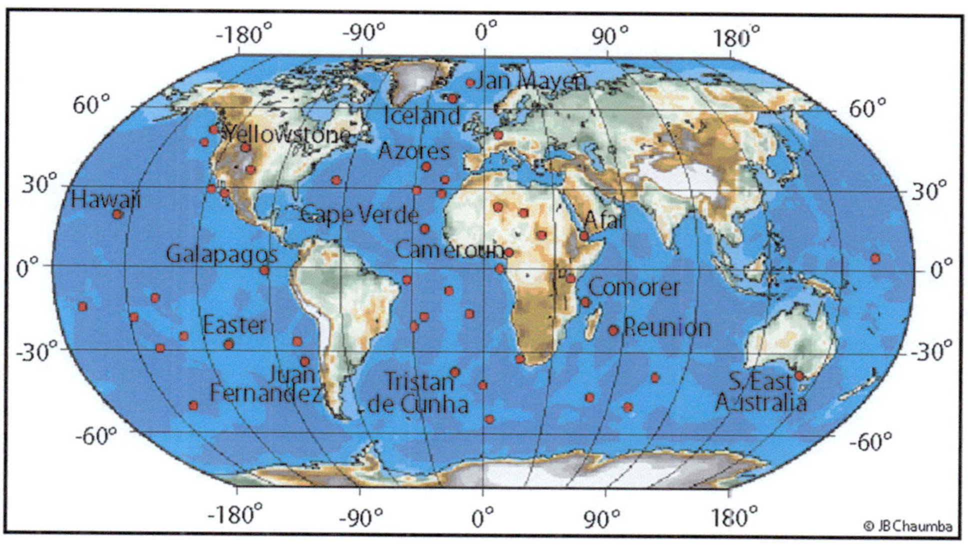

Fig. 6.14 Map of the world showing the distribution of hot spots. Modified after http://eqseis.geosc.psu.edu

We have already discussed that the mantle, which is the source of the heat for a hot spot, is the source of a column of hot rock that ascends from the core-mantle boundary to the base of the lithosphere (Figs. 6.13 and 6.14). As the overlying plate moves over the hot spot, a chain of volcanoes forms. Some of the volcanoes which have been carried from the hot spot in the direction of plate motion tend to become submerged under water as the lithosphere becomes colder and denser, and subsides to result in the formation of a **guyot** if their tops are flat (Fig. 6.15). A guyot is an undersea flat-topped volcano whose top part once rose above sea level but was subsequently flattened mainly due to wave erosion (Fig. 6.15). Evidence that guyots must have once been built to above sea level comes from fossils such as coral reefs, which are organisms that survive only in shallow waters. A **seamount** is a volcano that did not build up to sea level and it therefore remained submerged below sea level (Fig. 6.15). Both seamounts and guyots, however, build up from the ocean floor during the time that they would have been atop the hot spot.

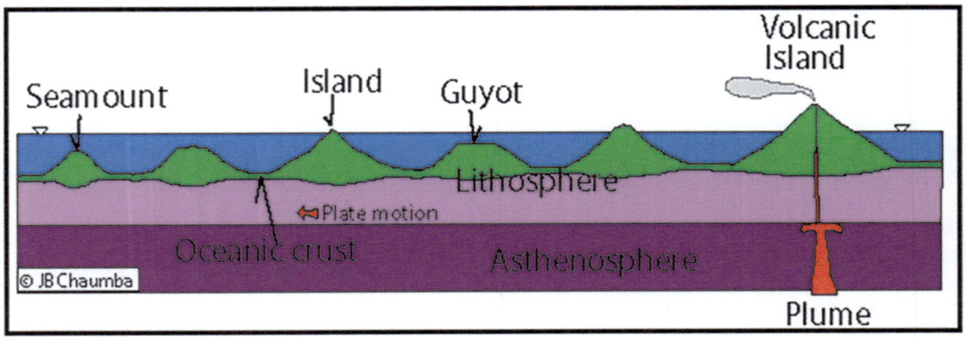

Fig. 6.15 The volcanic island to the extreme right is presently atop a hot spot where it experiences volcanism. With time, a volcano can be carries to the west away from the hot spot where the top is flattened by erosion to form guyots. Where the volcano does not build up to sea level and its top is not flattened, a seamount is formed

6.6 Plate Tectonics Theory

We have already discussed that according to the theory of plate tectonics, the cold and rigid lithosphere of the Earth, which overlies the weak and hot asthenosphere, is broken into eight [8] major plates (African, Antarctic, Eurasian, Indo-Australian, Nazca, North American, Pacific, and South American) and ten [~10] small plates (for example, Philippine, Scotia, Caribbean, and Cocos) that slowly move relative to one another. The Pacific plate is the largest of all plates. There are also numerous minor plates, which are plates that are typically grouped together with an adjacent major plate. A microplate, such as the Madagascar and Rovuma plates within the African plate or the Juan de Fuca and the Gorda plates within the North American Plate, may be defined as a plate with an area of less than 1 million km^2. Further, Earth scientists infer that as distance from the oceanic ridge increases, the older the oceanic crust becomes, and the older the sediments above the oceanic crust become (see Sect. 6.4.2). This observation implies that ocean ridges must be areas where plates are moving apart, and new oceanic crust is being created.

If new oceanic crust is being created at ridges, it must be destroyed somewhere since the size of the planet is constant. Thus, mountain ranges are areas where crust is being destroyed under three different scenarios. The first scenario involves both continental crusts being destroyed by crumbling such as in the case of the Himalayas. In the second scenario, only oceanic crust is destroyed by crumbling and subduction such as in the Aleutian chain of islands. In the third scenario, both continental and oceanic crusts can be destroyed such as what is occurring in the western margin of South America where oceanic crust subducts and continental crust are crumbled. The latter two scenarios involve sinking (subducting) of the oceanic plate beneath another (but older) oceanic plate, or subducting beneath another continental plate. Subduction in both scenarios leads to the formation of deep oceanic trenches and chains of volcanoes on the overriding (above) plate.

The majority of earthquakes on our planet by far occur along plate boundaries (Figs. 6.16 and 6.17). By plotting the location of these earthquakes, Earth scientists are able to map plate boundaries (Figs. 6.16 and 6.17) and the depths at which the earthquakes occur. Further, from paleomagnetic studies, Earth scientists are able to map even the location of past plate movements with great precision. Tectonic (or lithospheric) plates have been determined to migrate at an average rate of between ~0.60 cm/yr and 10 cm/yr (~0.2–3.9 in/yr). The Pacific plate migrates at relatively fast rates of 4 cm (1.6 in) annually, whereas the North Atlantic lithospheric plate migrates at very slow velocities of 1 cm (0.4 in) per year. On average, tectonic plates migrate at very slow velocities ranging from 5 to 10 cm/yr (2–3.9 in/yr).

A new tectonic map of the world incorporating a global compilation of active plate boundaries and geological provinces is shown in Fig. 6.18. Initial plate and province boundaries in this map were constructed from a combination of regional models that were refined using a variety of geoscientific constraints such as earthquakes, relative global position satellite motions, mapped faults, and radiometric ages, among others. The plate model shown in Fig. 6.18 shows improved correlation to observed earthquakes and occurrences of volcanoes within deformation zones and small plates called microplates.

6.7 Types of Plate Boundaries

Three major different types of plate boundaries have been identified (Fig. 6.19). A fourth plate boundary, which is not currently well understood, has recently been identified. Plate boundaries are interpreted as a consequence of plate motion (Table 6.1).

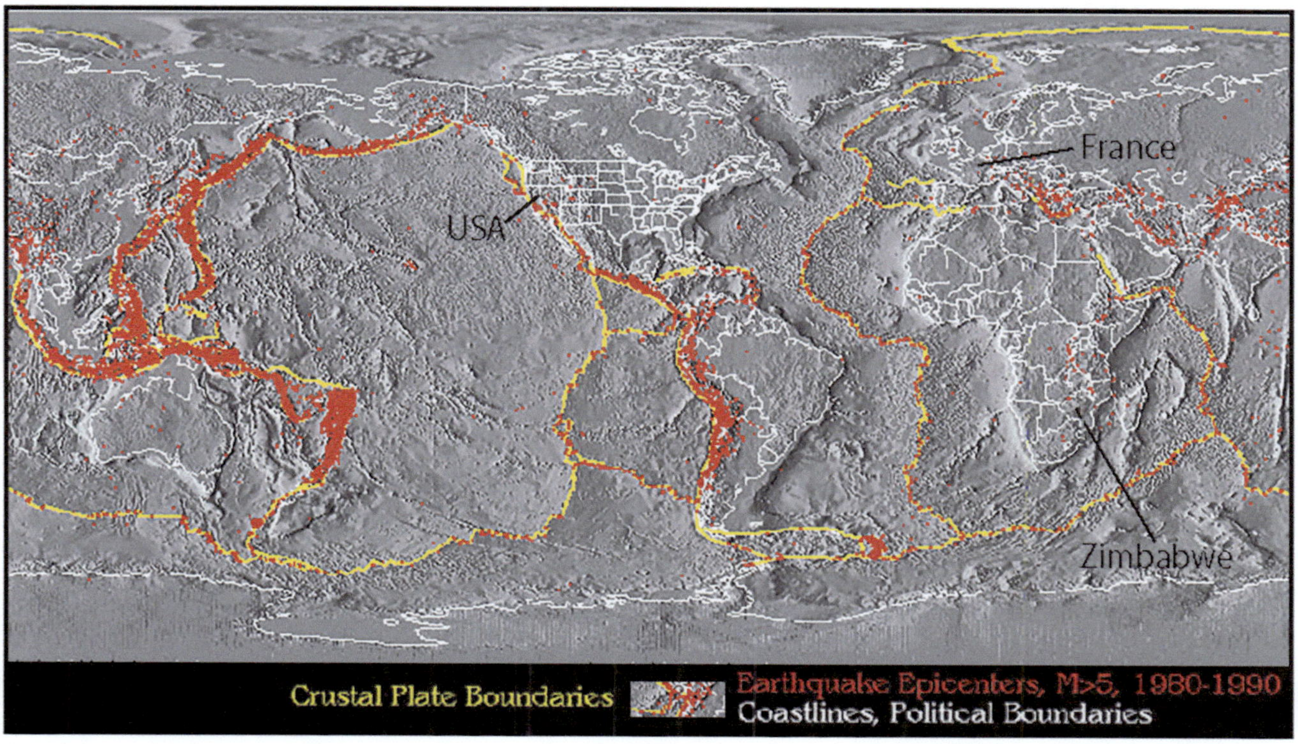

Fig. 6.16 Map of the world showing the distribution of earthquakes (red dots) of magnitude higher than 5 from 1980 to 1990. Most earthquakes are located along margins of plates whereas very few earthquakes are located within the interior of plates. Credit UC Berkeley

Fig. 6.17 Map showing world tectonic plates. Tectonic plates are of different shapes and sizes, and they migrate in different directions. Red arrows indicate direction of plate motion. Credit USGS

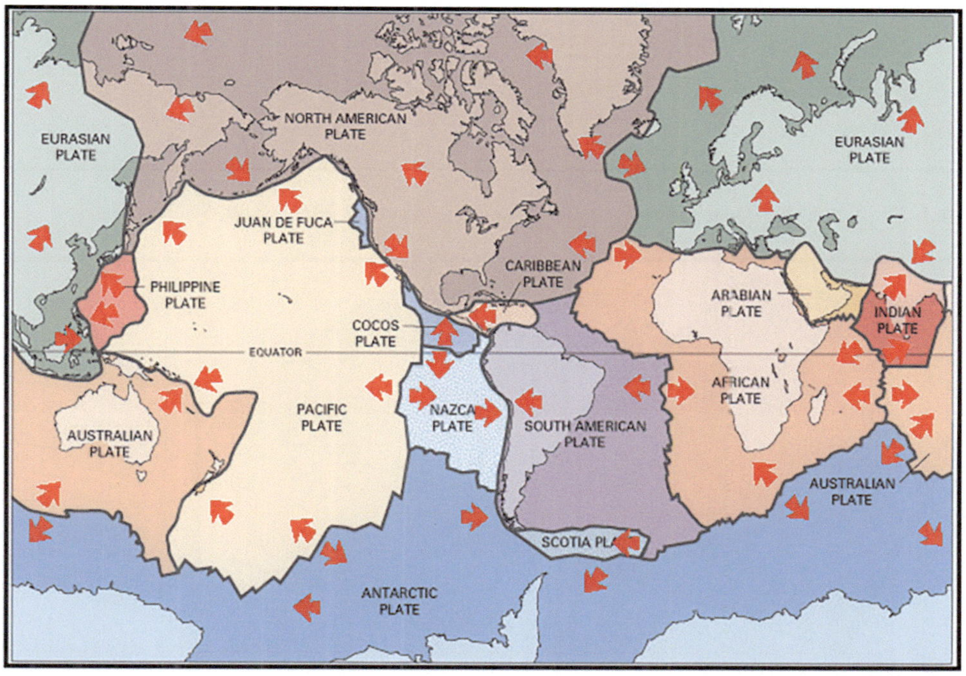

6.7.1 Divergent Plate Boundaries

Divergent plate boundaries (Fig. 6.19a; Table 6.1) are formed where two plates are moving away from a plate boundary. Any space that is created due to two plates moving apart is instantly filled with new mantle material in the form of magma that is generated below the divergent plate boundary due to decompression melting. The magma solidifies to form basalt on the ocean floor and gabbro beneath it. Divergent plate boundaries (derived from *di* meaning "apart" and the Latin verb *vergere* meaning "to move") are also known as constructive or spreading plate

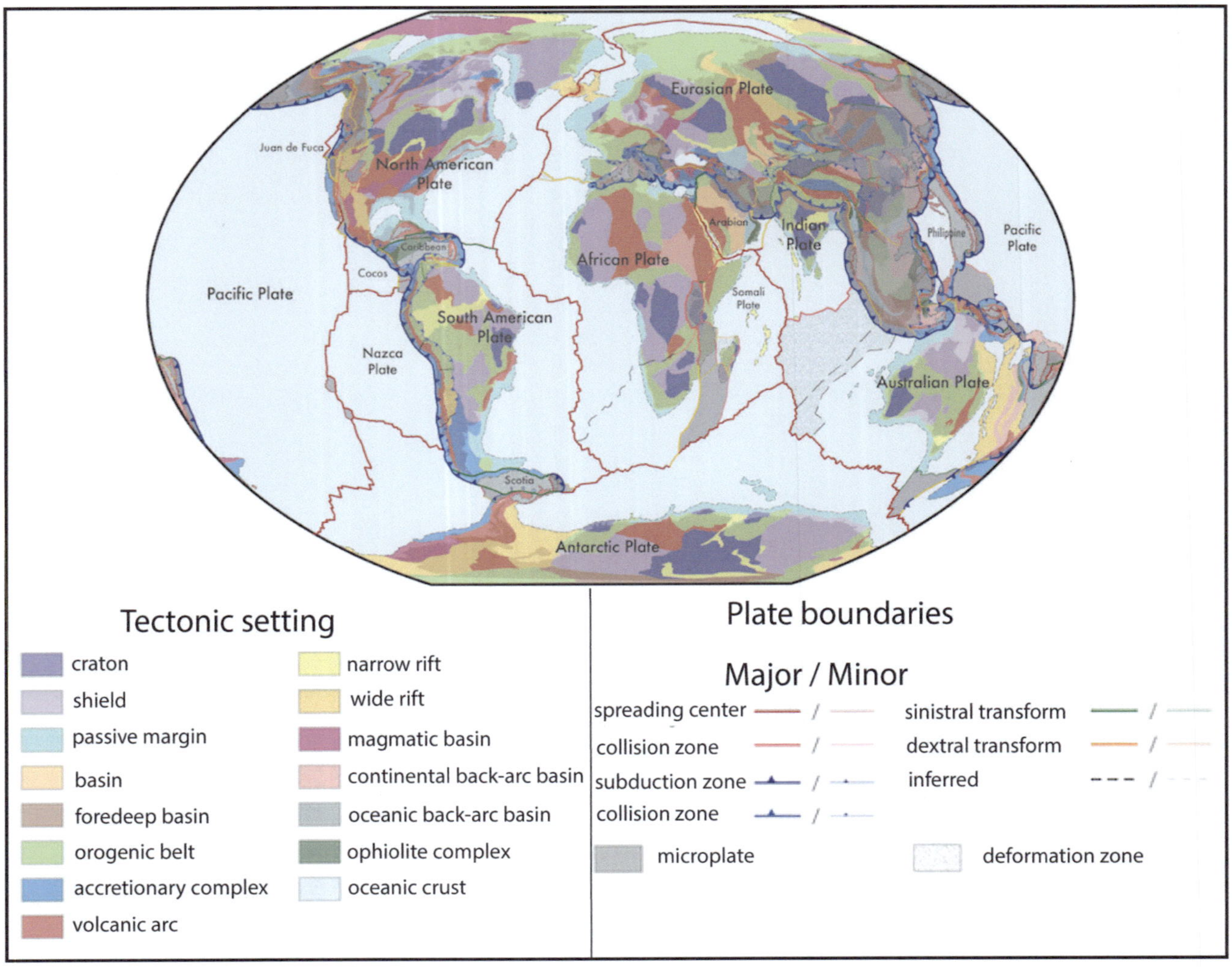

Fig. 6.18 New tectonic map of the world (after Hasterok and others 2022—Earth-Science Reviews)

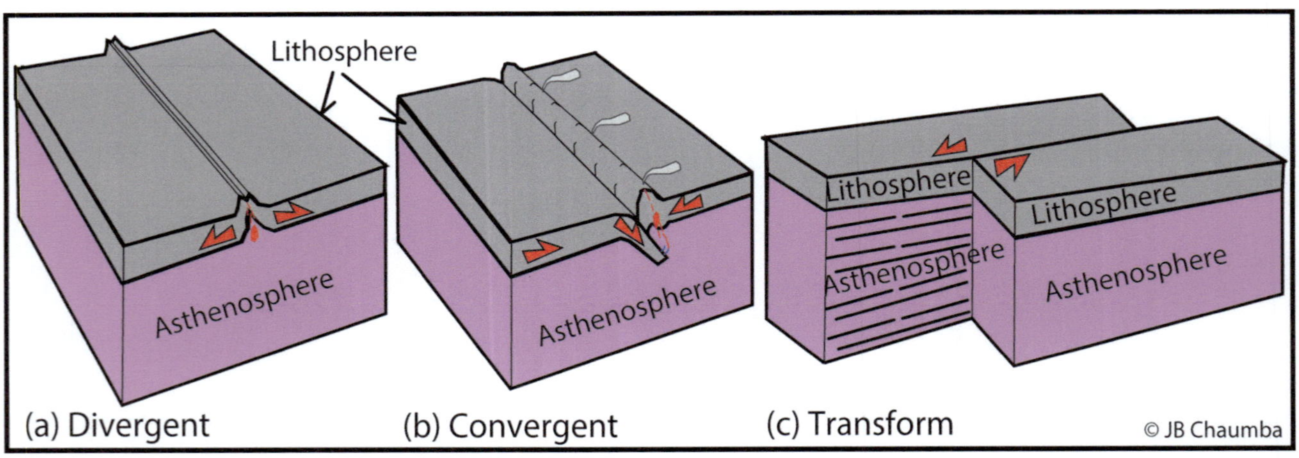

Fig. 6.19 Diagram showing divergent (**a**), convergent (**b**), and transform (**c**) plate boundaries. The red half arrows shown indicate direction of plate motion

Table 6.1 Summary of different types of plate boundaries and examples

Plate boundary type	Lithosphere type	Stress types and plate dynamics	Type of landforms and products	Common examples
1. Divergent	Dominantly oceanic	Tensional stresses and spreading; two plates moving away from a boundary	Mid-ocean ridge, rift valley, volcanoes, earthquakes; basalt and rhyolite	Mid-Atlantic Ridge (Africa-South American plates boundary); Central Indian Ridge (African and Indian plates); East African Rift; Red Sea Rift
2. Convergent	Oceanic-continental	Compressional stresses and subduction of oceanic plate	Ocean trench, continental volcanic arcs, earthquakes; andesite	Andes (Nazca and South American plates boundary); Cascades volcanoes (Juan de Fuca and North American plates); Peru-Chile trench
2. Convergent	Oceanic-oceanic	Compressional stresses and subduction of denser oceanic plate	Ocean trench, island arc volcanoes, earthquakes; andesite	Lesser Antilles Volcanic arc (Caribbean and South American plates boundary), Aleutian islands (Pacific-North American plates boundary). Aleutian trench
2. Convergent	Continental-continental	Compressional stresses and collision of the two plates	Mountain belt and earthquakes; none to minor andesite volcanism	Himalayas (Indian and Eurasian plates boundary), Appalachians (North American-African and Eurasian plates), Alps (Africa-Eurasian plates)
3. Transform	Oceanic-oceanic or continental-continental	Shear stresses and two plates slide past each other	Earthquakes; no volcanism	San Andreas fault (California, USA), North Anatolian fault (Türkiye), Alpine fault (New Zealand)
4. Plate boundary zones	Oceanic-oceanic or continental-continental?	Tensional or compression or shear stresses, deformation	Earthquakes, volcanoes, mountain ranges; andesite	New Zealand

margins since new oceanic lithosphere is generated here in the form of basalt and gabbro beneath it resulting in two adjacent plates spreading laterally. Divergent plate boundaries are dominantly located at mid-oceanic ridges although some can also form within continents when extensional forces cause thinning of the continental lithosphere. With continuous extension, continental lithosphere will eventually open up to become an ocean basin.

Divergent plate boundaries that are generated within continents initially produce rifts, which later become rift valleys. The most active divergent plate boundaries, however, are generated between oceanic plates and they are often referred to as mid-oceanic ridges. Divergent plate boundaries, thus, are generated as the plates pull away from each other resulting in the generation of new crust. One can envisage three stages in the development of a divergent plate boundary with the first and early stage involving rifting such as the East African Rift System discussed below. A second and intermediate stage would involve opening up of a narrow and elongated ocean basin such as the Red Sea. A third and final stage can be equated to an elongated and wide ocean such as the Atlantic Ocean.

It is along divergent plate boundaries where geysers spurt superheated water on the continents and where hydrothermal fluids gush out along hydrothermal vents to form **black smokers**. A black smoker can be defined as a vent on the seafloor where superheated acidic water containing dissolved matter, typically black sulfide minerals, is ejected (Fig. 6.20a, b). A black smoker, therefore, is a chimney that is formed from deposition of low-pH black-colored iron sulfides. Black smokers get their name from the suspended minerals they contain which are deposited on the ocean floor upon being chilled by cold seawater. In some instances, a **white smoker** can be produced, which is a vent where alkaline hydrothermal fluids are emitted on the ocean floor (Fig. 6.20c). The main difference between black smokers and white smokers is that hydrothermal fluids emitted by white smokers are cooler (260–300 °C) than those that are emitted by black smokers (360–375 °C). Further, white smokers are located further away from the mid-ocean ridges than black smokers. Lost City, a hydrothermal vent field in the central Atlantic Ocean, is a famous white smoker. Earth scientists believe that seawater which infiltrates downward interacts with mafic or ultramafic rocks to produce high-pH fluids that precipitate silica as well as barium, silicon, or calcium sulfates (which are white in color) upon mixing with seawater to produce the white color of white smokers.

Some Earth scientists think that the origin of life can be traced to the interface between alkaline hydrothermal fluids

Fig. 6.20 Black smokers and white smokers. **a** Black smoker—Sully Vent in the Main Endeavour Vent Field of the northeast Pacific Ocean showing a bed of tube worms that cover the base of the black smoker (credit: NOAA). **b** A black smoker vent at South Cleft located at the Juan de Fuca Ridge in the northeastern Pacific Ocean (Photo credit: NOAA). **c** Image of a white smoker on Kawio Barat submarine volcano on Sangihe arc, Sangihe Islands, Indonesia, spewing white smoke captured at a depth greater than 1,850 m (credit: NOAA Okeanos Explorer Program). **d** Tube worms have their longer tail ends firmly planted on the ocean floor, whereas the shorter and upper red plumes sway to resemble a field of poppies (credit: NOAA)

and seawater where the right conditions for life to emerge were met. Earth scientists discovered that chemical energy that is contained in hydrothermal fluids is responsible for the development of intense microbial activity. Consequently, there are specific animal communities such as giant **tube worms** (Fig. 6.20a, d), among others, that feed in and around the hydrothermal fluids that are fed by geothermal energy. The geysers and hydrothermal solutions are heated by magma beneath the divergent boundary, and minerals are often dissolved by the hydrothermal solutions which then precipitate black smokers.

When extensional forces stretch continental crust, thinning of the continental lithosphere occurs which leads to rifting which can ultimately break a continent apart. A narrow and linear zone of normal faulting called a rift can be generated. The Basin and Range Province of the western US and the East African Rift in East Africa are examples of active rifting currently taking place. In the East African Rift System (Fig. 6.21) which is a developing divergent plate boundary, a part of Eastern Africa which includes the Horn of Africa region, is, thus, a continental rift zone that is in the process of breaking apart from the rest of the African continent.

The East African Rift first formed 22–25 million years ago, and it is a narrow zone delineating where the African plate is in the process of splitting into the Nubian plate (the larger plate to the west) and the relatively smaller Somali plate to the east, at a rate of 6–7 mm (0.2–0.3 in) per year. With continued extension, Earth scientists estimate that this continental lithosphere can split up within 10 million years when the Somali plate finally breaks off from the Nubian plate leading to the formation of a new narrow ocean basin.

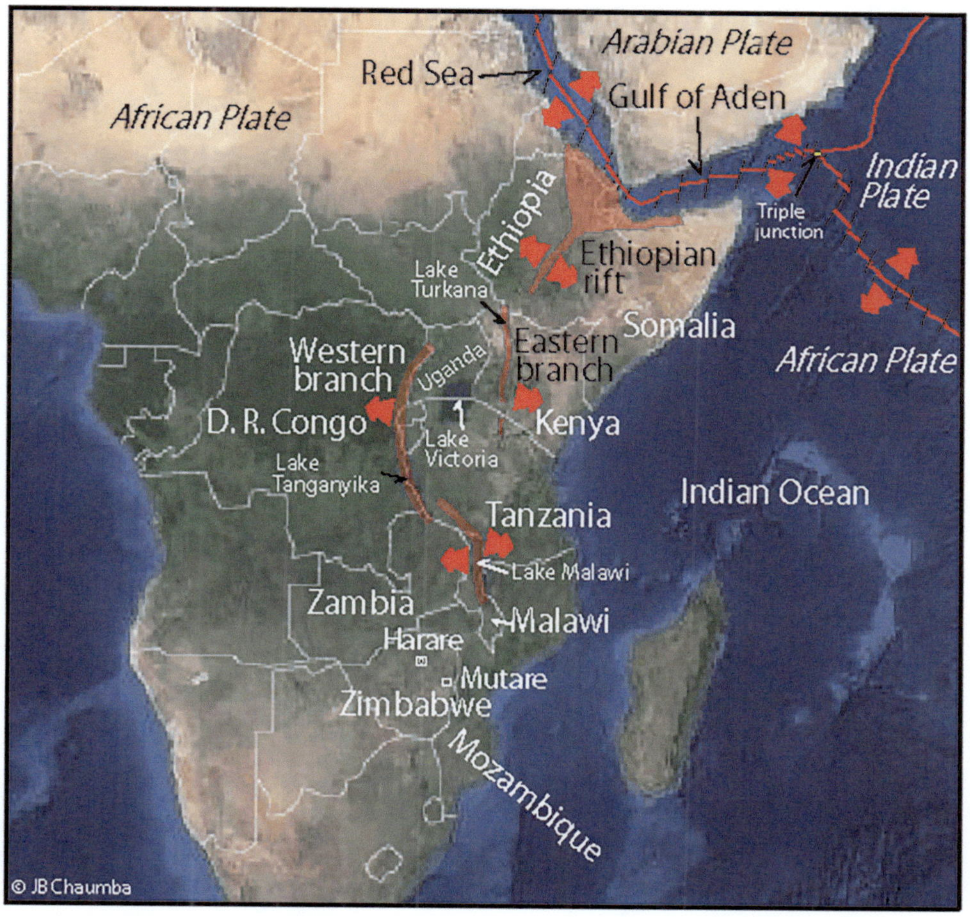

Fig. 6.21 Google Earth map showing the extent of, and rift segment names for, the East African Rift System which is growing. The names of major lakes within the rift system are also shown, as well as Lake Victoria which is not part of the rift system. Red arrows denote extensional forces

During movement of plates, each plate is assumed to largely remain intact and rigid in its interior, except at the margins where deformation, destruction, and growth of a plate can occur.

6.7.2 Convergent or Destructive Plate Boundaries

Convergent plate boundaries (Figs. 6.19b and 6.22, Table 6.1) are generated where two plates collide, resulting in one plate being destroyed due to subduction beneath another plate. When two plates collide, the impact from the collision results in buckling up of the edge of one or both plates into mountain ranges, and sometimes bends the other plate down to produce a deep oceanic trench marking the side of subduction. Oceanic trenches are narrow, elongated, and deep depressions in the ocean floor which are parallel to plate boundaries. Ocean trenches, therefore, mark zones of subduction. There are three types of convergent plate boundaries which we discuss next.

6.7.2.1 Oceanic-Continental Plate Convergence

Oceanic-continental convergence occurs when one of the plates is composed of continental lithosphere whereas the other plate is composed of oceanic lithosphere (Fig. 6.22a). The Andes Mountains in western South America and the Rocky Mountains in western North America are examples of this type of collision.

6.7.2.2 Oceanic-Oceanic Plate Convergence

Oceanic-oceanic plate convergence occurs when both of the colliding plates are composed of oceanic lithosphere (Fig. 6.22b; Table 6.1). The denser oceanic plate is subducted and heated up in the process leading to flux melting. The magma that is generated ascends and accumulates on the overriding oceanic lithosphere, often resulting in the formation of a chain of islands (i.e., an island arc) on the overriding oceanic lithosphere. Zones of active seafloor spreading can also occur behind the island arc to form a **back-arc basin** or a marginal sea (Fig. 6.23). A back-arc basin is a type of submarine basin that is gener-

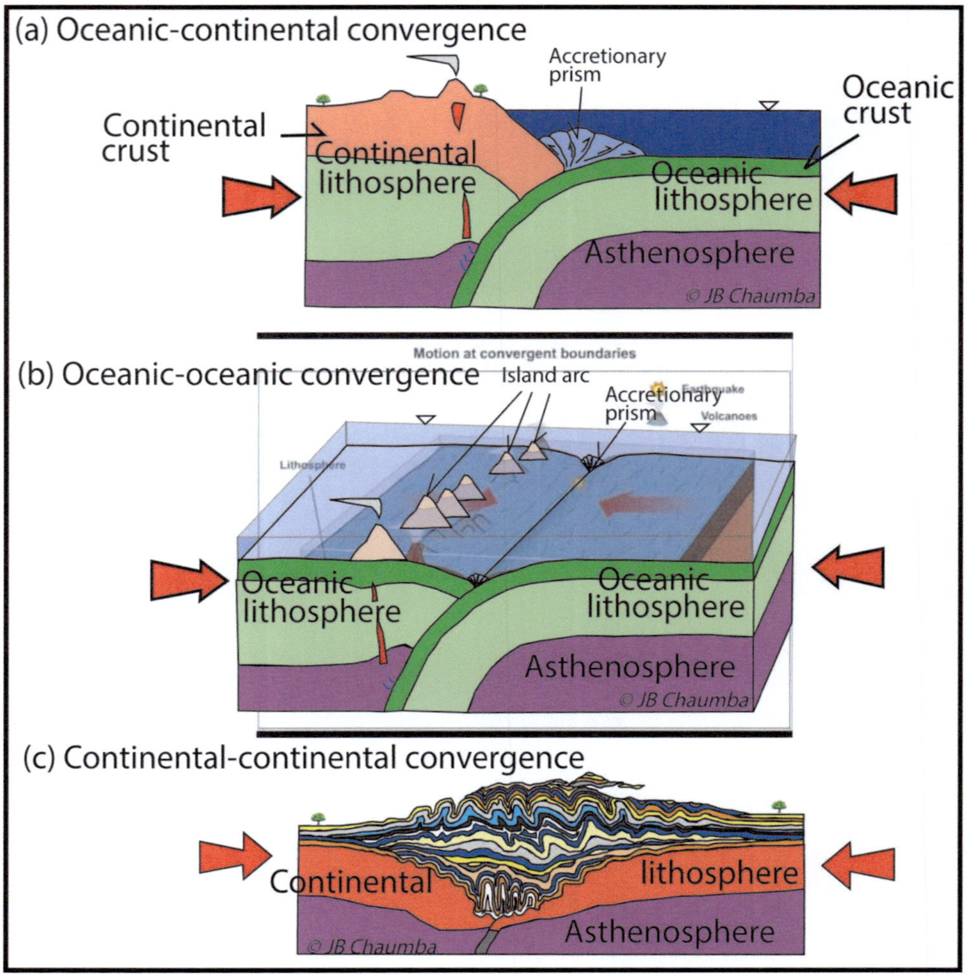

Fig. 6.22 Different types of convergent plate boundaries. **a** Oceanic-continental convergence occurs when an oceanic lithosphere is subducted beneath a continental lithosphere. **b** Oceanic-oceanic convergence occurs when one oceanic lithosphere (the older and denser one) is subducted beneath another oceanic lithosphere (the younger and less dense one). **c** Continental-continental convergence occurs when two continental lithospheric plates collide resulting in shortening and thickening of the crust since neither plate is subducted due to their low densities. Red arrows represent compressive stresses

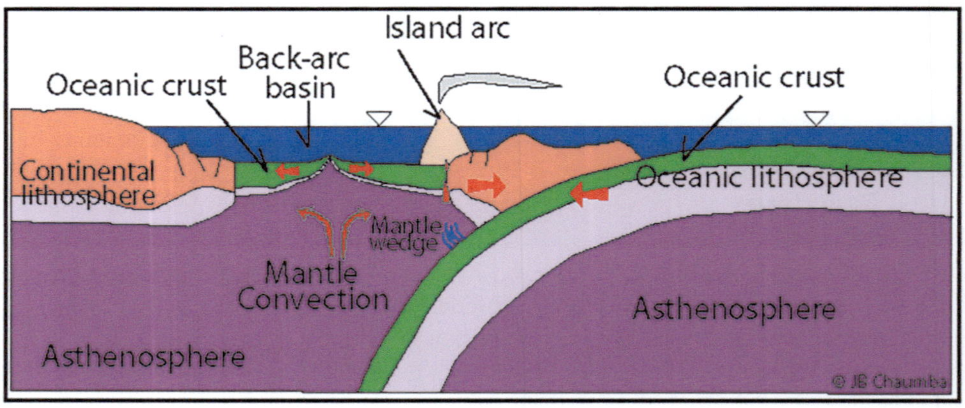

Fig. 6.23 A back-arc basin or marginal sea forms due to extension which occurs behind some island arcs

ated at some convergent plate boundaries which occur in association with island arcs and subduction zones due to tensional forces (Fig. 6.23). Several back-arc basins occur in the western Pacific Ocean. Most of them, back-arc basins are formed as a consequence of oceanic trench rollback whereby a subduction zone migrates towards the plate that is subducting.

6.7.2.3 Continental-Continental Plate Convergence

Any two continental lithospheric plates that collide are both too light to subduct, so when a continental-continental collision occurs, huge mountain ranges are created (Fig. 6.22c; Table 6.1). Examples of continental-continental convergence are the Himalayas in Asia, the Appalachians in eastern North America, and the Alps in Europe (Table 6.1).

Subduction, therefore, occurs when one of the two tectonic plates that are undergoing collision is composed of oceanic crust (Fig. 6.22a, b). In the case of two oceanic plates colliding, the denser plate is subducted underneath another oceanic plate which is younger and therefore of relatively lesser density. With continued subduction, the oceanic slab gets heated up, expelling water in the process. The expelled water induces flux melting to generate magma in the mantle wedge which ascends to form island arcs. If the overriding plate (the one that lies above the subducting slab) is continental crust, the volcanoes form a chain of islands termed a continental volcanic arc (Fig. 6.22a) whereas an island arc is formed when the overriding slab is also of oceanic crust (Fig. 6.22b).

Eventually, the subducted slab is melted and destroyed either at a depth of 660 km (the base of the Transition Zone) or at the core-mantle boundary. Powerful earthquakes (both intermediate and deep) shake wide areas on both sides of the plates at a convergent plate boundary. If one of the subducting oceanic plate leads to flux melt, magma that is generated rises into and through the overriding plate, solidifying into new crust on the overriding plate. The magma that is generated by flux melting solidifies to form andesite, an intermediate-colored, medium-density rock that makes up the continents and islands at convergent plate boundaries. Granite and basalt can also be generated at convergent plate boundaries. Thus, at convergent boundaries, continental crust, made of andesite, is created, and oceanic crust is destroyed.

6.7.3 Transform Plate Boundaries

Transform plate boundaries (Figs. 6.19c and 6.24, Table 6.1) develop where plates slide past each other, and this can involve either two oceanic plates or two continental plates. At transform plate boundaries (Figs. 6.19c and 6.24), crust is neither created nor destroyed since the plates only slide horizontally past each other resulting in cracking and breaking of the lithosphere. Due to the friction that is generated, lithospheric plates do not simply glide past each other, but stresses build up in both plates as plate motion occurs. With continued buildup of stress due to friction, the energy stored in the rocks can be released when it exceeds the threshold of the rocks leading to the occurrence of earthquakes. When earthquakes occur, any natural or human-made structures that cross a transform plate boundary can be offset, i.e., split into two parts that are then transported in opposite directions. Within the zone of friction where movement occurs, the rocks are pulverized as the plates move past each other, resulting in the formation of linear fault valleys on the continental lithosphere or canyons on oceanic lithosphere.

A location where three plate boundaries meet is referred to as a **triple junction** (Figs. 6.21 and 6.24). Triple junctions are named after the plate boundary types that intersect at any location. In the case of the intersection of the East African Rift with the two arms of the Red Sea rift system (Fig. 6.21), a ridge-ridge-ridge triple junction is formed. A further example is the ridge-ridge-ridge triple junction where the African, Indian and Arabian plates intersect (Fig. 6.21), as well as the trench-transform-transform triple junction where the North American, Gorda, and Pacific plates intersect (Fig. 6.24).

6.7.4 Plate Boundary Zones

In certain cases, convergent boundaries can lead to the formation of broad zones in which plate boundaries are not well defined and the effects of plate interaction are not yet fully understood. Plate boundary zones can be defined as the zones where interaction between adjacent plates occurs during collision, pulling apart or sliding past each other. Plate boundary zones range in width from a few kilometers to a few hundred kilometers. New Zealand is interpreted to be a deforming plate boundary zone that is characterized by continued steady deformation (Fig. 6.25; Table 6.1). Movement between the Australian and Pacific plates as well as the deformation occurring in the boundary zone between the two plates leads to the formation of mountain ranges of New Zealand, earthquakes, and volcanic eruptions that are characteristic of this country. Volcanic activity, earthquakes, mountain building, and oceanic trenches are all located along these plate boundaries (or faults). Along convergent boundaries, lithosphere that is destroyed by subduction or mountain building is roughly balanced by the generation of new oceanic lithosphere along divergent margins through seafloor spreading. The creation and destruction of the lithosphere ensures that the total volume of the lithosphere remains the same as our planet is neither expanding nor shrinking in size. This inference of the plate tectonics theory is also referred to as the principle of the conveyor belt.

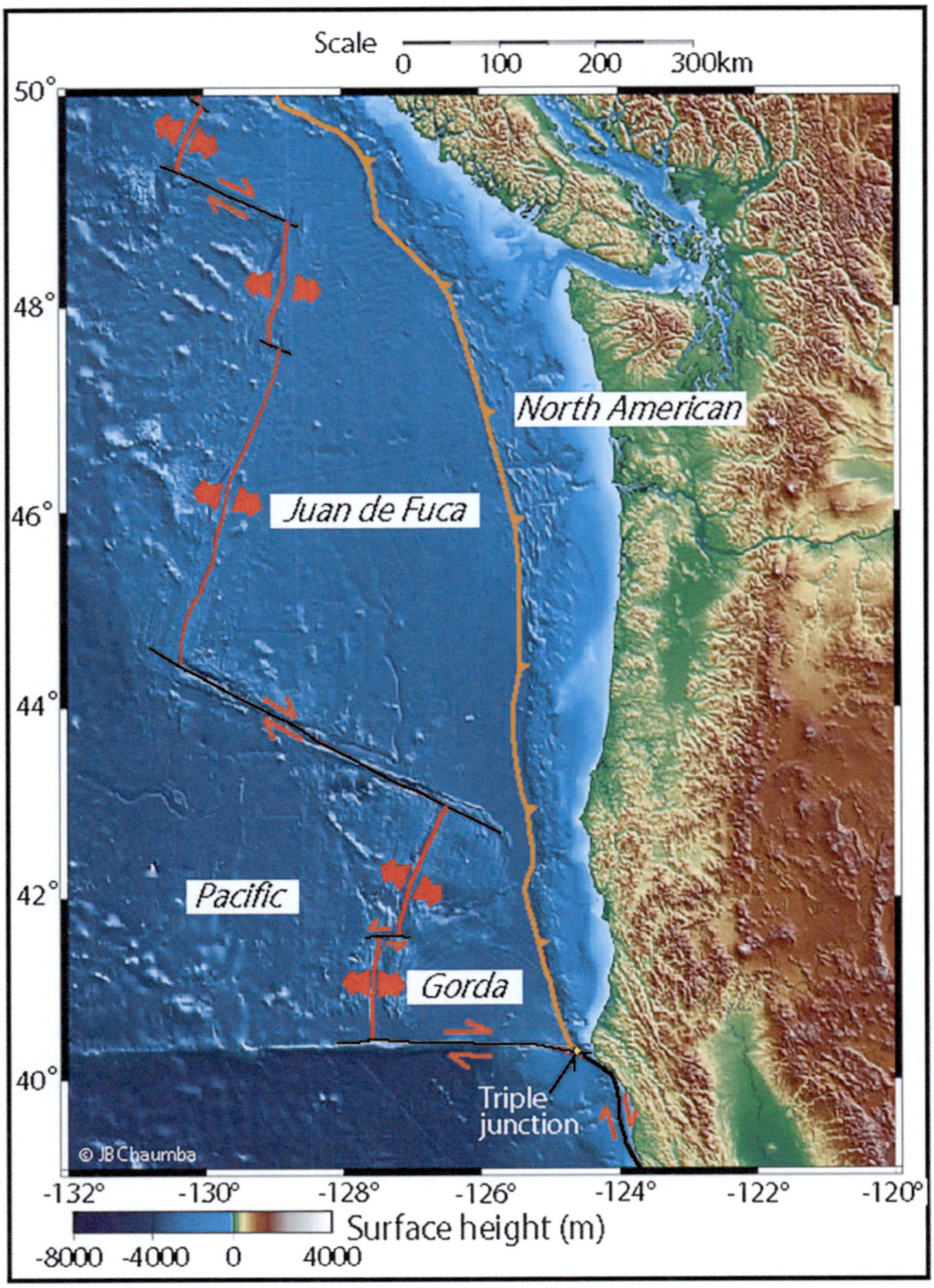

Fig. 6.24 Map of the Juan de Fuca and Gorda plates located offshore of the Pacific Northwest. The two plates migrate eastward from the mid-ocean ridge and spreading center located on the western edges of the two plates towards the trench on their eastern margins where they both subduct beneath the North American plate termed Cascadia subduction zone. Cascadia subduction zone can produce earthquakes of magnitude 9. Photo credit UC Berkeley

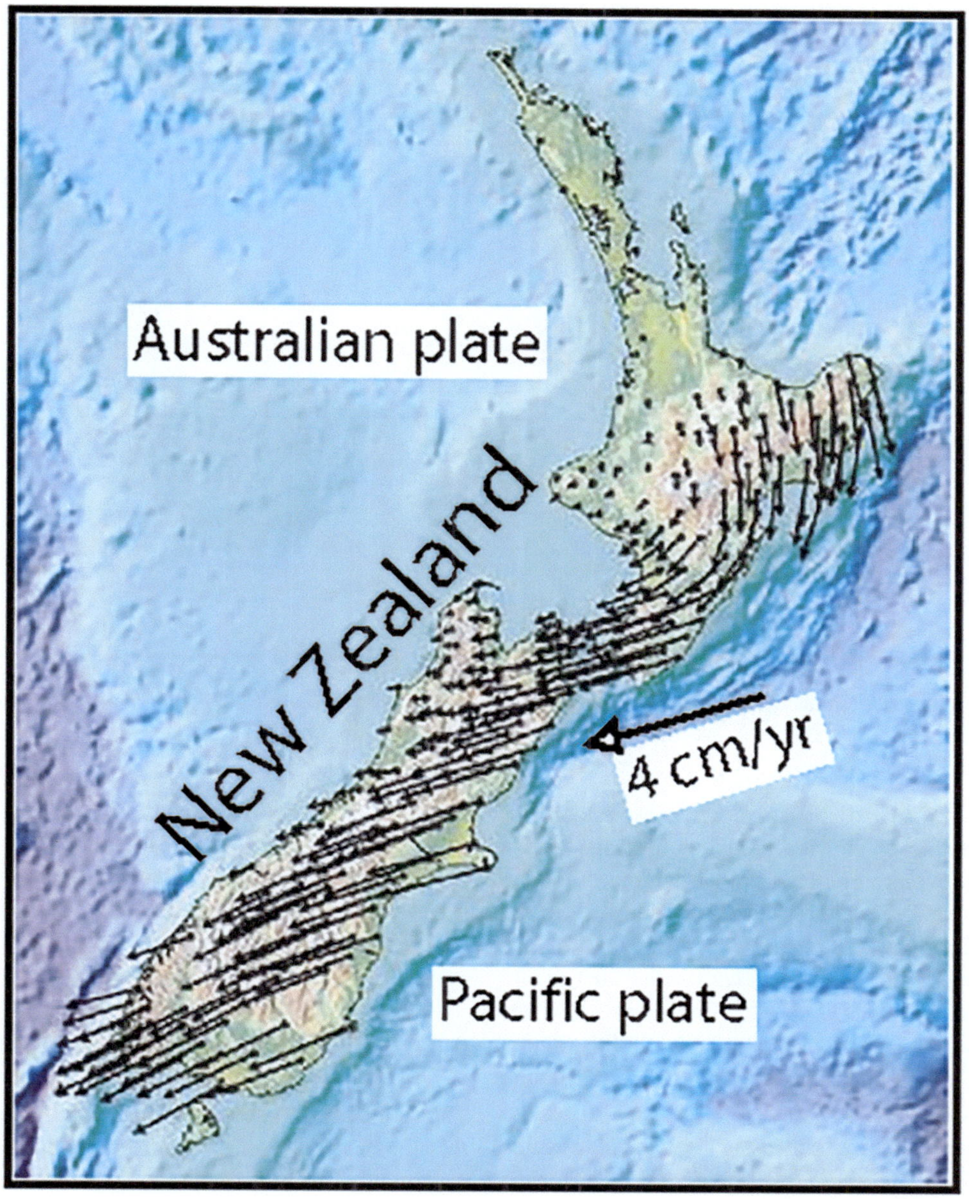

Fig. 6.25 Diagram illustrating a plate boundary zone such as New Zealand located between the Australian and Pacific plates where interaction between these two adjacent plates occurs during collision and sliding past each other. *Credit* GNS Science New Zealand

6.8 Determination of Rates of Plate Motion

Movement of lithospheric plates can be described in terms of either relative or absolute plate motion. **Relative plate velocity** is when the movement of one plate is described relative to that of another plate which is also in motion. **Absolute plate velocity** is when the motion of plates is described relative to a fixed location beneath the plates, such as in the mantle. Since velocity is defined as distance divided by time, the relative plate velocity is determined by measuring the distance between the axis of the ridge and a location on the plate whose age is known: the relative plate velocity is then obtained by dividing the measured distance from the axis of the ridge to the location of known age by the measured age of the ocean floor at the location.

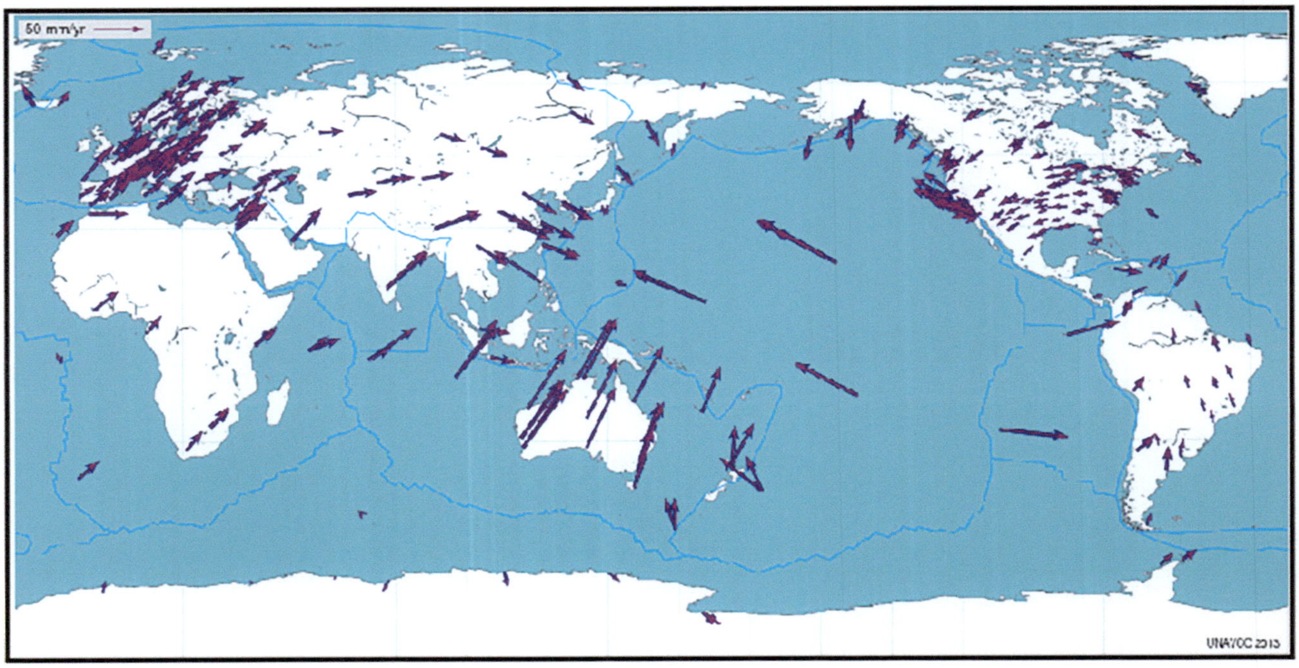

Fig. 6.26 Composite global positioning system (GPS) velocity map of the world (credit: UNAVCO). Purple colors denote the rate and direction of plate motion

To calculate the relative plate velocity, the velocity of the plate on one side of the ridge axis of the ridge is first calculated, and the value obtained is then doubled. A location whose rock has been dated to be 2 Ma, and it is 20 km from the axis of a ridge, for example, can have its relative velocity determined. One begins by converting 20 km into centimeters to get 2,000,000 cm. Next divides this distance by the age (2 Ma = 2,000,000 years), that is, 2,000,000 cm/2,000,000 years, which simplifies to a relative plate velocity of (1 cm/yr) $\times$ 2 = 2 cm/yr when it is doubled. To calculate absolute plate velocity, one assumes a fixed position for a mantle plume. The track of the hot spot volcanoes is interpreted to provide a record of the velocity of the plate which is then used to calculate the absolute plate velocity, as well as the direction of plate motion, by measuring the distance of hot spot volcanoes from the fixed mantle plume. The location of mantle plumes has been shown to vary slightly, however, implying that calculating the absolute plate velocity is not simple but complicated. Since lithospheric plates are known to move at very slow rates equivalent to the rate at which a person's fingernails grow, global positioning systems (GPS) are used to measure very slow plate velocities (Fig. 6.26).

By reconstructing the position and latitude of the plates making up Earth's lithosphere at various times in the past, it has been concluded that most, or all, continents have assembled to form a single landmass called a supercontinent, and broken apart, multiple times in the geologic past. However, the far back in time we venture, the accuracy of the track of the continents is reduced. During the past 250 Ma, the location and shape of the continents and how they have evolved and assembled to form the last supercontinent Pangea is now known with some reasonable accuracy (Fig. 6.27).

6.9 Driving Forces of Plate Tectonics

The pull of the deeply plunging lithosphere is one of the forces that Earth scientists believe may drive plate separation. Tectonic plates are interpreted to be able to migrate since the Earth's lithosphere is composed of greater mechanical strength than the underlying asthenosphere. Plate movement is thought to be driven by a combination of processes. The first of these processes is **mantle convection** (Fig. 6.28a). Density variations which occur laterally in the mantle are thought to result in convection cells developing in the mantle. Mantle convection is the slow motion of mantle which is inferred to be solid. **Slab pull** (Fig. 6.28b) is the second process responsible for driving plate tectonics. At subduction zones, the relatively cold and dense oceanic crust subducts or is "pulled" down into the mantle. The third of these forces is **ridge push** (Fig. 6.28c), which is the motion of the seafloor away from spreading ridges due to variations in topography (the ridge is a topographic high) and density changes in the crust (density increases as newly formed crust cools and moves away from the ridge).

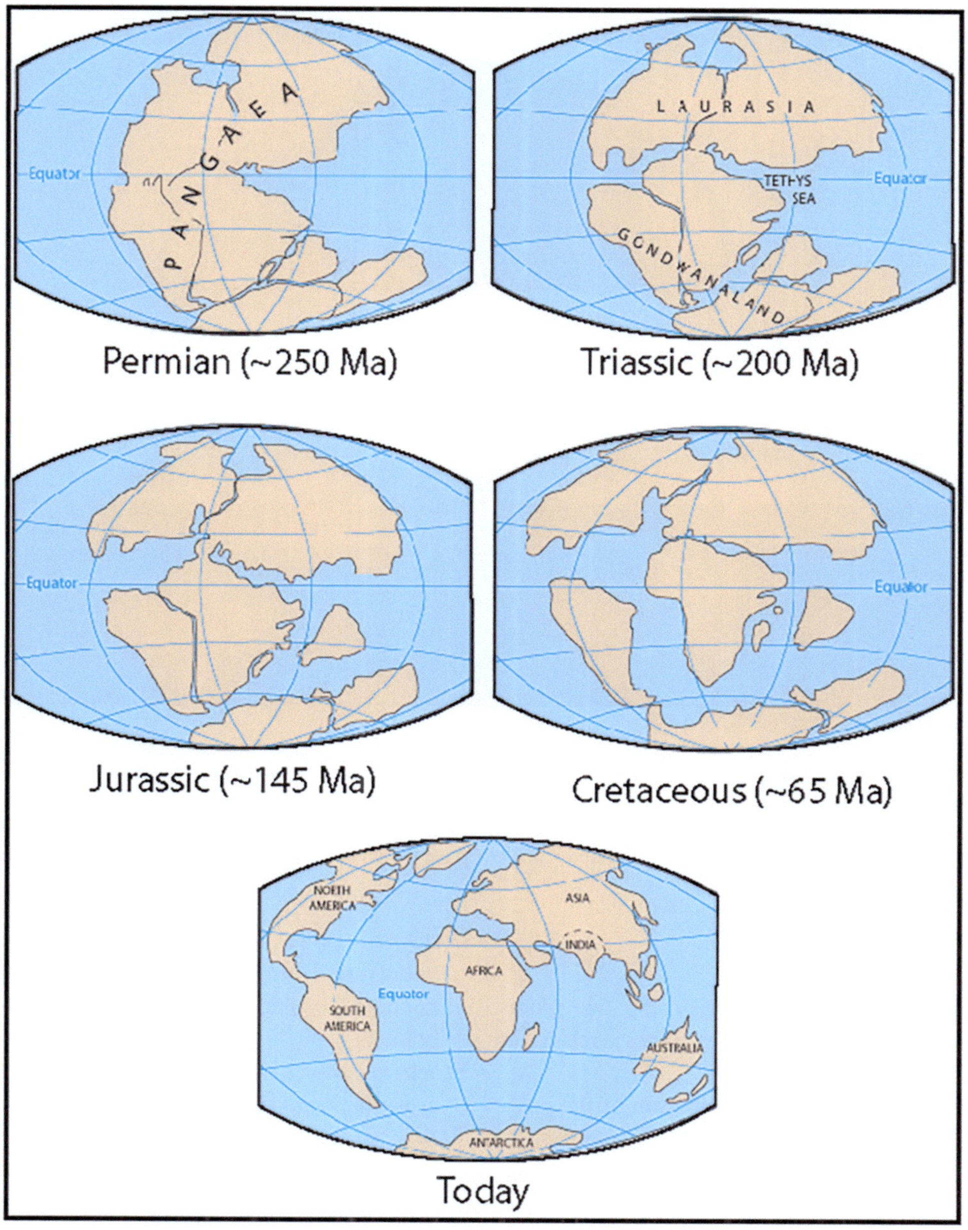

Fig. 6.27 The last time the continents assembled, they formed a supercontinent called Pangea which began to fragment 225–200 million years ago (Ma). Shown here are the fragmenting continents at different times until the present day. *Credit* USGS

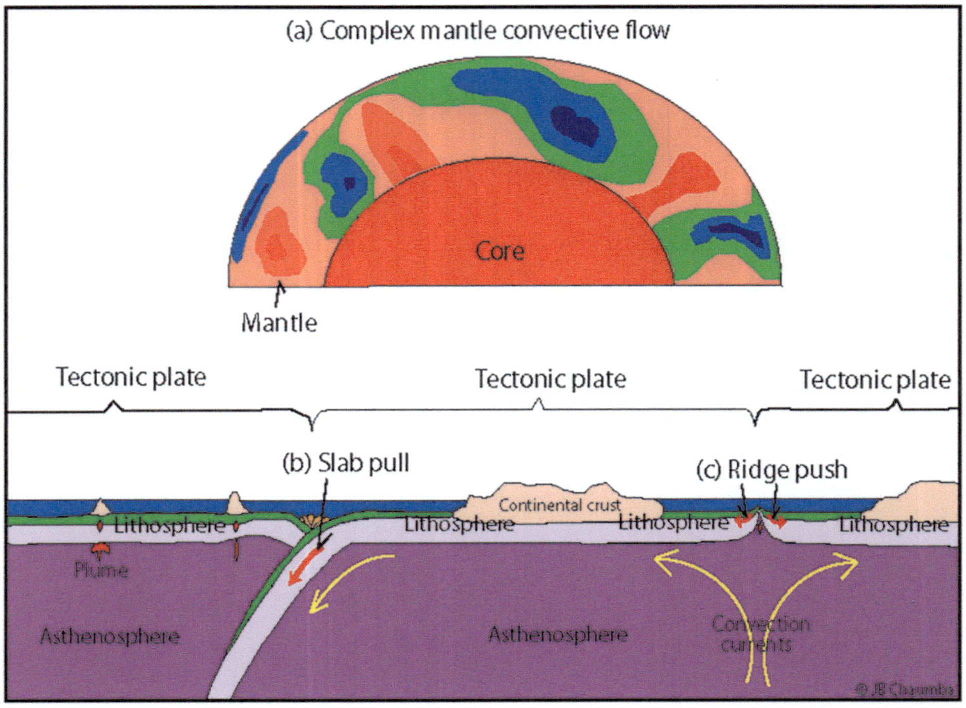

Fig. 6.28 Three different types of forces are inferred to drive motions of plates. **a** Convective floor in the mantle caused by hot rock (reddish colors) which rises after being heated and becoming less dense than surrounding rocks, whereas cold rock (bluish colors) sinks as it is denser, resulting in convection which carries the lithospheric plate above with it. **b** Denser lithosphere sinks or subducts into the less dense mantle beneath it, pulling the rest of the lithospheric plate in the process. **c** Higher elevation of the mid-oceanic ridge results if forces which are exerted laterally away from the ridge called ridge push which helps to drive the motions of plates

Glossary

Absolute plate velocity the movement of any two plates relative to a fixed location below the plates such as in the mantle

Apparent polar wander the perceived motion of the paleomagnetic poles of Earth relative to a continent while regarding the continents being studied as having fixed positions

Axial rift valley a small valley measuring approximately 2–10 km wide; it can be entirely absent where the ridge-spreading rate is fast

Back-arc basin a zone of extension located behind an island arc where sedimentation occurs. A spreading center within the back-arc basin is parallel to the plate boundary is thought to be a result of extension.

Black smoker a geothermal vent that occurs on oceanic ridges of the seafloor where superheated water at temperatures of 360–375 °C containing a lot of suspended matter such as black sulfide minerals is ejected

Convergent plate boundary a zone of compression where two plates collide, resulting in either one oceanic lithosphere plate being pushed into the mantle beneath, or both continental lithosphere plates deforming and shortening

Divergent plate boundary a zone of extension where either two oceanic lithosphere plates or two continental lithosphere plates are moving away from a plate boundary.

Fracture zone a linear and elongated feature on oceanic crust that forms from the action of offset mid-ocean ridge axis segments and extends from transform faults away from a ridge into a plate.

Gondwanaland an unusually large continental mass that was composed of the continents of Africa, Australia, Antarctica, South America, as well as India, which today lie in the southern hemisphere except India. Gondwanaland, or Gondwana, was not a supercontinent.

Guyot an undersea flat-topped volcano whose top part once rose above sea level but was subsequently flattened mainly due to wave erosion

Hot spot a volcanic region that derives its magma from a mantle material that is anomalously hot compared with its surroundings. The location of a hot spot is independent of plate margins

Hot spot track motion of a plate over a hot spot leads to the formation of a chain of volcanoes that increase in age with increasing distance away from the hot spot. Any active volcano marks the current location of the hot spot.

Large igneous province LIPs are voluminous accumulations of igneous rocks such as continental flood basalts and associated intrusive rocks, oceanic plateaus, seamount groups, submarine ridges, and ocean basin flood basalts. LIPs formed in massive volcanic events and formed independent of plate setting, are thought to have resulted from a mode of mantle convection that is different from that responsible for plate tectonics on Earth and they form on the continents, in ocean basins, and along margins between the two, and either completely within a plate or at boundaries of plates.

Magnetic declination is the angle between magnetic north and true north, and it varies over time as well as changing with location. Magnetic declination values east of true north are positive, whereas values west of true north are negative

Mantle convection is the slow movement of the solid silicate mantle of Earth due to convection currents which transport heat from its interior toward the surface. Mantle convection is the most efficient mechanism of transporting heat

Mantle plume a column of hot magma in the mantle that rises due to convection and is thought to cause volcanic activity away from plate margins such as in hot spots

Oceanic ridge a continuous submarine mountain rage extending, that is, 80,000 km (50,000 miles) long and stretches through all the world's oceans

Paleoclimate a climate that was prevalent at any particular time in the geological past

Paleomagnetism a branch of geophysics that is concerned with the magnetism in rocks that was induced by Earth's magnetic field at the time that they were formed

Paleopole the magnetic pole of Earth as it was situated at any particular time in the geological past

Pangea a supercontinent that incorporated almost all the landmasses the last time when almost all the continents assembled

Plate boundary zone are the regions of interaction between adjacent plates where they collide, pull apart and/or slide past each other, and they range in size from a few kilometers to a few hundred kilometers wide

Plate tectonics the lithosphere of Earth is composed of rigid plates that are constantly moving at velocities such that over millions of years, the shape of continents is changed, mountain ranges and ocean basins are created and destroyed, for example

Relative plate velocity movement that describes the motion of one tectonic plate relative to another plate that is also in motion

Ridge push a sliding force occurring at mid-ocean ridges due to the result of the rigid lithosphere sliding down the hot and raised asthenosphere below a mid-oceanic ridge

Slab pull plate motion that is caused by subduction caused by the weight of a cold and dense plate that is sinking into the mantle at oceanic trenches

Seamount an underwater mountain

Supercontinent the assembly of most or all of the continents on Earth to form one large landmass.

Tectonic plate an irregularly shaped block of solid rock that is composed of either continental or oceanic lithosphere, and/or both, varying greatly, from a few hundred to thousands of kilometers across. An example is the African plate.

Transform fault a strike-slip fault that is located at the boundary between any two lithospheric plates

Transform plate boundary a zone where any two lithospheric plates slide past each other

Triple junction a location where three lithospheric plates meet

Tube worm giant worms which occur at hydrothermal vents at divergent plate boundaries on the ocean floor. Where geothermal heat causes circulation of heated water upwards, and of cold water seeping downwards through faults and joints in rocks making up the ocean crust.

White smoker a hydrothermal vent that emits alkaline high-pH hydrothermal fluid on the ocean floor at temperatures of 260–300 °C and away from the mid-ocean ridges

7 Igneous Activity, Volcanoes, and Volcanic Hazards

Abstract

The chapter begins with defining a volcano, the manifestation of molten material reaching the earth's surface, and intrusive and extrusive features. Intrusive features such as batholiths, plutons, dikes, sills, lopoliths and laccoliths are defined and illustrated with the use of some diagrams and photographs. Similarly, extrusive features such as volcanoes, cinder cones, volcanic blocks and bombs, lava flows, pillow lavas, lapilli, and volcanic ash are similarly described. Types of lavas are also described. Continental and oceanic hot-spot volcanoes are discussed with the aid of their world distribution on both world and regional maps, as well as photographs of some of the features. A table is used to summarize the various types of volcanoes based on their size, tectonic setting, approximate size and shape, magma type, eruption characteristics, and tectonic setting. The chapter concludes with a discussion of active, dormant, and extinct volcanoes.

Learning Goals

Studying this chapter will enable you to:

- draw and label batholiths, laccoliths, lopoliths, stocks, plutons, dikes, and sills, and describe how each of these was formed,
- compare and contrast effusive and explosive eruptions,
- sketch and illustrate a typical volcanic cone,
- compare and contrast shield volcanoes, composite volcanoes, and cinder cones,
- describe, define, and distinguish materials emitted by an erupting volcano,
- list and describe the common volcanic hazards experienced in areas where volcanic eruptions occur,
- contrast fissure and vent eruptions,
- define flood basalts and list some common examples,
- describe the distribution of volcanoes on our planet in the context of plate tectonics,
- describe some methods that can be used to predict when volcanoes can erupt,
- list the hazards associated with volcanic eruptions, and
- recognize and differentiate among active, dormant, and extinct volcanoes.

Supplementary Information The online version contains supplementary material available at https://doi.org/10.1007/978-3-032-20231-4_7.

7.1 Introduction

Mount Agung, located on Bali island in Indonesia (Fig. 7.1a), is often referred to by the Balinese as the "Great Mountain" who consider it to be the most sacred place on the island. Mount Agung, however, is an active **volcano** and most recently it erupted intermittently from November 2017. Earthquakes from Mount Agung volcano, which occur as a consequence of magmatic ascent, or volcanic earthquakes, had been detected since August 10, 2017. In late November 2017, Mount Agung erupted five times causing thousands to evacuate, disrupting air travel and causing environmental damage. A small **phreatic eruption** occurred on November 21, 2017, and an ash cloud top reached a height of 3,842 m (12,605 ft.) above sea level. An eruption can be defined as a period during which lava and/or **pyroclastic materials** (*pyroclastic* or fire fragments, from the Greek work *pyro* which means "fire," and *clast* meaning fragment) are emitted through the vent of a volcano.

Prior to the 2017 eruption, Mount Agung had last erupted in 1963 when it killed nearly 1,600 people. If one considers the rapid population growth, Bali's population has almost doubled since the eruption of 1963. Consequently, Mount Agung poses a greater threat to the inhabitants of Bali Island

J. B. Chaumba, *Introductory Physical Geology*, Springer Textbooks in Earth Sciences, Geography and Environment,
https://doi.org/10.1007/978-3-032-20231-4_7

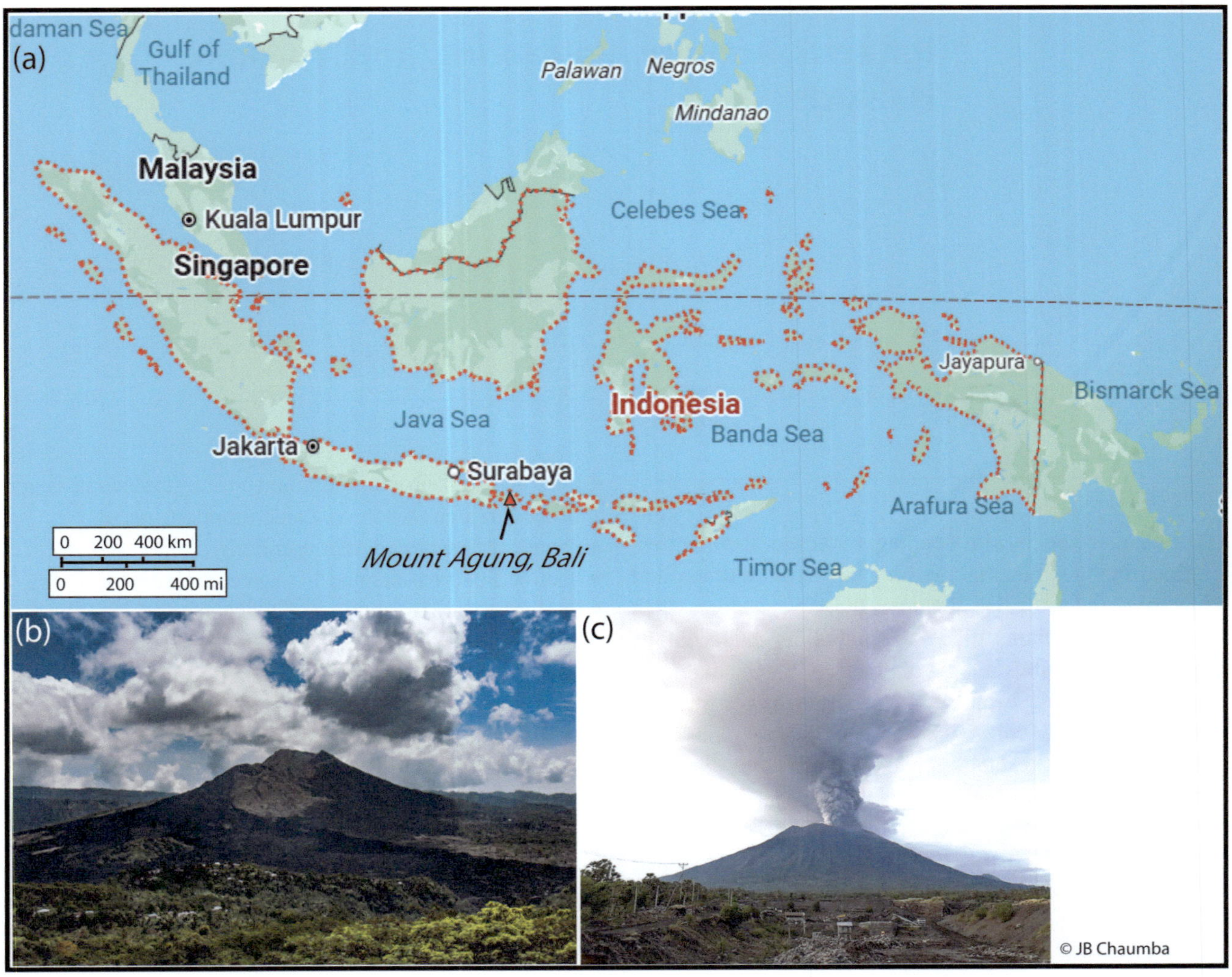

Fig. 7.1 **a** Map of Indonesia showing the location of Mount Agung. **b** Mount Agung in Bali, Indonesia as pictured in September 2017 before erupting (credit: Google maps). **c** Pyroclastic material from Mount Agung in Bali, Indonesia, during eruption in November 2017 (credit: Michael W. Ishak)

in the future. The rapid growth in Bali's population was also one of the reasons why in excess of 100,000 people had to be evacuated in late 2017 when tectonic activity increased due to magma rising towards the Earth's surface. Airports close to Bali were closed and flight cancellations were in excess of 400 during this recent eruption of Mount Agung, in addition to lost revenue from a decline in tourists of ~30%.

Thirteen percent of the world's volcanoes lie within Indonesia, a country which is also located on the Pacific rim of fire (see Chap. 6). Consequently, one would expect both volcanic eruptions and earthquakes (Chap. 8) to commonly occur in this nation. Since Indonesia is composed of several islands, the probability of tsunamis (Chap. 8) is also high. Against all this, news that this country's early tsunami warning system was not been in operation since 2012 surprised many. This was only revealed after at least 429 people perished (plus hundreds of others who were either injured or missing) when a tsunami occurred the Anak Krakatau (which means "Child of Krakatau") volcano triggered undersea landslides which resulted in a tsunami on Saturday, December 22, 2018. Tsunamis caused by landslides occur far less frequently than those caused by earthquakes occurring under water.

Several other volcanoes erupted in 2018. Examples are the Volcan de Fuego (or Chi'gag), an active stratovolcano located in Guatemala, Mount Sinabung in Indonesia, and Mount Etna, an active stratovolcano on Sicily, Italy. Although lives lost from each of these 2017–2018 volcanic eruptions have been in the few hundreds, some volcanic eruptions in the recent past have resulted in relatively high casualties of human lives (in the thousands and tens of thousands, for example) within the last 50 years. The 1985 Nevado del Ruiz volcanic eruption in Colombia resulted in the loss of 23,000 lives, and 3,500 lives were lost during the 1982 eruption of the El Chichón volcano in Mexico.

If one considers the loss of property as well as the impact on the economies in the affected areas, it is easy to deduce the devastating effects of volcanic eruptions. Volcanic eruptions are, therefore, one of our planet's most catastrophic natural disasters. Knowledge of when volcanic eruptions are most likely to occur, as well as where they are also likely to occur in the near future, and why they occur in those areas is important in reducing the loss of both lives and property. Further, if we can more accurately predict when a volcano is likely to occur, people can be evacuated in a timely manner which can prevent the loss of lives. In this chapter, we will examine these factors and many others, starting off by defining what a volcano is.

One deadly pyroclastic eruption includes the eruption of Mount Pelée 1902 on the island of Martinique in the Lesser Antilles of the West Indies on the eastern Caribbean. Pyroclastic flow almost flattened the town of Saint-Pierre, the most densely populated town of the island leaving ~29,500 dead within minutes. The eruption of Mount Pelée is among the deadliest volcanic disasters of the twentieth century. Only a handful of people escaped the eruption, with only two known survivors.

7.2 What is a Volcano?

A volcano can be defined as a mountain or hill (Fig. 7.1b, c) that is formed on the surface of a planet or moon by magma that is escaping from a planet or moon's interior. A volcano is typically conical in shape (Fig. 7.1b, c), and it is produced by an accumulation of extrusive volcanic material. The escaping volcanic material causes an eruption which can forcefully eject volcanic material high in the sky (explosive or violent eruption) or the volcanic material can gently ooze out (effusive or calm eruption). However, the vent through which magma spews out onto the surface of a planet or moon is also termed a volcano. Although volcanic eruptions are spectacular and draw thousands of visitors each year to watch them, most magma that is generated in the mantle and lower crust (see Chap. 3) does not make its way to the surface. Let us first discuss the features that are formed by magma that solidifies either deep within the Earth or near its surface, i.e., intrusive igneous activity, before returning to discuss further those features that are formed by magma that does make its way to the Earth's surface to form extrusive features. This approach allows us to trace the path taken by magmas from deep within a planet where they are generated to the surface of a planet where the journey of molten material ends. Although intrusive igneous features form below the Earth's surface and are only visible after millions of years of erosion that eventually expose them at the surface of a planet, studying them is useful in that it allows us to glean into subsurface processes which occur hidden from our inquisitive eyes.

7.3 Intrusive Igneous Features

Since any magma that is generated in the upper mantle and lower crust is less dense than its host rock, also referred to as **country rock**, the magma ascends towards the Earth's surface. As magma ascends, it tends to be trapped within the Earth's crust in a variety of crustal settings to form a wide spectrum of intrusive igneous features such as batholiths, plutons, stocks, lopoliths, laccoliths, dikes, and sills. The key characteristics in naming these diverse intrusive igneous features are the shape, the size of the resultant body, as well as the manner in which the magma intrudes the country rocks. Let us now briefly describe these intrusive igneous features, starting with the deeper and larger intrusive igneous features and then work our way to the shallower and relatively smaller igneous intrusive features.

A **batholith** can be defined as a massive and irregular intrusive body of considerable magnitude, generally composed of felsic or intermediate igneous rock (Fig. 7.2 and Table 7.1). Batholiths are the largest intrusive igneous features known, and the minimum size of a batholith is an exposed areal extent of 100 km^2 (40 mi^2). Classic examples of numerous batholiths of 2.6–2.7 billion years in age, such as the Chinhamhora and Hwedza batholiths, are exposed on the Zimbabwe craton (Fig. 7.3), as well as the Sierra Nevada and Idaho batholiths which crop out in western North America (Fig. 7.4). The Sierra Nevada batholith ranges in age from 210 to 88 million years old, whereas the age of the Idaho batholith, which ranges from 180 to 45 million year old, indicating that plutonism in the Idaho batholith began, and lasted later, than that in the Sierra Nevada batholith. Images of the Chinhamhora batholith (Fig. 7.5a) as well as the Half Dome, Sierra Nevada batholith (Fig. 7.5c) are also shown in Fig. 7.5. Other batholiths from other parts of the world, such as the 405-million-year-old Wologorong batholith of southern New South Wales, Australia (Fig. 7.5b), as well as the 1.4-billion-year-old granitic batholith of Rocky Mountain National Park in Colorado, USA (Fig. 7.5d) are also shown in the same figure.

A **laccolith** is a mushroom- or lens-shaped intrusive igneous feature that was emplaced between layers of rock resulting in an uplift of overlying rocks to form a dome but whose base is planar (Figs. 7.2 and 7.6a; Table 7.1). The upwarping of the rocks overlying a laccolith is interpreted to be caused by the high pressure of the injected magma. Examples of laccoliths include the Traprain Law in Scotland, UK (Fig. 7.6a) and the Henry Mountains in Southeastern Utah, USA. The Henry Mountains Laccolith can be best viewed along State Route 95, facing to the west. A **lopolith** is a saucer or basin-shaped and concordant intrusive igneous body with a convex downward bottom (Figs. 7.2 and 7.6b; Table 7.1). A lopolith is parallel to the layers or trend of the rocks that it intrudes. The sagging shape of a lopolith is interpreted to be due to a decrease in

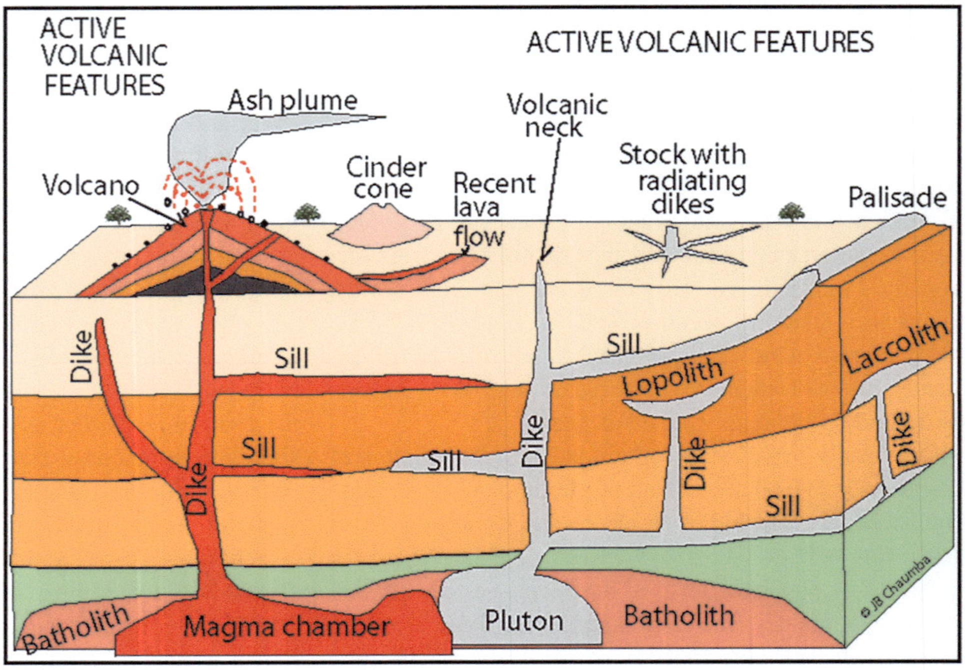

Fig. 7.2 Diagram illustrating some of the various intrusive igneous features described in the text. It can be seen in this diagram that plutonic igneous bodies are often connected to extrusive igneous features such as volcanoes

volume of the magma chamber that occurs as a consequence of crystallization of magma as the weight of the overlying rocks collapses into the volume that was previously occupied by the magma in the chamber. Lopoliths are, thus, composed of dense mafic magmas that permit depression of the overlying rocks upon cooling. The 2.06 billion year-old Bushveld Complex in South Africa (Fig. 7.6b) and the 2.58 billion year-old Great Dike of Zimbabwe are world-famous examples of lopoliths.

A **pluton** is an intrusive igneous feature with a blob shape that ranges in size from tens of meters (feet) to tens of kilometers (miles). An example of a pluton is the 3.2 billion year-old Dalmein pluton in Mpumalanga Province, South Africa. Several plutons constitute a batholith. A **stock**, which can be defined as a smaller pluton occurring in relative isolation, has a diameter of less than 20 km (12 miles). A **sill** is a tabular body or sheet intrusion that is intruded conformably into any pre-existing fabric or layering in the country rocks (Figs. 7.2 and 7.6c; Table 7.1). A **dike** can be defined as a tabular body or sheet intrusion that cuts across any pre-existing fabric or layering in the country rock (Figs. 7.2 and 7.6d; Table 7.1). Both dikes and sills are generally characterized by uniform thickness, are generally uniform in shape, and they tend to be planar bodies. Since not all rocks can have a fabric or layering, tabular bodies intruding massive rocks may be defined on the angle they make with the horizontal. Dikes tend to be vertical bodies whereas sills tend to be horizontal bodies. In general, sills tend to be parallel with the Earth's surface.

Table 7.1 Types and identifying characteristics of various intrusive features

Intrusive igneous feature	Description	Example
Batholith	Massive and irregular intrusive body of considerable magnitude	Idaho batholith, Idaho, USA
Stock	A relatively small pluton occurring in relative isolation, has a diameter of less than 20 km (12 miles)	Henry Mountains, Utah, USA and the Parashi stock in Colombia
Pluton	A blob-shape intrusive igneous feature that ranges in size from tens of meters (feet) to tens of kilometers (miles)	Clayton Peak stock, Utah, USA and Dalmein pluton in South Africa
Laccolith	Mushroom or lens-shaped intrusive igneous feature with a planar base	Traprain Law, Scotland; Henry Mountains, Utah, USA
Lopolith	Saucer-shaped and concordant intrusive igneous body with a convex downward bottom	Bushveld Complex, South Africa and Great Dike, Zimbabwe
Dike	Not conformable to pre-existing attitude of rocks	Mashonaland dikes and East Dike, Zimbabwe
Sill	Conformable to pre-existing attitude of rocks	Palisades Sill, New York and New Jersey, USA

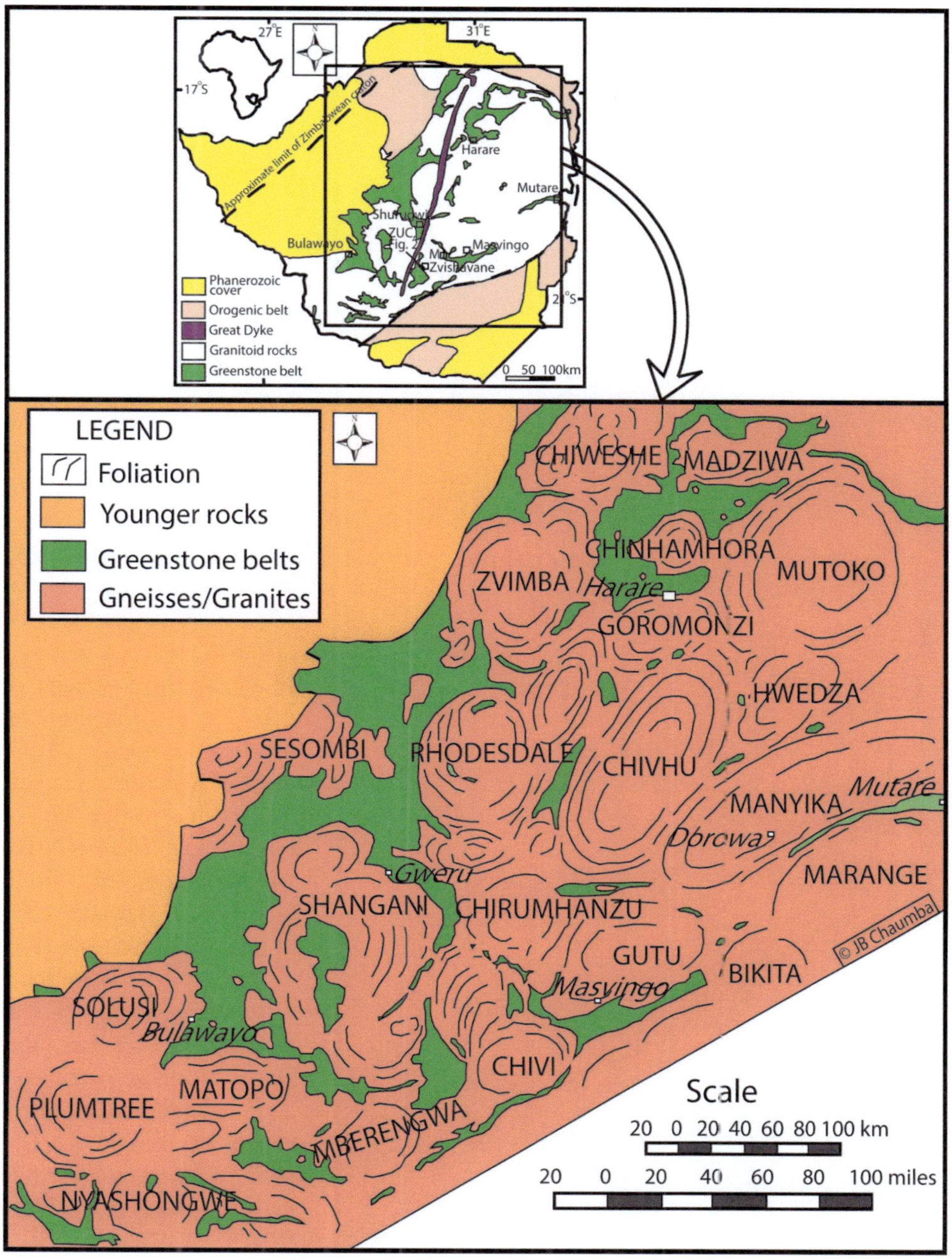

Fig. 7.3 Map of part of Zimbabwe showing the distribution of numerous batholiths which intruded the greenstone belts within the Zimbabwe craton

7.4 Causes and Types of Volcanic Eruptions

Volcanic eruptions occur due to magma moving upward to the surface of the Earth from the upper mantle and lower crust where it is generated since it is less dense than surrounding rocks inside the Earth. Upon reaching the Earth's surface, the magma can either erupt violently if the rate of upward migration is rapid, or oozes out if it does not contain a high proportion of volcanic gases. Of course, an eruption can only occur if magma gushes out of a volcano before it cools and solidifies inside the Earth to form the intrusive igneous features that we described above. Magma can only erupt if its viscosity and concentrations of volatiles (both described in Chap. 3) permit its rapid ascend toward the Earth's surface. Magmas with high viscosities (see Table 3.1) are not likely to ascend to the surface of a planet due to their high resistance to flow. When high-viscosity magmas reach the Earth's surface, they tend to erupt explosively (or violently). Low-viscosity and/or more fluid magmas are more likely to rapidly ascend and easily reach the Earth's surface.

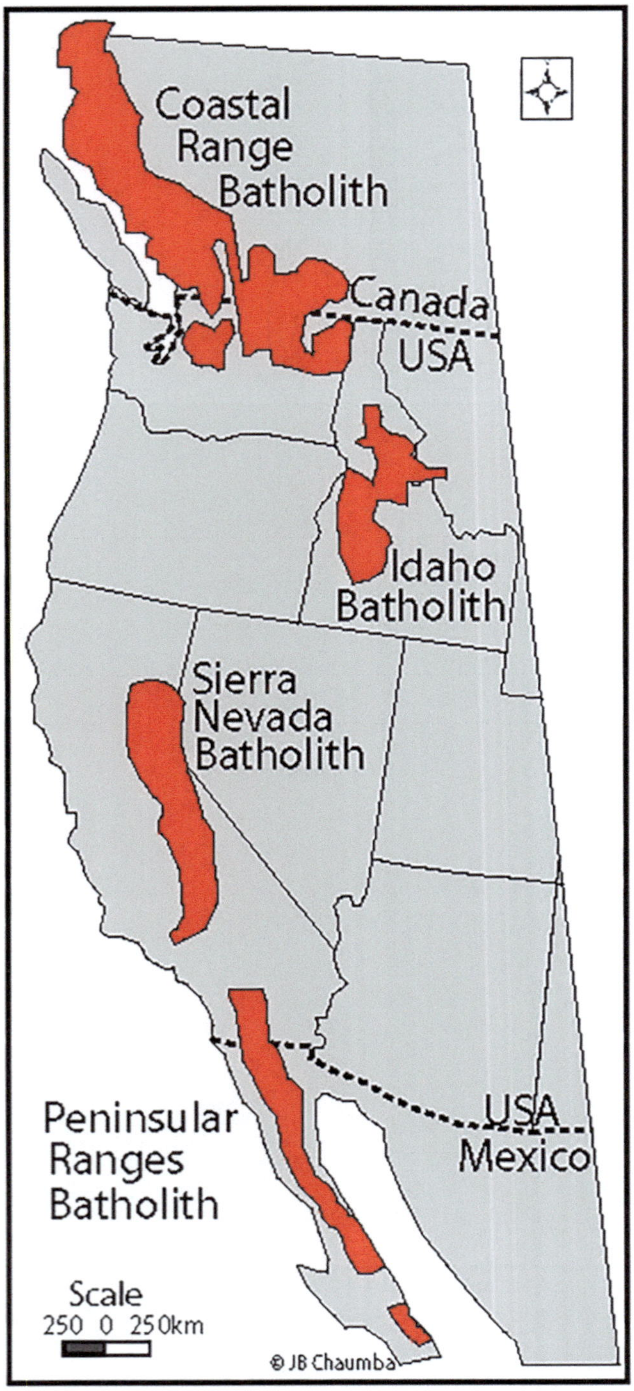

Fig. 7.4 Map showing the distribution and extent of some of the batholiths of western North America such as the Coastal Range, Idaho, Sierra Nevada, and Peninsular Ranges batholiths

Magmas can either contain almost no volatiles at all (dry magmas) or they can contain up to 15% dissolved volatiles (wet magmas) such as water (H_2O), carbon dioxide (CO_2), nitrogen (N_2), hydrogen (H_2), sulfur dioxide (SO_2), and chlorine (Cl). Water is the dominant volatile, often accounting for up to 50% of gases that erupt at a volcano. Consequently, magmas possess molecules that contribute to the atmosphere and hydrosphere of a planet. Deep within the mantle and lower crust, the gases occur as dissolved components within the magma due to the high pressure exerted on the magma by the country rocks. As the magma ascends towards the surface of a planet, the pressure exerted on the magma decreases, enabling unmixing of the volatiles and magma. The volatiles, due to their lower density in comparison to that of the magma, tend to migrate faster (and expand in the process) upwards through the magma to pond near the top of the ascending magma. The result of the faster ascent of volatiles is pressure that is applied against the overlying country rocks.

When the ascending magma meets an obstacle, increased pressure is applied on the overlying country rocks due to increased accumulation of the volatiles which can lead to bursting of the overlying country rocks or obstacles. Bursting of the overlying country rocks results in an **explosive eruption** as volatiles rapidly expand when volatile-rich felsic magmas are extruded. Pavlof volcano, located on the Aleutian chain of islands of Alaska, USA, is an example of an explosive volcano (Fig. 7.7a) which often results in debris and ash deposits, as well as lava, accumulating to produce the volcano shape. A violent or explosive eruption can last anywhere from hours to days until the trapped gases escape, after which the eruption can become a quiet eruption (also called an **effusive eruption**) where magma erupts without any violence. Kīlauea volcano, located on the Big Island of the Hawaiian chain of islands, USA, is an example of an effusive volcano (Fig. 7.7b).

The composition of the magma and its temperature determine the viscosity of magmas. As the silica content of the magma increases, the viscosity of the magma also increases. Felsic magmas, which are generally low temperature, oxygen ions occurring at the corners of the oxygen-silicon tetrahedral (see Chap. 2) that are not bonded tend to form temporary but relatively strong bonds which link the oxygen-silicon tetrahedra chain clusters (see Chap. 2). In general, felsic magmas have high viscosities and, thus, felsic magmas tend to move more slowly. Mafic magmas, however, are characterized by higher temperatures and low silica contents which result in low viscosities. Consequently, mafic magmas flow much more easily and they are more likely to ascend to the Earth's surface with ease. Mafic magmas, thus, tend to ascend and erupt more easily whereas felsic magmas have a tendency to easily accumulate in magma chambers within the Earth's crust where they slowly cool and form intrusive igneous features discussed above.

Where low-viscosity and more fluid mafic magmas ascend to the Earth's surface, volatiles can escape from the mafic magma much more easily as the volatiles encounter very little resistance to form effusive eruptions. Volatiles within felsic magmas, which are char-

Fig. 7.5 Photographs of some batholiths. **a** Part of the 2.6-billion-year-old Chinhamhora batholith near Domboshava, Zimbabwe (credit: Zimbabwe Flora). **b** Woolpack Rocks, located in the Cathedral Rock National Park, part of the 405-million-year-old Wologorong batholith, southern New South Wales, Australia (credit: New South Wales National Parks and Wildlife Service). **c** Half Dome, Yosemite National Park, part of the 210–45-million-year-old Sierra Nevada batholith of the Western USA (credit: Sierra Mountain Center). **d** Exfoliation dome in 1.4-billion-year-old granitic batholith of Rocky Mountain National Park, Colorado (credit: Colorado Geological Survey)

acterized by high viscosities, tend to encounter resistance resulting in the buildup of pressure, ultimately leading to explosive eruptions upon the magma reaching the Earth's surface. Intermediate magmas are characterized by relatively lower temperatures and relatively high silica concentrations which also result in explosive eruptions.

7.5 Extrusive Igneous Features

A pyroclastic flow, also called a pyroclastic density current or a pyroclastic cloud, is a current of hot gases accompanied by volcanic matter that moves very fast. **Tephra** (*Greek* for ash) is often used as a collective name for any airborne accumulation of pyroclastic materials. Tephra can race down the flanks of a volcano due to the force of gravity and spread laterally at average velocities of ~100 km/h (62 mph), although pyroclastic flows can reach speeds of up to 700 km/h (430 mph). The temperatures of tephra can be as high as 1,000 °C (1,830°F). Pyroclastic flows are common where volcanoes erupt explosively. The gradient or the slope of the volcano, the force with which the pyroclastic material is ejected with, as well as the density of currents determine the velocity of pyroclastic materials. Although tephra is not consolidated, coherent rock types that are formed from the consolidation of pyroclastic accumulations are called pyroclastic rocks. Tephra is classified on the basis of particle size (Table 7.2; Fig. 7.8).

Fig. 7.6 Some images of various intrusive igneous features. **a** View of Traprain Law, a laccolith near A1 Highway near East Lothian, Scotland, viewed from the southeast (credit: Undiscovered Scotland). **b** An image of volcanic rocks of the Bushveld Complex, South Africa, an example of a lopolith. View looking south along strike of Critical Zone and Main Zone of the Bushveld Complex, Olifants River, South Africa (credit: D.L. Reid). **c** Relatively narrow sills (brown) intruding black shales north of Segundo, Colorado and west of Trinidad, Colorado. (credit: Colorado Geological Survey). **d** A dike exposed on the wall of Hanson Aggregates' Bolingbroke Quarry, Bolingbroke, Georgia, USA (credit: S.D. Clark)

7.5.1 Types of Lava

In Chap. 3, we learned that we can classify the types of magma generated in the mantle and lower crust into ultramafic, mafic, intermediate, and felsic types (Table 3.1). It is estimated that mafic lava constitutes in excess of 90% of the volume of all lava erupted on Earth. The majority of this is mafic magma that is emitted at oceanic ridges. The remaining ~10% of lava is either of intermediate lava (constituting ~9%) or felsic lava which constitutes ~1%. Volcanic ash is dominantly erupted by felsic lavas.

Lava flows that are characterized by broken lava blocks with sharp edges and corners and rough surfaces are termed **'a'ā** flows (Fig. 7.9a, b). 'A'ā is a Hawaiian term for lava that is composed of loose, irregular, and spiny blocks. The irregular and spiny surface that is produced by solidified 'a'ā flow makes walking on it very difficult (Fig. 7.9a, b), as Hawaiians learned. The irregular and spiny surface hides a massive dense core beneath 'a'ā flows (Fig. 7.9a, b). This massive and dense core is the most active part of the 'a'ā flow. As pasty lava in the core travels downslope, the clinkers are carried along at the surface. The hardened fragments at the leading edge of an 'a'ā flow tumble down the steep front and they are buried quickly by the advancing flow to produce a layer of lava fragments located at both the bottom and the top of an 'a'ā flow.

Lava flow that is characterized by smooth and ropy surfaces is termed **pāhoehoe** flow (Fig. 7.9b, c). Pāhoehoe flows are associated with low rates of effusive eruptions at low flow rates volumetrically (2–5 m^3/s, or ~71–177 ft^3/

Fig. 7.7 **a** Pavlof volcano, Aleutian Islands, Alaska, USA, eruption March 28, 2016, showing an explosive eruption in which ash was dominantly ejected (credit: AVO of USGS). **b** Mafic lava erupting from Kupaianaha 100 m in diameter lava pond of "Pu'u 'O'o" volcano, Hawai'i in December 1986 (credit: soest.hawaii.edu)

Table 7.2 Classification of volcaniclastic fragments and volcaniclastic sediments (tephra) based on size of particles

Size of particle	Name of volcaniclastic fragments	Name of volcaniclastic sediments
<2 mm (0.08 in)	Volcanic ash	Tuff
2–64 mm (0.08–2.5 in)	Lapilli or cinders	Lapillistone
>64 mm (2.5 in)	Volcanic bombs	Agglomerate/volcanic breccia
>64 mm (2.5 in)	Volcanic blocks	

s). Front velocities (1–10 m/h or ~3–33 ft/h) of pāhoehoe flows are also slow. Pāhoehoe flows are generally thin, and they range from a few centimeters (inches) to a few meters (feet) in thickness. The skin of pāhoehoe flows forms due to cooling by air and it is not disrupted during its low-velocity flow and tends to maintain its smooth, unbroken, well-insulating surface (Fig. 7.9). Further, the temperature and viscosity of pāhoehoe flows do not change very much even at distances of tens of kilometers from the vent which ensures that the advancing front of a pāhoehoe flow is composed of hundreds or thousands of active toes. A pāhoehoe flow typically consists of thousands of small flow units termed toes (Fig. 7.9b, c), and a single toe is generally 30–50 cm (~12–20 ft) wide, <30 cm (12 ft) thick, and 1–2 m (~3–7 ft) long. An 'a'ā flow, in contrast, generally consists of only 1–2 large flow units.

Lava tubes (Fig. 7.9d) are natural conduits within solidified lava flows that can resemble caves, through which lava travels beneath the surface of a lava flow. Lava tubes form by the crusting over of lava channels and pāhoehoe flows. A broad lava-flow field often consists of a main lava tube and a series of smaller lava tubes that supply lava to the front of one or more separate flows. When the supply of lava stops at the end of an eruption or lava is diverted elsewhere, lava in the tube system drains downslope and leaves partially empty conduits beneath the ground. Such drained tubes commonly exhibit "high-lava" marks on their walls, generally flat floors, and many lava stalactites that hang from the roof. Lava can also erode downward, deepening the lava tube and leaving empty space above the flowing lava.

Pillow lavas are the most volumetrically abundant types of lava (Fig. 7.10a, b). They are characterized by elongated and interconnected flow lobes that display elliptical or circular cross-sections. Mid-ocean ridges are sites where the overwhelming majority of pillow lavas are erupted. However, pillow lavas are also encountered where submarine volcanic eruptions occur, such as at hot spots where lava is erupted subaqueous (underwater) (Fig. 7.10a) within intraplate oceanic settings. Such locations include the Hawaiian-Emperor seamount chains, Galapagos, and Tristan de Cunha, for example. When lava is erupted underwater, thin streams of lava are extruded in a form that resembles toothpaste squeezed out of a tube (Fig. 7.10a). The outer surface of the molten lava is quickly chilled by the water and solidifies, whereas the interior remains molten and continues to flow, lengthening the tubes. Lava flows with similar textures on land can be recognized as being uplifted submarine (or sublacustral) lavas. Ancient rocks displaying pillow structures, such as the ~3.4 billion year-old volcanic rocks from the Onverwacht Group of the Barberton greenstone belt in northeastern South Africa and Swaziland (Fig. 7.10b), are interpreted to indicate underwater eruption at the time they were formed.

Lahars are formed when pyroclastic materials mix with water to form a volcanic mudflow. Lahars are common

Fig. 7.8 Ranges of sizes of pyroclastic materials that are extruded by a violent eruption. **a** A duck carcass lies on the ground in Escuintla, Guatemala blanketed with volcanic ash from the Volcan de Fuego (Spanish for "Volcano of Fire") eruption of Sunday June 3, 2018 (credit: Luis Soto, Associated Press). **b** Image of lapilli (credit: USGS). **c** Image of a volcanic bomb (Photo by author). **d** The light-gray-colored block of rock is interpreted to be part of an ~830-year-old pyroclastic eruption deposit on the northwest rim of Moku ʻāweoweo at Mauna Loa summit caldera. See backpack for scale (credit: F.A. Trusdell—USGS)

where moist soils mix with materials that are ejected during explosive eruptions. In other instances, a volcano covered with snow can also form lahars when hot pyroclastic materials from an erupting volcano cause either melting of the snow or melting of a glacier.

7.5.2 Structure of a Volcano

During its upward movement from the lower crust and mantle, most magma may accumulate in magma chambers to form intrusive features we have just discussed above. However, some magma does escape from the magma chamber and erupt to form a volcano at the surface of a planet. Magma that rises towards the surface of a planet follows **pipes** or **conduits** (Fig. 7.11) and erupts through a **vent** (Fig. 7.11). A vent is the surface opening that is located on the top of a volcano, and it marks where volcanic material is ejected through from the interior of a planet. All volcanoes contain a central vent underlying the summit **crater** of the volcano. A crater is a bowl-shaped depression that is located at the summit of a volcano, with a diameter less than 1 km (0.6 miles). A crater can form by accumulation of pyroclastic materials around the rim of a vent, or it can form during explosive eruptions due to erosion of the wall of a crater by the ejected materials. A meteorite impact also causes a depression called a crater. A **caldera** is a very large, bowl-shaped depression greater than 1 km (0.6 miles) in diameter that is rimmed by in-facing scarps. Calderas generally form by the collapse of the top of a volcanic cone or group of cones due to removal of the support provided by an underlying magma. Crater Lake in Oregon, USA, is an example of a caldera. In some instances, subsequent minor eruptions

Fig. 7.9 **a** A Hawaiian Volcano Observatory (HVO) Earth scientist photographing the front part of a channelized ʻaʻā lava flow from the Kīlauea volcano eruption as it burns scrub and trees bordering an older flow on June 28, 2016 (credit: USGS). **b** An HVO Earth scientist sampling lava for chemical analysis from the margin of a lobe of a pāhoehoe flow on June 23, 2016. The lava erupted from Kīlauea volcano, Hawaii (credit: USGS). **c** An Earth scientist (Loren Antolik) writing down field notes while standing near a lava tube (credit: Tim Orr, USGS). **d** Lava tube photographed on the Galapagos (credit: D.T. Nelson)

may build small cones on the floor of the caldera, which may still later fill up with water, with the later eruptions forming islands, as in the case of Wizard Island in Oregon, USA.

Examples of stratovolcanoes, which are being formed as a consequence of the collision between the North American plate and the Juan de Fuca plate, are Mount Rainier and Mount Baker (Fig. 7.12a, b, respectively). Mount Bachelor (Fig. 7.12c), located in the U.S. state of Oregon (Fig. 7.12c), is another comparable example of an active stratovolcano. Mount Meru (Fig. 7.12d), a dormant stratovolcano located 70 km west of Kilimanjaro in southeast Arusha Region, Tanzania, has a height of 4,562 m (14,980′). Mt Meru is formed from decompression melting associated with an emerging divergent plate boundary associated with the East African Rift Valley.

7.5.3 Flood Basalt Volcanism

Typically, a conduit has the shape of a chimney or a pipe. In some instances, however, large volumes of basaltic lava can erupt at the surface of a planet via a linear feature such as a joint or a fracture, in which case it is referred to as a fissure. Flood basalts are thought to have been erupted through **fissure eruptions**. Fissure eruptions occur when lava is extruded via linear fractures in the crust called fissures (Figs. 7.13 and 7.14), which tend to be produced where plates can ultimately diverge. The erupting basaltic lava can then spread out laterally to form layer upon layer of lava flows termed **flood basalts**. Curtains of lava can be produced beneath continental rifts (Fig. 7.13), or when lava erupts from linear chains of cinder cones, or even when lava is emitted along linear fractures (Fig. 7.14). Fissure

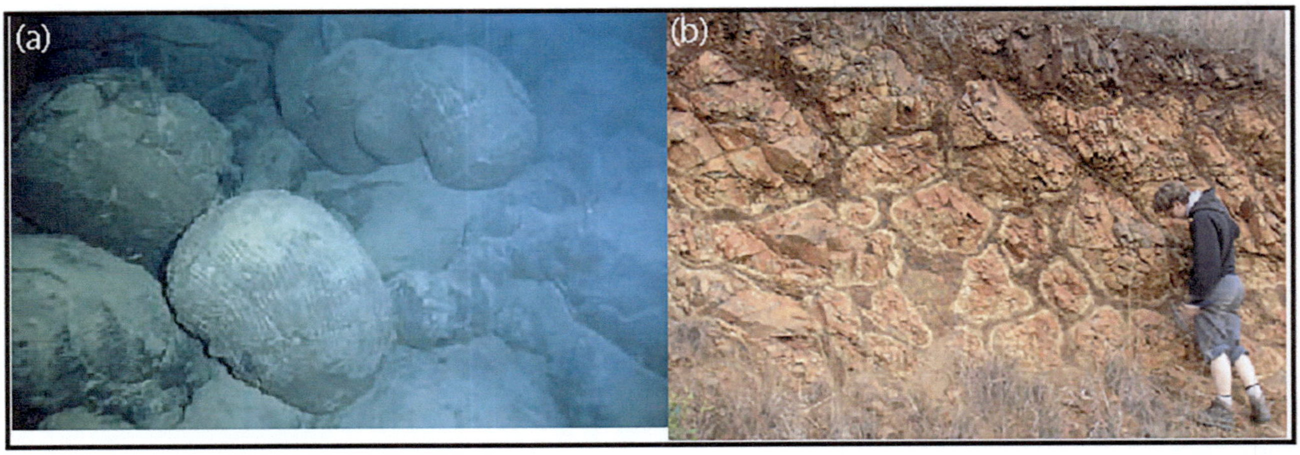

Fig. 7.10 **a** Pillow lavas on the Hawaiian island of Loihi, a seamount located to the southeast of the Big Island of Hawaii. Pillow lavas are produced due to submarine eruption of magma (credit: USGS). **b** Image of ~3.5 billion year old pillow lavas from the Onverwacht Group of the Barberton greenstone belt in northeastern South Africa and in Swaziland. *Credit* Chris Rowan

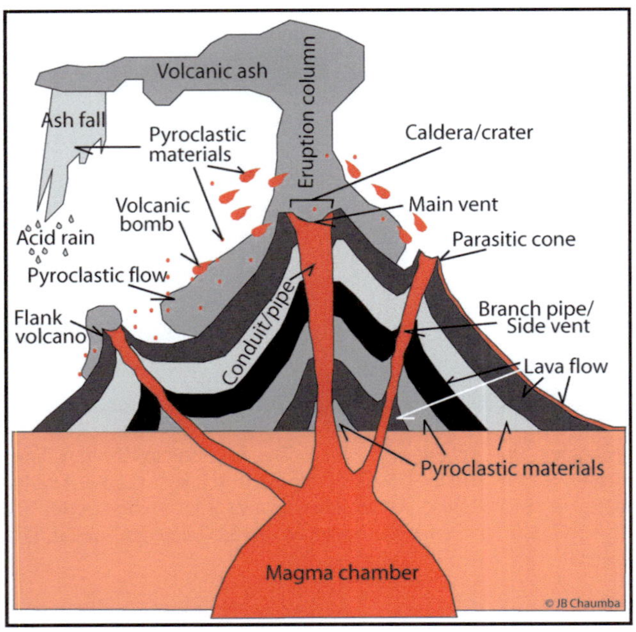

Fig. 7.11 Diagram illustrating the anatomy of a volcano

eruptions are products of huge outpourings of low-viscosity (fluid) basaltic magma, which tends to flood or cover extensive areas to form flood basalts or extensive lava plateaus (Figs. 7.13 and 7.14). The flood basalts which erupt along fissures can reach speeds of 40 km/h (~25 mi/h). Such large volumes of magma that have been extruded, or intruded, within continental crust, within oceanic crust, or at the margins of continents, are termed **large igneous provinces** (**LIPs**).

LIPs produced by mafic magmas are interpreted to form when the head of a mantle plume causes higher degrees of partial melting upon reaching the base of the lithosphere in comparison to the normal partial melting that can occur in the asthenosphere since the head of a mantle plume is characterized by higher temperatures. Consequently, an unusually hot and large volume of magma is produced which forms LIPs. Further, in the scenario when the head of a mantle plume is located beneath a rift setting, more decompression occurs due to thinning of the lithosphere which then triggers even more partial melting which produces flood basalts, such as at Iceland or at Afar in the East African Rift system.

In some instances, instead of basaltic lavas, huge volumes of pyroclastic ash have also been erupted on continental crust, such as by the Yellowstone hot spot, and these are also referred to as LIPs. Mantle plume-related magma that intrudes the continental crust induces melting of the crust which generates silicic magmas than erupt explosively resulting in pyroclastic ash that can cover extensive areas to form these LIPs. In addition to basaltic lavas being erupted at continental rift settings, silicic magmas which produce explosive eruptions can also be produced. Further, both basaltic lava and pyroclastic materials can be erupted at continental rift settings which form stratovolcanoes such as those occurring along the East African Rift system that we discussed in Chap. 6.

7.5.4 Continental and Oceanic Hot-Spot Volcanoes

Huge volumes of mostly basaltic magma were extruded or intruded by processes not considered to be "normal" ones such as sea-floor spreading and subduction during the history of Earth, particularly during the past 256 million years (called the Mesozoic (252–65 million years ago) and Cenozoic (66 million years ago–present)). LIPs that were formed earlier during the Proterozoic (2500–541 million

Fig. 7.12 Photos of examples of stratovolcanoes. **a** Mount Rainier, Washington State, USA (credit: L. Zeigler), **b** Mount Baker, Washington State, USA (credit: L. Zeigler), **c** Mount Bachelor, Oregon, USA (credit: R.H. Simoyi), and **d** Mount Meru, Tanzania (credit: R.H. Simoyi)

Fig. 7.13 Diagram illustrating the structure of a fissure eruption and the plume model for the formation of flood basalts

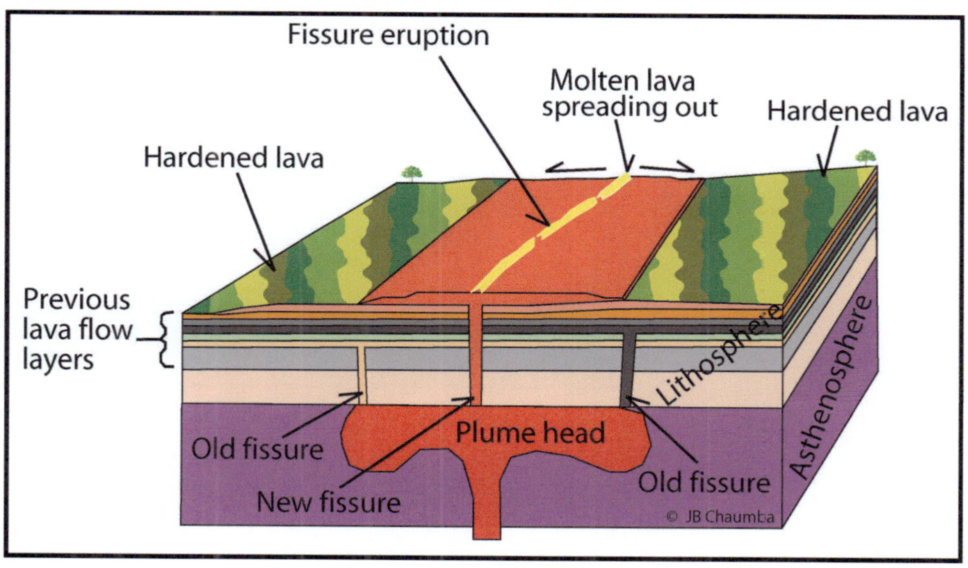

years ago) and Paleozoic (541–252 million years ago), but these have been largely eroded and are now only represented by deep-level conduits of magma such as layered intrusions (for example, the Great Dike of Zimbabwe, Bushveld Complex in South Africa, as well as the Stillwater Complex in Montana, USA), sills (for example, the Palisades sill in New York and New Jersey, USA), and dike swarms (for example, the Mashonaland dikes in Zimbabwe). LIPs occur in the form of lava plateaus or continental flood basalts on continents, or oceanic plateaus and ocean basin flood basalts on ocean basins (Fig. 7.15). It must be noted here that some LIPs are composed of silicic rocks (Fig. 7.15). Several LIPs were formed during rifting and break-up of continents as well as during uplift on a regional scale.

The largest eruptions of lava on earth are represented by voluminous continental flood basalts as well as ocean basin flood basalts (Fig. 7.15). Flood basalts are considered a type of LIP that is common on the surface of our planet (Fig. 7.15).

Fig. 7.14 **a** Fountains of lava erupting and spreading laterally from fissures of Pu ‘u ‘Ō ‘ō during the early phase of Episode 48, Kilauea Volcano, Hawaii, USA. The eruption, which began on January 3, 1983, continued to July 1986 (credit: USGS). **b** An aerial view of lava erupting and spreading laterally from a fissure to the east of Leilani Estates subdivision, Hawaii, USA, on May 13, 2018 (credit: Terray Sylvester/Reuters)

Some continental flood basalts often exceed 2000 km^3 (480 mi^3) in volume. Consequently, such voluminous eruptions of lava result in the formation of very thick stacks of basalt lava flows. As noted above, LIPs were erupted both on the continents (each continent contains continental flood basalts) and also in submarine environments to form oceanic plateaus at different times throughout the history of our planet. The 66 million-year-old Deccan Traps (Indian Sanskrit language meaning “step” for the step-like topography produced by the stacked lava layers) in India. The 249 million-year-old Siberian Traps in Russia, the Columbia River basalts in the US Pacific Northwest and southern British Columbia, Canada, and the 183 million-year-old Karoo flood basalts in southern Africa are typical examples of continental flood basalts. Some other typical examples of flood basalts are the 16 million-year-old Columbia River Plateau in the US Pacific Northwest (Fig. 7.16a), the 30 million-year-old Ethiopian Traps (Fig. 7.16b), the 8–1.3 million-year-old Akhurian (Arpaçay) River gorge of the Kars plateau in eastern Turkey (Fig. 7.16c), and the Paraná flood basalt province in Brazil (Fig. 7.16d).

Let us briefly describe some select continental flood basalts. The Columbia River flood basalts (CRFB) form a continental flood province which extends in excess of 210,000 km^2 mostly in the US states of Washington, Oregon and Idaho (Fig. 7.17), with limited exposures in parts of northern Nevada. Lava flows of the CRFB, which

Fig. 7.15 Diagram showing the distribution of extensive flood basalt plateaus of the world

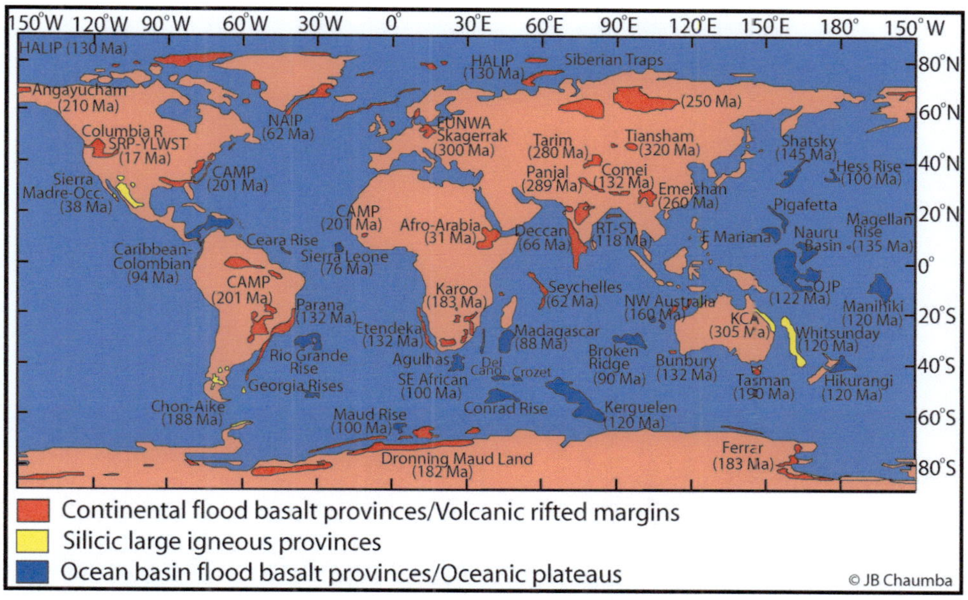

are layered and thick, were erupted as flood basalts and were formed from effusive eruptions. The CRFB were formed from about 16.7 –5.5 million years ago from north-northwest-trending linear fissures. Earth scientists think that upwelling resulting from a mantle plume situated above a hot spot was responsible for generating the magmas that formed the CRFB. Further, it is thought that since the eruption of the CRFB, the North American plate has been moving in a west-southwestwardly motion, and the hot spot that fed the CRFB now lies beneath the Yellowstone volcano located in the northwest part of the US state of Wyoming. The bulk of the greatest volume of the CRFB is thought to have been erupted in a time span of about 1.1 million years (around 16.7–15.6 million years ago).

The Deccan Traps, located in west-central India (Fig. 7.18), are one of the largest continental flood basalts on Earth. The Deccan Traps are composed of flat-lying basalt lava flows which are >2,000 m (6,500 feet) in thickness and cover an area of almost 500,000 km^2 (~200,000 mi^2). The original area which was covered by the Deccan Traps lava flows is estimated to been as high as 1.5 million km^2 (600,000 mi^2). Earth scientists think that the volume of basalt forming the Deccan Traps is 512,000 km^3 (~12,275 mi^3). Most of the Deccan Traps basalt were erupted from 67 to 64 million years ago. Toxic gases that were emitted by the eruption which produced the Deccan traps may have induced a change in the global climate which led to the demise of dinosaurs 65 million years ago.

The Karoo continental flood basalts (KCFB) of southern Africa (Fig. 7.19) are dominantly composed of tholeiitic basaltic rocks. The KCFB are thought to be related to the break-up of southern Gondwana (as referenced to present-day poles) and the opening of the Indian Ocean. The KCFB extend over Archean (Zimbabwe and Kaapvaal) cratons, the Proterozoic Limpopo orogenic belt, as well as Permian–Jurassic basins as lava flows, sills, and huge radiating dike swarms. The KCFB extends over an area of ~3 × 10^6 km^2. The age of the KCFB yields a time span of emplacement of the tholeiitic basaltic rocks of ~1.0 Ma, with a peak age at 184 Ma. Some Earth scientists, however, think that the emplacement of the KCFB started to the north and then migrated southwards, and the eruption of the tholeiitic basalts occurred over a period of at least 6 million years. Such a longer period of eruption is not consistent with mantle-plume-related partial melting, but it may have been caused by melting of a thick (that is, greater than 100 km in thickness) lithospheric mantle. Further, huge radiating dike swarms of some KCFB have also been interpreted by some Earth scientists to be consistent with a mantle plume that affected the lithosphere which allowed subsequent propagation of the magma, as well as feeding lava flows, over great distances located further away from the focal-impact-zone mantle plume.

It has been suggested that huge amounts of volatiles, aerosols, and dust that are emitted into the atmosphere during eruption of flood basalts, which do not erupt violently, may have been responsible for some mass extinctions. Cooling of global climates may have been triggered due to aerosols of sulfuric acid, warming due to greenhouse gases such as carbon dioxide and sulfur dioxide, as well as acid rain. Mafic magmas such as basalts are enriched in dissolved sulfur and aerosols of sulfuric acid produced from sulfur volatiles, chiefly sulfur dioxide, that are released into the stratosphere by convecting mantle plumes, which rise above fissures and vents of volcanoes. Changes in oceanic circulation as well as in the chemistry of the ocean, in addition to oxygenation mainly from

Fig. 7.16 Images of flood basalts from various selected parts of the world. **a** Columbia River flood basalts of the states of Washington, Oregon, and Idaho, USA. Resting on columnar basalts are the Twin Sisters, and both of these are part of the Columbia River flood basalts, viewed from the Wallula Gap (credit: P. Abongwa). **b** Aerial view of the 1500 m thick, 30 million-year-old Ethiopian Traps exposed between Addis Ababa and Lalibela, Ethiopia. Notice the step-like layers exposed on the upper part of the valley cliffs (credit: Hervé Bertrand). **c** Akhurian (Arpaçay) River gorge of the Kars plateau in eastern Turkey that was carved in basaltic sequence of Pliocene age. This location is near the ruins of the ancient Armenian city of Ani, Turkey. The gorge seen is ~80 m deep. These South Caucasus basalts are ~3 million-year-old and are exposed in Turkey, Georgia, and Armenia (credit: Vladimir A. Lebedev). **d** View of basaltic plateaus of the 135 million-year-old Paraná flood basalt province in Brazil, Uruguay, Paraguay, and Argentina. For scale, see the road and the houses in the valley (credit: Carla Barreto)

mafic magmas that were associated with submarine eruptions which formed oceanic plateaus are thought to have affected the environment, although indirectly.

Earth scientists also think that LIPs may have been responsible for climate changes due to volcanic gases such as carbon dioxide, sulfur dioxide, methane, and the burning of coal deposits which lead to acidification of the oceans and global warming. A link between the voluminous eruptions such as fissure eruptions of some flood basalts and mass extinctions has been proposed, but the possibility of the involvement of other factors in such extinctions cannot be ruled out. Rifting of the continents, changes in the spreading rates of ocean floor, variations in sea level, and an increase in tectonism and volcanism may also have been suggested to have been responsible for mass extinctions. The concurrence of both a huge impact event as well as the eruption of flood basalts has been suggested to have been responsible for austere mass extinctions. Some significant mass extinctions such as the one that occurred at the boundary of the Cretaceous and Tertiary (or K/T) event, however, are suggested by some Earth scientists to have been caused by meteorite or asteroid impacts. The unusually long duration of the Karoo flood basalts (of at least 6 million years), however, would thus imply that a lower eruption rate would only have had a minor influence on global climate and fauna. Further, a longer time span of eruption of the tholeiitic basalts would enable the atmosphere, as well as the ocean, to act as climate buffers which likely prevented any significant destabilization of the ecosystem.

One can envisage sills as lava flows that were subsequently buried by later deposits. Both sills and lava flows are tabular and sheet-like bodies which can display **columnar jointing**. Devil's Tower (Fig. 7.20a), located in northeastern Wyoming, USA, is comprised of a 40.5-million-year-old igneous rock called a phonolite porphyry. As the magma which formed the phonolite porphyry crystallized, hexagonal columns were produced. The sides of columns in rocks which display columnar jointing range from 3 to 8, with 6 being the most common number of sides. At Devil's Tower,

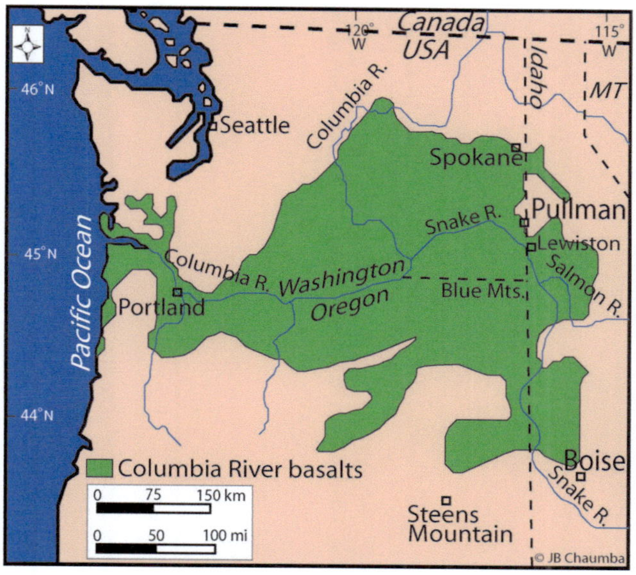

Fig. 7.17 Map showing the exposures of the Columbia River flood basalts in the USA states of Washington, Oregon, and Idaho

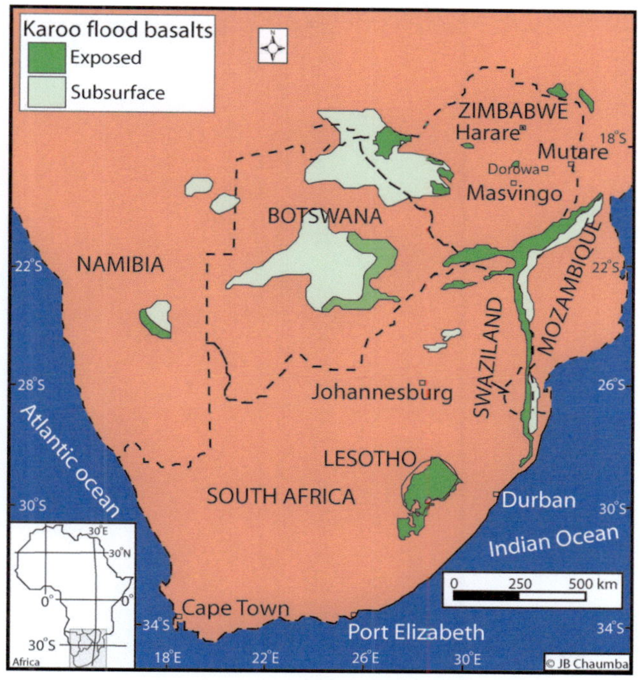

Fig. 7.19 Distribution of Karoo flood basalts of southern Africa

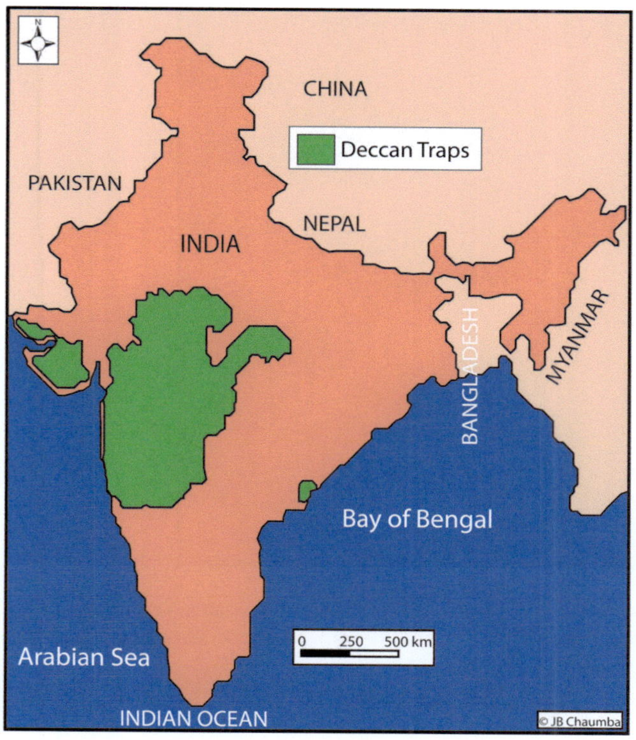

Fig. 7.18 Map showing the distribution of the Deccan Traps in India which were extruded dates from 67 to 64 million years ago

however, the hexagonal columns are typically 4-, 5-, and 7-sided, and the columns are approximately 1.8 m (6 ft.) in diameter. With continued cooling of the rock from the outside, the vertical columns shrank in cross-section (horizontally) which caused cracks to develop. The cracks occur at 120-degree angles which lead to the formation of roughly 6-sided columns. Devil's Tower is a bewildering feature that protrudes out of the prairie surrounding the Black Hills. It is considered sacred by the Northern Plains American Indians who consider Devil's Tower to be sacred. Numerous parallel cracks make Devil's Tower one of the finest crack climbing areas in North America, although only 1% of the 400,000 annual visitors visit its summit. Not far from Devil's Tower are located the Missouri Buttes which are composed of four separate summits which rise from an eroded platform of mesa called the Butte Divide. The elevation of the Butte Divide is 1,420 m (4,650 feet).

Before we discuss more extrusive igneous features, let us discuss an igneous feature which can be considered to be transitional between intrusive and extrusive environments. Located in northwestern New Mexico, USA, is Shiprock (Fig. 7.20b), which is interpreted to be a remnant of a volcanic eruption that occurred approximately 30 million years ago. The main part of the landform is 600 m high, and 500 m in diameter. To the local Navajo American Indians, Shiprock is known as Tse Bitai (meaning "the winged rock"). Shiprock is interpreted to be the central feeder pipe or a volcanic neck of a volcano which today has since been eroded away. The volcanic neck is composed of fractured volcanic rock (breccias) that is crosscut by numerous veins of solidified magma. The rock that forms Shiprock was formed from magma that was enriched in potassium termed a **minette**. Minettes are rocks that are formed when very small degrees of partial melting of the Earth's mantle occur. At the time it was formed, it is interpreted that Shiprock was probably 750–1000 m (2460–3281 ft.) below the land surface. The current high elevation of Shiprock in comparison to the surrounding landscape is a result of faster erosion rates of less

Fig. 7.20 **a** Columnar jointing displayed by the Devil's Tower at Devil's Tower National Monument. The image was obtained by the USGS with 3DR Solo Drone. Devil's Tower rises 265 m (867 ft.) above the surrounding landscape (credit: D.T. Nelson). **b** Photograph of Shiprock, a volcanic neck located in the state of New Mexico, USA. Shiprock towers 520 m (1700 ft.) above the surrounding landscape (credit: D.T. Nelson)

resistant surrounding landscape whereas Shiprock is relatively more resistant to erosion.

Associated with Shiprock are dikes, which are tabular sheets of igneous rocks radiating away from Shiprock; up to 6 dikes in total have been identified. The dikes are also interpreted to have been emplaced at some depth below the earth's surface at the same time that the Shiprock was also intruded. These dikes now form elongated ridges rising above the landscape as they are also similarly more resistant to weathering than the less resistant country rock of sandstone and sandstone.

Situated 5.6 km (3.5 miles) to the northwest of Devils Tower, the Missouri Buttes (Fig. 7.21a) are also composed of columnar phonolite comparable in age to the Devil's Tower. Further examples of rocks which display columnar joints are the Devil's Postpile National Monument in California as well as the Giant's Causeway in Northern Ireland. Columns at the Devil's Postpile National Monument and the Giant's Causeway, at approximately 0.61 m (2 ft.) in diameter, however, are much smaller in diameter in comparison to those at the Devil's Tower.

Fig. 7.21 **a** A photograph of Devils Postpile in California, USA, showing vertical columns of basalt that were formed from slow cooling of a thick lava flow. During cooling of the thick lava flow, shrinking occurred and, together with accompanying stresses, produced the vertical columns of the rocks (credit: USGS). **b** Giant's Causeway, near the village of Portballintrae, Northern Ireland (credit: E. Kepekepe)

Giant's Causeway (Fig. 7.21b), located in Northern Ireland, is a promontory of 50 to 60-million-year-old basalt columns extending for a distance of 6 km (4 miles) along the country's northern coast. It is estimated that there are approximately 40,000 stone pillars that compose Giant's Causeway, with each column being typically with five to seven sides of irregular dimensions. The columns jut out of the cliff faces to resemble steps that creep into the sea. Giant's Causeway is interpreted to have been produced from successive flows of lava which slowly extended toward the coast, cooling as it came into contact with the sea. Consequently, columns were formed from the layers of basalt. In addition, pressure between the columns helped shape them into polygonal "pillars" that are as long as 25 m (82 feet) in height and range in diameter from 38 to 51 cm (15–20 inches). The polygons form arrays along cliffs which average some 100 m (330 feet) in elevation.

7.6 Types of Volcanoes

In general, a volcano consists of a volcanic cone which is a mountain or hill that was formed by erupted materials, as well as a volcanic crater which is a bowl-shaped depression with steep walls located at the vent (see Fig. 7.11). Collapse of the volcanic materials surrounding the vent of a volcanic crater forms a depression due to the magma chamber beneath being emptied by eruption of the magma. A variety of shapes of volcanic cones and craters, as well as their sizes, occurs. The chemical composition of the erupting materials, as well as the style of volcanic eruption, determines the dimensions and shapes of volcanic cones and craters. Next, these different dimensions and shapes of volcanoes are described and illustrated.

7.6.1 Cinder Cones

Cinder cones, which resemble cone-shaped hills with a vent at the top, are dominantly composed of loose, vesicular, and low-density basalt called scoria. Cinder cones are composed of pyroclastic materials called cinders. Cinder cones typically have steep slopes due to pyroclastic materials piling up to generate conical hills whose slopes range from 30 to 40°. Cinder cones are the smallest types of volcanoes (Table 7.3, Fig. 7.22), typically less than 450 m (1500 ft.) in height, and they are circular in plan view. Cinder cones lack any lava flows which can act as glue to hold the pyroclastic materials together; hence cinder cones are easily eroded. Magma of any composition can form cinder cones, provided the magma contains enough volcanic gases to propel the lava into the atmosphere and form a cone of volcanic material around the vent (Table 7.3). Since pyroclastic materials are ejected high above the vent, they tend to cool down before landing which ensures that the pyroclastic fragments don't stick together.

Due to their simple nature, numerous examples of cinder cones exist on Earth. Examples of cinder cones include Koko Crater in Hawaii, Mt. Rukinear in the Democratic Republic of the Congo, Mount Elephant in Victoria, Australia, Puy de Pariou in Auvergne, France, and Flourmill Cone in British Columbia, Canada.

7.6.2 Stratovolcanoes or Composite Volcanoes

The majority of volcanoes on Earth (~60%) are **stratovolcanoes** or composite volcanoes. Stratovolcanoes (Table 7.3, Fig. 7.23) are formed by multiple eruptions, with the eruptions often recurring from over a few hundred to over hundreds of thousands of years. Although other magmas can be involved, the most common type of magma that is erupted in the formation of stratovolcanoes is andesite. Stratovolcanoes (Fig. 7.23a), which, as stated above, comprise the largest percentage of the Earth's individual volcanoes, are mostly characterized by eruptions of more viscous dacitic and andesitic magmas in comparison to basalt. Stratovolcanoes formed from andesitic magmas are built mostly of pyroclastic materials, but some of the andesitic magma may intrude fractures within the cones resulting in the formation of dike or sills. Stratovolcanoes often grow to great heights resulting in their slopes becoming unstable, leading to collapse of the stratovolcano from gravitational forces. Stratovolcanoes can rise to two to three thousand meters in height above their bases.

Table 7.3 Table illustrating different types of volcanoes

Type of volcano	Approximate size and shape	Magma type and eruption characteristics	Tectonic setting	Example
Cinder cone	Generally small in size, range from 10 to 100 s of m in diameter, steeply dipping (>20°)	Mostly mafic magmas; form from early stages of a shield- or rift-related eruption that are enriched in gas	Various; can form on flanks of larger volcanoes	Paricutin, Mexico; Lava Butte, Oregon, USA; Coffee Crater, British Columbia, Canada
Statovolcano/composite volcano	Medium sized, range from 1000 s of m in radius, moderate dips (10°–30°)	Magma composition ranges from felsic to mafic, and from explosive to effusive eruptions	Almost all are formed above subduction zones	Stromboli, Italy; Mount Mayon, Philippines; Mt. St. Helens, Washington, USA
Shield volcano	Immense size (up to several 1,000 m high and 200 km in diameter), gentle dips (~2° to 10°)	Dominantly mafic magmas, effusive eruptions, some cinder cones form on flanks of shield volcanoes	Associated with mantle plumes; some form at spreading ridges	Mauna Loa, Hawaii, USA; Skjaldbreiður, Iceland; Erta Ale, Ethiopia
Large igneous provinces	Enormous (up to millions of km^2) and 100 s of m in thickness	Magma is always mafic and individual flows can be 10 s of m in thickness	Associated with mantle plumes	Siberian Traps, Russia; Columbia River basalts, USA; Rajmahal Traps, India

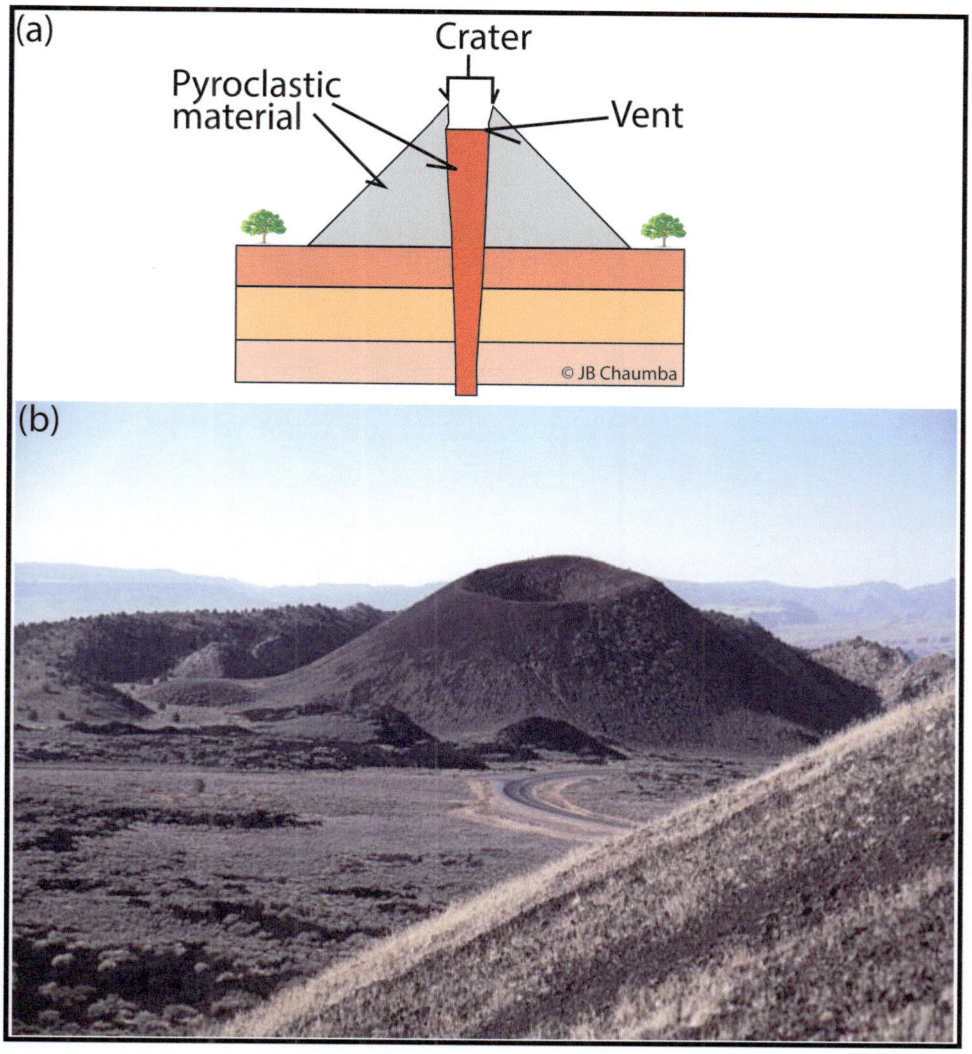

Fig. 7.22 **a** Diagram illustrating a cinder cone. **b** Photograph of Southern Santa Clara cinder cone in Utah, USA, as viewed from the flank of the Northern Santa Clara cinder cone. See road for scale. *Credit* Global Volcanism Program, Smithsonian Institution, Washington, DC, USA

Viscous magmas enable pressure of volcanic gases to build up by preventing their easy escape from the dacitic and andesitic magmas which lead to violent eruptions. Both lava and pyroclastic materials are involved in the formation of stratovolcanoes, a feature that imparts a layered structure in stratovolcanoes; hence, their other name is composite volcanoes. Stratovolcanoes are often encountered above subduction zones. Several stratovolcanoes occur in linear chains where they are often separated by several tens of kilometers. A number of stratovolcanic chains occur on Earth, chiefly around the Pacific Ocean where they form the "ring/rim of fire." Common examples of stratovolcanoes include Mt. Rainier in Washington State, USA; Mayon volcano in the Philippines; Mount Fuji in Japan; Cotopaxi volcano in Ecuador; Krakatoa in Indonesia; Mount Vesuvius near Naples in Italy; and Stromboli, also in Italy (Fig. 7.23b).

7.6.3 Shield Volcanoes

When basaltic magma erupts through a central vent, it spreads out laterally in all directions for great distances from the vent due to its low viscosity. With continued eruptions, successive lava flows often produce a broad, low, and gently dipping conical structure termed a **shield volcano** (Table 7.3, Fig. 7.24a, b). A shield volcano derives its name from its resemblance to the shield of a warrior that is lying on the ground (Fig. 7.24a, b).

The wide summit and gentle slope of a shield volcano are a consequence of the swift-flowing basaltic magma which has low viscosity. Lava tubes also provide pathways for the basaltic magma enabling the magma to flow great distances away from the vent while it is still hot. The largest volcanoes on Earth are shield volcanoes. Examples of shield volcanoes include Mauna Loa (Fig. 7.24b) and Mauna Kea in

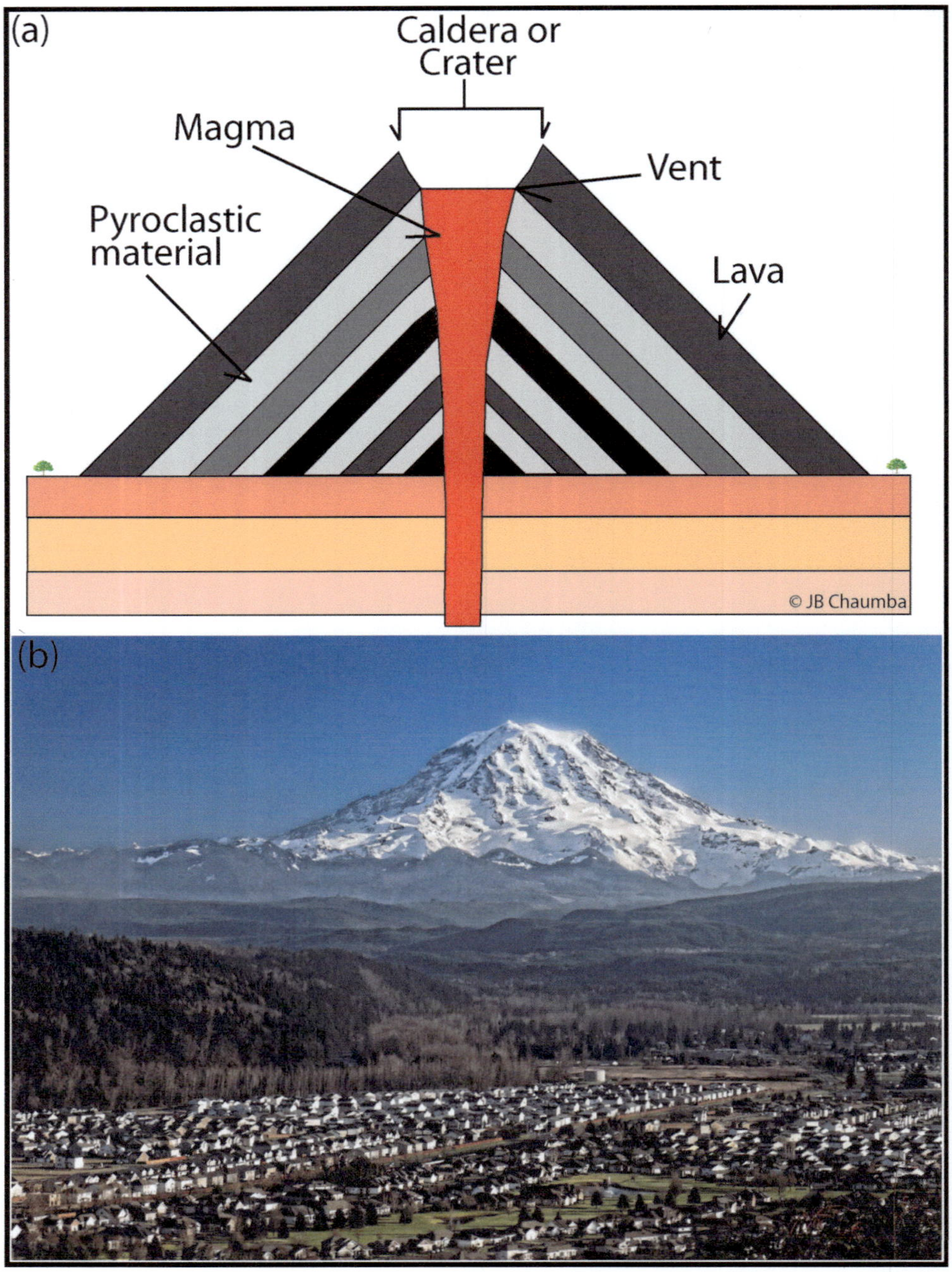

Fig. 7.23 **a** Diagram illustrating a stratovolcano composed of alternating layers of lava and pyroclastic materials. **b** A photograph of Mt. Rainier, Washington, USA, partly covered by thin snow. View of Mt. Rainier from South Hill with Puyallup Valley in the foreground, on January 18, 2014. *Credit* Ed Ruttledge, Cascades Volcano Observatory, USGS

Hawaii, USA; Erta Ale volcano located in the Afar region of northeastern Ethiopia; Masaya volcano in Nicaragua, Central America; Tolbachik volcano in Kamchatka, Russia; and Karthala volcano on the Comoros, Comoro islands.

7.7 Volcanic Hazards

In the United States, in excess of 50 volcanoes are known to have erupted at least once in the past 200 years, with Alaska, Hawaii, California, Oregon, and Washington the most volcanically active states of the country. A wide variety of hazards that can kill people and destroy property, however, are emitted when volcanoes erupt. If huge explosive eruptions occur, people and property hundreds of miles away can be endangered, as well as affecting global climate in some cases. Further, volcanic hazards such as landslides can occur even when no volcano is erupting.

7.7.1 Eruption Columns and Clouds

When a volcano erupts explosively, tephra (solid and molten rock fragments) as well as volcanic gases can be forcefully ejected into the atmosphere. Volcanic bombs, which are the largest rock fragments, tend to fall back to the

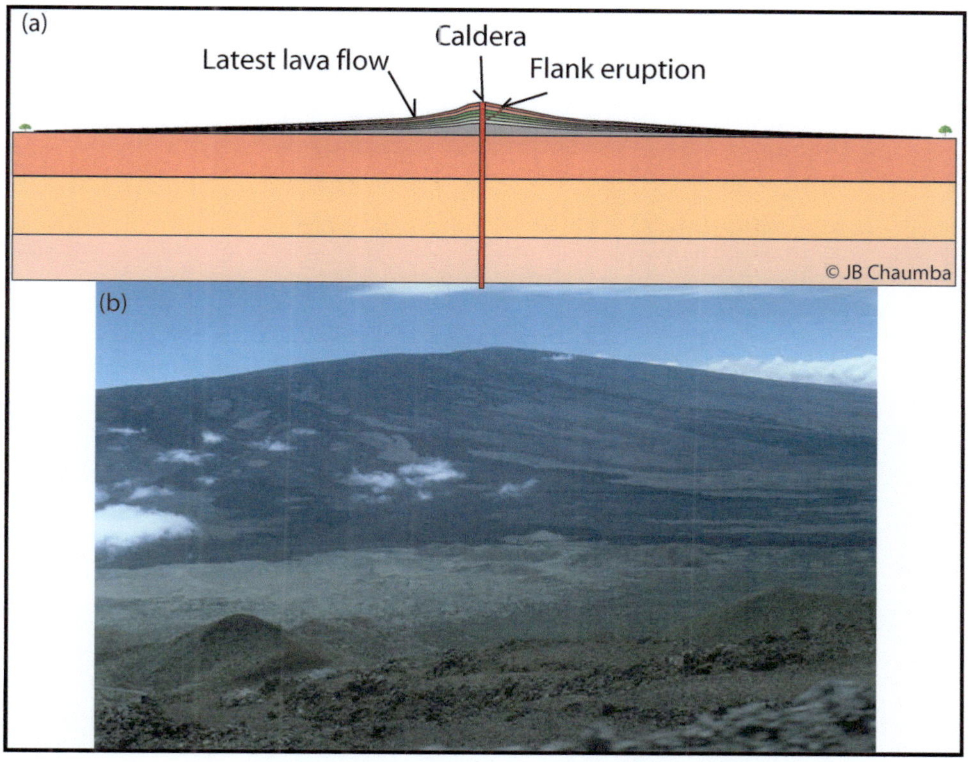

Fig. 7.24 **a** Diagram illustrating a shield volcano. **b** An image of part of Mauna Loa, an example of a shield volcano on the Big Island (or Hawaii), USA. *Credit* USGS

ground within a couple of miles of the vent. Fine-grained pyroclastic materials including volcanic ash, however, are ejected higher into the atmosphere to form an eruption column (Fig. 7.11). Eruption columns tend to grow rapidly, reaching in excess of 12 miles above a volcano in less than half an hour to produce a fan eruption cloud. The hazard from the volcanic ash in the eruption cloud is chiefly to aviation.

Commercial jets are frequently damaged when they inadvertently fly into volcanic ash clouds, where they risk crashing due to engine failure. Huge eruption clouds often extend for hundreds of kilometers (miles) downwind, which often results in ash fall covering enormous areas since wind carries the smallest volcanic ash particles the farthest distance from an erupting volcano. As an example, volcanic ash from the May 18, 1980, eruption of Mount St. Helens, Washington State, USA, fell as far east as Minnesota. Lastly, even minor ash fall can damage crops, electronics, and machinery whereas heavy ash fall can lead to roof collapse if it accumulates on top of buildings.

7.7.2 Volcanic Gases

Even before a volcano starts to erupt, any cracks in the ground can permit volcanic gases to easily migrate upwards and reach the Earth's surface through **fumaroles**. Fumaroles are small openings near volcanoes through which volcanic gases escape. During volcanic eruptions, a wide variety of volcanic gases (Fig. 7.25) can be emitted, with water vapor (steam) from heated groundwater being by far the dominant gas that is emitted by volcanoes. In addition to water vapor, other common volcanic gases emitted by erupting volcanoes are carbon dioxide (Fig. 7.25a), sulfur dioxide (Fig. 7.25b), hydrogen sulfide, hydrogen, and fluorine. The volcanic hazard posed by sulfur dioxide gas is that it can react with water droplets in the atmosphere to form acid rain, which is corrosive and harms vegetation. Carbon dioxide, which is colorless and odorless as well as being denser than air especially when it is cold, is often trapped in low areas in high concentrations that are often deadly to people, wildlife, and animals, as in the case of Lake Nyos in Cameroun, Africa (Fig. 7.25a). In regions where volcanic eruptions occur, or other areas where CO_2 emissions occur, it is highly recommended to avoid small depressions as low areas may often act as traps of CO_2.

In addition, fluorine gas, which in high concentrations is toxic, can be adsorbed onto volcanic ash particles that later fall to the ground. Fluorine gas accumulated on surface to form a thin film (i.e., adsorb) on volcanic ash particles tends to poison livestock that graze on volcanic ash-coated grass. Further, the adsorbed fluorine gas also contaminates domestic water supplies. Huge explosive eruptions, such as the eruption of Mount Pinatubo in the Philippines, on June 15, 1991, led to the injection of huge amounts of sulfur dioxide gas into the stratosphere where it can react with

Fig. 7.25 **a** Photograph of some of Nyos village compounds, Cameroun, surrounded by dead cattle that were killed by suffocation due to a cloud of carbon dioxide which inundated them on September 3, 1986, two weeks after the explosion of Lake Nyos (Photo credit: USGS). **b** George Kourounis investigating sulfur mounds and other volcanic gases being emitted by Papandayan volcano, Indonesia, in July 2008 (Credit: George Kourounis). **c** Earth scientists sampling volcanic gases emanating from the Geyser Bight Geothermal Area, Recheshnoi volcano, Umnak Island, Western Aleutian Islands, USA (credit: John Lyons, USGS)

water to form mist (an aerosol) of sulfuric acid. Aerosols of sulfuric acid have been shown to reflect solar radiation back into space which often leads to lowering of the Earth's mean surface temperatures by some degrees Celsius (°C) (several Fahrenheit, °F) for extended periods of time. Further, these sulfuric acid aerosols also contribute to the destruction of the ozone layer by altering chlorine and nitrogen. Sulfur dioxide, a colorless gas, is irritating to the skin, eyes, as well as to the respiratory system.

7.7.3 Lava Flows and Domes

Less viscous, low-silica basaltic lava can flow relatively fast at velocities of 16–48 km/h (10–30 mi/h), and spread out to form broad thin sheets that extend for several kilometers (miles) wide. As an example, basalt lava flows erupting from Kilauea Volcano on the Island of Hawaii, USA, since 1983, have led to the destruction of almost 200 houses, mostly due to incineration by the hot lava (Fig. 7.26a). The coastal highway near Kilauea volcano has also frequently been engulfed by this volcano.

7.7.4 Volcano Mass Wasting

A type of mass wasting that occurs when rapid downhill movements of rocky materials, snow, and (or) ice on volcanoes is called a landslide or debris avalanche. Landslides associated with volcanoes often range in size from small movements of loose debris on the surface of a volcano to massive collapses of the entire summit or sides of a volcano. Steep volcanoes tend to be very susceptible to landslides since they are constructed partly of loose volcanic rock fragments layers. Some rocks that form volcanoes are also easily hydrothermally altered by

Fig. 7.26 **a** Lava flow incinerates house in rural Pahoa, Hawaii, on Monday, November 10, 2015 (credit: County of Hawaii/AP). **b** Pyroclastic flow from Volcán de Fuego (which means "volcano of fire" in Spanish), Guatemala, on Sunday, June 3, 2018, cascades down the flank of the volcano (credit: AFP/ Getty images). At least 159 people were killed during this eruption

circulating hot and acidic ground water to produce soft and slippery clays. Landslides which occur on volcano slopes are often triggered either by magma moving upwards towards the Earth's surface, or by heavy rainfall. Relatively large earthquakes can also cause these materials forming the slopes of volcanoes to break free and move downslope.

7.7.5 Pyroclastic Flows

A **pyroclastic flow** occurs when rock fragments, hot ash, together with gas move swiftly down the flanks of a volcano (avalanche) due to the occurrence of explosive volcanic eruptions. A pyroclastic flow can also occur when the steep side of a growing lava dome disintegrates. A pyroclastic flow (Fig. 7.26b) is also referred to as a pyroclastic cloud or a pyroclastic density current. The temperature

Fig. 7.27 **a** Mt. Ruapehu, New Zealand, Lahar originating from its summit (credit: AAP/John Cowpland, NZPA). **b** View of lahar from Mt. Ruapehu, New Zealand, volcano under a bridge (credit: Phil Walter, Getty Images). **c** Lahar flows from Mt. St. Helens, Washington State, USA (Credit: Fabio Alves, Panoramio). **d** View of the city of Armero, Colombia, after the November 1985 lahar (credit: USGS)

of a pyroclastic flow can exceed ~800 °C (1,830°F) and move at speeds that can reach 700 km/h (430 mph). Pyroclastic flows tend to follow valleys, and they often destroy and burn everything in their paths. Pyroclastic surges, which are lower-density pyroclastic flows, commonly overflow ridges that are hundreds of meters (tens of feet) in height.

7.7.6 Lahars

Lahars are debris flows or mudflows that are dominantly composed of volcanic materials on the flanks of volcanoes. Flows of mud, rock, together with water comprise lahars, and they can swiftly rush down channels of streams and valleys at speeds of 32–65 km/h (20–40 mi/h), often covering distances exceeding 80 km (50 miles) (Fig. 7.27b, c). Some lahars can be equated to fast-moving rivers of wet concrete since they contain so much rock debris (60–90% by weight). Close to the source (Fig. 7.26a) of lahars, the flows are capable of ripping up and carrying houses, huge boulders, and trees, for miles downstream whereas further downstream, lahars entomb everything in their path in mud (Fig. 7.27b–d).

Lahars are one of the deadliest volcanic hazards. Due to their potential to be formed both during an eruption as well as when a volcano is "quietly" erupting. The water that helps generate lahars can originate from heavy precipitation, or from melting of snow and ice, or even from the blowing off of a crater lake at the summit of a volcano. Large lahars are a potential hazard to many communities downstream from glacier-clad volcanoes, such as Mount Rainier in Washington State, USA, as lahars can easily destroy any settlement along their paths.

When Mount Pinatubo erupted on June 15, 1991, a lahar occurred when the eruption coincided with the arrival of a typhoon which caused a deadly mixture of rain and pyroclastic materials to rain down on cities and towns that are close to the volcano. The Nevado del Ruiz in the Andes mountains of Colombia (Fig. 7.27d) is a well-documented lahar which resulted in the loss of thousands of people and caused extensive damage to infrastructure. The Armero tragedy was one of the major consequences of the eruption of the Nevado del Ruiz in Tolima, Colombia, on November 13, 1985, leading to the death of more than 20,000 people. The Colombian government had received warnings from multiple volcanological organizations to evacuate the area after the detection of volcanic activity two months earlier, but written records state that it did not take any action. As pyroclastic flows erupted from the volcano's crater, they melted the mountain's glaciers, sending four enormous lahars (volcanically induced debris flows, landslides, and mudslides) down its slopes at 50 km per hour (30 miles per hour). The lahars picked up speed in gullies and coursed into the six major rivers at the base of the volcano; they engulfed the town of Armero, killing more than 20,000 of its almost ~29,000 inhabitants. Casualties in other towns, particularly Chinchina, brought the overall death toll to ~23,000.

Stratovolcanoes are known to have caused the highest casualties of any volcano type even though they do not erupt as explosively as large silicic caldera complexes. One of the reasons why stratovolcanoes have killed so many people is that they are the most common type of all volcanoes. Consequently, more people would settle close to stratovolcanoes than any other type of volcano. Further, mass movements that are associated with stratovolcanoes can cause further fatalities.

To mitigate volcanic hazards, it is encouraged to avoid settling close to areas that are prone to volcanic eruptions through the construction of volcanic hazard assessment maps which delineate areas occurring in areas susceptible to volcanic hazards. Lava flows can also be diverted to prevent them from incinerating settlements. Powerful pumps have been successfully used by firefighters in Iceland to spray cold seawater onto an approaching lava flow with the aim of freezing it.

7.8 Active, Dormant, and Extinct Volcanoes

An **active volcano** (Fig. 7.28a) is one that has erupted since the last ice age, that is, in the last ~10,000 years. An active volcano is characterized by a smooth conical shape (Fig. 7.28a). An **erupting volcano**, which is a volcano in eruption, is an example of an active volcano. A **dormant volcano** (Figs. 7.28b and 7.29a) is one that has not erupted in the last ~10,000 years, but it has the potential to erupt again. The shape of a dormant volcano tends to be characterized by the presence of gullies on its flanks due to erosion. Further, the conical shape, although still present, is generally no longer smooth, but the volcano is still close to being perfectly conical (Fig. 7.28b). An **extinct volcano** (Figs. 7.28c and 7.29a) is one that is not expected to erupt again. An extinct volcano tends to be deeply eroded, with only remnant hills remaining (Fig. 7.28c). Mt. Bachelor, Oregon (Fig. 7.29), which is inferred to have last erupted in 2800 BC, is an example of a dormant shield volcano within the Cascade Range in the state of region, USA. Mount Bachelor is also located in the state of Oregon, USA, relatively close to the Three Sisters.

Where the volcano is composed of soft pyroclastic materials which easily get eroded away, only the more resistant plug of the volcano (such as Shiprock, New Mexico) and

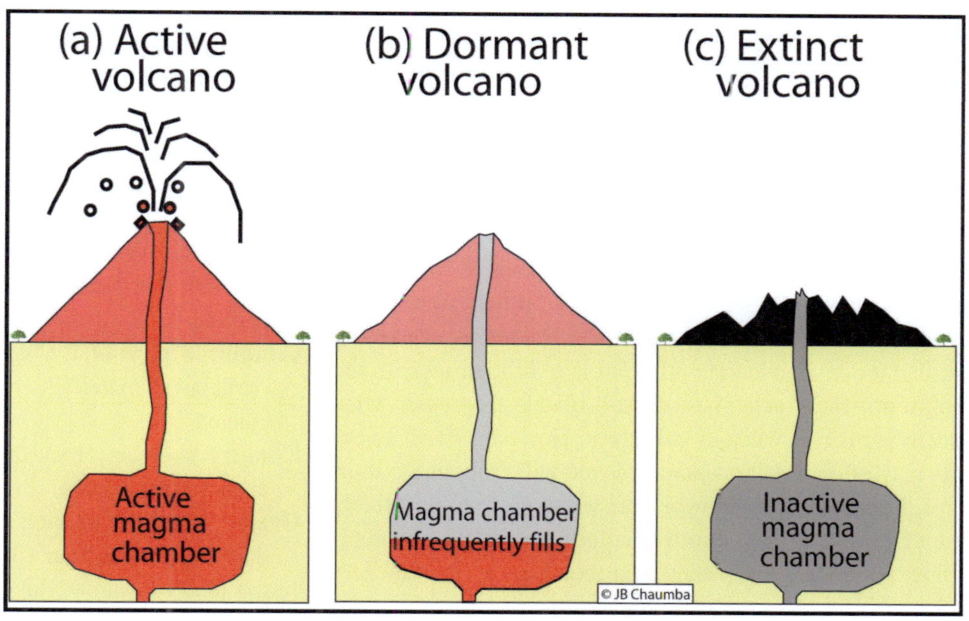

Fig. 7.28 Image illustrating active (**a**), dormant (**b**), and extinct (**c**) volcanoes

Fig. 7.29 **a** Mount Edgecumbe at the southern end of Kruzof Island, west of Sitka in Alaska, is an example of a dormant volcano (credit: USGS). **b** Black Butte, Oregon, USA, is an example of an extinct volcano. Note the extensive erosion in the form of channels on the flanks of the volcano (credit: USGS)

dikes that are both composed of solidified magma remain after erosion. It must be borne in mind, however, that these definitions of active, dormant, and extinct volcanoes are not set in stone as there have been some volcanoes thought to be "extinct" that erupted. Further, the rate of erosion of a volcano depends on the climate where the volcano is located, as well as the presence or absence of agents of erosion such as rivers or glaciers. If a volcano lies within a tectonically active region, the chances that the volcano is still active are very high, as opposed to a volcano that is located in an area that is no longer active tectonically which would more likely be a dormant or an extinct volcano.

7.9 Predicting When a Volcano Will Erupt

The best way to avoid being a victim of hazards from volcanoes is to avoid living close to them. Unfortunately, people settled in areas near active volcanoes long before Earth scientists studied them and knew as much as they do today. Since preventing volcanic eruptions is not an option, preventing fatalities due to volcanic eruptions, therefore, depends on predicting when a volcano can erupt. Can Earth scientists predict when volcanic eruptions will occur? The benefits to any successful prediction of an active volcano are crucial to saving lives through issuing of timely warnings to inhabitants who live near volcanoes. Predicting when any particular active volcano will erupt is a difficult task. In the long term, Earth scientists are still unable to predict when exactly an active volcano will erupt. However, Earth scientists may often predict when a volcano will erupt in the short term. In order to predict when an active volcano will likely erupt in the short term, Earth scientists rely on closely monitoring a volcano. Any changes in the shape of the volcano, any gases emitted, and activities such as earthquake activity as well as heat flow, are useful in predicting when a volcano will likely erupt.

Glossary

'a'ā a Hawaiian denoting a type of lava flow with a rough rubbly surface that is composed of broken lava blocks called clinkers which makes walking on it very difficult and slow

Active volcano a volcano that has erupted at least once during the past 10,000 years

Caldera a large depression with a diameter exceeding 1 km (0.6 mi) that formed when the summit of a volcano collapses due to inflation of a magma chamber

Crater bowl-shaped depression occurring at the top of a volcano formed from pyroclastic material of pyroclastic material on rim of vent or by erosion of walls of crater due to rapidly ejected particles

Cinder cone a cone that occurs around a volcanic vent and is formed by fragments of lava and pyroclastic materials that are ejected during volcanic eruptions

Columnar jointing a structure that develops in basalt rocks and it is composed of columns typically hexagonal in shape and they are separated by joints or fractures in the rock that form due to contraction of the rock during cooling

Conduit a pipe of a volcano through which lava and/or pyroclastic materials from the interior of a planet are ejected

Country rock any rock that encloses an igneous intrusion or any other feature

Crater a large depression at the summit of a volcano with a diameter of less than 1 km caused explosion blowing the top of a volcano of the crater walls by ejected materials

Dike any igneous rock that cuts across any earlier formed rocks or cuts across any pre-existing horizontal structures such as layering or bedding

Dormant volcano a volcano that, although not erupting, it is an expected to erupt again

Effusive eruption a type of volcanic eruption whereby magma quietly and steadily flows out of a onto the surface of the Earth as lava

Explosive eruption a type of volcanic eruption whereby volatile-rich magma and/or pyroclastic materials are violently and rapidly ejected from the interior of a planet out of a onto its surface

Erupting volcano a volcano that, although it is active, it is not having an eruption

Extinct volcano a volcano that has not erupted for at least 10,000 years and it is not expected to erupt again

Fissure eruption is a linear volcanic vent through which lava generally quietly erupts and it is also known as a volcanic fissure or fissure vent or eruption fissure

Flood basalt voluminous eruptions that inundate vast areas of a planet, often covering broad regions of the land or ocean floor with flat-lying lava surfaces

Fumarole a small opening near volcano through which volcanic gases can escape

Laccolith a typically lens-shaped body of igneous rock that was emplaced between layers of rock leading to uplift to form a dome-shaped feature

Lahar a fluid mass of volcanic fragments that rapidly flows downhill and it usually contains a mixture of ash, breccia, and boulders with rainwater, or river water, or lake water displaced by the lava flow from the erupting volcano

Large igneous province a very large accumulation of igneous rocks composed of sills, lava flows, and even tephra deposits formed when huge volumes of magma are erupted and accumulate on the surface of a planet, either on the continent or on the ocean floor

Lava tube a tunnel that if formed within a solidified lava flow that was previously occupied by flowing molten lava when it formed a subsurface opening

Minette a dark-colored lamprophyre dominantly composed of very coarse-grained biotite and groundmass or fine-grained alkaline feldspar minerals

Lopolith a large igneous intrusion which is lenticular in shape with a depressed central region and is generally concordant with the surrounding rock layers. A dike or funnel-shaped body that occurs beneath the lopolith acted as the feeder

Batholith the largest intrusive igneous body with an area of at least 100 km^2 (40 square miles) that formed deep within the crust but later exposed at the surface due to erosion

Pāhoehoe smooth and unbroken mafic lava composed of an undulating, smooth, ropy, or billowy surface

Phreatic eruption a type of volcanic eruption produced by water percolating into the magma chamber where it is heated and turns into steam

Pillow lava lava that is characterized by pillow-shaped structures that are formed due to the eruption of the lava under water

Pipe a conduit occurring beneath a volcano through which magma and/or pyroclastic materials passed through and that is composed of volcanic breccia, fragments of older rock, and solidified magma

Pluton a large body of plutonic rock that is formed when magma slowly crystallizes below the surface of the Earth

Pyroclastic flow dense mixture of lava, pyroclastic fragments, hot ash, and volatiles that are erupted explosively from a volcano and swiftly flows downslope at great speeds, destroying everything in its path

Pyroclastic material/fragment a cloud of extremely hot ash, lava fragments carried through the air, and volatiles, and it moves swiftly due to the buoyancy provided by the volatiles

Shield volcano a domed and broad volcano composed of gently sloping sides that is formed from the eruption of fluid, mafic lava of low viscosity

Sill a concordant, tabular and sheeted intrusion that has been emplaced between either layers of sedimentary rock, or parallel to fabric or foliation in metamorphic rock

Stock a discordant rock that was formed from slow cooling within the crust and it is smaller than a batholith and it has a surface exposure of less than 100 km^2 (40 mi^2)

Stratovolcano a steeply dipping and tall conical volcano that is generally composed of several layers of hardened viscous lava, tephra, and volcanic ash that is formed from periodic, explosive eruptions

Tephra a general term for all pyroclastic materials that is ejected into the air by an erupting volcano such as ash, dust, and cinders, that is ejected, regardless of composition, fragment size, or emplacement mechanism

Vent an opening in the earth's crust from which magma, pyroclastic materials, and volatiles are ejected from the interior of a planet

Volcano a generally conical mountain or hill with a vent or a crater or vent through which lava, pyroclastic materials, and volatiles are erupted from the interior of the earth

8 Earthquakes

Abstract

After defining an earthquake, a table of number of earthquakes of magnitude 5 and above, and estimated fatalities, which occurred in the last decade on a global scale based on USGS data, follows. Definitions of earthquake terminology accompanied by illustrations follow, as well illustration of the elastic rebound theory in both extensional and compressive settings. Types of seismic waves are explained and illustrated. How to locate the epicenter of an earthquake with data from three recording stations is explained by using an example. Similarly for the determination of size of an earthquake. The modified Mercalli intensity scale and moment magnitude scale of measuring earthquake intensity are described. Earthquake depths, high-frequency earthquake zones, and earthquake prediction are all described, together with various methods of mitigating earthquake casualties and damage.

Learning Goals

Studying this chapter will enable you to:

- Define an earthquake
- List the triggering mechanisms of earthquakes
- Contrast earthquake intensity and magnitude scales
- Differentiate the epicenter of an earthquake from its hypocenter or focus
- Describe the use of seismographs in locating the epicenter of an earthquake
- Describe how the size of an earthquake is determined
- Locate the major belts on Earth that are prone to earthquakes
- Explain some of the challenges of predicting earthquake occurrences
- List and describe steps that can prevent damage of earthquakes
- Explain why earthquakes are concentrated only in some zones of our planet and not everywhere.

8.1 Introduction

We will begin this chapter by briefly mentioning a couple of notable **earthquakes** (episodes that cause the ground to shake) that occurred in different parts of the world in recent history. On February 6, 2023, a **moment magnitude scale**, M 7.8, earthquake, called the Kahramanmaraş earthquake of 2023 Türkiye, struck both central and southern Türkiye at 04:17 local Türkiye time (01:17 UTC). The earthquake **epicenter** was located 37 km (23 mi) west-northwest of Gaziantep. The earthquake occurred at a depth of 10 km (6.2 mi). The modified Mercalli intensity scale of the earthquake was XII (which is the highest value—Extreme) surrounding the epicenter as well as in Antakya, Türkiye. Another earthquake (M 7.7) with an epicenter located 95 km (59 mi) north-northeast from the first one followed at 13:24 local time. The death toll in Türkiye was 53,537, and between 5,951 and 8,476 in Syria. The earthquake was the deadliest earthquake in what is now present-day Türkiye since the 526 Antioch earthquake caused extensive damage and fatalities of tens of thousands. Northern Syria was seriously affected by these earthquakes.

On September 5, 2022, at 04:52:19 (UTC), an M 6.6 earthquake struck approximately 43 km (27 mi) southeast of Kangding in Luding County in Sichuan Province, China. The earthquake occurred at a depth of ~10 km. The earthquake led to the death of 93 people. Another 423 were injured and 25 remained missing. A strong earthquake (an episode that causes the ground to shake)—the 2016 Central Italy earthquake, measuring 6.2 on the moment magnitude scale, struck in the Apennine Mountains of Central Italy

---

Supplementary Information The online version contains supplementary material available at https://doi.org/10.1007/978-3-032-20231-4_8.

J. B. Chaumba, *Introductory Physical Geology*, Springer Textbooks in Earth Sciences, Geography and Environment,
https://doi.org/10.1007/978-3-032-20231-4_8

Fig. 8.1 Images from the Amatrice, Italy earthquake on August 24, 2016. **a** Rescuers at work in the village of Pescara del Tronto, Italy (credit: Gregorio Borgia/Associated Press). **b** The rubble in Amatrice, Italy, on August 24, 2016. View of Main Street, Amatrice, Italy, before (**c**) and after (**d**) the earthquake (credit: Google/Associated Press). Images from the village of Pescara del Tronto, Italy before (**e**) and after (**f**) the earthquake (credit: Google/EPA)

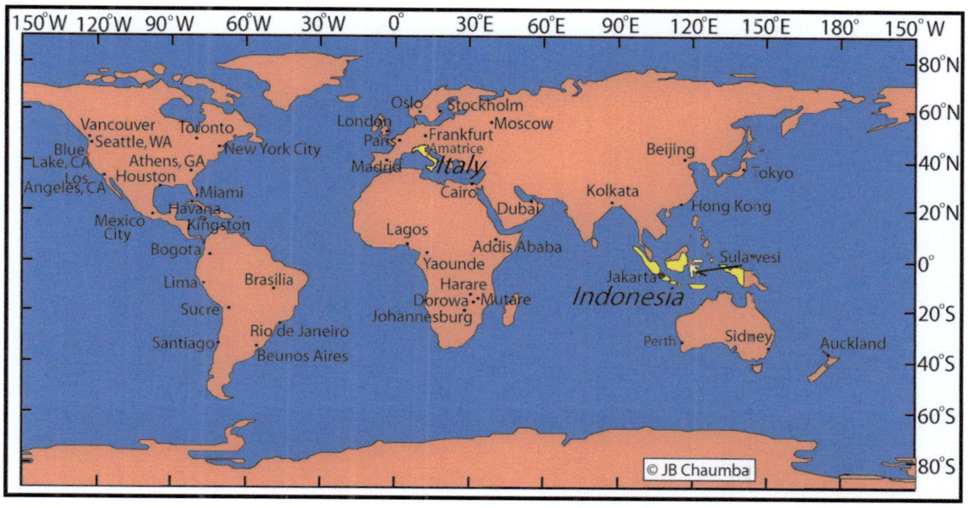

Fig. 8.2 Map showing the location of the 2016 Central Italy earthquake and the 2018 Sulawesi earthquake and tsunami in Indonesia. The location of a light earthquake that occurred close to Blue Lake, California, on February 6, 2019, is also shown

on August 24, 2016, at 01:36 (UTC) causing lots of deaths and extensive destruction of buildings (Figs. 8.1 and 8.2). The focus of this Central Italy earthquake was at a depth of ~4 km and its epicenter was located near the town of Accumoli. This earthquake led to the deaths of 299 people. The mayor of Amatrice, Italy, lamented that "half the town no longer exists." It has been suggested that stone churches and other buildings that were constructed centuries before the earthquake struck when little was known about earthquakes contributed to the high fatalities. Why did these earthquakes occur in those part of the world and not in Nancy, France, or in Toronto, Ontario, Canada, or in Masvingo, Zimbabwe, for example? In this chapter, we will learn more about the answer to this question, as well as to several other questions you may have in mind about earthquakes.

Each year, thousands of earthquakes of magnitude 5 and higher occur worldwide (Table 8.1). The number of earthquakes per year, or earthquake **frequency**, decreases as the earthquake magnitude increases (Table 8.1). Earthquakes of magnitude 8 and higher have the lowest frequency, with a single earthquake on average being recorded yearly (Table 8.1). The lower the earthquake magnitude, however, the higher the earthquake frequency. Millions of earthquakes of magnitude 3 and lower occur each year, but such earthquakes tend to occur along spreading centers mostly in oceans and are not felt by most people.

Another notable, although still a shallow but large earthquake, called the 2018 Sulawesi earthquake and tsunami, struck in the neck of the Minahasa Peninsula, Indonesia, on September 28, 2018. The epicenter of the major earthquake (magnitude 7.5 on the moment magnitude scale) was located 77 km (48 mi) away from the provincial capital of Palu in Donggala, Central Sulawesi, Indonesia (Figs. 8.2, 8.3). A series of foreshocks, the largest one of magnitude 6.1 on the moment magnitude scale, which occurred earlier in the day preceded the large earthquake. A localized tsunami caused by the earthquake struck Palu, Indonesia, sweeping away houses, buildings, and other structures near the coast (Fig. 8.3). At least 2,256 people died due to both the earthquake and the triggered tsunami, making it the deadliest earthquake worldwide in 2018. The tsunami had an estimated maximum height of 4–7 m (13–23 ft.), striking the towns of Palu, Donggala, and Mamuju, along its path. Further, liquefaction of the soil in areas in and around Palu occurred, triggering mudflows which submerged several buildings contributing to the high fatalities.

Not all earthquakes cause destruction to property and death. A light earthquake that occurred at 18:23:01 (UTC) on Wednesday, February 6, 2019, had a magnitude of 4.1 on the moment magnitude scale and its epicenter was located 20 km (12 miles) SE of Blue Lake, CA (Fig. 8.2), based on data from the USGS. This light earthquake had a hypocenter

Table 8.1 Worldwide earthquakes 2000–2021 and number of earthquakes and fatalities (data from USGS)

Magnitude	2010	2011	2012	2013	2014	2015	2016	2017	2018	2019	2020	2021
8.0 +	1	1	2	2	1	1	0	1	1	1	09	3
7–7.9	23	19	12	17	11	18	16	6	16	9		16
6–6.9	150	185	108	123	143	127	130	104	117	135	112	140
5–5.9	2209	2276	1401	1453	1574	1419	1550	1455	1674	1492	1312	2047
Estimated deaths	226,050	21,942	689	1,572	756	9624	1297	1012	4535	244	NYA	NYA

NYA—Not yet available.

Fig. 8.3 Images from the September 28, 2018, Palu, Indonesia earthquake. **a** The ruin of a mosque damaged by earthquake and tsunami is seen in Palu, Central Sulawesi. Indonesia, Friday, September 28, 2018. The earthquake struck the island of Sulawesi and triggered a 3-m (10 ft.) high tsunami that swept away houses in Palu (credit: AP Photo/Rifki). **b** The tsunami swept a ship ashore in Palu, Central Sulawesi, Indonesia, (credit: AP Photo/Rio Mario). **c** A department store building is seen heavily damaged by earthquake in Palu, Central Sulawesi, Indonesia (credit: AP Photo/Rifki). **d** People survey outside the shopping mall which was damaged following earthquakes and a tsunami in Palu, Central Sulawesi, Indonesia, Sunday, September 30, 2018. Rescuers try to reach trapped victims in collapsed buildings after hundreds of people are confirmed dead in a tsunami that hit two central Indonesian cities, sweeping away buildings with massive waves (AP Photo/Tatan Syuflana). **e** The Ponulele Bridge before the earthquake (credit: Google Earth), and the same bridge after the earthquake **f** (credit: Carl Court/Getty Images)

(focus) of 22 km (14 mi), and it resulted in neither destruction to property nor to death. Later in this chapter we will discuss the reasons why some earthquakes, even though they are relatively moderate, can result in extensive destruction to property and high fatalities whereas some great earthquakes may similarly cause neither destruction to property nor death. Again, in Indonesia, a gigantic tsunami occurred on December 26, 2004; the product of a series of huge waves that were triggered by a magnitude 9.2 underwater earthquake, which killed approximately 228,000 people mostly in Indonesia and in 12 other countries.

8.2 Fatalities from Earthquakes

Although numerous earthquakes occur on Earth each day, the good news is that the majority of them are small in magnitude and therefore result in neither loss of life nor destruction to property and infrastructure. When earthquakes occur in uninhabited areas with no infrastructure, such as in remote locations, the ground just shakes and causes damage to buildings causing loss of lives. Within the oceans, particularly at divergent spreading boundaries such as the mid-Atlantic Ridge, the bulk of small to minor earthquakes that occur here daily result in no loss of life or destruction to any property. However, whenever moderate to large earthquakes, or even minor earthquakes, occur in densely populated areas such as Los Angeles, California, they are usually accompanied by massive destruction to property and/or multiple loss of lives. Indeed, lots of lives have been lost from earthquakes throughout history. In 2004, the magnitude 9.1 earthquake that struck off the West Coast of Northern Sumatra, Indonesia (Table 8.1), led to the loss of ~298,000 people. That earthquake played a huge role in making 2004 record the highest number of fatalities on the graph shown in Fig. 8.4. The year with the second highest number of fatalities, ~226,000 people, resulted from a number of earthquakes which was topped by the 2010 Haiti earthquake of magnitude 7.0 (Fig. 8.4) on the moment magnitude scale. The 2010 Haiti earthquake alone claimed the lives of ~230,000. There are some differences, however, as we will discuss later in this chapter, which can play a role on whether multiple fatalities and destruction of property occurs.

Although large magnitude earthquakes shake the ground more than relatively lower magnitude ones, relatively lower magnitude earthquakes occur more commonly than large magnitude earthquakes (Fig. 8.4b). The higher the earthquake magnitude, it appears, the lower the frequency of the earthquake (Fig. 8.4b). It can be observed that 5–5.9 magnitude earthquakes occur much more frequently than the 6–6.9 magnitude earthquakes, but both magnitudes occur more frequently than the 7–7.9 magnitude earthquakes (Fig. 8.4b). Consequently, earthquakes with magnitudes of 8 and higher, thus, have the lowest frequencies (Fig. 8.4b).

8.3 Definition of an Earthquake and How It is Triggered

The shaking of the ground that results from rapid movement(s) of any adjacent blocks of the upper part of the Earth (lithosphere—see Chap. 6) as they slip past one another along a fault(s) is called an **earthquake**. Consider any two blocks within the Earth's crust where one is displaced relative another (Fig. 8.5a). Although a fault is represented as a line in 2-D diagrams (Fig. 8.5a), in reality faults are represented by surfaces as shown by the 3-D diagram in Fig. 8.5b. The line or surface along which the displacement occurs is termed a fault, and the point at which the displacement occurs is termed the **focus** or **hypocenter** of an earthquake (Fig. 8 5a, b). The point at the Earth's surface situated directly above the focus of an earthquake is termed the **epicenter** (Fig. 8.5a, b). When an earthquake occurs, **seismic waves** travel concentrically away from the earthquake focus (Fig. 8.5b). The epicenter represents the location where seismic waves first reach the Earth's surface on a horizontal surface. At the Earth's surface, seismic waves propagate concentrically away from the epicenter (Fig. 8.5b). The **fault trace**, or trace of a fault (Fig. 8.5b), is the intersection of a fault with the Earth's surface. A fault trace represents the line that is commonly plotted on a geologic map to represent a fault as shown in Fig. 8.5b.

8.4 Elastic Rebound Theory

As tectonic plates (see Chap. 6) slowly shift themselves, strain builds in the lithosphere. Consider an intact crustal block that has not yet been affected by deformation which later at Time 0 develops a joint from tectonic motion (Fig. 8.6). Continued stresses being applied to this block results in increasing strain or rock deformation from Time 1 to Time 2. The bending of rocks due to continued applied stress on rocks can occur on time scales of decades to hundreds of years. At Time 3, the stress applied to the rocks substantially increased to overcome the frictional resistance which led to faulting. During the springing back, or faulting, the layers of rocks slip back to their original undeformed state (Time 3 in Fig. 8.6). Thus, it is the rapid movement of one block of rock past another one on a fault (Time 3 on Fig. 8.6). The snapping back of the rock layers to their original undeformed state is synonymous with an earthquake occurrence and it is accompanied by the generation of seismic waves (earthquakes) that radiate away from the point of rupture. This process of springing back of rock layers to their undeformed state upon release of stresses built up in rocks is referred to as the elastic rebound. The process of gradual accumulation and release of stress and strain is what is termed the **elastic rebound theory** of earth-

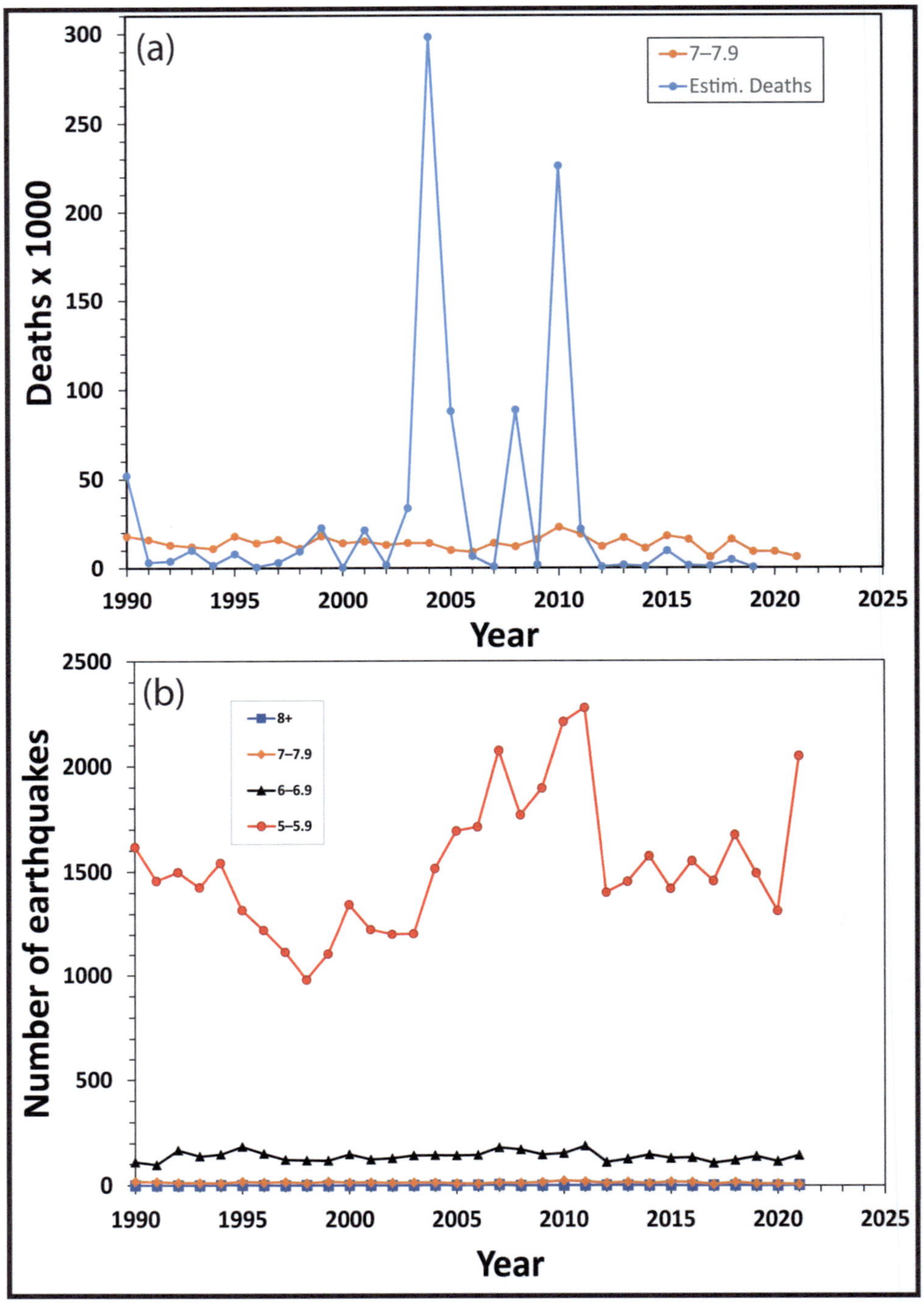

Fig. 8.4 **a** Graph showing the deaths (in thousands) from earthquakes by year from 1990 to 2021. **b** Graph showing the number of earthquakes of different magnitude ranges per year from 1990 until 2021. Data in both plots from USGS

quakes. It was proposed by the Henry F. Reid who investigated the 1906 San Andreas earthquake, and it ascribes earthquakes to be a product of sudden elastic rebound of previously stored energy in rocks.

The elastic rebound theory can also be illustrated at convergent plate boundaries. Here, convergence leads to stress being applied to two colliding lithospheric plates (Fig. 8.7). With continued convergence, the overriding plate is dragged downward by the subducting oceanic slab, storing up energy in the overriding plate. Eventually, increased stresses will lead to the overriding plate snapping back to its original undeformed state, causing earthquakes to occur in the process (Time 3 in Fig. 8.7). The process restarts (Time 4) and repeats itself. A comparable process also occurs at convergent plate boundaries (Fig. 8.8). At convergent plate boundaries, a subducting slab (Fig. 8.8 time 1) drags the overlying plate downward, causing strain on the latter (Time 2). With continued convergence, slip eventually occurs (Time 3), resulting in the occurrence of an earthquake as the overriding plate snaps back to its orig-

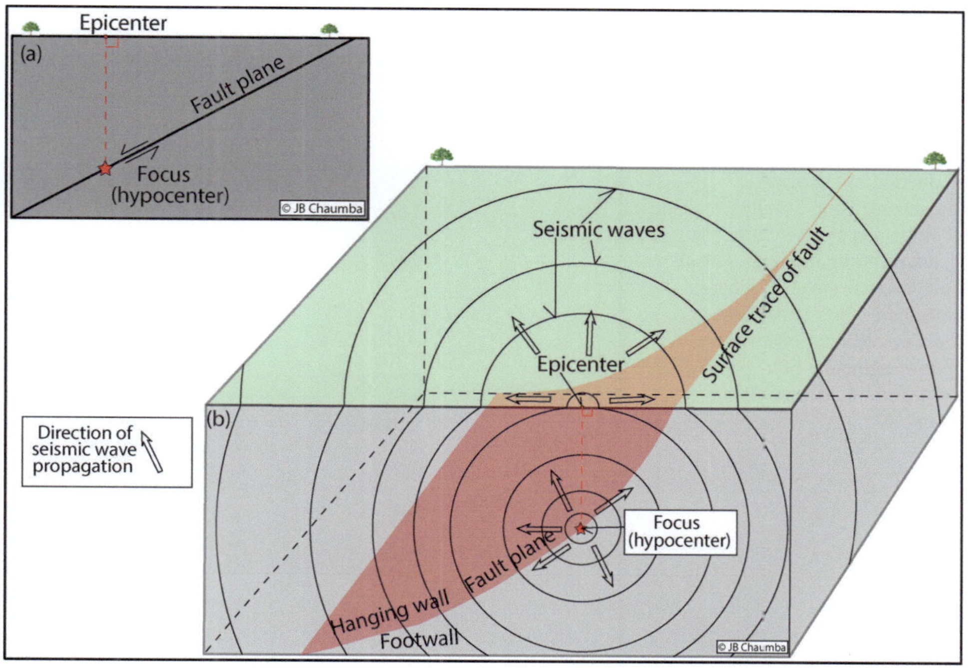

Fig. 8.5 Diagram showing the location along a fault where displacement occurs (focus or hypocenter) generating seismic waves. Seismic waves travel outward in all directions from the earthquake focus. The epicenter of an earthquake is the ground location directly above the earthquake focus

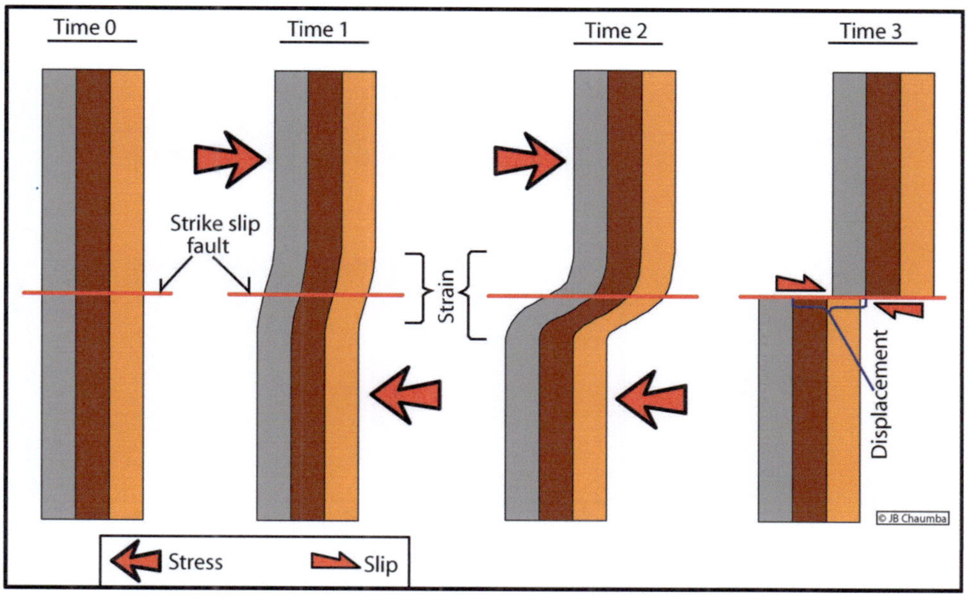

Fig. 8.6 Illustration of the elastic rebound theory. Time 0—starting position on rock layers with a pre-existing join. Time 1—blocks on opposite sides of a fault with stresses exerted in opposite directions lead to rock deformation. Time 2—increased strain results from continued stressed. Time 3—slip accompanied by and earthquake occurs resulting in displacement. The process (Time 0–Time 3) is repeated reusIting in displacements of hundreds of kilometers of tens of millions of years

inal undeformed state (Time 4). The process restarts with continued convergence.

8.5 Ring of Fire and the 20 Largest Earthquakes in the World

The majority of earthquakes (together with the majority of volcanic eruptions) tend not to be randomly distributed but they occur in specific areas, such as along plate boundaries. The major zone of earthquake and volcanic eruptions is the circum-Pacific Ring of Fire which marks the boundary between the Pacific plate and several other surrounding plates (Fig. 8.8). This zone is aptly termed the Ring of Fire as it is the most seismically and volcanically active belt on Earth. It is largely within this Ring of Fire where the bulk of the world's largest earthquakes have been recorded.

Twenty of the largest earthquakes ever recorded are located largely around the Pacific Ocean, as shown in Fig. 8.9. Ten of the largest recorded earthquakes are listed in Table 8.2. The magnitude of the 10 largest earthquakes ranges from moment magnitude scale of 9.6 (Bio-Bio, Chile) to a number of earthquakes with a moment magni-

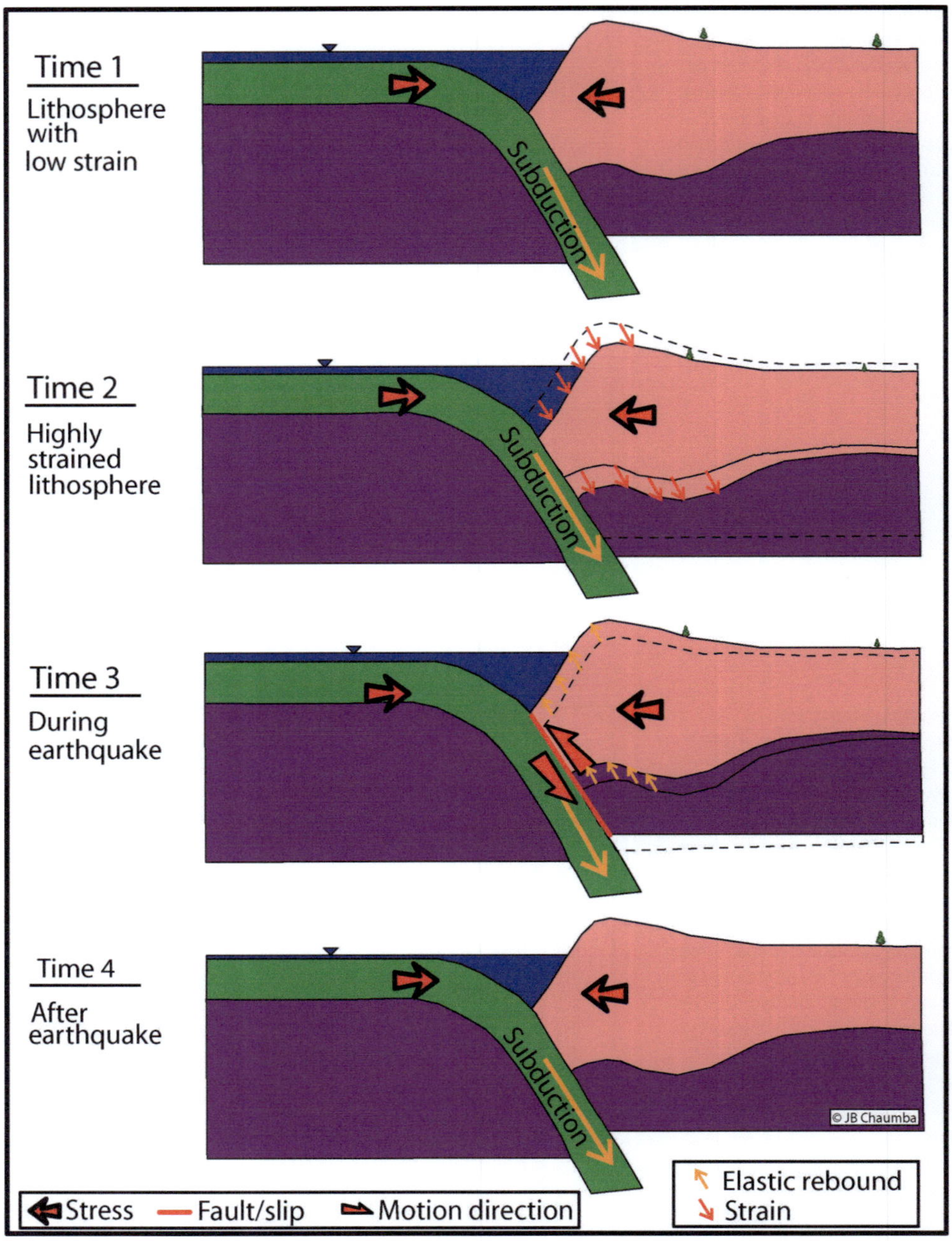

Fig. 8.7 Elastic rebound theory at a convergent plate boundary. Time 1—oceanic lithosphere and continental lithosphere collision occurs. Time 2—increased strain results from continued subduction of the oceanic lithosphere. Time 3—slip accompanied by and earthquake occurs resulting in displacement. Time 4—unstrained continental lithosphere starts to be subjected to stresses leading onset of strain on the continental lithosphere and the entire process (Time 0–Time 3)

tude scale of 8.6 (Table 8.2). Only the earthquake of moment magnitude scale 8.6 from Eastern Xizang (China)-India border is not located in the Ring of Fire (Fig. 8.9), lending support to the high seismic activity of the Ring of Fire in terms of both the frequency and high magnitude of this zone.

During an earthquake, the major rupturing produces the **mainshock**. However, prior to the mainshock, one or more smaller earthquakes called **foreshocks** tend to occur. The formation of cracks within the region where the mainshock later causes the foreshocks. After the mainshock, it is also common for one or more smaller earthquakes termed **aftershocks**; with the largest aftershock being a tenth of magnitude compared to the mainshock. Relaxation of the elastic deformation after the occurrence of the mainshock is interpreted to be responsible for aftershocks. Generally, the larger the mainshock, the larger and more numerous the aftershocks. Furthermore, the larger the mainshock, the longer the aftershock will continue.

Displacement along a fault tends to be greatest in the vicinity of where faulting originates (focus), decreasing in amount as one moves away from the focus of an earthquake. The amount displacement during the 2011 Tohoku, Japan earthquakes was as much as 30 m (100 ft). During the 1906 California earthquake, earth scientists estimate that displacement on the San Andreas Fault was up to 7 m

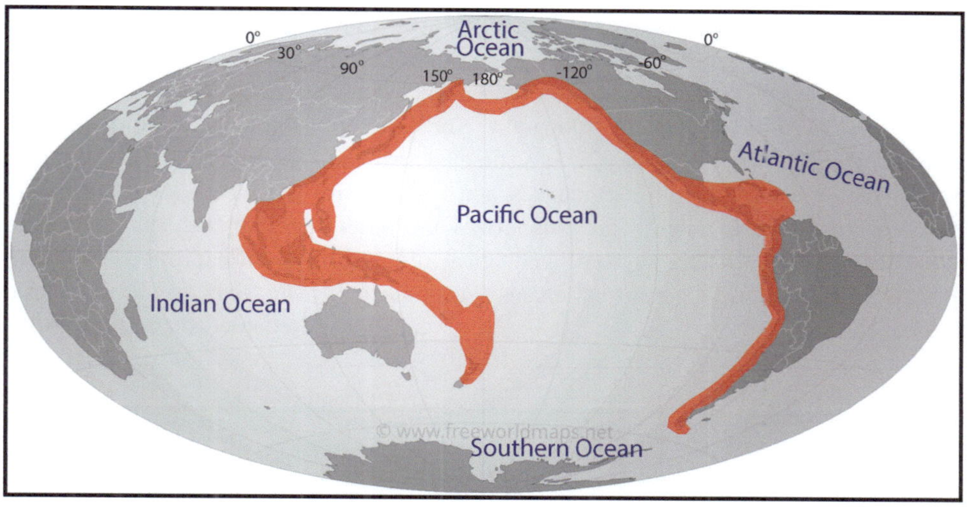

Fig. 8.8 Pacific centered world map showing the Ring of Fire. Base map obtained from https://www.freeworldmaps.net/world/pacific-centered/pacific-centered-worldmap-hd.jpg

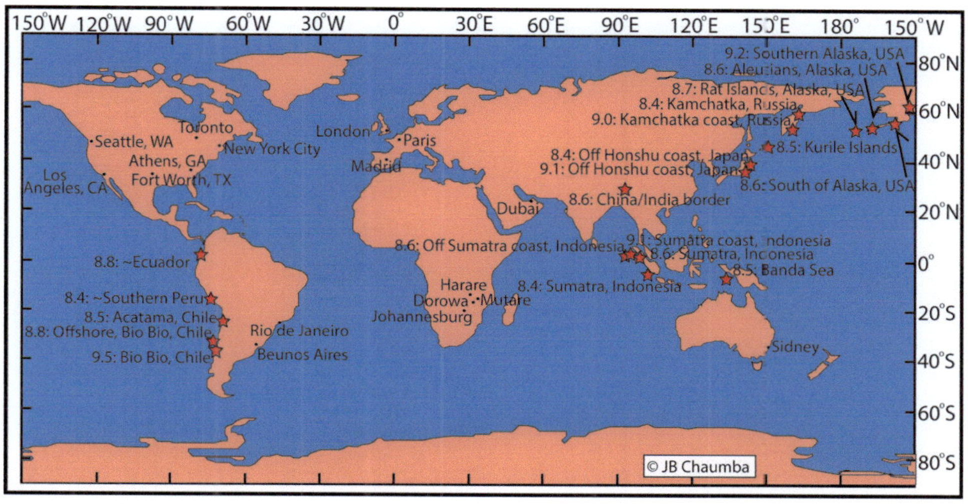

Fig. 8.9 Location map of the world's 20 largest earhquakes and their magnitudes. Data from USGS

Table 8.2 The 10 largest earthquakes in the world (Data from USGS)

	Mag	Location	Alternative name	Date (UTC)	Time (UTC)	Latitude	Longitude
1	9.5	Bio-Bio, Chile	Valdivia earthquake	1960-05-22	19:11	38.14°S	73.41°W
2	9.2	Southern Alaska	1964 Great Alaska Earthquake, Prince William Sound Earthquake, Good Friday Earthquake	1964-03-28	03:36	60.91°N	147.34°W
3	9.1	Off the West Coast of Northern Sumatra	Sumatra-Andaman Islands Earthquake, 2004 Sumatra Earthquake and Tsunami, Indian Ocean Earthquake	2004-12-26	00:58	3.30°N	95.98°E
4	9.1	Near the East Coast of Honshu, Japan	Tohoku Earthquake	2011-03-11	05:46	38.30°N	142.37°E
5	9.0	Off the East Coast of the Kamchatka Peninsula, Russia	Kamchatka, Russia	1952-11-04	16:58	52.62°N	159.78°E
6	8.8	Offshore Bio-Bio, Chile	Maule Earthquake	2010-02-27	06:34	36.12°S	72.90°W
7	8.8	Near the Coast of Ecuador	1906 Ecuador–Colombia Earthquake	1906-01-31	15:36	0.96°N	79.37°W
8	8.7	Rat Islands, Aleutian Islands, Alaska	Rat Islands Earthquake	1965-02-04	05:01	51.25°N	178.72°E
9	8.6	Eastern Xizang-India border region	Assam, Tibet	1950-08-15	14:09	28.36°N	96.45°E
10	8.6	Off the West Coast of Northern Sumatra		2012-04-11	08:39	2.33°N	93.06°E

(23 ft). Small magnitude earthquakes, however, can only have small displacements that can be a few mm (inches) in magnitude. Since the San Andreas Fault developed approximately 15–20 million years ago along the North American and Pacific plates boundary, it is estimated that the total displacement from both earthquakes and creep is at least a whopping 350 miles! This emphasizes the observation that the slow nature of some geological processes ultimately lead to very significant changes of planetary landscapes.

8.5.1 Types of Seismic Waves

The study of earthquakes and seismic waves that move through and around planet Earth is called **seismology**, and Earth scientists who study earthquakes and seismic waves are called **seismologists**. Waves of energy that are generated by the sudden breaking of rock within the Earth's crust or an explosion are called **seismic waves**. Seismic waves are, thus, the energy that travels through the Earth and are recorded on **seismographs** (Fig. 8.10). A seismograph, in

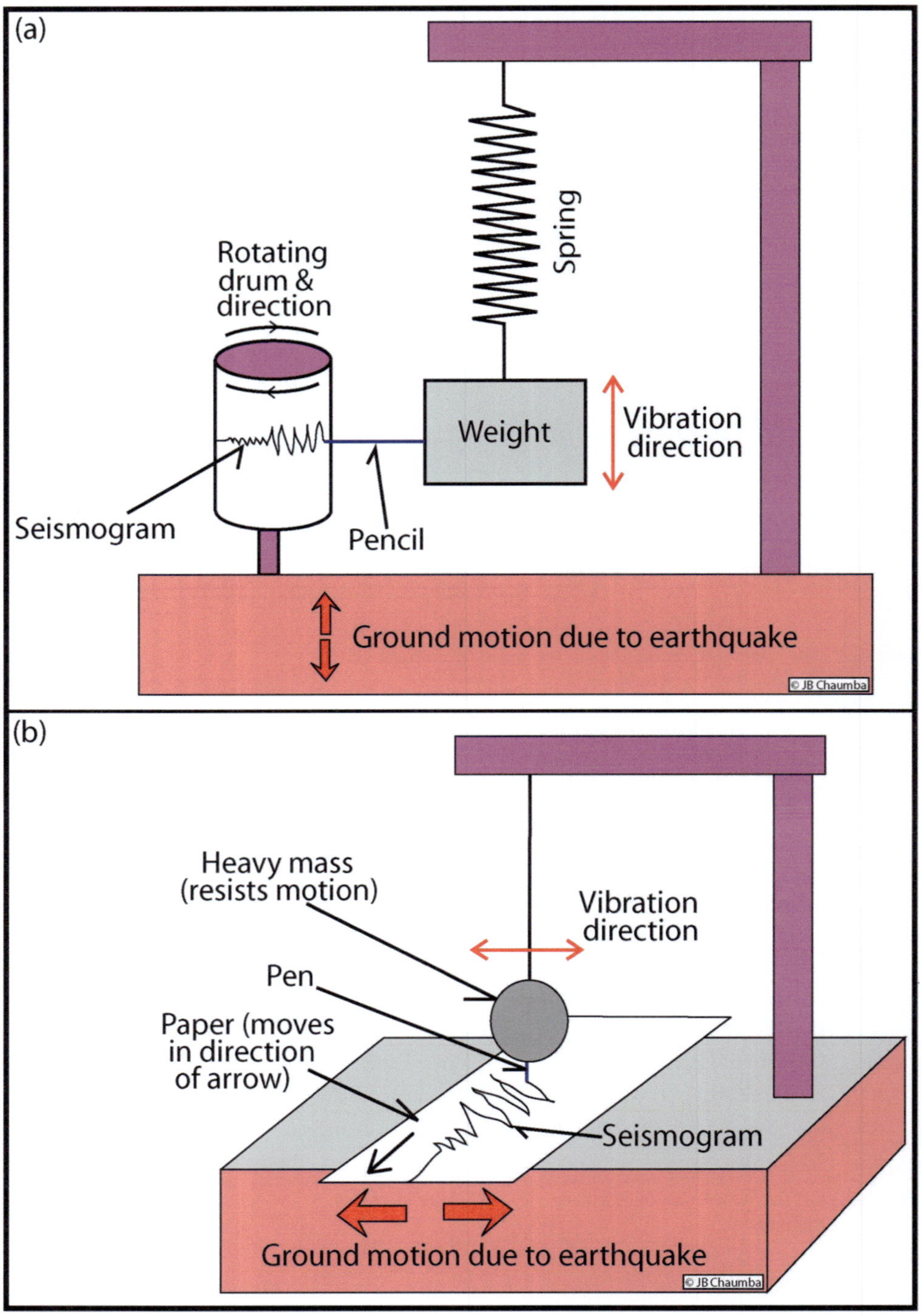

Fig. 8.10 Seismometers. **a** Up-and-down motion is recorded by a vertical-motion seismometer and **b** back-and-forth horizontal motion is recorded by a horizontal-motion seismometer

general, refers to the seismometer and its recording device as a single unit. An instrument that is used to detect and record earthquakes is referred to as a seismograph, or **seismometer**. The internal part of the seismograph is called a seismometer. A seismograph may be a pendulum or a mass mounted on a spring (Fig. 8.10). Seismometers, however, often tend to be used synonymously with seismographs.

A seismograph is an instrument that is used to record the motion of the ground during an earthquake. Seismographs are installed in the ground where they monitor ground shaking. Numerous seismographs are now installed throughout the world, and they operate as part of a seismographic network. Seismographs are securely mounted onto the surface of the earth so that when the earth shakes, the entire unit shakes, with the exception of the mass on the spring. The mass has inertia, and it remains in the same place during shaking of the ground (Fig. 8.10a, b). The recording device on the mass records the relative motion between itself and the rest of the instrument such that as the seismograph shakes under the mass, ground motion is recorded. The mechanisms of seismographs are no longer manual, but they work by measuring electronic changes that are produced by the motion of the ground with respect to the mass. A seismograph generally consists of a mass attached to a fixed base such that during an earthquake, the base moves whereas the mass does not. The motion of the base with respect to the mass is then transformed into an electrical voltage which is recorded on paper (Fig. 8.10). The record made on paper is proportional to the motion of the seismometer mass relative to the Earth, and it is typically converted, mathematically, to a record of the absolute motion of the ground.

The recording of the ground shaking at any specific location of a seimograph is called a **seismogram**. The horizontal axis on a seismogram represents time (Fig. 8.11), which is measured in seconds, whereas the vertical axis represents ground displacement and it is usually measured in millimeters. A straight line (Fig. 8.11), with the exception of small wiggles that are caused by local disturbances or "noise" and the time markers, is recorded when there is no earthquake. However, seismograms are now digital, and there are no more recordings on paper.

When an earthquake occurs, seismic waves produced by the slip travel in all directions away from the point of rupture. There are several different kinds of seismic waves, and they all move in different ways. These waves can be classified into two categories. Two main types of waves are distinguished on the basis of where they move. These are **body waves** and **surface waves**. Body waves travel through the inner layers of the Earth, whereas surface waves can only travel along the surface of the Earth in a manner akin to ripples on water.

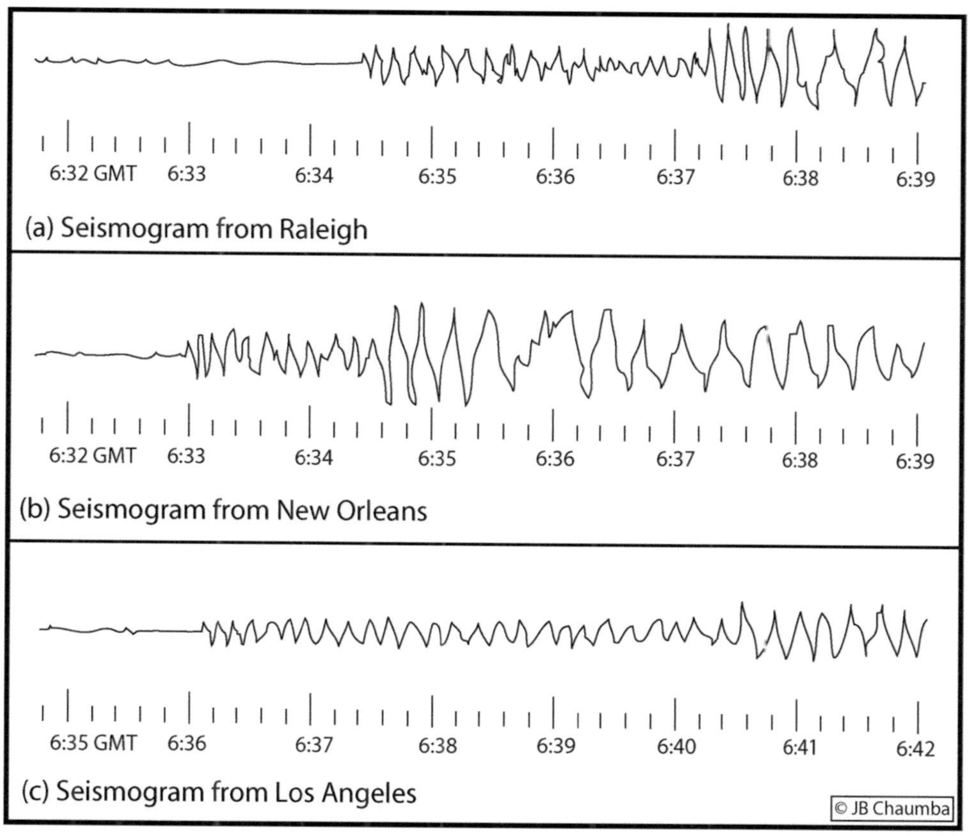

Fig. 8.11 Examples of hyothetical seismograms for seismographs located in Raleigh, North Carolina (**a**), New Orleans, Louisiana (**b**), and Los Angeles, California (**c**)

8.5.2 Body Waves

Body waves emitted by an earthquake travel through the interior of the Earth and arrive at a recording station before surface waves. Body waves are of a higher frequency than surface waves. Body waves are subdivided into **primary waves** (or **P-waves**) and **secondary waves** (or S-waves or shear waves). Primary waves are the compressional waves that cause the back-and-forth vibrations parallel to the direction of wave propagation. These waves are the fastest kind of seismic waves and, consequently, they are the first to reach a seismic station. Primary waves are capable of moving through both solid rocks and fluids (such as the liquid outer core layers of the Earth). Primary waves push and pull the rock that they move through in a manner akin to sound waves pushing and pull the air (Fig. 8.12a). The pushing and pulling of primary waves are responsible for these waves being also referred to as **compressional waves**. When particles are subjected to a P-wave, they move in the same direction that the wave is propagating in, or the direction that the energy is traveling in.

Secondary waves (or **S-waves**) are shear waves, and they are the second types of body waves. Secondary waves cause up-and-down vibrations perpendicular to the direction of wave propagation, and they are the second wave one feels soon after an earthquake. Secondary waves are capable of moving only through solid rock, not through any fluids. The incapability of surface waves to move through fluids helped seismologists to the interpretation that the outer core of the Earth's is liquid. Secondary waves are slower than primary waves. In addition, surface waves make rock particles vibrate up and down or side to side, perpendicular to the direction in which the wave is propagating. Secondary waves travel at ~60% of the velocity of P-waves.

On a secondary wave form, a crest is the maximum amount of positive (or upward) displacement from the rest position, whereas a trough is the point on the medium that exhibits the maximum amount of negative (or downward) displacement from the rest position (Fig. 8.12b). The amplitude is the maximum amount of displacement of a particle on a medium from its rest position. The amplitude is, thus, the distance from rest to crest. The amplitude can also be thought of as the measured from the rest position to the trough position. The wavelength is the length of one complete wave cycle.

8.5.3 Surface Waves

Surface waves are much lower frequency waves that travel only through the Earth's crust. Surface waves travel at the slowest velocities in comparison to body waves (P- and S-waves). Surface waves, thus, reach a recording station after body waves and they are easily distinguished on a seismogram. However, surface waves are responsible for the bulk of the damage and destruction that are associated with earthquakes. This damage and the strength of the surface waves are, however, reduced in earthquakes that occur deeper in the Earth's crust.

Two types of surface waves exist. **Rayleigh waves**, which are named for John William Strutt, Lord Rayleigh, who mathematically predicted their existence. Rayleigh waves propagate by rolling along the ground just like waves roll across an ocean or a lake. Rayleigh waves (Fig. 8.13) move the ground up and down and side to side in the same direction that the wave is propagating. Rayleigh waves are responsible for most of the shaking that is felt from an earthquake. Shaking of the ground caused by Rayleigh waves tends to be much larger than that caused by other waves.

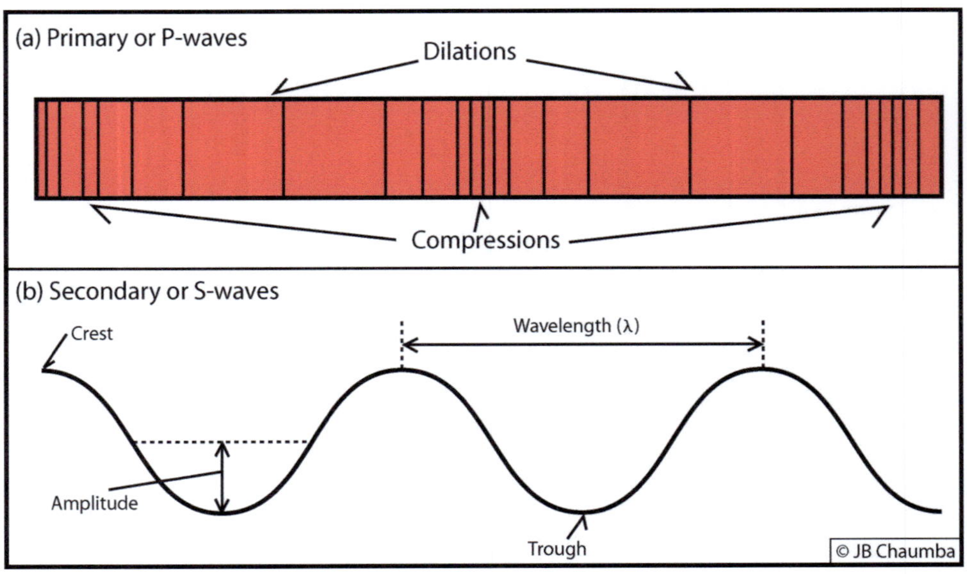

Fig. 8.12 **A** Illustration of primary wave's expansions (dilations) and contractions (compressions) produced propagation of P-waves. **b** Illustration of crest, trough, wavelength, and amplitude of secondary waves

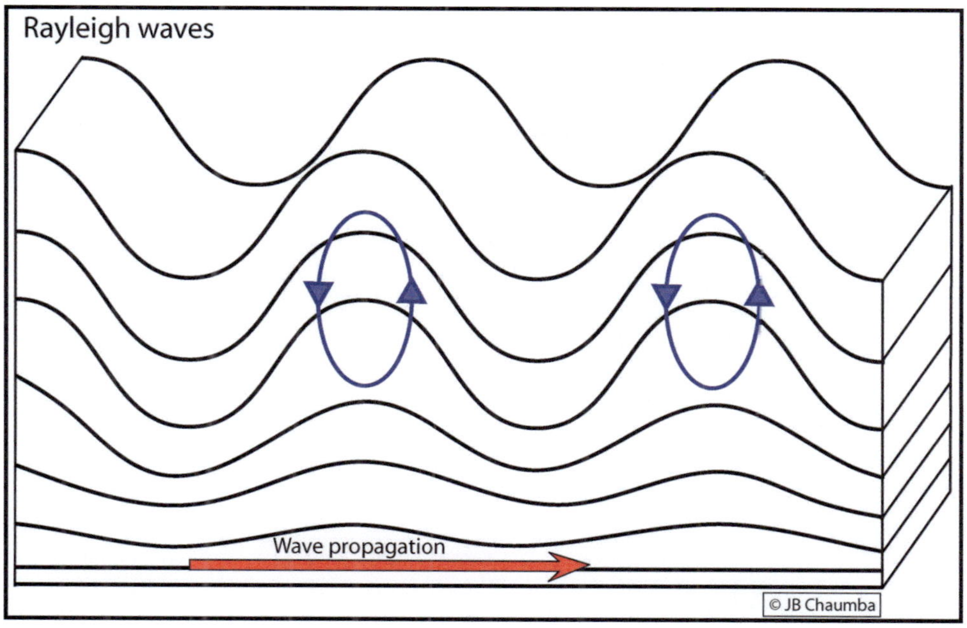

Fig. 8.13 Illustration of a Rayleigh wave showing how particles in the crust follow circular paths as the wave propagates from left to right. With increasing depth, surface waves die out

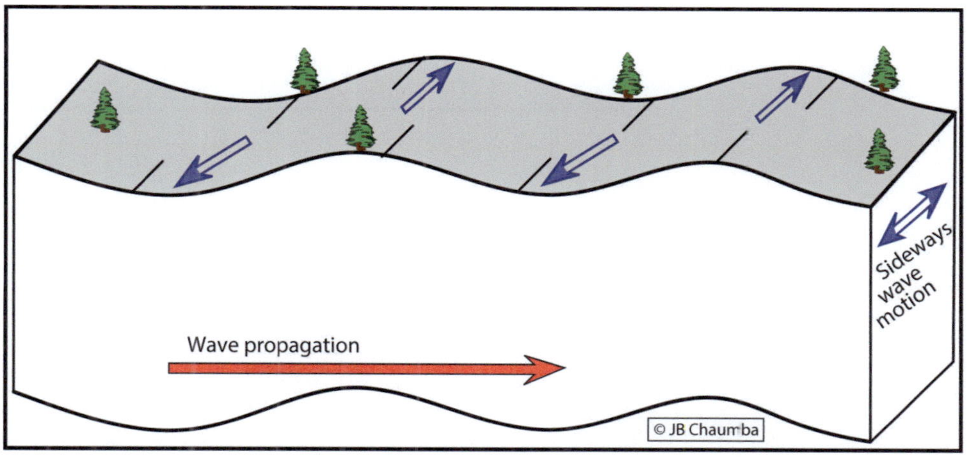

Fig. 8.14 Illustration of a Love wave showing how particles on the Earth's upper crust move back and forth like a writhing snake as the wave propagates from left to right. With increrasing depth, surface waves die out

Love waves, the other type of surface waves, named after A.E.H. Love, a British mathematician who worked out the mathematical model for these types of waves, are the fastest surface waves. Love waves move the ground from side to side (Fig. 8.14) and they are confined to the surface of the Earth's crust. Love waves produce entirely horizontal motion.

8.6 Locating the Epicenter of an Earthquake

If an earthquake occurs at any point in the Earth's crust such as a location that is shown in Fig. 15, seismic waves travel in all directions away from the focus of the earthquake. Suppose we have three seismometers at three stations (A, B, and C; Fig. 8.15) which record the arrival times of P, S, and surface waves. To locate the epicenter of an earthquake, one first determines the difference between arrival times of the P- and S-waves recorded by a seismograph, or the time delay. Examination of any seismogram which shows the signals that are generated by the P-wave, S-wave, and surface waves (Figs. 8.15, 8.16). From the signals at each seismometer, one can calculate only the distance from the epicenter to that seismogram from the time delay, but not the direction from the epicenter to the seismometer station.

To determine the epicenter of the earthquake, one needs first to calculate the distance from the epicenter to three seismometer stations. A circle, at the scale of the map, with a radius corresponding to the distance of that seismoeter station to the epicenter is drawn on a map from each of the three seismometer stations (Fig. 8.17). All the three circles will intersect at one point, which denotes the epicenter of the earthquake (Fig. 8.17).

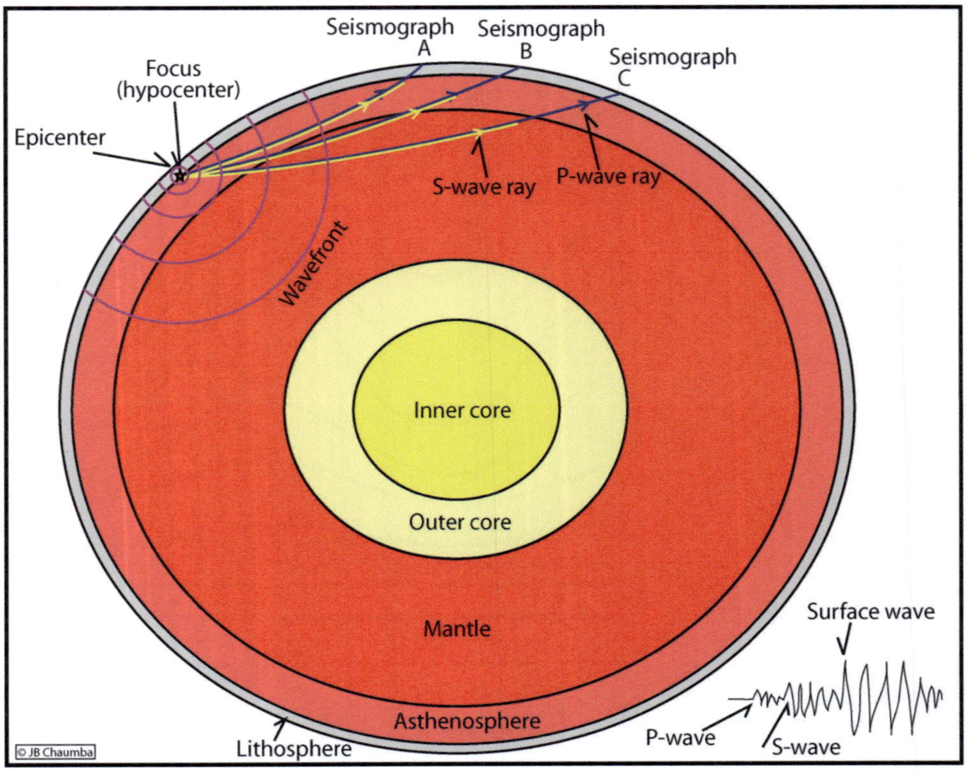

Fig. 8.15 Diagram showing Earth's different layers with different types of seismic waves traveling at different velocities from the focus of an earthquake. Seismographs at stations A, B, and C each record seismograms with different signals generated by different types of seismic waves

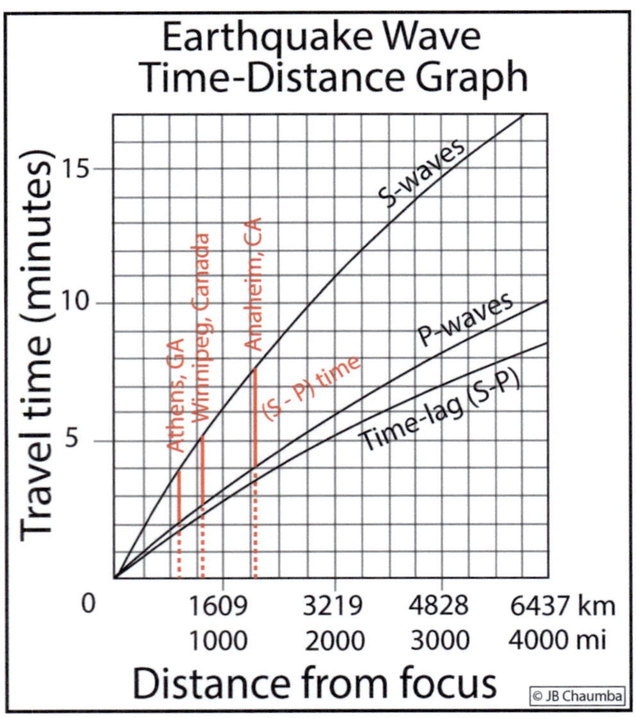

Fig. 8.16 Plot of travel times of secondary and primary waves versus distance from the focus of an earthquake. The time lag (arrival time of secondary wave minus that of primary wave) is also plotted

8.7 Determining the Size of Earthquake

A number of different ways have been devised to determine different aspects of an earthquake. Earthquake intensity is a measure of ground shaking and damage caused by an earthquake at a locality; thus, earthquake intensity values change from location to location. As body waves traveling from the focus of an earthquake reach the Earth's surface, ground shaking occurs. Additionally, ground shaking occurs as surface waves move along the Earth's surface. Determining the amount of ground shaking is of critical importance to Earth scientists in determining how "big or small" an earthquake is. Giuseppe Mercalli, an Italian scientist, assessed both the damage from an earthquake and perceptions of people of the ground shaking in the region affected by the earthquake. The **modified Mercalli intensity scale**, named after him, is shown in Table 8.3. Roman numerals are used to denote earthquake intensities with the highest value of XII coinciding with the epicenter of an earthquake, and progressively smaller values being assigned to zones further away from the epicenter. Since energy is lost by earthquake waves as they travel away from the epicenter, modified Mercalli intensity values progressively decrease with increasing distance from the epicenter (Fig. 8.18). An

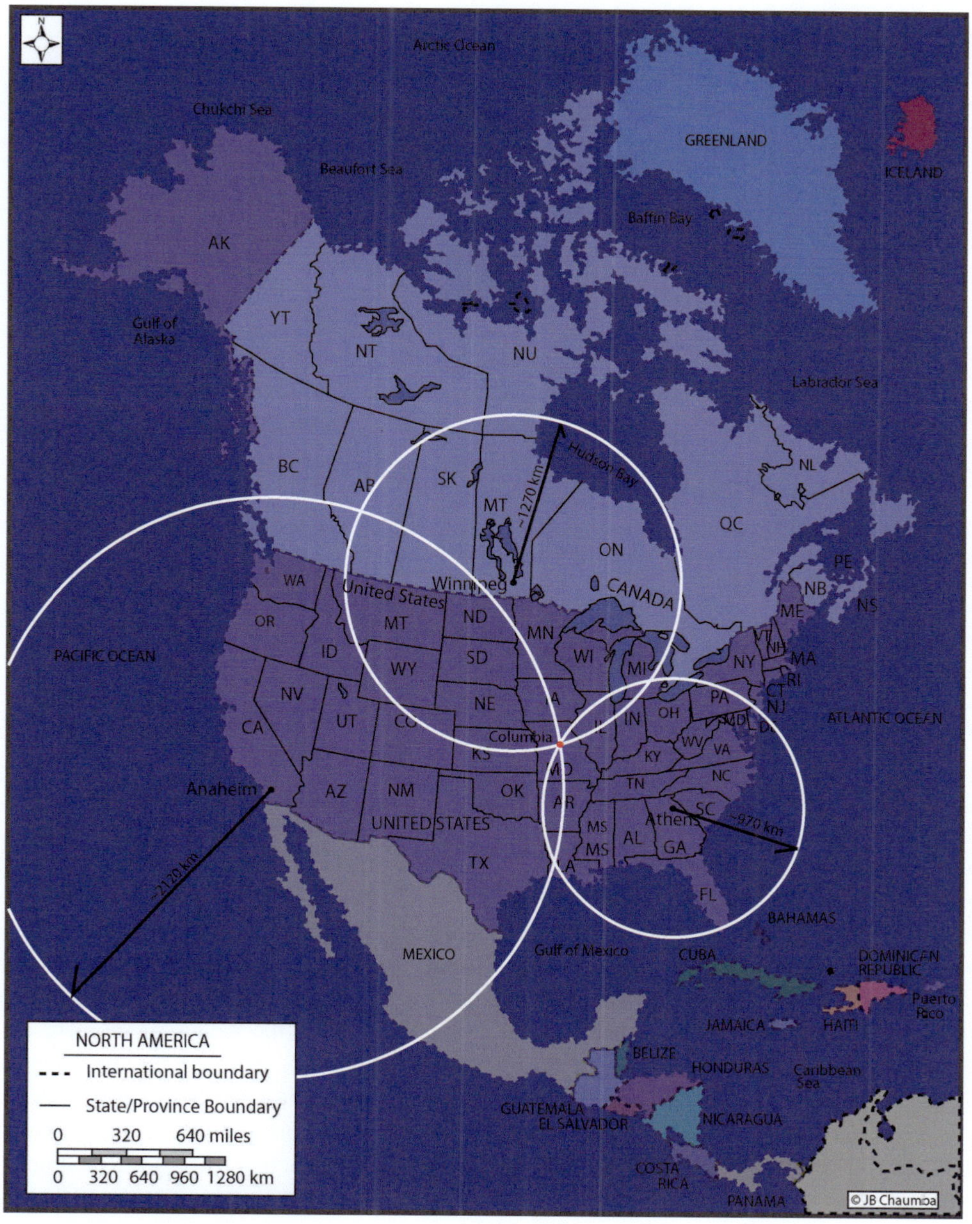

Fig. 8.17 Diagram illustrating triangulation that is used to locate the epicenter of an earthquake. For an earthquake measured in Athens, Georgia, Winnipwg, Canada, and Anaheim, California at the respective distaces shown, its hypothetical epicenter would be in Columbia, Missouri. Columbia, Missouri marks the intersection of three circles drawn from the three recording stations in the three cities at the scale of the map

example of a modified Mercalli intensity scale map of the 2016 Armatrice, Italy earthquake is shown in Fig. 8.19.

The magnitude of an earthquake measures its size or size of the earthquake source. The earthquake magnitude records the same number irrespective of where the seismometer is located or what the ground shaking feels like. The **Richter scale**, named after its inventor Charles F. Richter in 1935, measures the largest amplitude (wiggle) on the recording and is not accurate particularly for larger earthquakes. The Richter scale is logarithmic one in that an increase of one unit magnitude represents an increase of ten times in the maximum amplitude of ground motion. An earthquake of magnitude 5, thus, results in ground motion that is 10 times greater than a magnitude 4 earthquake. An earthquake of magnitude 5 shakes the ground 1000 times greater than a magnitude 2 earthquake. A chart that illustrates how to adjust the distance of a seismometer station from an epicenter is shown in Fig. 8.20. The Richter scale is no longer used, particularly for earthquakes larger than about 5.5. Earthquake magnitudes are now routinely measured by the **moment magnitude scale**, which was defined by Thomas C. Hanks and Hiroo Kanamori in 1979. The moment magnitude scale measures the size of an earthquake based on how

Table 8.3 Modified Mercalli intensity scale

Shaking	Modified Mercalli intensity	Destructiveness (extent of damage and shaking)	Richter scale
Not felt	I	Not felt except by only by very few people under favorable conditions but detected by seismometers	1.0–3.0
Weak	II	Felt by people at rest especially those on upper levels of buildings	3.0–3.9
Weak	III	Easily felt by people inside buildings, especially those in upper levels of buildings, standing vehicles can rock slightly, vibrations comparable to those of passing big trucks	
Light	IV	Felt by people indoors and some outdoors, walls make cracking sounds and doors, dishes and windows disturbed. Stationary vehicles rocked noticeably	4.0–4.9
Moderate	V	Felt by almost everyone, unstable objects are overturned and windows are broken	
Strong	VI	Felt by everyone but damage is negligible. Some heavy furniture can move and some plaster falls	5.0–5.9
Very strong	VII	Poorly built and designed building suffer heavy damage, but in strong buildings of good design damage is negligible	
Severe	VIII	Ordinary buildings damaged considerably with partial collapse but slight damage in specially designed structures. Walls, chimneys, and monuments fall	6.0–6.9
Violent	IX	Specially designed structures suffer extensive damage, buildings shifted off their foundations	
Extreme	X	Frame structures and most masonry with foundations. Rails bent. Strongly built wooden structures are destroyed	7.0 and higher
Extreme	XI	Almost all masonry structures are destroyed. Bridges are destroyed and rails are greatly bent	
Extreme	XII	Total destruction of all buildings and bridges. Objects are thrown into the air and lines of sight and level are distorted	

much energy is released. The moment magnitude scale relates to the distance of movement along a fault or fracture, or slip amount, and the area of the fault. A seismographic network now records earthquakes whereby each seismic station in the network measures ground motion at that site.

Earthquakes are also classified in categories, depending on their magnitude, which range from great to minor, as shown in Table 8.4. The larger the magnitude, the lower the frequency of the earthquake (Table 8.5). Earthquakes of magnitude 8 and greater are classified as great earthquakes and earthquakes of magnitude 7.0–7.9 are classified as major earthquakes on the moment magnitude scale (Table 8.4). Minor earthquakes are those with magnitudes of 3.0–3.9 on the moment magnitude scale (Table 8.4). Close to a million earthquakes of magnitude 2.5 or less occur each year (Table 8.5); however, these earthquakes are not felt by human beings as the majority of them occur along mid-ocean ridges and other divergent boundaries.

8.8 Energy Released

An alternative way that the size of an earthquake can be measured is to determine how much energy they released. The amount of energy that an earthquake radiates is a measure of the potential for damage to structures made constructed by humans. Since an earthquake releases energy at a variety of frequencies, seismologists include all frequencies of ground shaking for the earthquake in order to determine an accurate value of energy released. Whereas each whole number increase in magnitude represents a tenfold increase in the measured amplitude, each whole number represents 32 times increase in energy release. The energy release can commonly be converted into another type of magnitude called the Energy Magnitude (M_e). It must be pointed out that the values of moment magnitude and the energy magnitude are not the same since the energy magnitude and moment magnitude measure two different properties of the earthquake (Fig. 8.21). By converting the moment magnitude to energy using the equation $\log E = 5.24 + 1.44\ M$, where M is the energy magnitude, one can roughly estimate the energy release.

8.9 Prediction of Earthquakes

Worldwide maps of earthquake zones have been assembled (Fig. 8.22) as well as regional and national maps that show seismic hazards. Such maps are designed to help the world, and some nations, particularly those with high risks of being affected by earthquakes, prepare for future earthquakes and put plans in place to mitigate fatalities and potential damage from earthquakes.

In the Americas, the major earthquake zones stretch from Alaska's southern and western coastline extending Alaska to the southern tip of Chile's western coastline in South America (Fig. 8.22). In 1964, one of the most powerful earthquakes in modern history, measuring 9.2

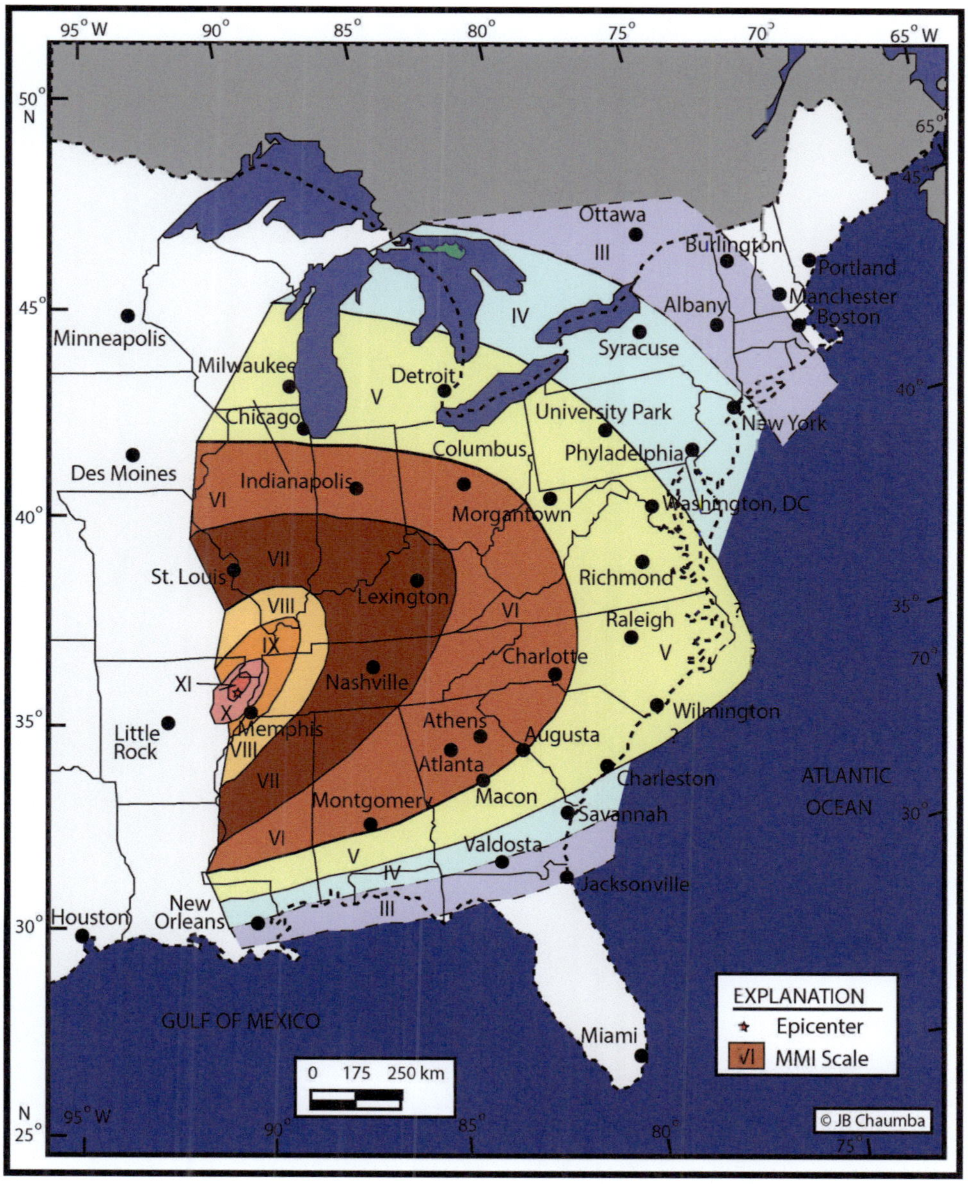

Fig. 8.18 Modified Mercalli intensity scale for the New Madrid, Missouri, USA 1812 earthquake (after U.S. Geological Survey). Only the eastern half-segment is shown

on the moment magnitude scale, struck Alaska's Prince William Sound, through the San Andreas and its associated fault system and the California Peninsula, the Pacific Ocean's coastline of both Central and South America, to the southern tip of Chile (Fig. 8.22). Major earthquakes such as the 1964 Anchorage, with a magnitude of 9.2, and the magnitude 7.7 of San Francisco 1906 earthquake also affected the seismically active zone.

In the USA, besides the west coast and its high-frequency earthquake, the other high-frequency earthquake zones occur in the New Madrid area in southeast Missouri and adjacent states (Fig. 8.23). Additionally, minor high-frequency earthquake zones occur in Charleston, South Carolina and southeastern Tennessee (Fig. 8.23). Seismic hazards are high on the west coast along the Andreas fault and the southern coast of Alaska. In Hawaii, high seismic hazards are largely confined to the Big Island (Hawai'i) due to volcanism that is associated with the hot spot situated beneath this island. A close-up of the seismic hazard map for California is shown in Fig. 8.24. The San Andreas Fault is a large strike-slip fault, also called a transform fault, that separates the North American and Pacific Plates which can generate destructive earthquakes.

8.10 Earthquake Depths and High-Frequency Earthquake Zones

The majority of earthquakes on Earth are characterized by foci that lie within a few tens of kilometers of its surface. Based on the foci depth, earthquakes with less than

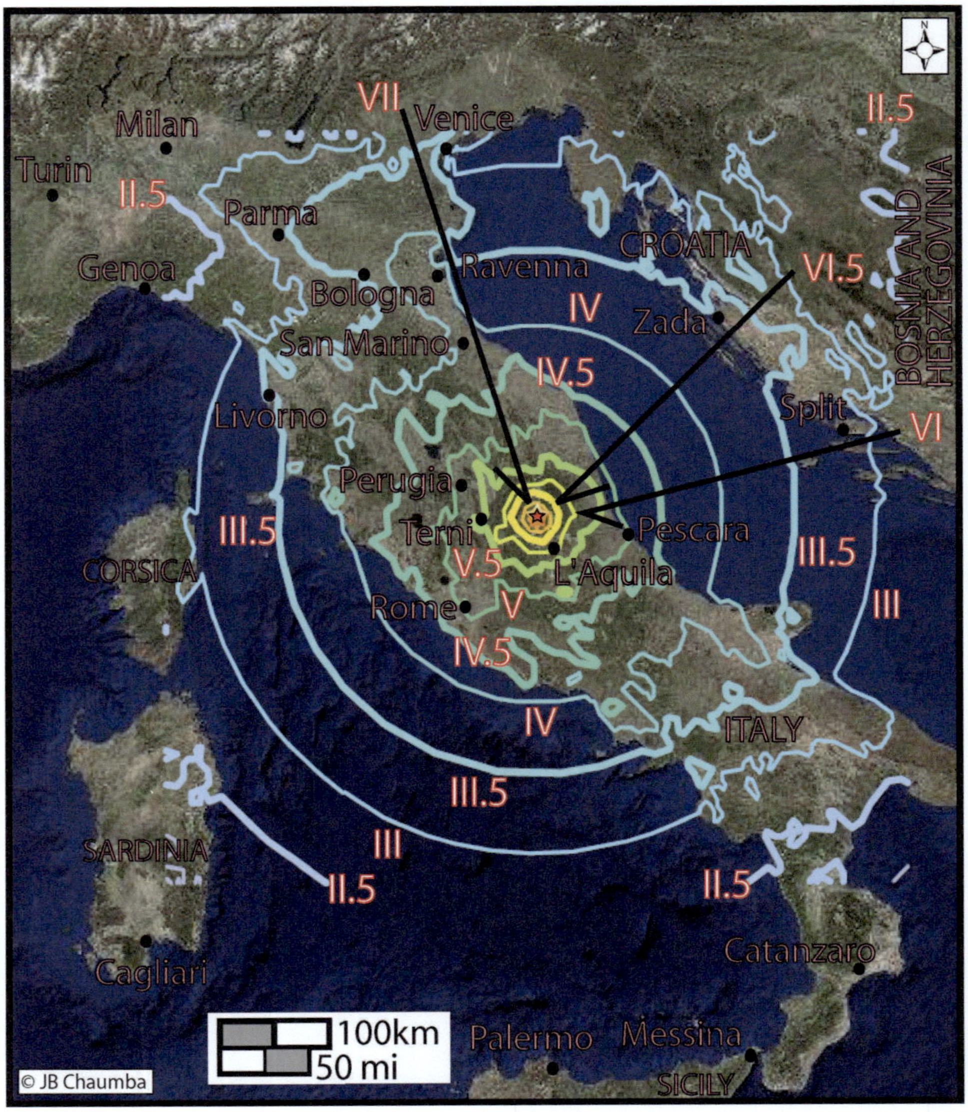

Fig. 8.19 Modified Mercalli intensity scale map of the 2016 Armatrice, Italy earthquake (credit: USGS)

70 km depth are classified as *shallow-focus* earthquakes (Fig. 8.21). Earthquakes that are characterized by foci with depths of 70–300 km are classified as *intermediate-focus* earthquakes. The third and final classification is for those whose foci at depth of > 300–700 km which are termed *deep-focus* earthquakes (Fig. 8.21). In terms of the frequency of earthquakes, shallow-focus earthquakes have the highest frequencies. Approximately two-thirds (75%) of the total energy released from earthquakes on Earth is contributed by shallow-focus earthquakes in comparison to deeper ones. Scientists have observed that intermediate and deep-focus earthquakes are rarely destructive and that the shallower an earthquake is, the more damage it produces at the surface. Below a depth of ~700 km from the Earth's surface, there have been no earthquakes observed to occur. The reason for the lack of earthquakes at greater depths is that the rocks at those depths are very hot and are subjected to high pressures such they undergo plastic deformation of flowage rather than brittle deformation of faulting and breaking.

The distribution of earthquake foci on our planet is not random. When earthquake foci are plotted on a map of our planet (Fig. 8.25), it can be observed that they are largely confined to specific areas of our planet. These "zones" where earthquakes are concentrated are termed **seismic zones** (from *seismos*, the Greek word for "shaking"). Large portions of our planet are relatively free of earthquakes, particularly the large earthquakes. Small earthquakes, however, have foci which also plot in some parts of our planet (Fig. 8.25).

8.11 Earthquake Prediction

Is it possible for Earth Scientists to accurately predict when and where a big earthquake will strike? Let us first

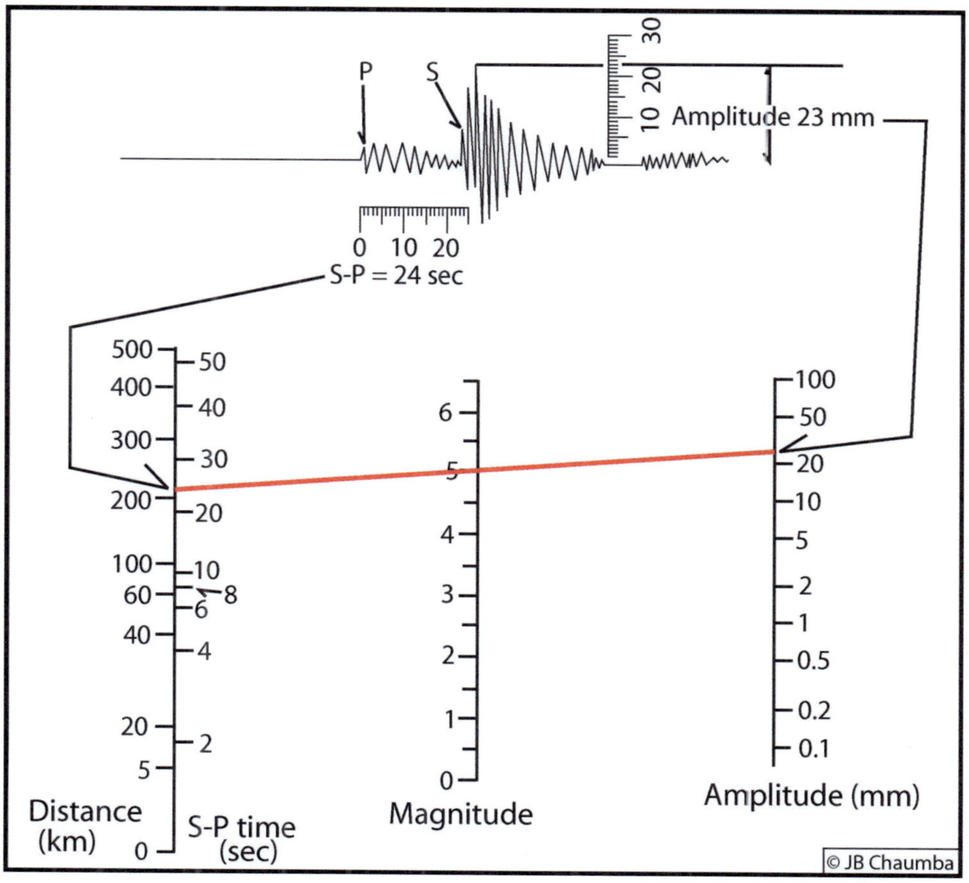

Fig. 8.20 Calculating earthquake magnitude (after Richter's (1958) *Elementary Seismology* book). First measure amplitude of largest wave (23 mm in this case), then calculate the difference between arrival time of S-wave and P-wave (S–P) (24 s here) to determine the distance of to the epicenter. The next stage involves drawing a line from the distance (km)/(S-P sec) column to the point on amplitude (23 mm) column. The Richter magnitude can then be read off the magntitude (middle) column (magnitude 5 in this case)

Table 8.4 Earthquake magnitude classes

Magnitude	Intensity at epicenter	Class
≥ 8	X–XII	Great
7.0–7.9	IX–X	Major
6.0–6.9	VII–VIII	Strong
5.0–5.9	VI–VII	Moderate
4.0–4.9	IV–V	Light
3.0–3.9	≤ III	Minor

define the **recurrence interval**, which is the average time between consecutive earthquakes of a certain magnitude in one region. Although earthquakes are known to happen periodically, the precise time between any successive earthquakes does not equate to the exact time between successive earthquakes. On the scale of decades or centuries, Earth scientists can make some long-term predictions. The likelihood that an earthquake will occur in a given year is called the annual probability and it is calculated by.

$$\text{Annual probability} = 1/\text{recurrence interval}$$

As an example, if a magnitude 6 earthquake on the San Andreas fault, for example, is 200 years, then its annual probability = 1/200, or 0.5%. Thus, there exists a 0.5 in 100 chance of a magnitude 6 earthquake occurring on the San Andreas Fault in any given year.

Thus, the simple answer to the first question of predicting the time scale of years or hours (or short-term prediction)

Table 8.5 Earthquake magnitude scale

Magnitude	Estimated number each year	Earthquake effects
8.0 or greater	One every 5 to 10 years	**Great earthquake**. Has potential to totally destroy communities near the epicenter
7.0 to 7.9	20	**Major earthquake**. Serious damage
6.1 to 6.9	100	May cause a lot of damage in very populated areas
5.5 to 6.0	500	Slight damage to buildings and other structures
2.5 to 5.4	30,000	Often felt, but only causes minor damage
2.5 or less	900,000	Usually not felt, but can be recorded by seismograph

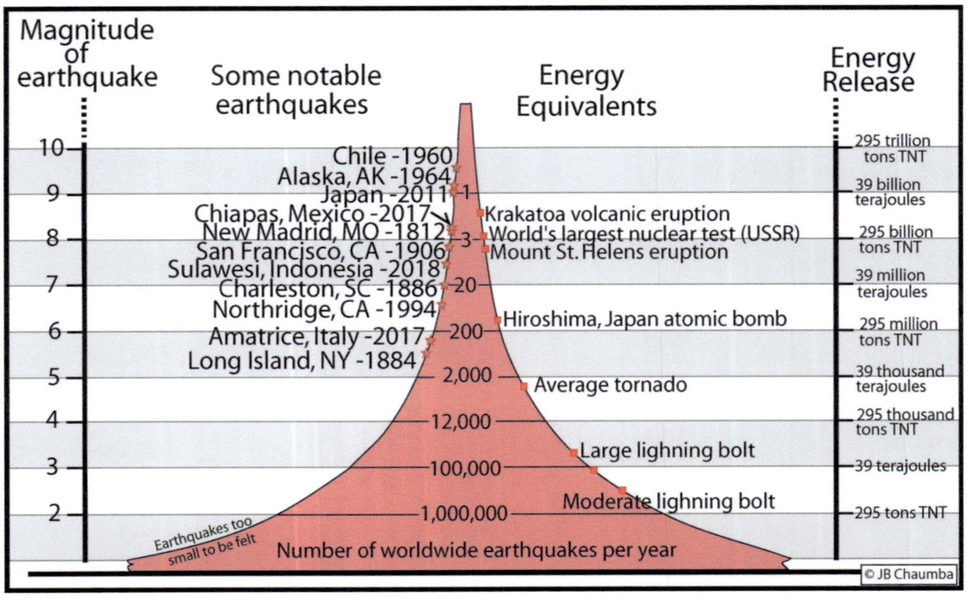

Fig. 8.21 Diagram showing the magnitudes of some earthquake and energy release, and their comparison with other natural and man-made events

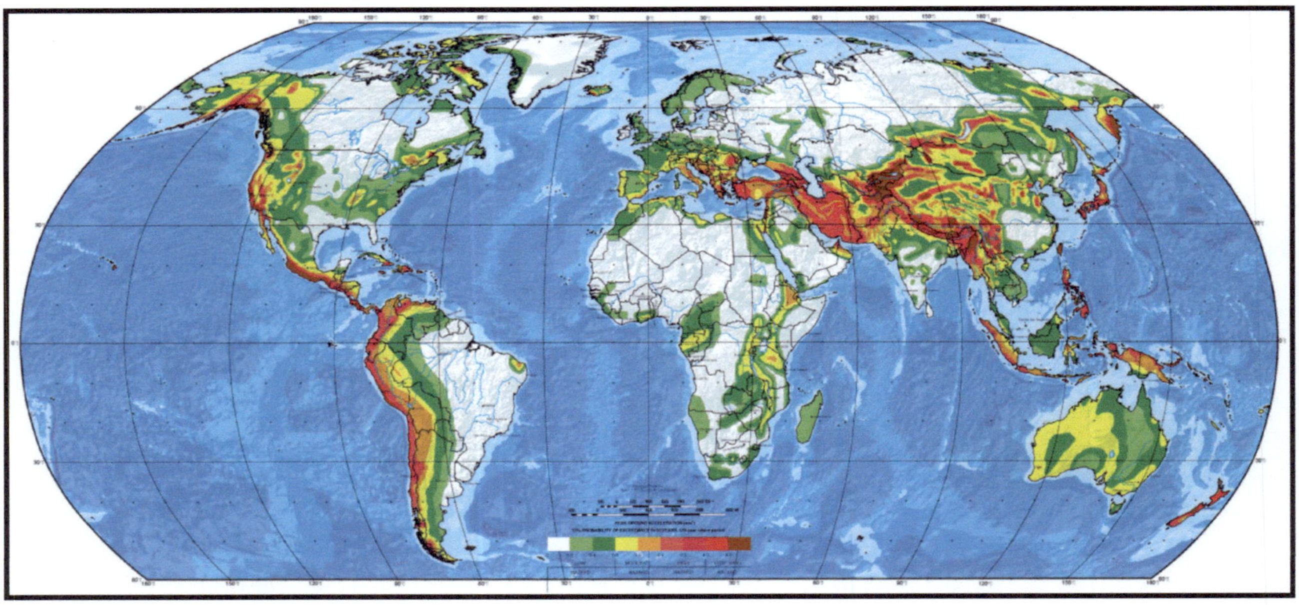

Fig. 8.22 Map showing global seismic hazard map. Orange, reddish, and brown colors denote areas with the highest earthquake hazards (after European Space Agency)

an earthquake will occur is not yet! Since strain builds up in rocks that are composed of different kinds of minerals with different responses to applied tectonic stresses, we cannot yet tell when slip is likely to occur on any fault.

The answer to the "where?" a big earthquake is likely to occur can be partly answered in zones that have not experienced any earthquakes in hundreds of years, yet they are part of plate boundaries or faults where relatively high rates of movements occur. Such fault zones where slip has not recently occurred, called seismic gaps, can be delineated on maps and are areas that are potentially susceptible to experiencing the next big earthquake. However, given the complexities of different proportions of minerals with different yield strengths that compose rocks, it is not an easy task to precisely pinpoint where exactly, and when, the next big earthquake will strike.

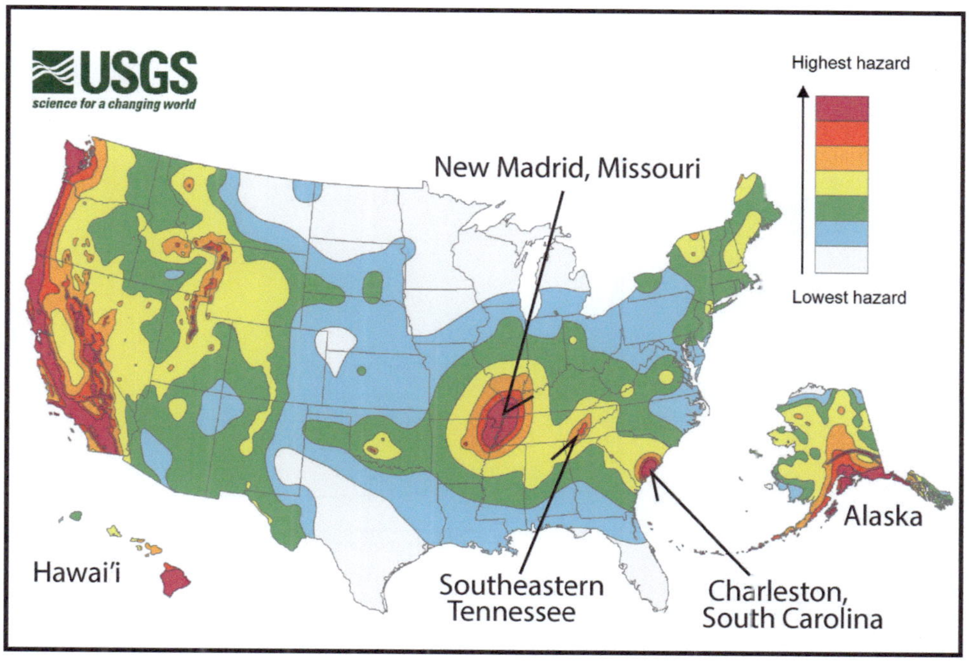

Fig. 8.23 Seismic hazard map of the United States (image credit: U.S. Geological Survey). Seismic hazards are high on the west coast along the Andreas fault and the southern coast of Alaska. New madrid, Missouri, Charleston, South Carolina, and Southeastern Tennessee also have high seismic hazards. In Hawaii, high seismic hazards are largely confined to the Big Island (Hawai'i) mainly from volcanism associated with the hot spot situated beneath this island

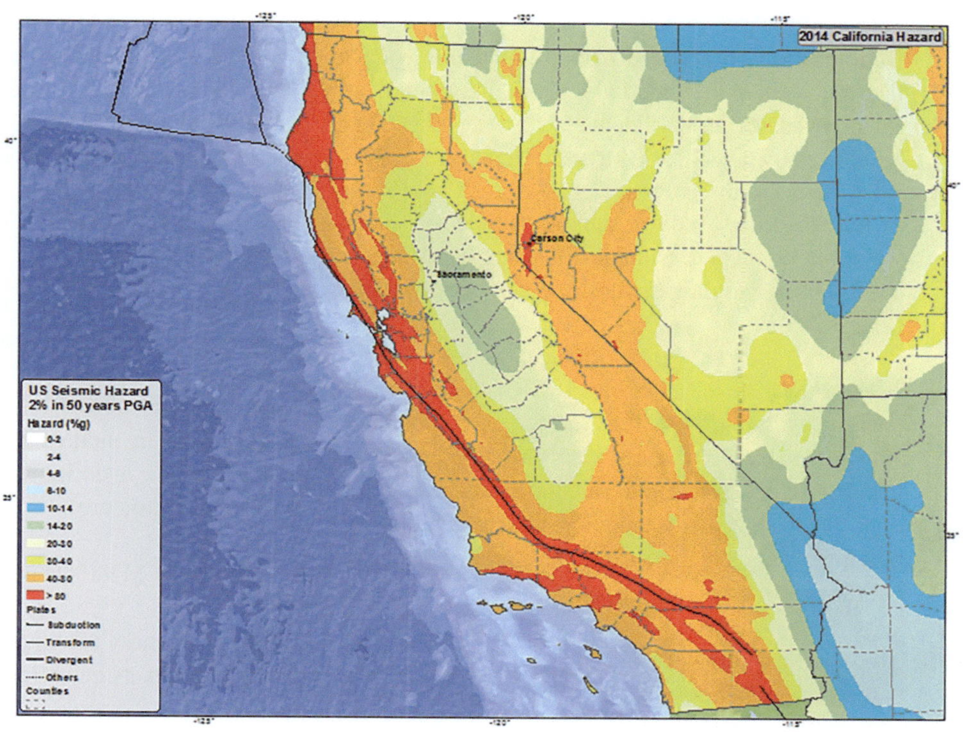

Fig. 8.24 Seismic hazard map of state of California, USA (credit: USGS)

8.12 Mitigating Earthquake Casualties and Damage

On September 19, 2017, at 13:14 CDT (18:14 UTC), the moment magnitude 7.1 Puebla earthquake struck Mexico City, Mexico, strongly shaking the ground for about 20 s and claimed the lives of 370 people, including 228 in Mexico's capital city. Similar to other earthquakes, the damage was not caused by the earthquake itself but by the collapse of more than 40 buildings with people inside, pointing to the necessity of **earthquake engineering**, or the construction of earthquake-proof buildings.

Earthquakes create horizontal pressure on buildings, causing them to collapse. On average, collapsing buildings cause \$2.1B in damage and 10,000 deaths a year. Listed below are some ways to help mitigate the effect of earthquakes.

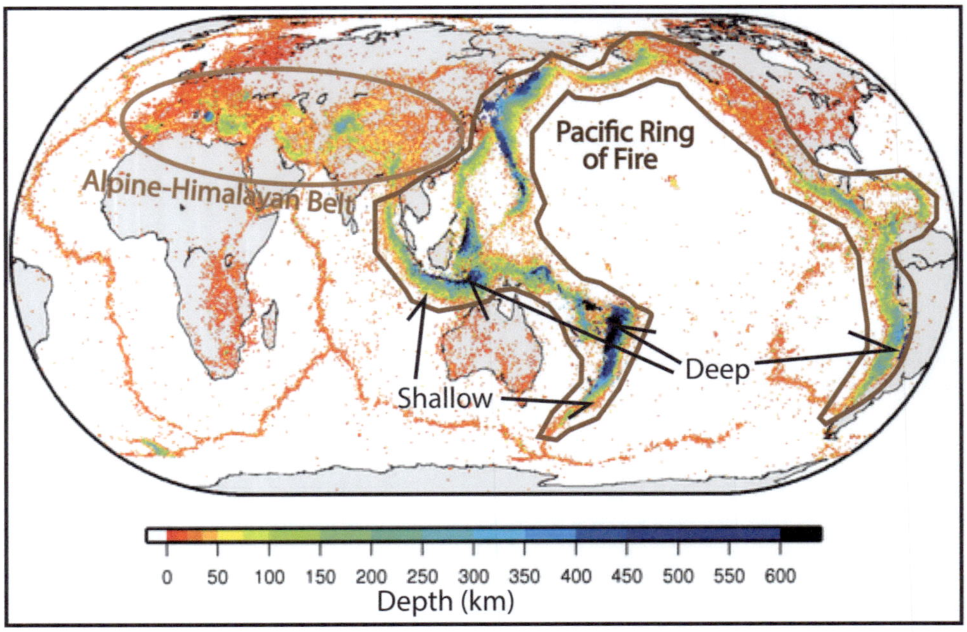

Fig. 8.25 International Seismological Center—EHB locations color coded by depth of located events from 1964 to January 2025. Earth's two major seismic belts are also shown, as well the shallow and deep earthquakes associated with subduction zones (after: International Seismological Center)

1. Construct Flexible Foundation
   i. Buildings are constructed on top of flexible pads so that they isolate the foundation from the ground.
   ii. In the event of an earthquake, only the base of the building moves while the structure remains steady.
2. Counter Forces with Damping
   i. Shock absorbers reduce the magnitude of shock-waves in a manner similar to shock absorbers in cars, helping buildings reduce the amount of shaking (Fig. 8.26a).
   ii. Pendulum power, used chiefly in skyscrapers, involves suspension of a huge ball with steel cables using a hydraulic system at the top of the building. When the building starts to sway, the huge ball behaves like a pendulum and moves in a direction opposite to that of the building which stabilizes the building, thus counteracting the frequency of the building during an earthquake.

3. Reinforcing Structures of Buildings
   i. Cross braces/beams and shear braces act to transfer movement away from the building to its foundation (Fig. 8.26b).
   ii. Diaphragms, or horizontal frames, tend to distribute forces from the floor and transfer them to the columns and walls of buildings (Fig. 8.26c, d).
   iii. Moment-resisting frames keep joints rigid while permitting structures to bend.
4. Shielding Buildings from Vibrations

Termed the "seismic invisibility cloak", rerouting and deflection of energy from an earthquake involves placing concrete and plastic rings under buildings so that earthquake waves are channeled around the building (Fig. 8.26e, f). The method involves generation of concentric plastic and rings of concrete at least 1 m (3 ft.) below the building's foundation, and then covering it.

5. Earthquake-Resistant Materials
   i. Steel and Wood

High ductility materials that possess an ability to undergo large deformations and tension used in construction help buildings to resist stress and vibration. For example, structural steel that allows buildings to bend without breaking is now used in modern buildings. In addition, wood also acts as ductile material due to its high strength despite its lightweight structure.

ii. Innovative Materials of Possible Future Use

Shape memory alloys possess the ability to endure heavy strain as well as reverting to their original shape. Fiber-reinforced plastic wrap that is made from polymers and may be wrapped around columns helps provide greater strength and ductility to structures. Earthquake zoning involves regulation encouraging settlement in areas underlain by unjointed crystalline rocks while discouraging settlement in areas underlain by loose, unconsolidated sediments and/or sedimentary rocks and jointed crystalline rocks. Additionally, settlement should be discouraged in areas composed of wet sand or mud that are susceptible to liquefaction. Construction of houses on steep to moderate sloping terrain should also be discouraged, whereby an assessment is conducted of land considered stable or unstable in areas prone to earthquakes.

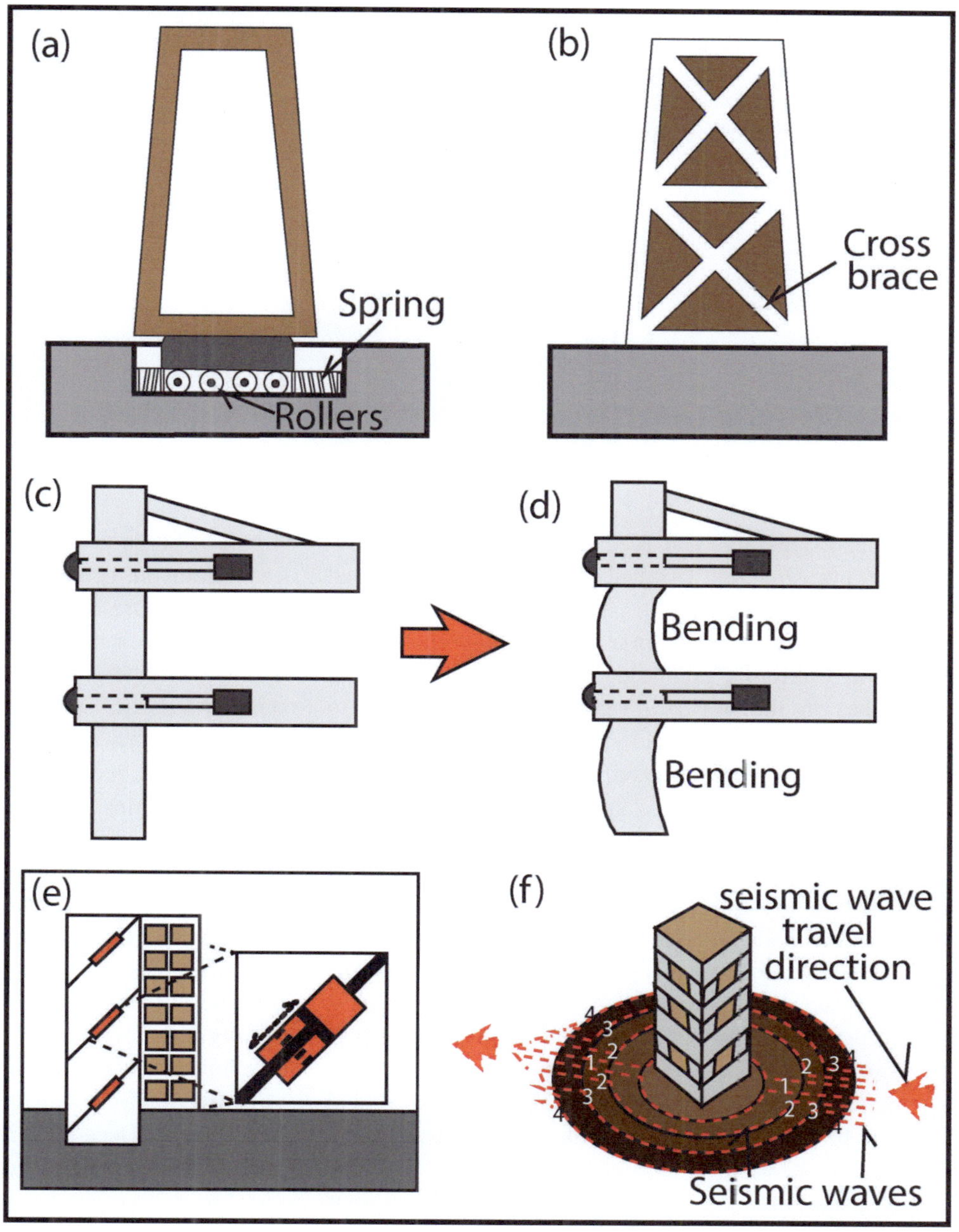

Fig. 8.26 Earthquake engineering. **a** Rollers or shock absorbers help lessen the severity of earthquake vibrations with wider bases. **b** Cross braces/beams on buildings with wider bases provide added strength during an earthquake. **c**, **d** Diaphragms, or horizontal frames, tend to distribute forces from the floor and transfer them to the columns and walls of buildings. **e**, **f** Rerouting and deflection of energy from an earthquake with concrete and plastic rings placed under buildings to channel earthquake waves around the building

During an earthquake, it is also recommended that employing the "Drop, Cover, Hold on" strategy can protect you in the event of an earthquake. A person should immediately i. drop down onto one's hands and knees; ii. cover one's head, neck, and entire body underneath a sturdy desk or table. One should get next to low-lying furniture that will not fall on you or down near an interior wall and cover one's head and neck with arms and hands if there is no shelter nearby; and.

iii. hold on to one's shelter until the shaking ceases. If the shaking shifts the shelter around, one should move around with the shelter.

8.12.1 Mitigating Damage and Casualties from Earthquakes

Destructive from earthquakes is in the form rupturing of faults, collapsing of buildings and other structures, tsunamis, landslides, liquefaction, and fires. Ground shaking (vibration of the ground during an earthquake) produces the greatest loss of property and personal injuries. Surface rupture is a consequence of vertical or horizontal movement on either side of a ruptured fault. Displacement of the ground, which tends to affect large areas, often produces severe damage to structures, pipelines, railways, and roads. Most earthquake damage is caused by seismic waves which

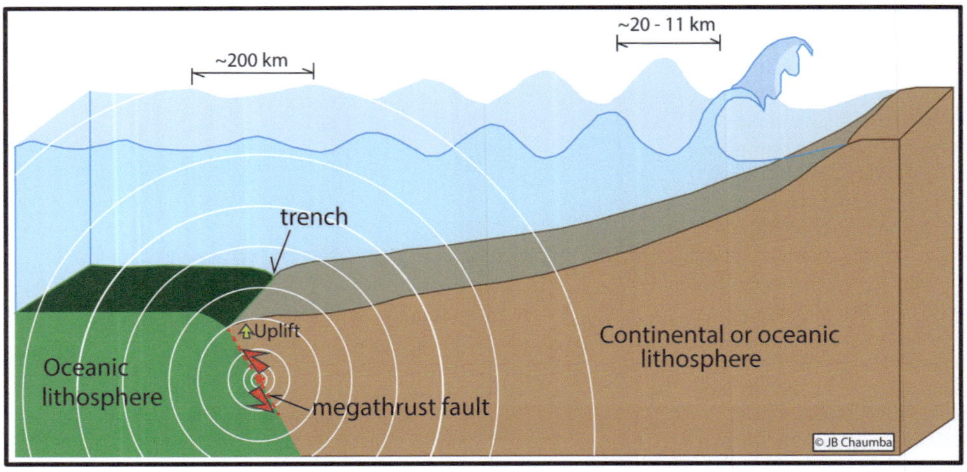

Fig. 8.27 Diagram illustrating the formation of a tsunami, an earthquake occurs underwater due to collision of an oceanic plate and a continental plate leading to forceful vertical motion that causes the water column to suddenly fall or rise

pass beneath buildings and roads, as well as other structures and infrastructure. Downed power lines and ruptured gas pipes contribute to fires.

It is estimated that collapse of buildings and infrastructure in California produced eight of the ten costliest earthquakes in the last 100 years. Earthquakes tend to trigger mudslides and landslides, particularly in areas that are composed of water-soaked soils. Rocks and debris from the landslides can cause fatalities, as well as destroy animals, buildings, vehicles, and trees. Landslides can also block transportation routes such as roads and rail roads and disrupt utility lines. Ground shaking from an earthquake can result in **liquefaction**. Liquefaction takes place when loosely packed, water-logged sediments at or near the ground surface lose their strength due to ground shaking. Liquefaction occurring beneath buildings and other structures can cause major damage during earthquakes. Foundations and supports of buildings, bridges, pipelines, and roads can be undermined during liquefaction, causing them to collapse or dissolve or sink into the ground.

Earthquakes that originate within the Pacific Ocean floor often generate **tsunamis**, which are a series of waves that are caused by earthquakes (Fig. 8.27) or undersea volcanic eruptions. Large tsunamis with waves that affect the entire ocean from the ocean floor to the surface are dangerous to property, infrastructure, and human health, with effects of destruction from tsunamis that can affect areas further beyond coastlines. Fires that result from earthquakes are common hazards. Earthquake fires are caused by electrical and gas lines that are dislodged due to the ground shaking. Gas that is set free from broken gas lines can spark, often starting firestorms. Vibrations from earthquakes can result in ground displacement as well as surface rupture. Surface rupture often causes damage to roads and buildings, often leading to loss of life and injuries, or impede ground transportation. The shaking from a major earthquake can shake and shift almost everything inside your home. During the 1994 Northridge, California earthquake, for example, the majority of the injuries were from heavy furniture and household objects falling on people.

Destruction from tsunamis is often extensive. As a consequence of the December 26, 2004, Banda Aceh, Malaysia strong waves which crashed on Patong Beach, Phuket, Thailand (Fig. 8.28a). The cities of Rikuzentakata, Iwate Prefecture, Honshu in Japan were flooded from the tsunami of March 11, 2011 (Fig. 8.28b).

Fig. 8.28 **a** Photograph of tsunami waves encroaching Patong Beach, Phuket, Thailand, from the 2004 Banda Aceh earthquake (S009/ Gamma-Rapho/Getty Images). **b** The flooded city of Rikuzentakata, Iwate Prefecture, Honshu, Japan. Flooding occurred as a consequence of an earthquake-triggered tsunami on March 11, 2011 (Japan Times)

Glossary

Aftershocks smaller earthquakes than the mainshock that occur after the largest shock of an earthquake sequence. Aftershocks may continue over for weeks, months, or even years.

Amplitude the size of the wiggles on an earthquake recording measured from the position of equilibrium.

Body wave a seismic wave capable of moving through the Earth interior. Examples of body waves are P- and S-waves.

Earthquake the rapid slip on a fault, leading to ground shaking and radiation of seismic energy. Volcanic or magmatic activity or other sudden changes in stress in the crust of the Earth can also result in earthquakes.

Earthquake engineering a process with several phases that range from the description of sources of earthquakes to the characterization of site effects and structural response, and to the description of measures of seismic protection.

Earthquake hazard any risk such as faulting, shaking of the ground, tsunamis, liquefaction, and landslide that is associated with an earthquake that can affect people's normal activities of people.

Earthquake risk the probable building damage and number of people expected to be killed or hurt if a likely earthquake on a particular fault occurs. Occasionally, earthquake risk and earthquake hazard are incorrectly used interchangeably.

Elastic rebound theory the slow accumulation and release of stress and strain which is then released during an earthquake due to the rapid elastic rebound of previously stored energy.

Epicenter the point on the Earth's surface located vertically above the focus (or hypocenter) within the crust where a seismic rupture occurs.

Fault a fracture along which blocks of Earth's crust on either side move relative to one another parallel to the fracture.

Fault gouge ground-up and crushed rock that is produced by friction between two sides from movement of a fault.

Fault plane the planar, or flat, surface along which slip occurs during an earthquake.

Fault scarp a feature on Earth's surface resembles a step and it is formed by slip on the fault.

Fault trace the intersection of a fault with the Earth's surface. It is also the line that is commonly plotted on geologic maps to represent a fault.

Focus the point within the crust where an earthquake rupture occurs. The focus is also commonly termed the hypocenter.

Foreshocks relatively smaller earthquakes preceding the largest earthquake in a series (mainshock). Not all mainshocks have foreshocks.

Ground motion the movement of the Earth's surface from earthquakes or explosions produced by waves that are generated by rapid slip on a fault or sudden pressure at the explosive source. The seismic waves travel through the earth and along its surface.

Hypocenter the point within the Earth where an earthquake rupture originates.

Intensity intensity is a number that is written as a Roman numeral to describe the severity of an earthquake based on its effects on humans and their structures and on the Earth's surface. Several intensity scales exist with the most commonly in the United States being the modified Mercalli scale.

Intraplate pertains to earthquakes and processes within the plates.

Left-lateral if one were to stand on the fault and look along its length, one observes a type of strike-slip fault where the left block moves toward you, and the right block moves away.

Liquefaction a process whereby water-saturated sediment temporarily loses strength and acts as a fluid, often caused by shaking caused by an earthquake.

Locked fault a fault that is not slipping due to the frictional resistance on the fault being greater than the shear stress across the fault (it is stuck). Such faults may store strain for extended periods until it is eventually released in an earthquake when the frictional resistance is overcome.

Love wave a surface wave having a horizontal motion that is transverse (or perpendicular) to the direction along which the wave is traveling.

Magnitude a number that characterizes the relative size of an earthquake and it is based on measurement of the maximum motion recorded by a seismogram.

Mainshock the largest earthquake in a sequence which may be preceded by one or more foreshocks, and almost always followed by many aftershocks.

Intensity a number written as a Roman numeral which describes the severity of an earthquake in terms of its effects on the Earth's surface and on humans and their structures. The modified Mercalli scale is an example of an intensity scale.

Magnitude a number characterizing the relative size of an earthquake. Magnitude is based on measurement of the maximum motion recorded by a seismograph.

Moment magnitude (Mw) scale scale based on the concept of seismic moment that is uniformly applicable to all sizes of earthquakes but it is difficult to compute.

P-wave also termed **compressional wave**—is a seismic body wave that shakes the ground back and forth in the same direction and the opposite direction as the direction the wave is propagating.

Rayleigh wave a seismic surface wave that causes the ground to shake in an elliptical motion, with neither transverse nor perpendicular movement.

Recurrence interval or return period the average time span between earthquake occurrences on a fault or in a source zone.

Richter scale or Richter magnitude scale a mathematical device developed in 1935 by Charles F. Richter to compare the size of earthquakes. The magnitude of an earthquake is determined from the logarithm of the amplitude of waves recorded by seismographs. Adjustments are included for the variation in the distance between the various seismographs and the epicenter of the earthquakes. On the Richter scale, magnitude is expressed in whole numbers and decimal fractions. Because of the logarithmic basis of the scale, each whole number increase in magnitude represents a tenfold increase in measured amplitude; as an estimate of energy, each whole number step in the magnitude scale corresponds to the release of about 31 times more energy than the amount associated with the preceding whole number value.

Right-lateral if one were to stand on the fault and look along its length, one observes a type of strike-slip fault where the right block moves toward you, and the left block moves away.

Secondary wave also called **shear wave**—is a seismic body wave that makes the ground vibrate back and forth perpendicular to the direction that the wave is propagating.

Seismic gap a seismic gap is a section of a fault that has produced earthquakes in the past but is now quiet. For some seismic gaps, no earthquakes have been observed historically, but it is believed that the fault segment is capable of producing earthquakes on some other basis, such as plate-motion information or strain measurements.

Seismicity the geographic and historical distribution of earthquakes.

Seismic wave an elastic wave that is generated by an impulse such as an earthquake or an explosion. Some seismic waves may travel either along or near the Earth's surface (Rayleigh and Love waves) or through the Earth's interior (P- and S-waves).

Seismic zone an area of seismicity probably that shares a common cause. An example is the San Andreas Fault.

Seismograph or seismometer an instrument that is used to detect and record earthquakes. In general, it consists of a mass attached to a fixed base. During an earthquake, the base moves and the mass does not. The motion of the base with respect to the mass is commonly transformed into an electrical voltage. The electrical voltage is recorded on paper, magnetic tape, or another recording medium. This record is proportional to the motion of the seismometer mass relative to the Earth, and it can be mathematically converted to a record of the absolute motion of the ground.

Seismogram a record written by a seismograph in response to ground motions produced by an earthquake, explosion, or other ground sources.

Seismograph also called a seismometer—is an instrument that consists of a mass attached to a fixed base and it is used to detect and record earthquakes.

Seismology seismology is the study of earthquakes and the structure of the Earth, by both naturally and artificially generated seismic waves.

Seismologist an Earth scientist who specialized in geophysics studying the origin and the propagation of seismic waves in geological materials.

Seismometer see seismograph.

Shadow zone the area of angular distances of 104–140° from a given earthquake that does not receive any direct P-waves. The shadow zone results from S-waves being stopped entirely by the liquid core and P-waves being bent (refracted) by the liquid core.

Shear wave or S-wave a seismic body wave that shakes the ground back and forth perpendicular to the direction the wave is moving.

Tsunami a giant wave that is caused by an earthquake or a volcanic eruption occurring under a huge body of water.

Wavelength the distance between any successive points of equal amplitude and phase on a wave such as crest to crest or trough to trough.

9 Crustal Deformation and Mountain Building

Abstract

Discussions of both stress and strain acting on rocks begin this chapter, followed by explanation of elastic and plastic deformation, as well as fracture, clarified by diagrams. Compressional, tensional, and shear stresses result in rocks undergoing deformation by distortion, displacement, and rotation. Diagrams and photographs are utilized to explain these stresses and the types of deformation they cause. Both diagrams and photographs are utilized to distinguish between brittle deformation and elastic deformation, and examples of each. Description of faults and their classification, including diagrams and photographs, are also discussed. The chapter ends with a discussion of the varied landscapes on our planet being ascribed to isostacy.

Learning Goals

Studying this chapter will enable you to:

- Contrast stress and strain.
- Describe the types of stresses and deformation.
- Recognize deformed rocks.
- Sketch and recognize normal, reverse, and thrust faults.
- Distinguish between reverse and thrust faults.
- Differentiate brittle and ductile deformation.
- Recognize and differentiate structures formed due to compression and extensional forces.
- Distinguish the different common types of folds.
- Identify Earth's major mountain belts and how they were formed.
- Define and recognize accreted terranes.
- Sketch and define isostacy.

Supplementary Information The online version contains supplementary material available at https://doi.org/10.1007/978-3-032-20231-4_9.

9.1 Introduction

We learned in Chap. 6 that several tectonic plates form the upper part of the Earth and that these plates are continually moving. In this chapter, we will examine the types of deformation and the resulting features such as mountains, faults, folds, the changing continents, and several other tectonic features that are formed. However, before we do this, we must first discuss factors that cause deformation and how rocks respond to applied tectonic stresses.

9.2 Stress and Strain Acting on Rocks

Stress is a force applied over an area on rocks which produces shape. Stress has units of force/area (such as kg/m^2 or lb/in^2). Pressure is a stress where the forces act equally from all directions. **Pascal's law**, named after Blaise Pascal who first enunciated it, states that in a fluid at rest in a closed container, any change in pressure in one part is transmitted without loss to every portion of the fluid and to the walls of the container. If stress is not equal from all directions, as is the case in the Earth's crust, it is termed **differential stress**. Four kinds of differential stress can act on rocks:

i. **Confining Stress**: The weight of all the material above a rock which pushes down and buries rocks.
ii. **Tensional Stress**: It is also referred to as extensional stress, which stretches rocks. Tension is the major type of stress at divergent plate boundaries.

J. B. Chaumba, *Introductory Physical Geology*, Springer Textbooks in Earth Sciences, Geography and Environment,
https://doi.org/10.1007/978-3-032-20231-4_9

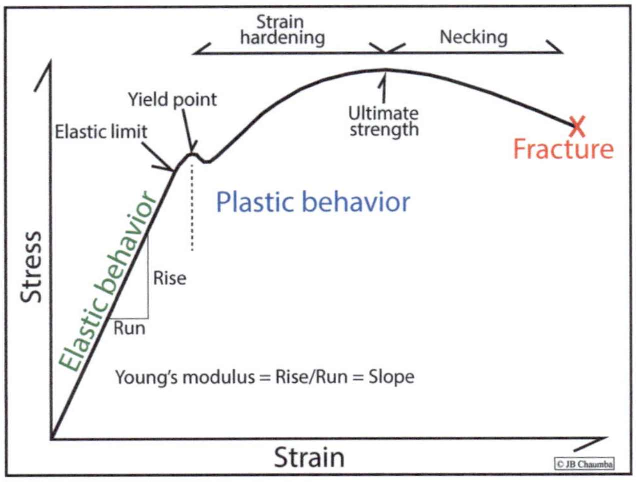

Fig. 9.1 Stress versus strain relationship

iii. **Compressional Stress**: It is also called shortening, which squeezes rocks together causing folding or fracturing of rocks. Compression is the major type of stress at convergent plate boundaries.
iv. **Shear Stress**: Shear stress occurs where parallel forces act in opposite directions. Shear stresses are the major stress types at transform plate boundaries.

When stress causes a material to change shape, it has undergone **deformation** or **strain**. Deformed rocks are common in geologically active areas. How a rock responds to stress depends on the rock type, the surrounding temperature, the pressure conditions that the rock is subjected to, the length of time the rock is under stress, as well as the type of stress applied to the rock. Rocks have three successive stages of deformation (Fig. 9.1):

i. **Elastic Deformation**: The strain in the rock is reversible, i.e., it returns to its original shape when the stress is removed. Elastic behavior is exemplified.
ii. **Plastic Deformation**: The strain in the rock is irreversible, i.e., the rock does not return to its original shape when the stress is removed, exemplifying plastic deformation.
iii. **Fracture**: The rock undergoes irreversible strain, i.e., wherein the rock breaks.

9.3 Deformation of Rocks

After their formation, some rocks may undergo changes. Rocks that have been deformed are those that have been affected by at least one of the following changes:

i. **Distortion**—Change in shape. Distortion occurs when the shape of a body of rock is altered, such as through folding of a layer of rock during deformation (Fig. 9.2a).
ii. **Displacement**—Change in their location. Displacement occurs when a body of rocks, for example, is moved during deformation to another location (Fig. 9.2b).
iii. **Rotation**—Change in their orientation. Rotation occurs when layers of horizontal rocks are tilted during deformation (Fig. 9.2c). Examples of deformed rocks are shown in Fig. 9.3. In addition to folding, the distorted rocks from Bighorns, Wyoming shown in Fig. 9.3a were also subjected to faulting.

The rocks exposed on a roadcut along Interstate 40 in southwestern Arizona shown in Fig. 9.3b underwent deformation whereby blocks on either side of the fault were displaced in the direction indicated by the arrows. Finally, the originally horizontal rock layers from a cliff face in Spanish Fork Canyon, Utah, which display tilted beds (Fig. 9.3c), were deformed by rotation in the direction indicated by the arrows.

9.4 Types of Stresses

Confining stress occurs when equal stress is applied to crustal rocks from all directions of the crust (Fig. 9.4a) leading to compaction. **Tensional stress** causes crustal rocks to be stretched and pulled apart (Fig. 9.4b). When two plates move farther away from one another, tension occurs. Tension also occurs when the ends of a single object move away from each other. Tensional stresses may have been the major cause of the breakup of the supercontinent Pangaea to form our present continents. **Compressional stress** (Fig. 9.4d) causes rocks to push or squeeze against one another. Crustal thickening or shortening is a consequence of horizontal compressional stresses whereas crustal thinning or even breaking off is caused by vertical compressional stresses. Mountains are formed due to high-impact compressional stresses that result when two plates collide. Compressional stresses can either push rocks together or cause the edges of one of the plates to rise over the other. **Shear stress** (Fig. 9.4e) pushes parts of the crust in different directions often resulting in part of the crust breaking off, reducing the size of a tectonic plate. Shear stress commonly occurs where two adjacent plates move in opposite directions. Friction generated at the edges of plates undergoing shear stresses tend to cause earthquakes.

In crustal rocks on Earth, examples of these types of stresses abound. Tensional stresses are common in areas

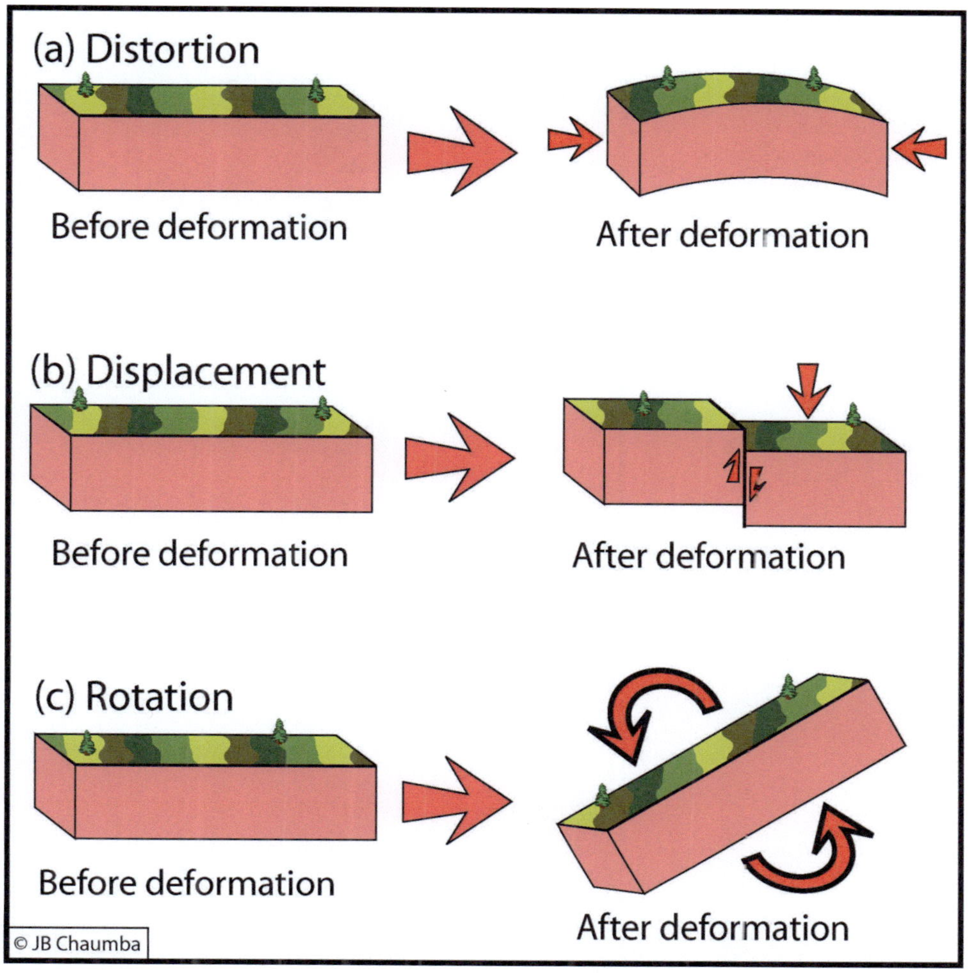

Fig. 9.2 Deformation includes distortion (**a**), displacement (**b**), and rotation (**c**)

where ranges and half-grabens (Fig. 9.5a) and basin and range, or horst and graben (Fig. 9.5b), are displayed. The Basin and Range Province in Nevada, USA, the East African Rift Valley in Eastern Africa, and oceanic spreading centers such as the Mid-Atlantic Ridge and the East Pacific Rise are examples of features formed by tensional stresses.

9.5 Types of Deformation

We have already discussed **elastic deformation** in Sect. 9.1, whereby a deformed object resumes its original shape once the applied stress is removed. When an object deforms by shattering into smaller fragments, it undergoes **brittle deformation**. A clay plate or dinner plate that one accidentally drops and it breaks into small pieces is a good example of brittle deformation. Play doll or dough, when deformed, easily bends as it does not break into small pieces. Deformation whereby an object bends or is twisted but does not break is termed **plastic deformation**. Some common examples of brittle deformation include **fractures**, **joints** (Fig. 9.6a, b), and **faults**. Fracture is a general which includes any break in rocks. Fractures and joints form in rocks that have been subjected to brittle deformation. A joint is a fracture without any noticeable movement on rocks on one side relative to those on the other side. Joints typically occur in sets, although single joints are also encountered. A fault is feature along which rocks on one side have moved relative to the rocks on the other side. Faults develop at relatively shallow crustal depths. Let us now discuss folding which rocks experience when they are deformed at relatively deeper crustal levels. Shearing and stretching also occur in rocks subjected to deformation at deep crustal levels. Rocks subjected to extensional stresses get thinner as a consequence of stretching.

9.6 Folds

A **fold** is a wavelike feature observed in deformed rocks. Common examples of ductile deformation include folds, domes, and basins. Folds form when one or more layers of originally horizontal and planar rock strata are bent or curved due to ductile deformation. Some folds, termed syn-

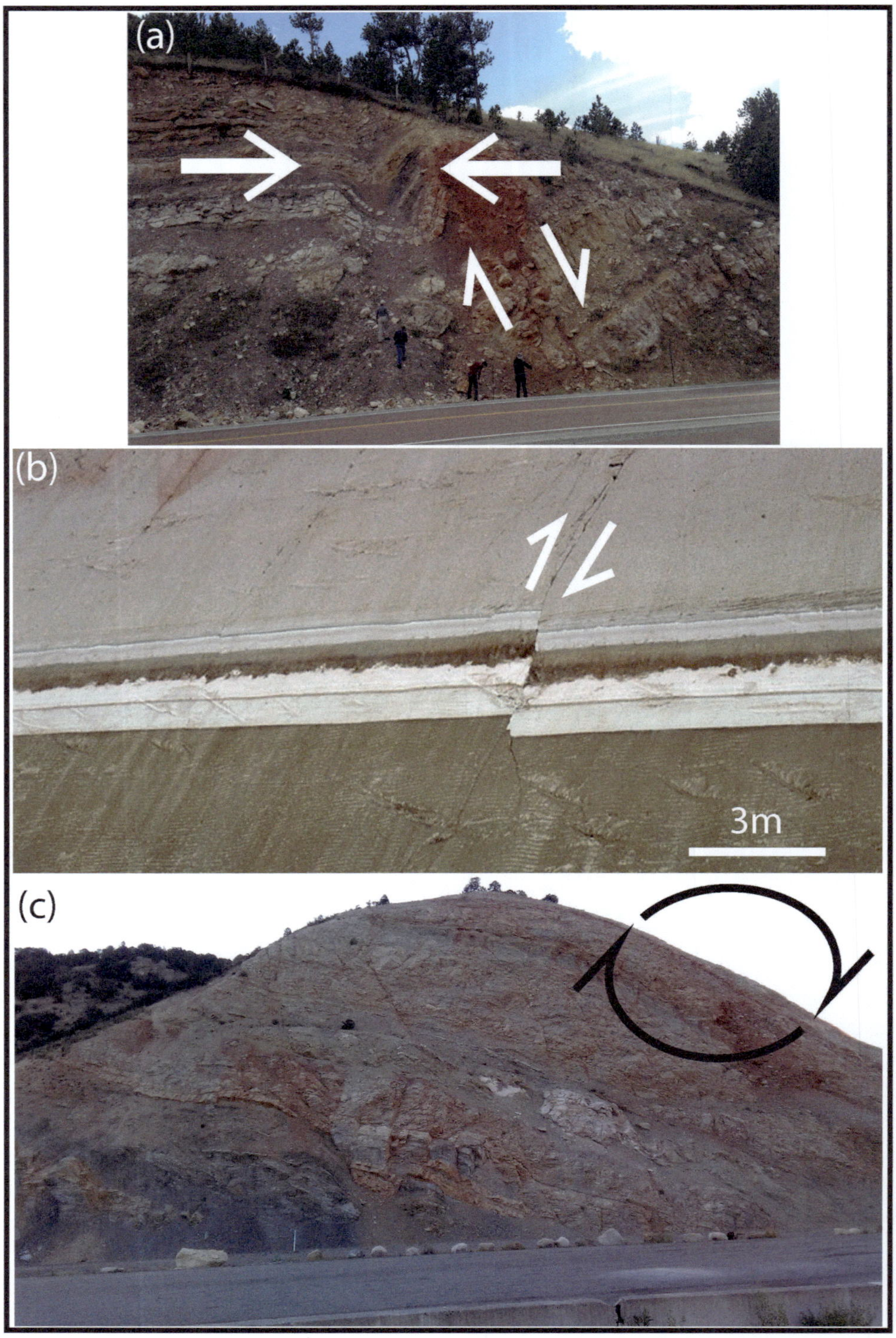

Fig. 9.3 Examples of deformed rocks. **a** Distorted rocks from Bighorns, Wyoming. In addition to folding, these rocks have also been faulted (credit: A. Gesualdo). **b** Deformed rocks exposed on a roadcut along Interstate 40 near Kingman, Arizona with blocks on either side of the fault displaced in the direction indicated by the arrows (credit: D. T. Nelson). **c** Originally horizontal beds were deformed by rotation in the direction indicated by the arrows on this cliff with tilted beds in Spanish Fork Canyon, Utah (credit: D. T. Nelson)

sedimentary folds, are formed from slumping of sedimentary material prior to lithification. Folds widely vary in size from microscopic crinkles to those occurring on the scale of mountains. Folds can occur as isolated features or as part of fold trains of different sizes (Fig. 9.7). Folds form different features which we discuss next.

Folds are classified based on the attitude of their fold axes as well as how they appear in cross sections that are perpendicular to the fold trend. The surface or plane that divides the fold as symmetrically as possible axial plane of a fold is the axial plane (Fig. 9.7). The axial plane varies from being vertical or inclined to horizontal, or at any intermediate angle (Fig. 9.7). A fold axis is the intersection of the axial plane with one of the strata that forms the fold. The axis can be horizontal or gently inclined in simple folds. However, the fold axis may be steeply inclined

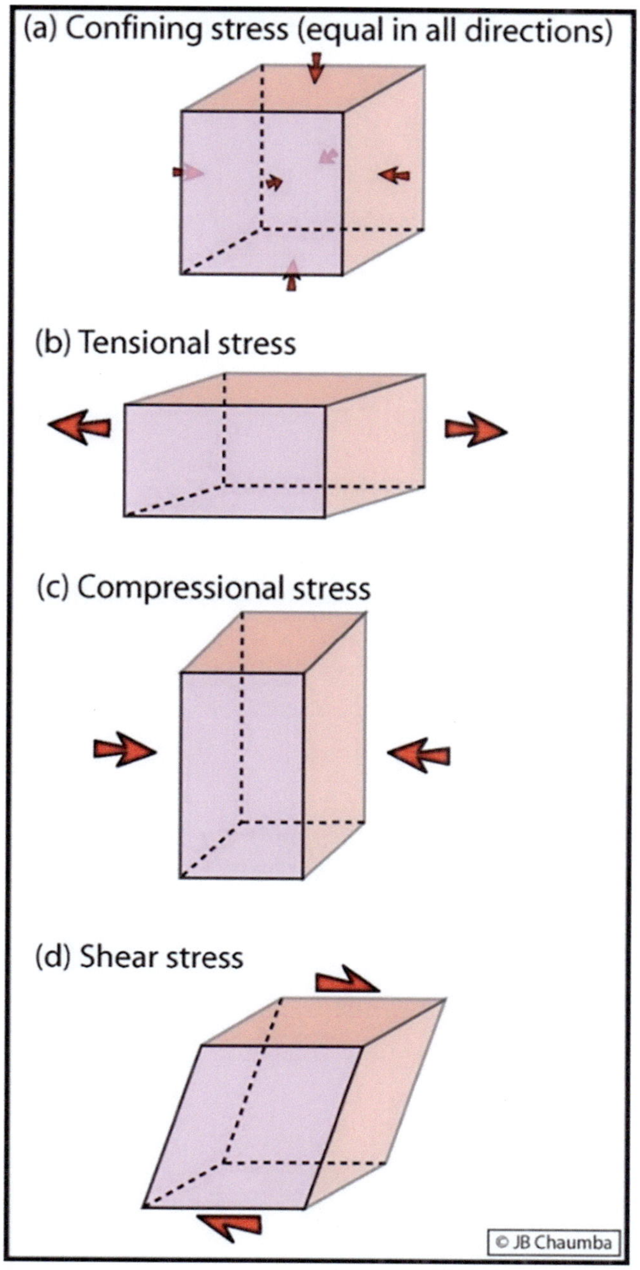

Fig. 9.4 Illustrations of: **a** confining pressure which is generated when an object experiences the same stress in all direction; **b** tensional stress, which is formed when opposite ends of an object are pulled away from each other; **c** compressional stresses, which form due to squeezing of an object; and **d** shear stress, which is generated when plates slide relative to one another

or even vertical (Fig. 9.7). The plunge of a fold in the angle of inclination of the axis measured from the horizontal. The limbs or flanks of the fold are the portion of the fold between two adjacent axes or slopes or limbs of a fold.

A fold anticline is convex upward part of a fold whereas a fold syncline is concave upward. A large anticline on which minor folds are superimposed is an anticlinorium, whereas a synclinorium is a large syncline on which minor folds are superimposed. A fold in which the axial plane is vertical is called a symmetrical fold, whereas a fold in which the axial plane is inclined is referred to as an asymmetrical fold (Fig. 9.7). In an overturned fold, which can also be referred to as an overfold, the axial plane inclined such that strata on one limb of the fold are overturned (Fig. 9.7). A **recumbent fold** possesses a linear, fold axial plane that is oriented at a low angle, that is, with a horizontal axial plane. The horizontal axial plane results in the formation of overturned layers in one limb of the fold (Fig. 9.7). The linear, fold axial plane that is oriented at a low angle in a recumbent fold results in overturned strata in one limb of the fold. Thus, when the axial plane of a fold is essentially horizontal, the fold is termed a recumbent (Fig. 9.7).

Some folds formed in highly deformed rocks can be complex (Fig. 9.8a). A **chevron fold** (Fig. 9.8b) is a structural feature that is characterized by repeated folded beds possessing straight limbs and sharp hinges. Chevron folds form in response to regional or local compressive stress and they tend to develop from repeated sets of layers with a V-shape (Fig. 9.8b). Chevron folds are typically encountered in rocks with regular alternating layers of contrasting competences. **Isoclinal fold**, derived from "iso" meaning "the same" (symmetrical), and "cline" means "angle" have the same angle and are parallel to each other. Isoclinal folds, therefore, imply folds with the "same angle" and these folds are both symmetrical and aligned in a parallel fashion. An example of an isoclinal fold is shown in Fig. 9.8c.

A **monocline** (derived from *mono* meaning one, and *kleinen* meaning incline) is a type of fold with only one limb that is bent. A monocline (Fig. 9.9a) forms when layers of rocks bend abruptly and resume the original attitude at lower or higher levels. A monocline is a step-like fold in rock strata consisting of a zone of steeper dip within an otherwise horizontal or gently dipping sequence. Thus, a monoline is a consequence of local warping in horizontal beds. On a monocline, steep inclined limbs separate rocks that lie at two levels (Fig. 9.9a). The East Kaibab monocline in Arizona, USA, is an example of a monocline.

Folds are classified into several types. An **anticline** (Fig. 9.9b) is a fold that is convex up and its oldest beds are encountered at its core, and it is formed by arching or upfolding of rocks. It is worth stating here that if age relationships between various strata are unknown, one should use term **antiform**. An antiform is a descriptive term denoting any fold that is convex up. The opposite of antiform is **synform**, which is a fold whose limbs close downward in strata for which the sequence is unknown. Limbs in an anticline dip away from each other. A **syncline** (Fig. 9.9c) is a fold formed by downfolding with younger rocks/layers located at the center of the fold. A syncline is generally a downward fold, referred to as a synformal

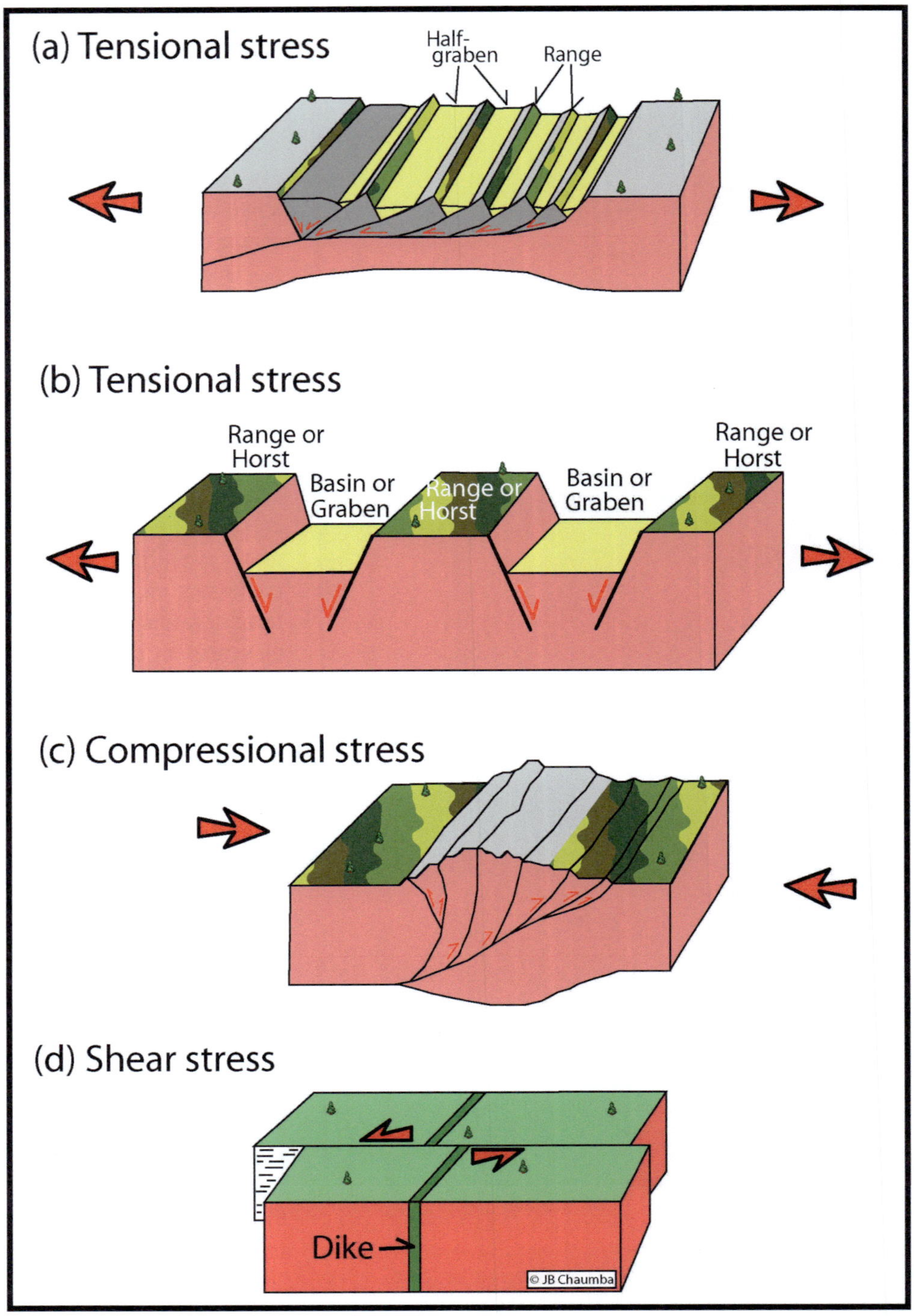

Fig. 9.5 Images showing ranges and half-grabens (**a**) and horsts and grabens or basins and ranges both formed from extensional stresses (**b**), collisional orogenies formed from compressional stresses (**c**), and displaced features such as dikes formed due to shear stresses (**d**). Compressional stresses are the chief cause of collisional mountain ranges (Fig. 9.5c) such as the Himalayas, the Alps, the Appalachians, the Andes, and the Rockies. Shear stresses (Fig. 9.5d) are responsible for the dislocation of once continuous features such as dikes and even man-made features such as communication lines

syncline, or a trough. Limbs of syncline dipping toward each other. A syncline that points upward, or perched, is formed where strata have been overturned and folded, referred to as an antiformal syncline. A syncline, thus, is a downward fold. Shown in Fig. 9.8a is an image of the Ovens anticline in Nova Scotia, Canada. A classic example of a syncline is displayed on the road cut on Interstate 68 in western Maryland (Fig. 9.10b, c).

Folds that are chaotic, random, and typically disconnected are termed **ptygmatic folds** (Fig. 9.11). Ptygmatic folds are common in sedimentary rocks that have undergone slump folding, as well as in migmatites and **décollement** (derived from French word "*décoller*," meaning "to detach from") detachment zones. Ptygmatic folds generally develop where the folded material is composed of greater viscosity than that of the surrounding rocks. Ptygmatic folds are also common in deformed sedimentary rocks in which slump folds are displayed.

Folds, which are bends or warps that are formed when one or more layers of originally planar surfaces such as

Fig. 9.6 Joints. **a** Joints, which are shallowly dipping sandstone to the right exposed which are part of the Henderson gneiss, western North Carolina, USA with joints are responsible for making the waterfall. **b** A closer view of joints of the Henderson gneiss in western North Carolina

sedimentary beds, are bent due to **ductile deformation** that occurs in response to compressive stresses. Folds form at great depths in the Earth's crust where the rock layers are subjected to high temperatures and pressures which permit rocks to crinkle. Folds occur on a variety of scales ranging in size from microscopic crinkles to those with amplitudes of kilometer (miles). In general, folds form in series. Folds can occur either as isolated folds or as extensive fold trains of different sizes.

Broadly similar to symmetrical folds are **isoclinal folds**, which are derived from the words "iso," meaning "the same" (symmetrical), and "cline," meaning "angle." Isoclinal folds, however, are parallel to each other and they possess the same angle. Isoclinal folds thus literally mean folds with

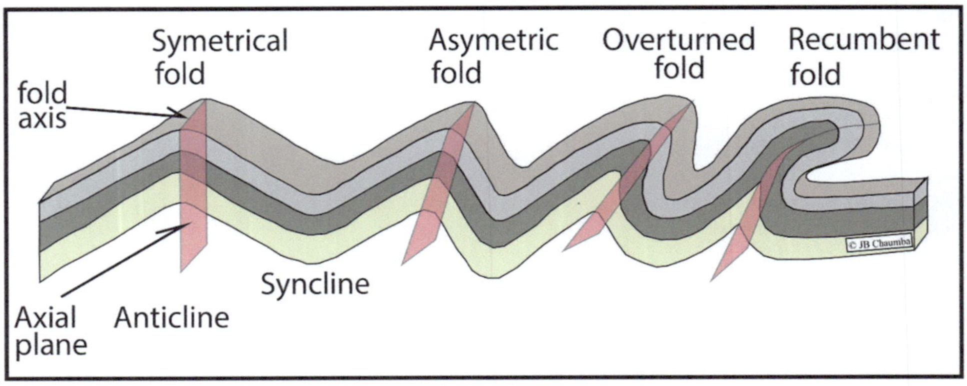

Fig. 9.7 Diagram illustrating different types of folds that form in deformed rocks

the "same angle." Isoclinal folds, thus, are symmetrical and they are aligned in a parallel manner.

A **non-plunging anticline** is one whose hinge line is horizontal, or parallel to sea level, as shown in Fig. 9.12a. A fold whose hinge line is not horizontal (not parallel to sea level) is termed a **plunging anticline** (Fig. 9.12b). The direction in which the axis is inclined denotes the direction of plunge. Sheep Mountain in Wyoming, USA is an example of a plunging anticline.

Where strata dip away from center in all directions, **domes** are developed (Fig. 9.13a). The oldest strata in domes occur in the center. Domes are generally circular to slightly elongated structures that are generated either as a consequence of upwarping or due to intrusion of magma which slowly crystallizes to form laccoliths. The Black Hills, located in western South Dakota, are some classic examples of a dome. Where strata dip toward the center in all directions, **basins** are formed (Fig. 9.13b). The relationships between the oldest and youngest rocks are key in differentiating between domes and basins (Fig. 9.13a, b). In basins, the youngest strata occur in the center. Domes, thus, can be envisaged as inverted bowls. Basins may develop above a lopolith or they can form in rocks deformed by huge meteorite impacts. Most basins are generated as a consequence of subsidence or down warping. Sedimentary rocks typically have domes and basins developed in them.

The ages of sedimentary layers in most sedimentary basins are useful in the identification of domes and basins. The youngest strata in domes occur furthest from the center, whereas in basins the youngest strata occur at the center (Fig. 9.13a, b). An example of a basin exposed in the Sahara Desert in Mauritania is shown in Fig. 9.14. Large basins and domes are commonly faulted, and basins and domes are widespread in the United States as illustrated in Fig. 9.15.

9.7 Faults

A **fault** is a planar fracture or discontinuity along which there has been displacement. Faults, which are like joints in that they also form in rocks that have been subjected to brittle deformation, range in scale from centimeters (inches) to kilometers (miles). For a fault that occurs at an angle to the horizontal, the displacement along a fault can be best described in terms of **hanging wall** and **footwall** (Fig. 9.16). Where a dip-slip fault is encountered, the hanging wall is the block of rock located *above* the fault, whereas the footwall is the block of rock located *below* the fault line (Fig. 9.16). The terminology can be illustrated using miners who hang their coats and lamps from the hanging wall as if it were a ceiling, and they would be standing on the footwall as if it is the floor below them (Fig. 9.16).

9.8 Classification of Faults

A dip-slip fault is one whose angle of dip is inclined to the horizontal and in which movement occurred parallel to slope of the fault as shown in Fig. 9.16. The block of rock occurring immediately above a dip-sip fault is termed hanging wall, whereas the block of rock occurring immediately below the fault is termed footwall (Fig. 9.15). Miners excavating ore deposits devised this terminology, which can also be illustrated diagrammatically with a lamp being hung from the hanging wall with a miner's feet resting on the footwall (Fig. 9.16). Where movement along the fault involves the hanging wall moving downward (down dip) and the footwall moving upward (up dip), as shown in Figs. 9.15 and 9.16a, a **normal fault** is produced. Where movement along the fault inclined at an angle of >45° to the horizontal and the footwall moves downward (downdip dip) and the hanging wall moves upward (up dip), as shown in Fig. 9.17b, a **reverse fault** is formed. If the angle between the fault and the horizontal is ≤45° and the hanging wall also moves upward relative to the footwall, a **thrust fault** is produced (Fig. 9.17c).

A **strike-slip fault** (Fig. 9.18) is formed when two tectonic plates or two crustal blocks slide past each other during an earthquake. Stress caused by two crustal blocks moving in opposite directions induces shearing along the

Fig. 9.8 Complex deformation of the ancient older than 3.35 billion-year-old gray gneisses of the Tokwe segment near Runde River, Zimbabwe (credit: Geological Society of Zimbabwe). **b** Chevron folds displayed on Millook Haven Beach, Cornwall, England (credit: Paul Berry). **c** Isoclinal folds to the right and parasitic folds on one limb of the fold to the left of hammer on outcrop (credit: Geological Society of Zimbabwe)

strike-slip fault. The friction between the tectonic plates is overcome during an earthquake which is triggered by the motion along the strike-slip fault. Strike-slip faults can be classified depending on the sense of motion along the fault. To determine the sense of movement on a strike-slip fault, imagine standing on one block and face the block on the other side of the fault. If the movement of the block on the other side is to the right, i.e., the ground opposite the fault moved to the right, then the fault is a **right-lateral fault** (Fig. 9.18a). However, if the movement of the ground of the block on the other side is to the left, i.e., the ground opposite the fault moved to the left, then the fault is a **left-lateral fault** (Fig. 9.18b).

A **fault scarp** is an offset, or a step, on the ground surface where the block on one side of a fault has moved vertically with respect to the other block. A fault scarp, thus,

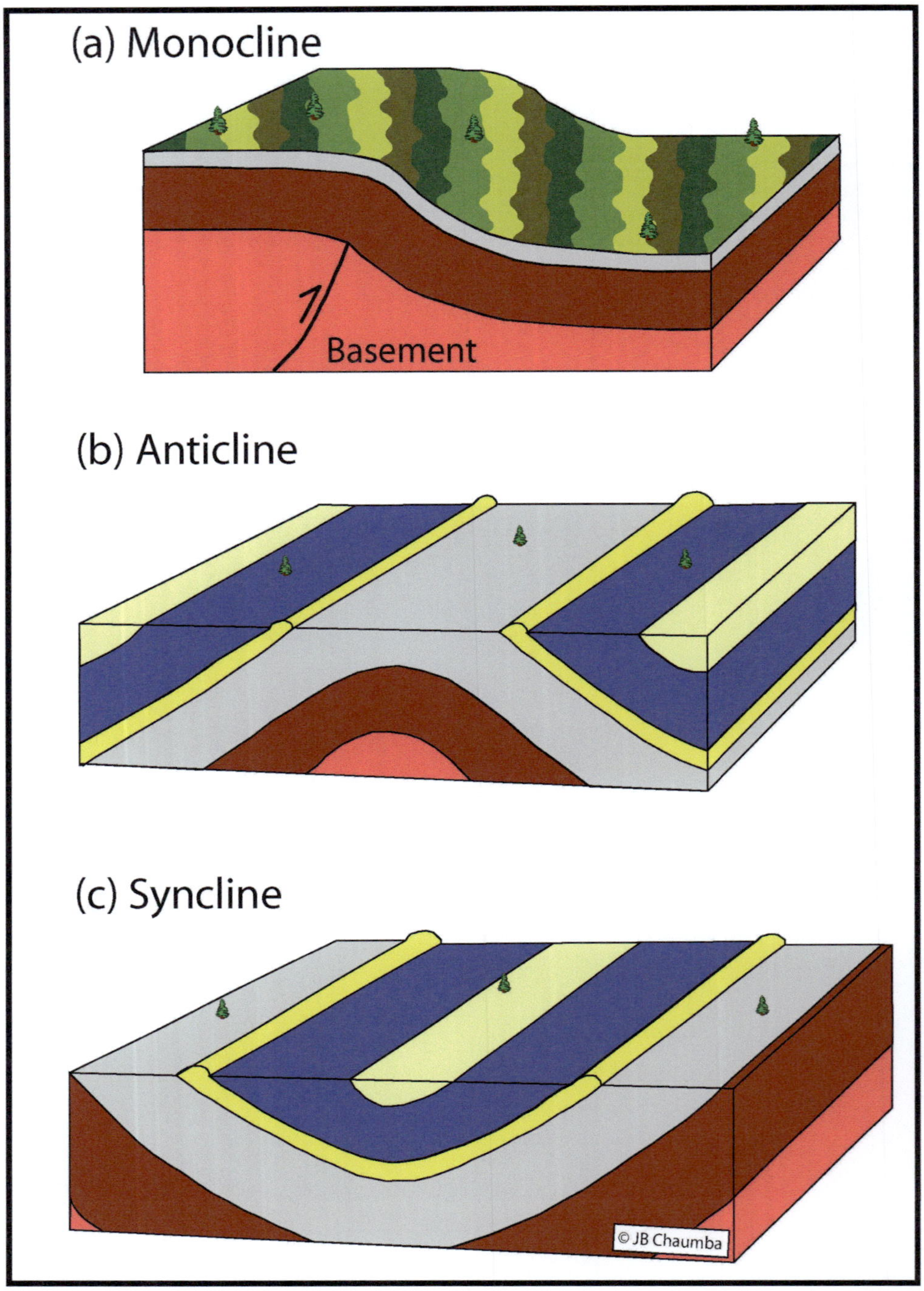

Fig. 9.9 Diagrams illustrating a monocline (**a**), an anticline (**b**), and a syncline (**c**)

is the topographic expression of a fault the land surface was displaced by movement along the fault. An example of a fault scarp is shown in Fig. 9.19 which was generated by the Leader Fault of the magnitude 7.8 Kaikōura earthquake of 2016 in South Island, New Zealand. This fault scarp measured approximately 200 m long and 3.5 m high.

A combination of both shear stresses and extensional (Fig. 9.20a) or compressional (Fig. 9.20b) stresses results in the formation of **oblique-slip faults**. Any two reference points that were joined before dip-slip faulting can be identified and connected after the faulting, such as the two corners indicated by black arrows in Fig. 9.20a, b. Most faults tend to have oblique-slip components, but most of them are characterized by very minor amounts of dislocations for the oblique-slip components to be recognizable and even be measured.

9.9 Deformation and Mountain Building

Most mountain ranges on Earth form because of the interaction of plates. Most mountains are produced at convergent plate boundaries and at divergent plate boundaries. Convergent plate boundaries leading to the formation of mountains occur at oceanic-oceanic plate boundaries, oceanic-continental plate

Fig. 9.10 **a** Image of the Ovens anticline, Nova Scotia, Canada (credit: Michael Rygel). **b**, **c** are images of the impressive syncline exposed through construction of Interstate 68 in western Maryland. **b** Aerial view (credit: T. W. Kammer). **c** View from the suspension bridge looking west across I-68 by the author. May 2019

boundaries, and continental-continental plate boundaries. The largest and longest mountain belt on Earth is the mid-ocean ridge which largely occurs beneath the ocean floor. The mid-oceanic ridge is almost 60,000 km (~37,000 miles) in length, and it has a depth of ~2,600 m (~8,500 ft) and it rises to a height of ~2 km (~1.24) above the deepest portion of an ocean basin. This feature is where seafloor spreading takes place along a divergent plate boundary.

At either an oceanic-oceanic plate boundary or an oceanic-continental plate boundary (Fig. 9.21), the subducting plate gets heated, leading to releasing of water from the rock which triggers partial melting in the mantle wedge above, generating magmas. Some mountains which form the Aleutians in Southwestern Alaska and some mountains which form the Caribbean Islands such as La Soufrière, St. Vincent & Grenadines, and Mont Pelée, Martinique of the Lesser Antilles are an example of mountains formed due to oceanic-oceanic plate collision. The magma generated in the mantle wedge is erupted and/or intrudes into the overlying plate, generating mountains (Fig. 9.21). The Andes Mountains of western South America are an example of a mountain chain formed due to oceanic-continental plate collision. An accretionary wedge (Fig. 9.21) is formed where the oceanic slab subducts due to scrapping off of sediments, and the compressional forces are transmitted to the overriding plate, forming folds and thrusts above a décollement fault (Fig. 9.21).

A décollement fault is a basal detachment fault which acts as a gliding plane between different masses of rock. Décollement faults form due to deformation, producing different deformation styles of rocks above and below the fault. Décollement faults form in both compressional and extensional environments. Compressional environments are characterized by folds and thrusts, whereas extensional environments are characterized by basins and ranges horsts and grabens. Volcanic arcs, which form at the boundary

Fig. 9.11 Ptygmatic folds, Himalayas Mountains of northern India (credit: Martin Schmidt)

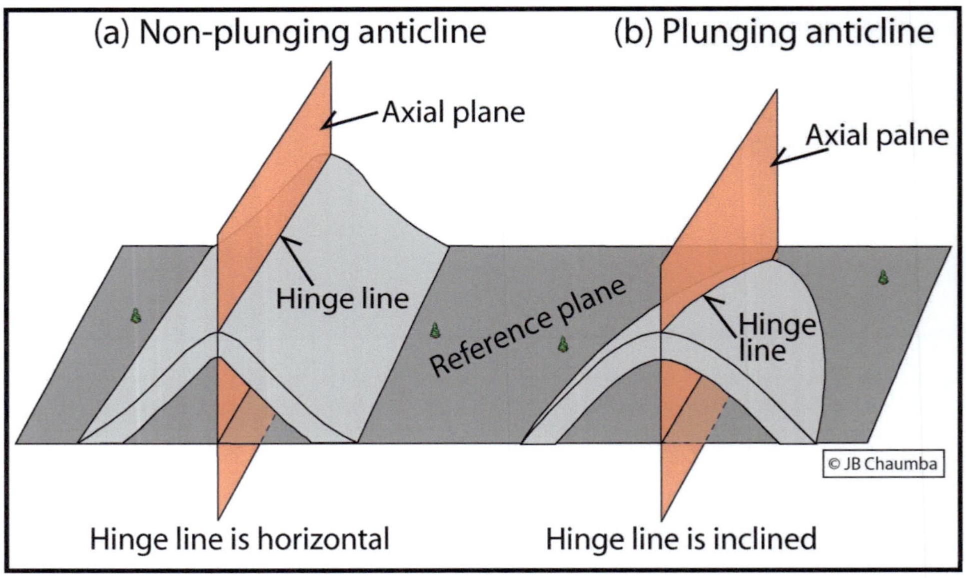

Fig. 9.12 Diagrams illustrating a non-plunging anticline in which the hinge line is horizontal (**a**) and a plunging anticline with a hinge line that is at an angle to the horizontal (**b**)

of two oceanic slabs through the subduction of one of them, often get joined at a **suture** zone to the margin of continents through **accretion** (Fig. 9.21). A zone where a crustal block was accreted to a continent is denoted by a suture (Fig. 9.21). Where folded rocks occur above the décollement fault, **fold-and-thrust belts** (Figs. 9.21 and 9.22). Fold-and-thrust belts are produced due to deformation of relatively thick sequences of shallow marine sediments. Accretion leads to the growth of continents as islands crush onto the margins of continents and are then added to the continents resulting in the formation of a larger continent. Slivers of oceanic crust that once separated either the sutured continents or a continent and an island are termed **ophiolites**, and ophiolites are common in convergent mountain belts. Finally, some mountains form due to rifting at such as the Basin and Range Province in Nevada, Utah, and Arizona in the western United States.

Examples of mountain ranges formed due to collision of continental crustal blocks are the Himalayas in Asia and the Alps in Europe. Thrust faults and reverse faults characterize collisional orogenies, with metamorphism, folding and foliation being intense at the center and becoming less deformed toward the edges of the mountains. Crustal thickening down to depths of ~70 km (43 miles) accompanied by uplifting of the continental crust to elevations of up to 8,848 m (29,031 ft), as in the case of the Himalayas, occurs when continents collide.

Fig. 9.13 Diagrams illustrating domes (**a**) and basins (**b**). Domes are formed from upwarping, whereas basins form as a consequence of down warping

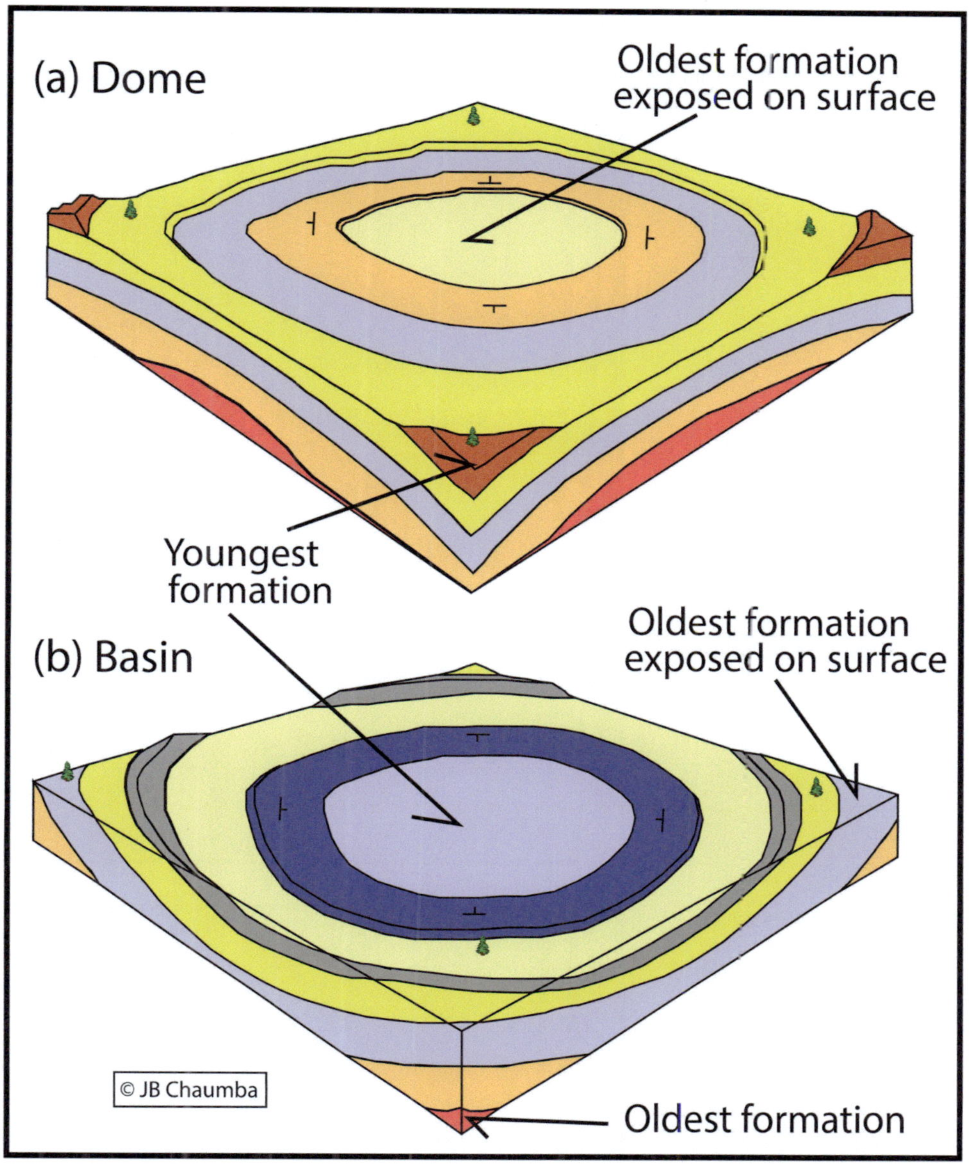

An **accretionary wedge** forms from sediments and ocean crust fragments that are scraped off the subducting slab and thrust against the developing volcanic arc (Fig. 9.21). When the oceanic plate on the leading edge of a continent is consumed due to subduction, the continental lithosphere eventually collides with the other continental lithosphere on the other plate, leading to the formation of a continent–continent orogeny (Fig. 9.22). The Himalayas are an example of a mountain chain formed as a consequence of continent–continent collision. Since continental crust is buoyant, it only subducts to limited depths which results in the continental crustal blocks colliding into each other to produce a collisional orogen (Fig. 9.22). Collisional orogeny results in the formation of huge volumes of metamorphic rocks, décollement faults, as well as fold-and-thrust belts (Fig. 9.22). The margin of the two continental blocks sutured at a continent–continent collision is also delineated by a suture zone (Fig. 9.22). A thrust fault that is hidden under the uppermost rock layers of the crust such that it is not exposed at the Earth's surface is called a **blind thrust fault** (Fig. 9.23).

Mountain ranges can form large portions of margins of some continents such as North America (Fig. 9.24) where younger rocks formed from deformation partially rim the margin of the continent. Interaction of plates leads to collisional mountains forming at convergent plate margins. Convergent margin plate interactions lead to the formation of subduction-related mountain ranges such as the Andes Mountains in western South America and the Rockies in North America. This subduction-related collisional orogen occurs where denser oceanic lithosphere is subducted beneath the buoyant continental lithosphere.

Fig. 9.14 Image of the Richat structure exposed in the Oudane area in Mauritania in the Sahara Desert. The structure is a dome with rock strata dipping away from the center (credit: NASA)

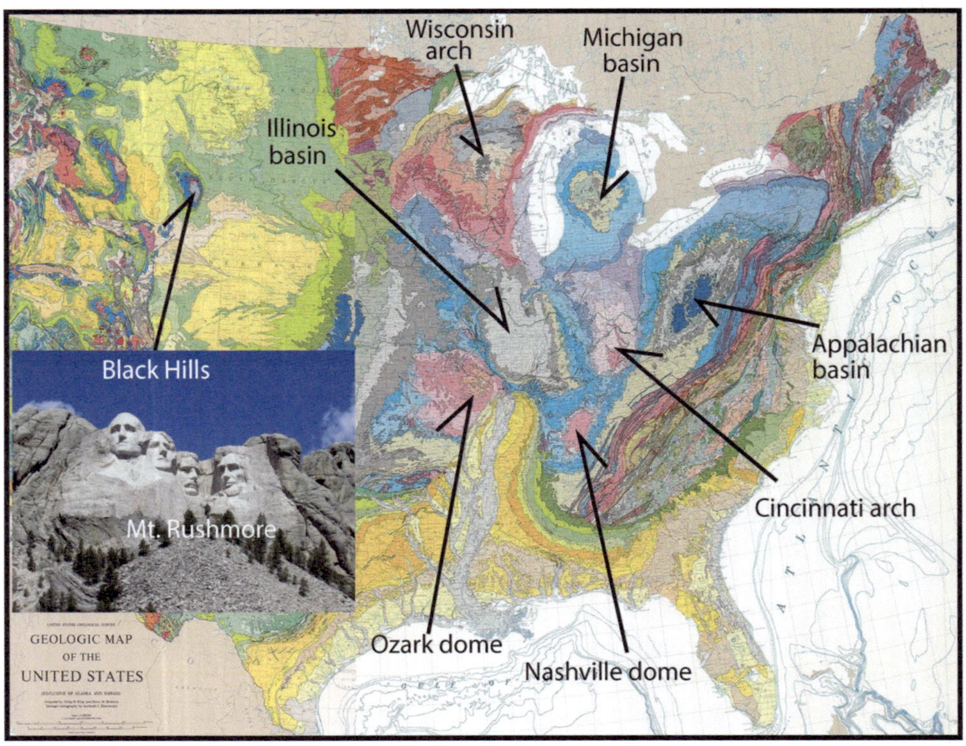

Fig. 9.15 May showing the geology of the eastern ~ two-thirds of the United States showing select domes and basins (geologic map of the United States: USGS)

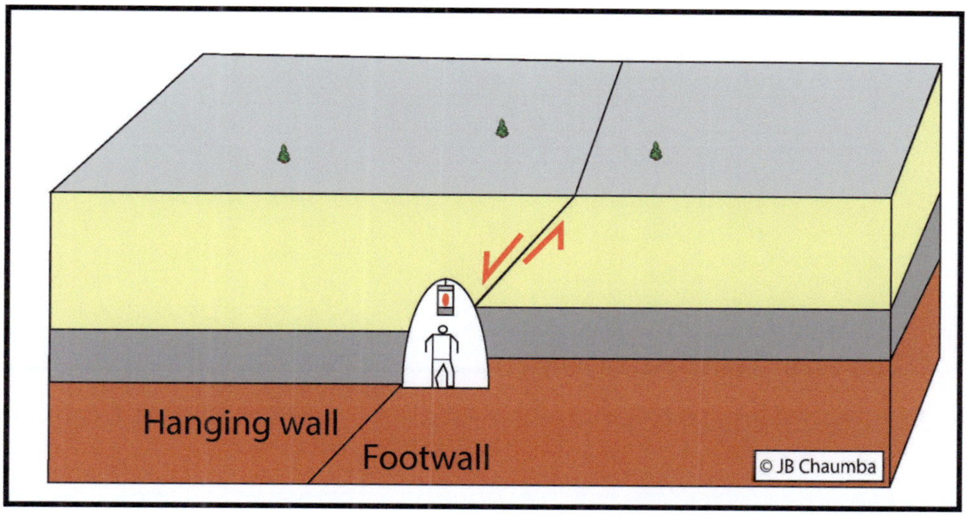

Fig. 9.16 The hanging wall is the block of crust located above an inclined fault (above the miner's head in the image), and the footwall is the block of crust below the fault (below the miner's feet)

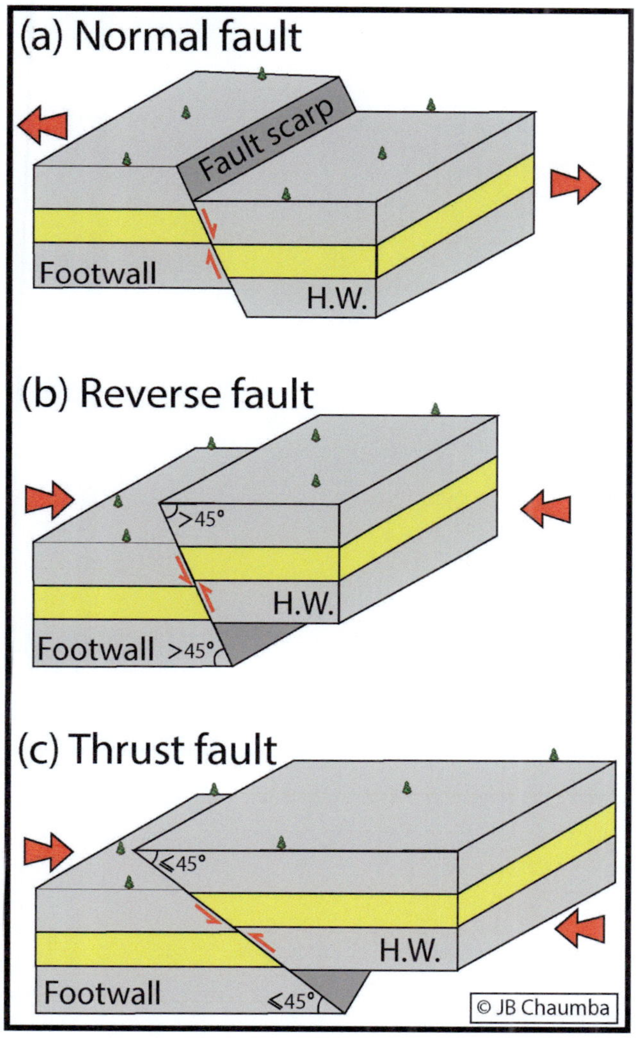

Fig. 9.17 Images illustrating a normal fault (**a**), a reverse fault (**b**), and a thrust fault (**c**). The angle of the fault with the horizontal is the key determinant in distinguishing between a reverse fault (>45°) and a thrust (≤45°)

Continental collision, however, also involves relatively small continental blocks. One consequence of plate collision involving smaller continental blocks either with continents or with each other is tectonostratigraphic terranes, or simply, **terranes**. A terrane is a package of rock sequence with distinctive geologic history and of regional extent that is bounded by faults. Some relatively large crustal blocks, large enough to form relatively small continents, termed **microcontinent**, comparable in size to the island of Madagascar are thought to have comprised some terranes. Geologists think that throughout geologic history, suturing of the smaller continental to the continents occurred at convergent plate margins. This suturing occurs to the margins of the overriding plate. Examples of terranes are the Wrangelia terrane on the western part of North America, the Meguma terrane and the Carolina terrane exposed in the northeastern and eastern part of North America, respectively, as well as Armorican terrane in Europe. In North America, the margin of the continent is composed of rocks generally younger than 600 million years (Fig. 9.24) that are thought to have been formed outboard of the continent but have since been accreted to the continent. The physiographic provinces of the eastern United States are also shown in Fig. 9.23, and most accreted terranes today form part of the Piedmont and Coastal Plain physiographic provinces.

Another example of an accretionary orogen, the Central Asian Orogenic Belt, is exposed in Asia (Fig. 9.25). Like other orogenic belts elsewhere, the Central Asian Orogenic belt is long and wide, consisting of rocks that have been folded, faulted, and eroded. This orogenic belt covers roughly 30% of the land surface area of the continent of Asia. The Central Asian Orogenic Belt is located within the boundaries of six countries of China, Kaza-

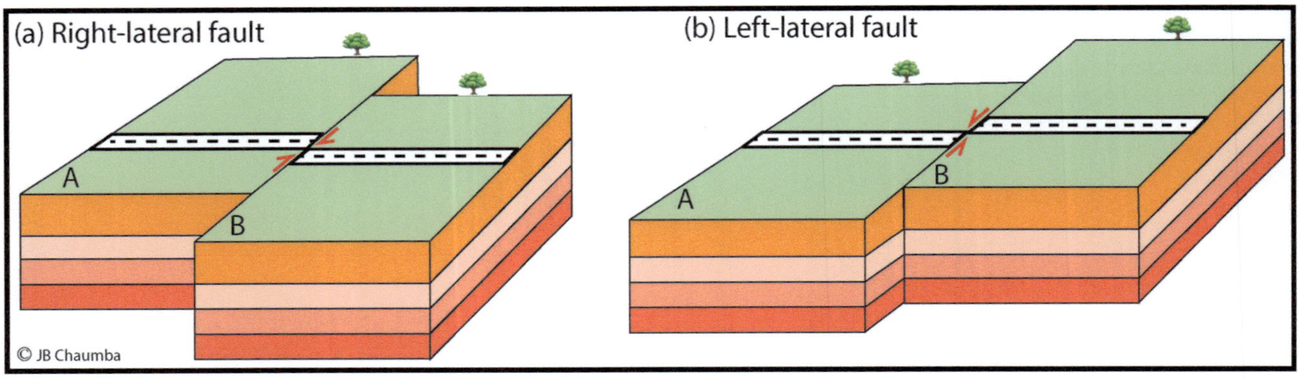

Fig. 9.18 Diagram illustrating the sense of movement along a right-lateral fault (**a**) and a left-lateral fault (**b**)

Fig. 9.19 The fault scarp that was formed on a section of the Leader Fault during the November 14, 2016, Kaikōura earthquake, New Zealand (credit: Katherine Pedley)

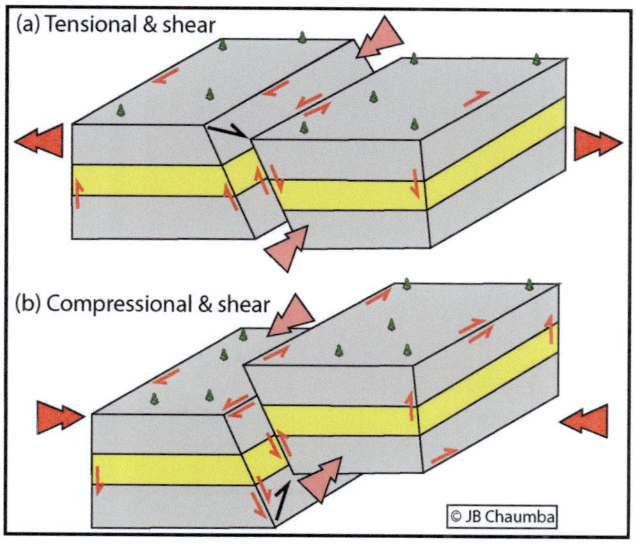

Fig. 9.20 Diagram illustrating the sense of movement on an oblique-slip fault. The red arrows indicate tensional and compressional arrows where the pink arrows indicate shear stresses. The black arrows in both arrows connect parts that were joint before slip-slip faulting

khstan, Kyrgyzstan, Mongolia, Russia, and Uzbekistan. In China, for example, crustal blocks such as the Tarim, Jungar, Qaidam, and Sulu (Fig. 9.25) are inferred to have been accreted to form part of Asia.

9.10 Origin of Earth's Varied Topography

The varied topography that we see on our planet can be explained using **isostacy**. The state of gravitational equilibrium that exists between lithosphere or crust of a planet and the mantle, enabling the crust to "float" at an elevation that is dependent on its thickness and density, is called isostacy. When different sizes of wooden blocks made of the same type of wood are inserted in the same bucket of water with the same volume of water (Fig. 9.26a), the water rises to different heights depending on the thickness and density of the wooden blocks (Fig. 9.26b–d). When a wooden block of pine is inserted in a jug, the block displaces water, and the top surface of the block is dependent on its thickness and its density (Fig. 9.26b). Oak, which is denser than pine, sinks to a deeper level than pine and its top level is lower (Fig. 9.26c) than that of pine (Fig. 9.26b). When a thicker pine block is now inserted in the same jug, it displaces more water than that displaced by the thinner block of pine (Fig. 9.26b).

Both oceanic and continental lithosphere, like the wooden blocks shown in Fig. 9.26, being less dense in comparison to the underlying mantle such as water in Fig. 9.26, "float" on underlying denser mantle. Conse-

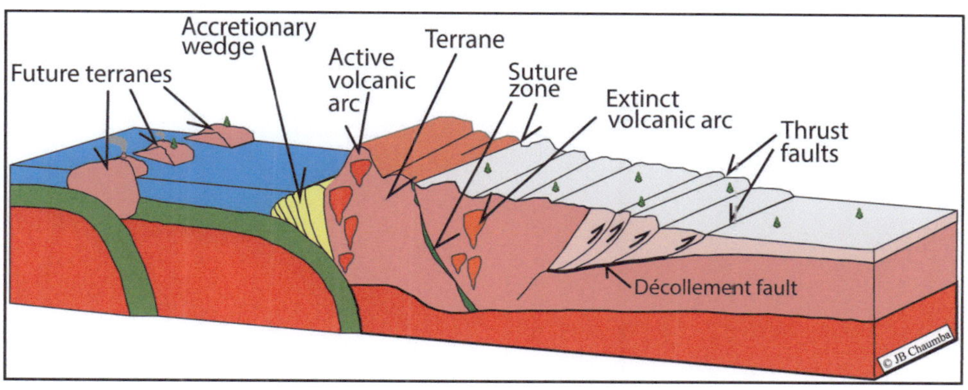

Fig. 9.21 Typical structures formed at an oceanic-continental plate boundary

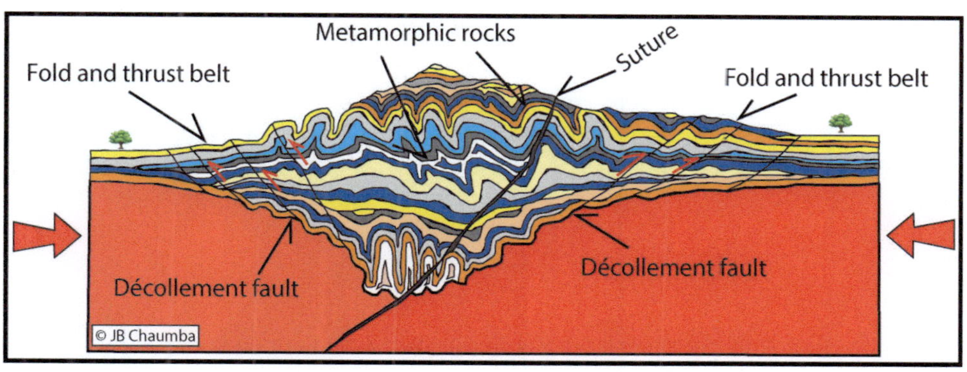

Fig. 9.22 Diagram illustrating features of collisional orogeny resulting from continent–continent collision. The red arrows indicate tensional stresses

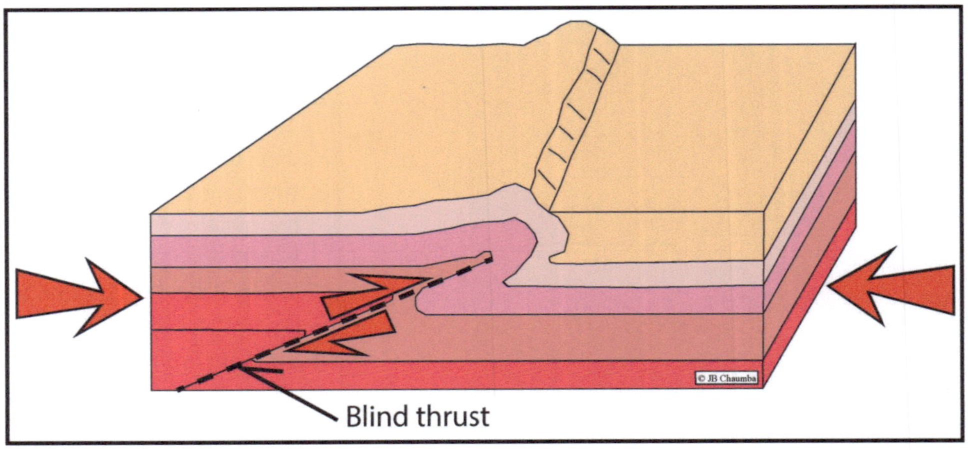

Fig. 9.23 Diagram illustrating a blind thrust fault which occurs below the Earth's surface

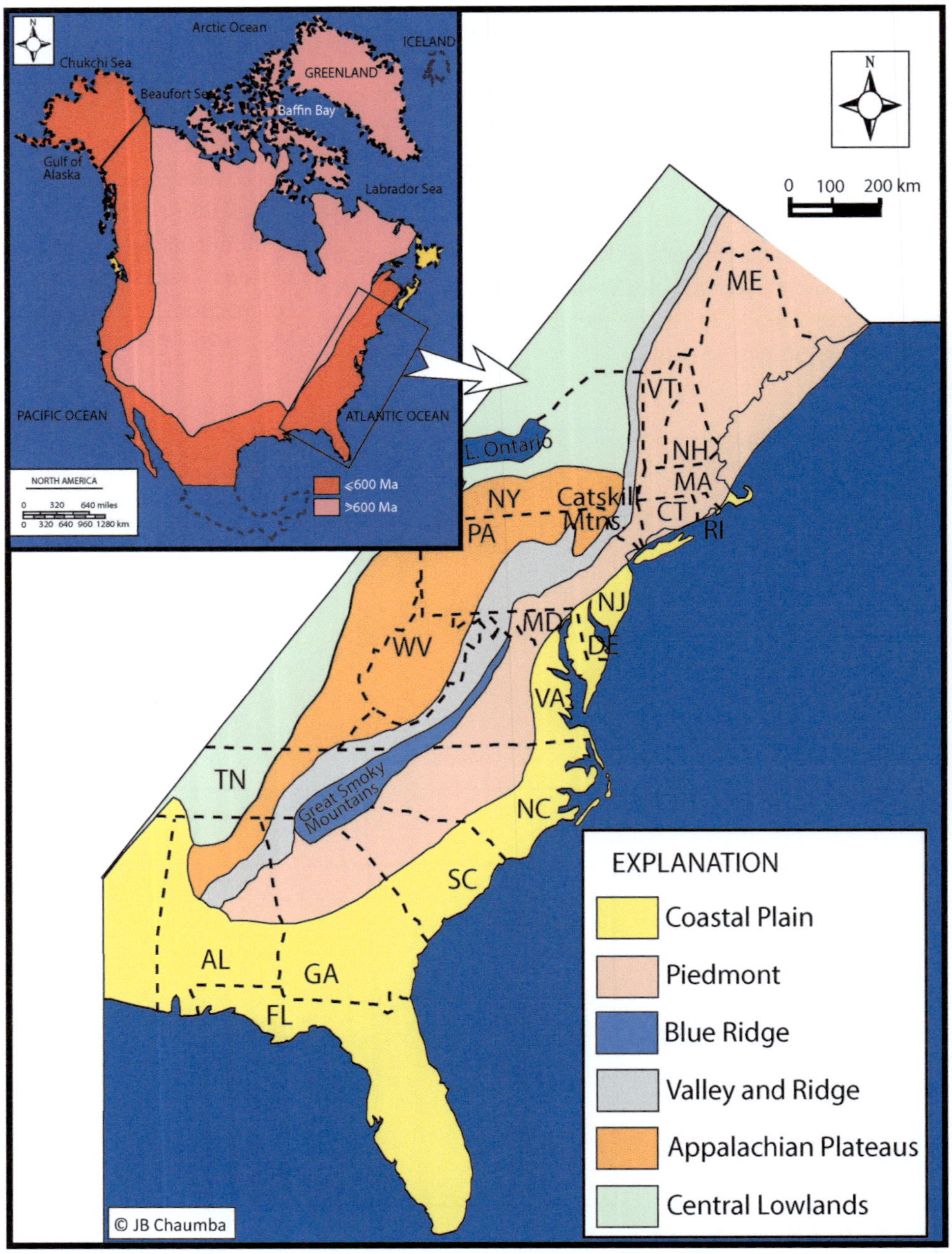

Fig. 9.24 Map showing accreted terranes of North America. The close-up shows the physiographic provinces of the eastern part of North America. The Piedmont physiographic province is composed of several accreted terranes

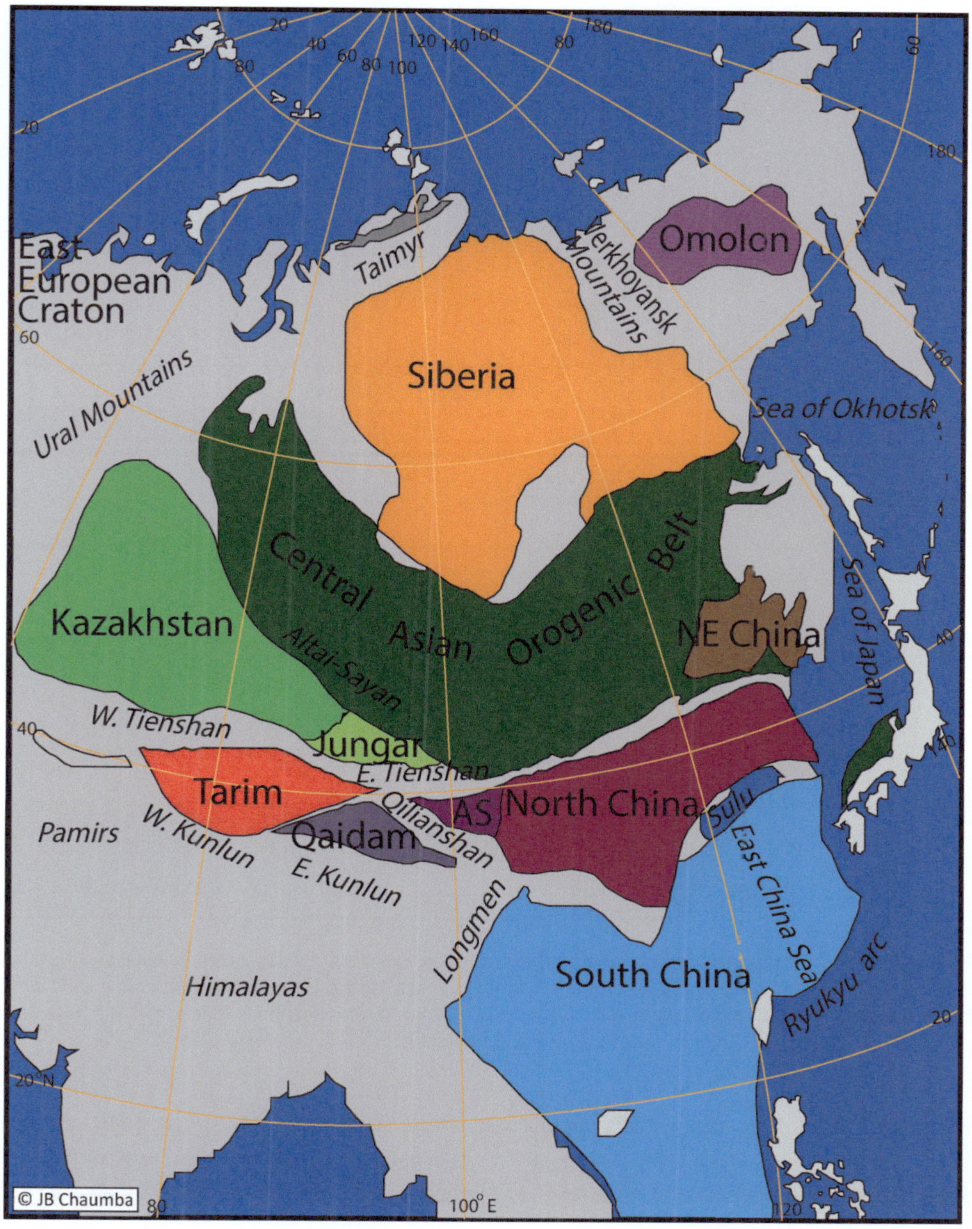

Fig. 9.25 Tectonic map of Asia showing the main terranes and the Central Asian Organic Belt

quently, where more weight is added to the crust, such as at Earth's mountain ranges, the depth to the asthenosphere increases, such as that shown in Fig. 9.26d. On average continental crust, the depth to the asthenosphere is shallow, such as denoted by Fig. 9.26b. **Isostatic adjustment** is this gravitational balance caused by loading and unloading of the crust.

Isostatic adjustment is illustrated in Fig. 9.27. Beneath mountain ranges, such as the Alps and Himalayas, a deep crustal root occurs. As the mountains grow vertically and horizontally due to continued collision, the crustal root also increases in size (Fig. 9.27a). However, when collision stops, erosion lowers the height of the mountains, accompanied by a corresponding decrease in the crustal root leading to rebounding of the crust (Fig. 9.27b). A similar process occurs when ice sheets build up on continents causing the crust to develop a root. When the ice sheets melt, crustal rebound occurs.

Compressional stresses that are experienced at continent–continent collisions lead to the formation of very high mountain ranges with elevations of up to 8,848 m (29,031 ft), as is the case in the Himalayas. Erosion is constantly acting on mountains even as they continue to be formed. However, as the mountains rise higher and higher, mass-wasting processes become more pronounced due to an increase in slope and weight. Thus, the forces acting as the base of high mountains also increase with increasing elevation of the mountains. This ultimately leads to rocks at the core of such mountains experiencing more weight such that they become hotter and weaker, often

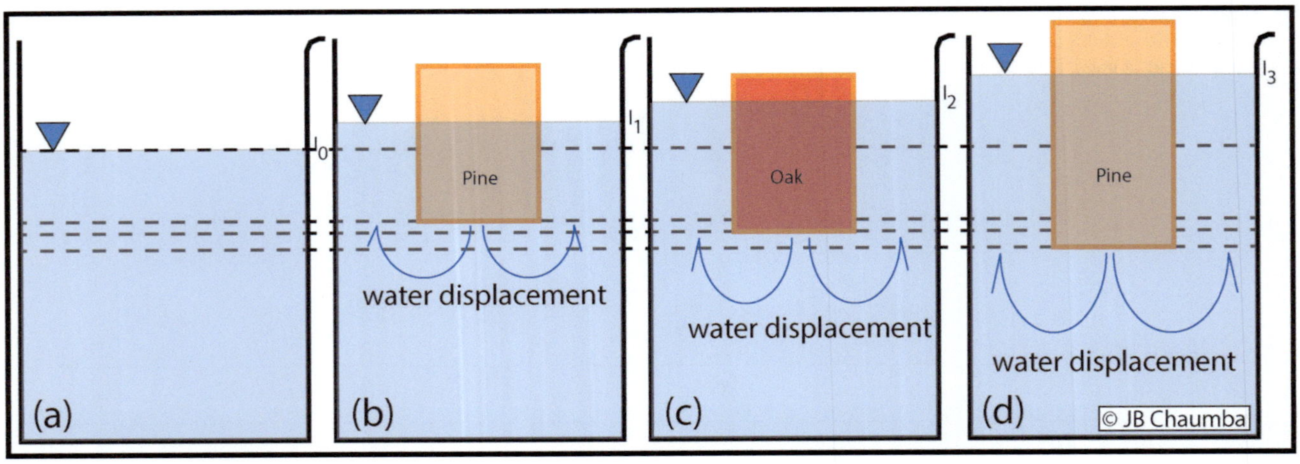

Fig. 9.26 Sketches illustrating the principle of isostacy

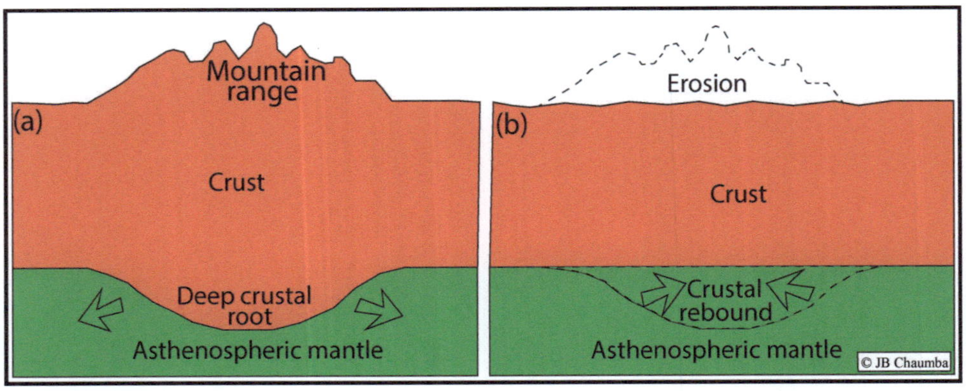

Fig. 9.27 Plots illustrating isostatic relationships between the crust and the lithospheric mantle. After a mountain building episode, mass has been added to a part of the crust to form thickened crust due to slow mantle displacement (**a**). The new mountain chain is eventually eroded away accompanied by crustal rebounding (**b**). The arrows represent slow asthenospheric mantle flow to produce the root (**b**), and asthenospheric mantle flowing back to its original position as erosion removes the mountain chain (**b**)

initiating lateral flow (Fig. 9.27a). Flow occurring at the bases of very high mountain ranges and in the asthenosphere triggers **gravitational collapse**. Faulting, in addition to ductile spreading at depth, together with subsidence in the crustal rocks is involved in gravitational collapse.

Glossary

Accretion A process by which crust is added to a continent at subduction zones at its edge due to subduction.

Accretionary wedge Also called an accretionary prism—is formed from sediments that are accreted onto the non-subducting tectonic plate at a convergent plate boundary.

Anticline A fold that is convex upward with older layers closer to the center of the structure.

Antiform A fold that is convex up.

Basin Strata dip toward the center in all directions and the youngest rocks occur at the center.

Blind thrust fault A type of thrust fault that is hidden under the uppermost rock layers of the crust.

Chevron fold Feature characterized by repeated folded beds with straight limbs and sharp hinges forming well-developed repeated set of V-shaped beds as a consequence of compressional stresses.

Compressional stress (Or shortening) squeezes rock together causing rocks to fold or fracture it and is common and convergent plate boundaries.

Confining stress The weight of all the material above a rock which pushes down and buries rocks.

Décollement Fault surface parallel that detaches or separates deformed rocks above from undeformed or differently deformed rocks beneath.

Differential stress Stress that is not equal from all directions, as stress in the Earth's crust.

Deformation Tectonic process which affects the size, shape, or volume of part of the Earth's crust.

Dip-slip faults Inclined fractures where the blocks have mostly shifted vertically.

Dome an Elliptical structure formed by upwarping of rock strata resulting in strata dipping away from the center where oldest strata in domes occur.

Elastic deformation Deformation in which a rock returns to its original shape when the stress is removed.

Fault scarp Feature on Earth's surface that resembles a step caused by slip on the fault.

Fold Undulation in rock strata caused by compressional forces.

Footwall The block of rock that lies below an inclined fault or a mineral deposit.

Fault Fracture between two blocks of rock along which movement has occurred relative to each other during an earthquake.

Fold-and-thrust belt A series of mountains forming foothills adjacent to an orogenic belt which formed as a result of compressional stresses.

Fracture A general which includes any break in rocks.

Gravitational collapse The decay of forces that support the deformed and thickened lithosphere due to the weakness of the lithosphere causing mountains to experience orogenic collapse.

Hanging wall The upper block or overhanging wall of an inclined fault or vein.

Horst An upthrown block that is located next to a graben in an extensional environment.

Isoclinal fold Very tight fold in which the fold limbs are parallel or almost parallel to one another.

Isostacy The Archimedes principle of hydrostatic balance between floating bodies such as the lithosphere "floating" on the asthenosphere.

Isostatic adjustment Nonelastic response of Earth's lithosphere to loading and unloading caused by erosion, deposition, ice accumulation, and deglaciation.

Joint A break in a layer or body of rock in which no visible or measurable movement occurred.

Left-lateral fault A type of strike-slip fault where the left block moves toward you and the right block moves away if you were to stand on the fault and look along its length.

Microcontinent Isolated pieces of continental crust, such as the size of Madagascar, that are surrounded by oceanic crust as well as seamounts and oceanic plateaus.

Moho, or Mohorovičić discontinuity Is the boundary between the Earth's crust and the mantle.

Monocline A step-like fold in rock strata consists of a zone of steeper dip within horizontal or gently dipping strata.

Non-plunging anticline An anticline whose hinge line is horizontal or parallel to sea level.

Normal fault A fault along which movement of the hanging wall is downward (down dip) and the footwall moves upward (up dip).

Normal stress Stress that is perpendicular to a given plane. Normal stress can either be compressional or tensional.

Oblique-slip fault A fault with a combination of both shear stress and extensional or compressional stresses.

Pascal's law In a fluid at rest in a closed container, a change in pressure in one part is transmitted without loss to every portion of the fluid and to the walls of the container.

Plastic deformation Irreversible strain imparted in the rock during deformation.

Plunging anticline A fold whose hinge line is not horizontal (not parallel to sea level).

Ptygmatic fold Folds that are chaotic, random, and often disconnected.

Recumbent fold Fold with a linear horizontal axial plane oriented at a low angle.

Reverse fault A fault inclined at an angle of >45° to the horizontal in which the footwall moves down dip and the hanging wall moves up dip.

Right-lateral fault A type of strike-slip fault in which the right block moves toward you and the left block moves away if you were to stand on the fault and look along its length.

Shear stress A type of stress which occurs where parallel forces act in opposite directions and it is common at transform plate boundaries.

Strain Change in shape or size resulting from applied stresses.

Strike-slip fault A fault along which two blocks slide past one another

Suture A zone where a continent or a crustal block was accreted to a continent.

Syncline A down folded structure with younger layers closer to its center

Tensional stress Stress perpendicular to a given surface that pulls rocks apart.

Terrane A package of rock sequence of regional extent with distinctive geologic history that is bounded by faults.

Thrust fault A shallow (≤45°) dip-slip fault in which the upper block, above the fault plane, moves up and over the lower block.

10 Fossil Evolution and Geologic Time

Abstract

The geologic history of our planet can be constructed from various techniques that are discussed in this chapter, such as the law of faunal succession, relative and radiometric dating, and cross-cutting relationships. Although Earth was formed ~4.54 billion years ago, currently known oldest dated rocks are 4.16 billion years old. The various ways through which fossil preservation occurs are discussed with the aid of illustrations. Diagrams and photographs, together with fossils, fossil assemblages, and index fossils that are used in deciphering Earth history are explained. Unconformity is defined, and various types of unconformities are discussed and illustrated using diagrams. The importance of radiometric dating in constraining events in Earth history, half-life and the three types of radioactive mechanisms, as well as the geologic time scale, are discussed and illustrated using diagrams.

Learning Goals

Studying this chapter will enable you to:

- Recognize the immensity of geologic time.
- Describe what evidence proves the immensity of geologic time.
- Recognize and differentiate types of unconformities.
- Describe relative dating.
- Deduce the geologic history of an area.
- Define principle of faunal succession.
- Define radiometric dating and provide examples of common dating techniques.
- Decipher Earth history utilizing a combination of relative dating and radiometric ages.
- Recognize the major subdivisions of, and utilize, the geologic time scale.

Supplementary Information The online version contains supplementary material available at https://doi.org/10.1007/978-3-032-20231-4_10.

10.1 Introduction

Almost every child starts going to school at a certain age, which is in general around the age of 5 or 6 years, no matter where on Earth that child lives. Consequently, governments can plan ahead for schools, colleges, and jobs for their citizens. Getting to know when a person was born, thus, is of importance. When it comes to our planet, human beings have always been fascinated by when Earth was "born," when life on Earth began, and when human beings first lived on Earth. No doubt attempts to determine how old our planet is date back centuries. In 1650, a high-ranking member of the clergy, the Irish Archbishop James Ussher, calculated that Earth was created in 4004 BC, suggesting that Earth was 5,650 years old. In the 1700s, **catastrophism**, or the belief that the varied landscapes such as canyons and mountains of our planet was a consequence of huge but sudden catastrophes, dominated. The causes of the catastrophes were not known, but they were short-lived since Earth was believed to be only less than 6000 years old.

The birth of modern geology is regarded as having commenced in the late 1700s, coinciding with James Hutton's *Theory of the Earth* book. In this book, he presented the principle of **uniformitarianism**, which states that the biological, physical, and chemical laws operating today have also been operating in the past. An understanding of processes and forces that happened on Earth in the geologic past can be gained by studying processes and forces that are operating on our planet today. Thus, "the present is the key to the past," is this idea of understanding current processes occurring on our planet to understand processes that shaped it in the past. James Hutton helped Earth scientists realize that over long periods of time,

J. B. Chaumba, *Introductory Physical Geology*, Springer Textbooks in Earth Sciences, Geography and Environment, https://doi.org/10.1007/978-3-032-20231-4_10

geologic processes produce huge amounts of change and form gigantic features such as the Himalayan Mountains and the Grand Canyon.

10.2 Time Scale of Geologic Events

The time that has elapsed since the formation of our planet billions of years ago is termed **geologic time**. Earth is now known to have been formed a very long time ago. Age determination of samples of meteorites (see Chap. 1) thought to have been formed at the same time as our planet has yielded ages 4.55 billion years old (Ga). The oldest known dated rocks on Earth, from the Nuvvuagittuq Greenstone Belt in Canada, are 4.16 billion years old, in the Hadean eon. These oldest known rocks are exposed along the eastern shore of Hudson Bay in northeastern Quebec near the Inuit municipality of Inukjuak. They are volcanic rocks composed of dark and light green-colored bands as well as pink and black flecks. Prior to this recent discovery, the Acasta gneiss found which forms part of the Canadian Shield in the Northwest Territories, Canada, with an age of 4.03 billion years old, were the oldest known rocks on Earth. As discussed in Chap. 1, the earliest-formed rocks on Earth are thought to have been reworked by plate tectonic processes. Later in this chapter, we will discuss how you can determine which rock formations are younger or older based on relative dating, as well as precise ages of when some rocks solidified from magma. Thus, our planet has had a very long history, on the scale of up to several billions of years, which enabled its various landforms and features to form.

In addition to the Canadian Shield, other shields exposed in Australia (Yilgarn and Pilbara cratons), South Africa (Kaapvaal craton), and Zimbabwe (Zimbabwean craton) also contain some of the oldest known rocks on Earth (Figs. 10.1a, b, c). As shown in Fig. 10.1 from the Kaapvaal craton, some of Earth's oldest rocks form part of these shields. The long expanse of time that has elapsed since some rocks were formed can be exposed where rivers incise into earlier-formed rock formations, such as in the Grand Canyon in the southwestern United States (Fig. 10.2a). Here, the Colorado River has exposed the 1.75-billion-year-old Vishnu Schist, which forms the basement onto which overlying strata were deposited. Although just above the Vishnu Schist, there is a period in which erosion removed the Vishnu Schist and thus no deposition occurred, multiple layers of the Grand Canyon record sedimentation that occurred. The 270-million-year-old Kaibab Formation, which forms the rims of the canyon and is a mere, are encountered at the topmost part of the Grand Canyon.

Some sedimentary layers represent relatively younger rocks of Earth history. Within the Runde River valley in Southern Africa, the Chilojo cliffs (Fig. 10.2b) were formed by red sandstones of Cretaceous age (we will discuss the geologic time scale later in this chapter). The Chilojo Cliffs (Fig. 10.2b), which crop out in the extreme south-eastern corner of Zimbabwe, are composed of the coarse-grained sandstones and conglomerates of Jurassic/Cretaceous age.

10.3 Fossil Preservation

A **fossil** can be defined as the remains, traces, or impressions of a once-living animal and/or plant. Consequently, sedimentary rocks contain remnants of once-living organisms such as leaves, shells, bones, and imprints of once-living organisms called **trace fossils** or ichnofossils. For any animal or plant to be turned into a fossil, suitable conditions must first be obtained. During **fossilization**, mineral and/or bone constituents are preserved into hard and/or petrified forms when suitable conditions are met. The study of fossils, including their age, method of formation, and their importance in evolution, is called **paleontology**. Sedimentary rocks commonly contain beds containing fossils.

Some parts of the Earth where sedimentary basins existed coinciding with the time when unique animals and organisms roamed our planet preserved those animals and organisms. The Morrison Formation (Fig. 10.3a), in which numerous dinosaur fossils once roamed and are preserved, is one such example. Several fossils of different dinosaur species of *Stegosaurus, Allosaurus*, and *Apatosaurus* can be viewed at the National Park Service in northwestern Colorado (Fig. 10.3a, b). Exhibits, including an 80-foot-long mural and other exhibits which reveal the history of dinosaurs (~1,500 dinosaur bones) and many other organisms that once roamed the Morrison environment during the Late Jurassic, can also be viewed in the Quarry Exhibit Hall. Dinosaurs have fascinated humans due, perhaps, to their large sizes, numerous varieties, and inferred way of life. However, it is important to note that several different organisms must have roamed our planet before, during, and after dinosaurs existed. We must also bear in mind that fossils likely represent only a small percentage of life that has existed on Earth since its formation. In order for the remains of once-living organisms to die and be preserved, suitable conditions must exist. Let us now discuss how animals and plants can be preserved to form fossils.

Various dinosaur fossils are also preserved in Dinosaur Ridge, Colorado (Fig. 10.4) as bones (Allosaurus, *Stegosaurus*, and *Apatosaurus*) as well as tracks (*Ornithominus* in Green Mountain, Colorado, *Acrocanthosaur*—a small meat eating dinosaur, *Eolambia*). *Eolambia* tracks, however, haven't yet been found in Colorado but Utah.

The process of a once-living organism becoming a fossil is called fossilization. Fossilization is a very rare process: of all the organisms that have lived on Earth, only a tiny percentage of them ever become fossils. If you were to

Fig. 10.1 **a** Lower and Upper Zones of the Stolzburg complex, Mawelawela Valley, view looking northeast. Fig Tree and Moodies Groups rocks are exposed in the distance. **b** Pillow basalts preserved in some amphibolites of the Barberton greenstone belt. **c** Pre-3.5-billion-year-old Theespruit formation, Barberton greenstone belt, showing folded barite-bearing felsic schists (*credit* C.R. Anhaeusser)

imagine, for example, the likely large number of gazelles that die in the Hwange National Park in Southern Africa per year. Their bodies are quickly eaten by hyenas, vultures, and other scavengers such that any remaining flesh is consumed by bacteria and insects and only bones are left behind. Even those bones are later fragmented into small pieces and are eventually destroyed themselves such that their nutrients are recycled in the soil. Thus, it is rare for any of the remains of the gazelle to be preserved as a fossil. In rare instances, however, a gazelle might be trapped and quickly buried in mud, and it is that gazelle that might be preserved and form a fossil that can be "found" millions of years from today. Now let's discuss the possibility that few organisms can be preserved to form fossils.

Fossilization can occur in several different ways. Some fossils can be perfectly preserved remains. An example of fossil preservation as preserved remains occurs when an original skeletal material and soft tissue such as insects is trapped in ancient tree sap called **amber** (Fig. 10.5a). Soft tissues such as insects, animal fur, and Neanderthal hunters can be well preserved in amber or can be frozen in glaciers, and their DNA can be studied and compared to those of animals now inhabiting our planet. **Permineralization**, another common fossilization method, occurs when wood fragments, shells, or bones are subjected to mineral-rich water that percolates through them while replacing some of the organic tissues with silica, pyrite, or carbonates. For example, petrified wood (Fig. 10.5b), many marine fossils,

Fig. 10.2 Images showing some of the sedimentary beds that reveal Earth history. **a** View of the Colorado River flowing in part of the Grand Canyon, Western USA (*credit* D.T. Nelson). **b** Image of sedimentary rocks forming the Chilojo Cliffs exposed along the banks of the Runde River, Gonarezhou National Park, Southeastern Zimbabwe. Elephant for scale (*credit* Tourism Zimbabwe)

Fig. 10.3 View of part of the Morrison Formation in which numerous dinosaur fossils are located. a Some dinosaur fossils that can be observed at the Quarry Exhibit Hall, Colorado (*credit* D.T. Nelson)

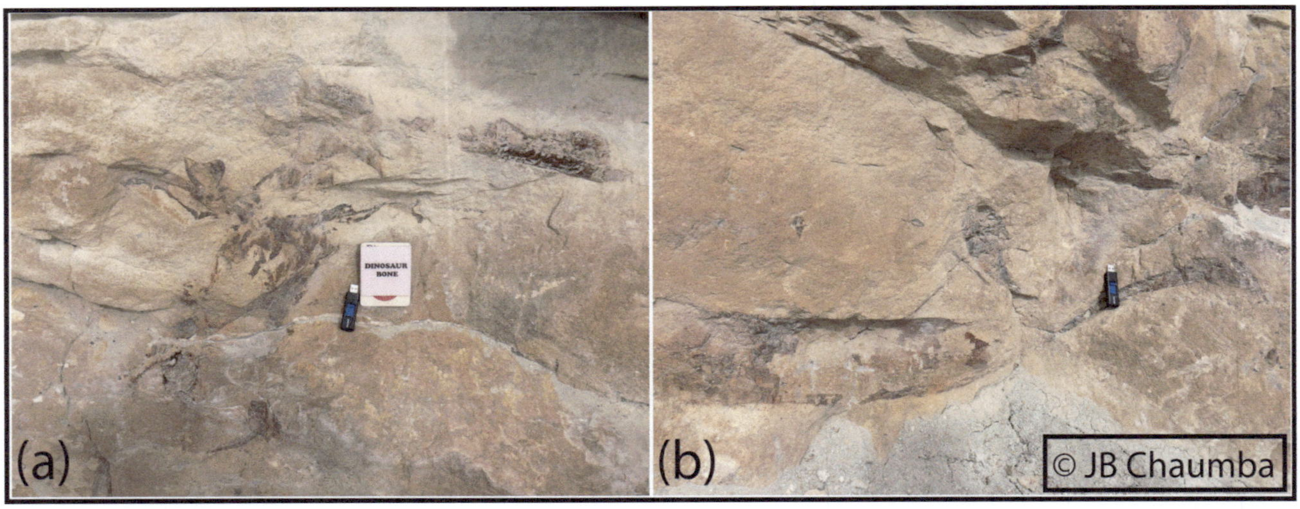

Fig. 10.4 **a** and **b** Some images of dinosaur bones exposed at Dinosaur Ridge in Colorado

and some fossil dinosaur bones were fossilized through permineralization.

Physical evidence of the way-of-life activities of extinct organisms is called trace fossils. Trace fossils are fossils that are not of the animal's actual body. Examples of trace fossils are footprints/tracks, trails, burrows, feeding marks, and resting marks. Dinosaurs, for example, left footprints along ancient rivers (Fig. 10.5c). Worms which burrowed in soft mud in some ancient oceans also formed some hollow tubes.

Exposed around Florissant in Teller County in central Colorado, in the Western United States is the Florissant Formation. The sedimentary rocks in the Florissant Formation are renowned for the abundant, diverse, and well-preserved fossil deposits of plants and insects (Fig. 10.5d) that are hosted in the mudstones and shales. An **impression** fossil, which is a type of trace fossil, forms when a leaf, skin, foot, or even a shell produces an imprint in mud or soft ground which later hardens to form a fossil preserving the shape of the original plant or animal (Fig. 10.5d). **Carbonization** can preserve leaves (Fig. 10.5e) by forming a carbon film impression of tree leaves through encasing an organism's remains by fine-grained sediments.

Fossil dung (Fig. 10.5f), or **coprolite**, are fossilized feces of animals that lived millions of years ago. Coprolites are other types of trace fossils, and they portray information about the stomach contents and feeding habits of the ancient animal. Shapes of coprolites can be studied by comparing them to those of modern analogues.

In some cases of fossilization, the original bone or shell can dissolve away to leave behind an empty space in the shape of the shell or bone. The original shell can also be replaced by a different mineral altogether such as pyrite or quartz. The depression that is formed is called a **mold** (Fig. 10.5g) and it only reflects the shape of an organism or its surface marking. A mold, however, does not portray information about the organism's internal structure. Hardened sediment that fills a **cast** is in the actual shape of the bones of the animal. The space may eventually be filled with other sediments to produce a cast (Fig. 10.5g) that can match the shape of the original animal's bones. Since their shells easily dissolve away, mollusks tend to occur as molds and casts.

Entirely different minerals can replace the original shell in some cases, or a bone can dissolve away altogether. For example, shells that were originally composed of calcite may be replaced by dolomite, quartz, or pyrite. If quartz fossils are surrounded by a calcite matrix, the calcite can be dissolved away by acid, leaving behind an exquisitely preserved quartz fossil. In other examples of fossilization, some fossils may form when their remains are compressed after being subjected to high pressures. When this happens, a dark imprint of the fossil can be left behind. Compression commonly affects fossils of ferns and leaves. The fossil record is important in understanding Earth history as it permits Earth scientists to learn about what the environment on Earth was like dating back to several hundreds of millions of years ago.

Although various types of animal and plant fossils have been discovered and described in the rock record, it is envisaged that these fossils only represent a very small percentage of the organisms and plants that have existed on Earth. Preservation of fossils is known to be possible in areas where rapid burial of dead organisms can occur. Since soft parts are easily destroyed by scavengers and decomposition by bacteria, organisms with hard parts are more likely to be preserved in the fossil record than those with soft parts.

Fig. 10.5 Types of fossils. **a** Scorpion in amber (*credit* Oregon State University), **b** Photograph of petrified wood, Petrified Forest National Park, Arizona, USA (*credit* D.T. Nelson), **c** Dinosaur footprint, Chewore inlier, Zimbabwe (*credit* Geological Society of Zimbabwe), **d** *Prodryas Persephone* butterfly fossil, Florissant Fossil Beds National Monument, Charlotte Hill, Colorado, USA (*credit* U.S. National Park Service), **e** Carbon film impression of Redwood tree leaf (*credit* Natural History Museum, Humboldt State University), **f** Coprolite from Salmon Creek, Lewis County, Washington, USA (*credit* Mindat), **g** Mold and cast (*credit* Nayturr.com)

10.4 Relative Dating

Relative dating is the determination of which rock unit is older or younger than another in any sequence of rocks. Relative dating is akin to stating that one classmate is older than another student without mentioning their dates of birth. Some principles discussed below are useful in determining the relative ages of geologic units and past geologic events. These principles were developed in the seventeenth century by the Danish scientist Nicholas Steno.

The *principle of superposition* states that within a sequence of layers of sedimentary beds, the oldest bed occurs at the base and that the layers progressively become younger with ascending order in the sequence (Fig. 10.6a–c). Any sedimentary bed, therefore, must be younger than the one that underlies it. The oldest and youngest layers must be, respectively, located at the bottom and top of a sequence of sedimentary rocks.

The *principle of original horizontality*, which states that, when sedimentary beds are horizontal when they are originally deposited (Fig. 10.6a–c). As sediments are being deposited, the force of gravity enables the sediments to accumulate in beds that are oriented horizontally. At the margins of a sedimentary basin, however, the beds may be slightly inclined as they conform to the slope of the underlying basement rocks. Subsequent to their formation, however, plate tectonic processes may lead to tilting the originally horizontally deposited beds such that they lie at an angle to the horizontal.

The *principle of lateral continuity* states that when sedimentary beds are formed, they are laterally continuous in

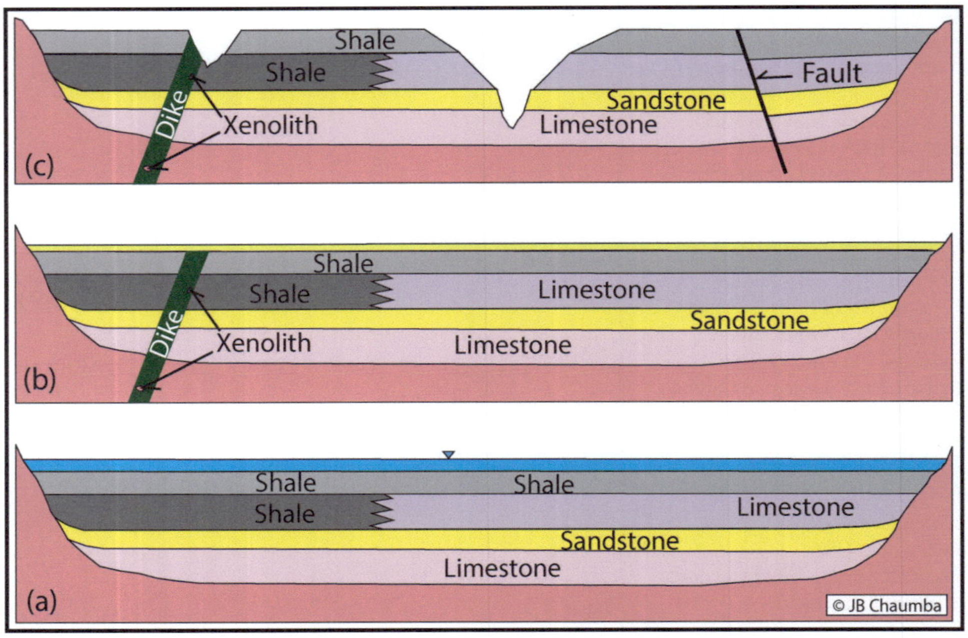

Fig. 10.6 Diagrams illustrating the lateral original horizontality and continuity of sedimentary beds (a, b, and c). The laterally continuous sedimentary beds can also grade into one another. **a** is a depiction of beds under water, **b** is a depiction of a time when the water has evaporated, and **c** is a depiction of the beds after erosion has affected some of the layers through incision by rivers to form a valley (to the left) and a canyon (to the right)

all directions although they may grade into different rock types or thin out at the margins of the sedimentary basin (Fig. 10.6a–c).

The *principle of cross-cutting relationships* states that a geologic feature that intrudes or cuts across another is younger than the feature that it intrudes or cuts across. Thus, both the dike and fault that cut across the sedimentary beds and basement rocks in Fig. 10.5b and c are younger than the rocks they intrude or cut across. Even the valleys in Fig. 10.10c are younger than the sedimentary rocks and dike they were incised into. The fault which cuts and displaces the rocks in Fig. 10.6b must be younger than the rocks it displaces.

The *principle of inclusions* states that a rock that contains fragments of another rock, called an inclusion, must be younger than the inclusion (Figs. Fig. 10.6b, c). The inclusions of the basement rock and shale contained by the dike (Figs. Fig. 10.6b, c) must be older than the dike which hosts them.

The *principle of baked contacts* states that any igneous intrusion that induces contact or thermal metamorphism to the rocks it intrudes is younger than the thermally metamorphosed rock or aureole. Thus, a rock containing the aureole is older than the intruding rock.

The principles that have been described are used by Earth scientists to determine the **geologic history** of an area. This involves determination of relative ages of sequence of events such as deposition, erosion, intrusion, faulting, folding that occurred in an area under study. If one were to encounter an area with the geology depicted in Fig. 10.7, an application of these principles comes in handy in unraveling its geologic history. In describing the geologic history of an area Fig. 10.7, gneisses denote the oldest rocks. Granitic rocks, which are the second oldest rocks in the section after the gneisses, later intruded the gneisses and in so doing incorporated xenoliths of the gneiss. The area was then subjected to erosion which affected the crystalline rocks to form a nonconformity. Above the nonconformity, a phase of deposition of sedimentary beds started, with the lowermost sedimentary beds containing inclusions of the underlying crystalline rocks. A dike, which contains xenoliths of them, was then intruded. Intrusion of the dike marked the final activity in the geologic history of the area shown in Fig. 10.7.

Dikes and xenoliths are commonly encountered on outcrops and in quarries. Shown in Fig. 10.8a is an inclined dike on a face of Bolingbroke Quarry in Monroe County, Georgia. The diabase dike, which cuts across the lithologies in the quarry, must be younger than all the other rock types that it intrudes. The mafic xenoliths (Figs. Fig. 10.8b, c), which are fragments of country rocks incorporated in the porphyritic intrusive rock exposed in Anson County, North Carolina, must be older than the Lilesville granite containing them.

10.5 Deciphering Earth History from Fossils

A group of species of fossils contained within a bed of sedimentary rocks is called a **fossil assemblage**. Due to extinction of fossils, a limited sequence of beds tends to contain a category of fossils which is not found in rock strata stratigraphically below or above that sequence of beds. Each

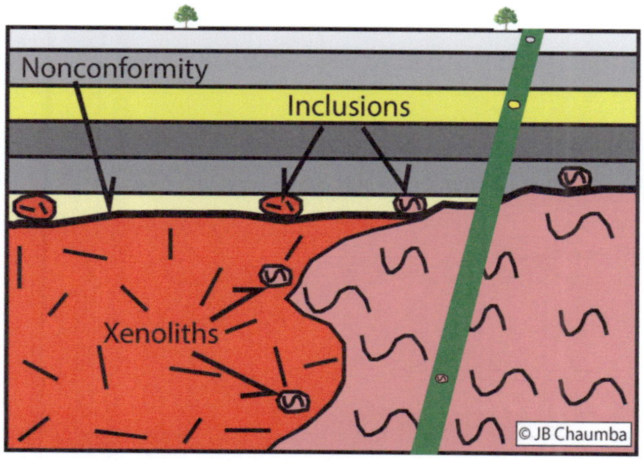

Fig. 10.7 Geologic cross section of an area. Gneisses, which are the oldest rocks in the section, were intruded by an igneous rock which incorporated gneiss xenoliths followed by erosion of the crystalline rocks to form a nonconformity. A period of deposition of sedimentary beds of various lithologies followed, with the lowest beds containing inclusions of underlying crystalline rocks. Intrusion of a dike with xenoliths of intruded rocks marked the final phase of the geologic history of the area

sedimentary bed is characterized by a unique fossil assemblage irrespective of where it was formed since life forms on Earth have changed through geologic history. The kinds of fossils occurring in rocks of different ages are, thus, also different. The *principle of fossil succession*, or *the law of faunal succession*, states that fossil species succeed each other in a definitive, recognizable order as once a species becomes extinct, it disappears and does not reappear in younger rocks. Consequently, strata from different areas that are characterized by the same types of fossils are of the same age. These ideas are ascribed to the British engineer William Smith.

An **index fossil** is any plant or animal that was preserved in the rock record such that it is characteristic of a particular short span of time or the environment in which it lived. An index fossil that is useful is abundant, easily recognizable or distinctive, and of a wide geographic distribution. Consequently, index fossils are critical in the correlation of strata.

Shown in Fig. 10.9 are two outcrops of fossil-bearing sedimentary beds. The brachiopod, which is only observed at outcrop A but not at outcrop B, won't be found in younger rocks. From the fossil assemblages observed at these two outcrops, an interval in the sequence in which the fossil species, called a **range**, can be determined (Fig. 10.9).

Based on the fossils observed at the two outcrops, the brachiopod is the oldest fossil, and the gastropod is the youngest fossil. The range from the two outcrops, as shown in Fig. 10.9, overlaps. Relative ages of beds in which the fossils occur are determined from the fossils. Although the sedimentary bed containing the gastropod at outcrop B does not contain a brachiopod fossil, it can be determined that this bed is younger than the brachiopod-bearing bed at outcrop A. Stratigraphic ranges and origins of some major groups of animals and plants are shown on the Geologic time scale in Fig. 10.10.

The Geologic time scale was developed in the nineteenth century using relative dating as a basis. Subdivisions of the Geologic time scale are into eons, eras, periods, and epochs (Fig. 10.10). The greatest expanse of time on the Geologic time scale is represented by eons—the Hadean, Archean, Proterozoic (these three are part of the Precambrian) and

Fig. 10.8 **a** An inclined dike exposed on the wall of Bolingbroke Quarry near Juliette, Monroe County, Georgia, USA. Steve Clark for scale. **b**, **c** Mafic xenolith in an outcrop of the Lilesville granite near Morven, Anson County, North Carolina

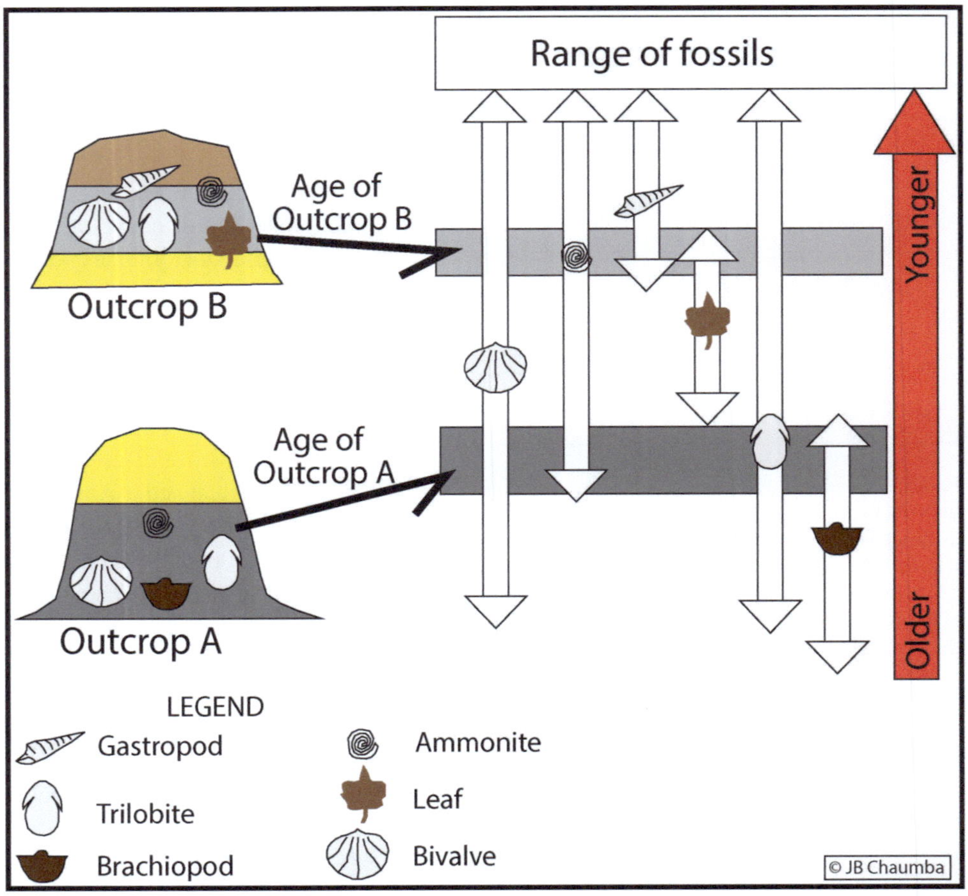

Fig. 10.9 Illustration of the principle of faunal succession based on fossil assemblages observed at outcrops A and B

the Phanerozoic. The ~4 billion years before the Cambrian is termed the Precambrian. The largest time expanses on the Geologic time scale are represented by eons (Fig. 10.10). The Hadean denotes the earliest eon in our planet's history prior to oldest known rocks. The Archean and Proterozoic eons are part of the Precambrian. Eras, which are subdivisions of eons, include the Paleozoic, Mesozoic, and Cenozoic (Fig. 10.10). The Precambrian, for example, is subdivided into seven eras (Fig. 10.10). Not that 88% of geologic time is represented by the Precambrian. The Precambrian is also used to denote eons and eras before the Phanerozoic. Profound worldwide changes in forms of life bound eras. Eras are subdivided into Periods, and Periods are subdivided into Epochs (Fig. 10.10). Period changes to the Geologic time scale are necessary to keep up to date with new radiometric ages for the boundaries as well as any changes in names of units.

Rock sequences in any region are subdivided into **stratigraphic formation**, or a formation for simplicity. A stratigraphic formation is a sequence of sedimentary layers that are traceable over large regions that represents the product of deposition over a definable time interval. A formation may either be composed of a single lithology or several lithologies.

10.6 Unconformities

Due to plate tectonic forces operating on Earth, change is always occurring such that here is neither one area where mountains have been forming since the formation nor an area where sediments have always been accumulating. Consequently, an area where sediments accumulated in the past is likely to have been uplifted and experience erosion in future. Areas with gaps in the sedimentary record occur either due to erosion or lack of deposition, or both. In such areas, where a continuous deposition is missing, gaps in the sedimentary record occur. An interval of time during which earlier-deposited rocks were eroded away or there was a lack of deposition is called an **unconformity**. There are different kinds of unconformities that are used to refer to this hiatus in sedimentation (Fig. 10.11).

A **paraunconformity** is a type of unconformity where sedimentary bends on either side of the unconformity are parallel (Fig. 10.11a). In a paraunconformity, very little erosion occurred prior to deposition of the sedimentary beds above the unconformity.

A **disconformity** is a type of unconformity where a surface of erosion occurs between two packages of sediment where no tilting affected the lower package of sediments prior to

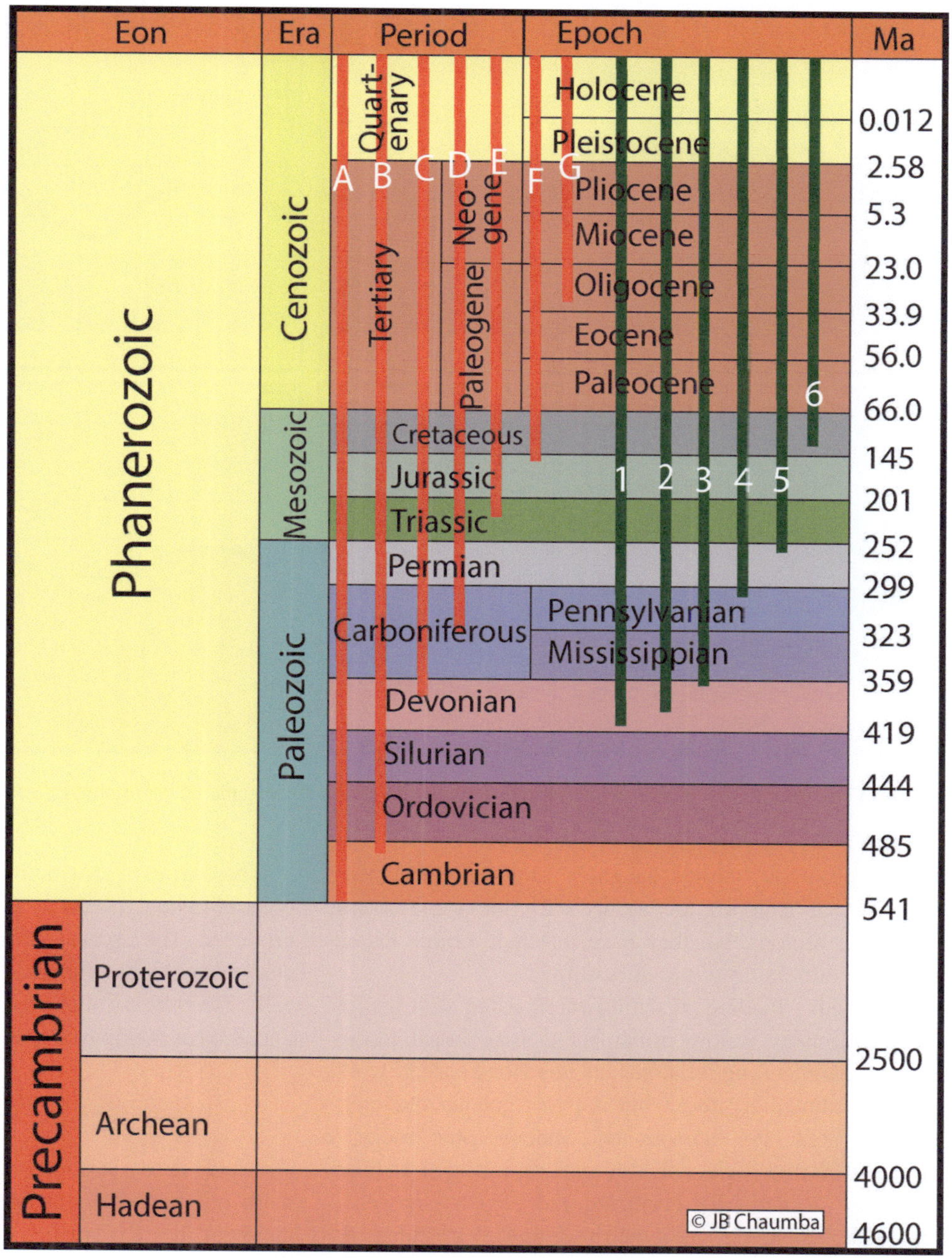

Fig. 10.10 Stratigraphic ranges and origins of some major groups of animals and plants are shown on the Geologic time scale. Key: For Animals shown in green bars: A—Animals with shells, B—Fishes, C—Amphibians, D—Reptiles, E—Birds, F—Humans. For Plants shown in green bars: 1—Club mosses, 2—Horsetail rushes, 3—Ferns, 4—Pines, 5—Ginkgos, 6—Flowering plants

deposition of the upper package of sediments (Fig. 10.11b). An **angular uniformity** occurs where tilting has affected an older package of sediments followed by truncation of the sediments by erosion. Finally, a younger package of sediments was then deposited above the surface of erosion (Fig. 10.11c). A **nonconformity** is a type of unconformity that separates underlying igneous and/or metamorphic rocks from overlying packages of sedimentary rocks (Fig. 10.11d). In a nonconformity, a long period of erosion usually occurs prior to deposition of overlying sediments.

Unconformities help Earth scientists to appreciate that the geological record in any one location is hardly complete, but that gaps in sedimentation are typically encountered. Events in Earth history such as mountain-building episodes where uplifting, erosion, and tilting of earlier-deposited beds are represented by unconformities. Shown in Fig. 10.12a, b, and c are examples of a disconformity, an angular unconformity, and a nonconformity, respectively.

10.7 Radiometric Dating

Radioactive dating is a method of dating rocks and minerals using radioactive isotopes.

More than 300 isotopes which occur naturally have been identified. The more common radioactive isotopes decay

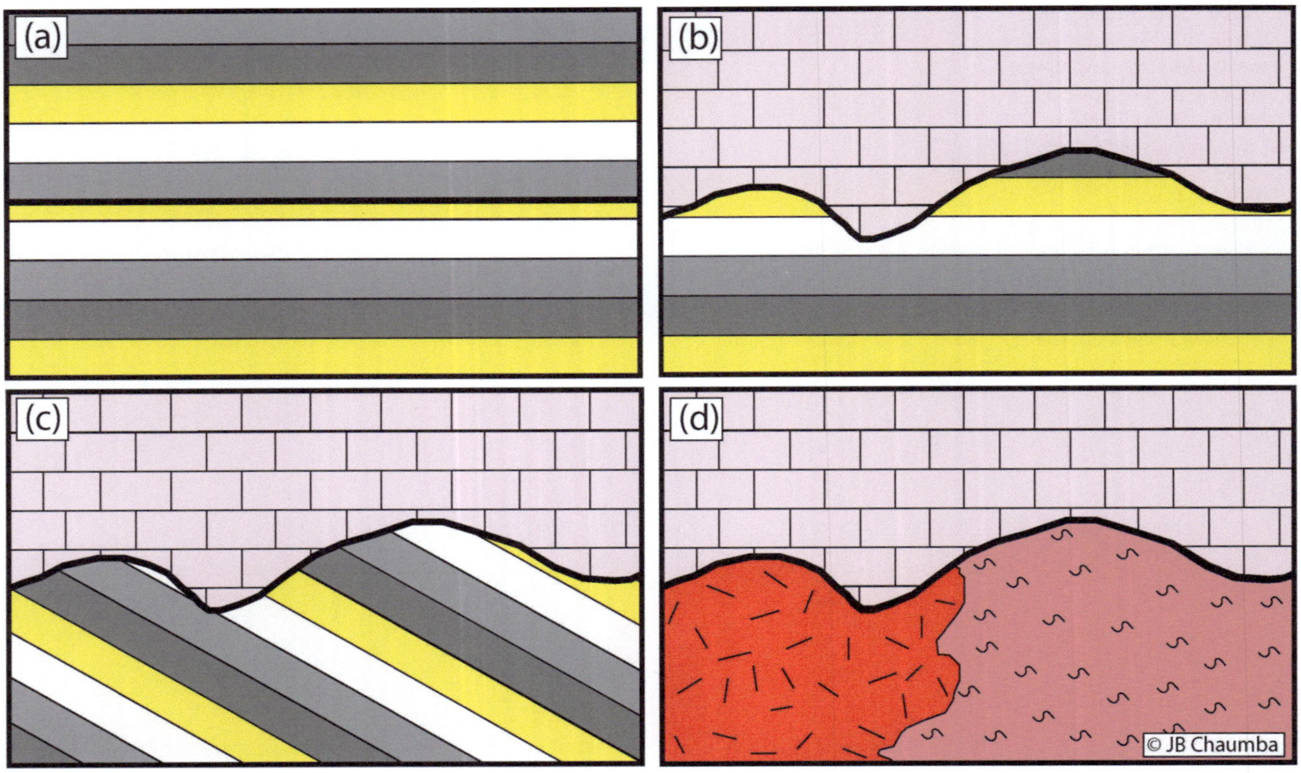

Fig. 10.11 Diagrams illustrating paraunconformity (**a**), disconformity (**b**), angular unconformity (**c**), and nonconformity (**d**)

into other isotopes termed daughter products. Recall from Chap. 2 that isotopes are atoms with the same atomic number, or protons, but they have different atomic masses due to the different number of neutrons.

The relative method of dating rocks using stratigraphic correlation, using various principles such as faunal succession and cross-cutting relationships described earlier, works well for sedimentary rocks but not for igneous and metamorphic rocks. For igneous and metamorphic rocks, a method of dating rocks and minerals that utilizes radioactive decay of naturally occurring radioactive isotopes is employed. Employing radiometric dating to calculate the age, in years, for rocks and materials involves the determination of either a short-life radioactive element or several long-life radioactive elements and their decay products. Carbon-14, denoted ^{14}C, is the short-lived isotope which decays to ^{14}N with a half-life of 5,730 years (Table 10.1). A **half-life** is the length of time it takes for one-half of the original, or parent, atom of the radioactive isotope to decay into an atom of a new, or daughter, isotope. Long-lived isotopes, such as the length of time it takes for the 238-uranium parent atom to decay to the 206-lead daughter atom (or simply $^{238}U \rightarrow {}^{206}Pb$), is in the billions of years (Table 10.1). An application of the suitable radiometric age-dating method (Table 10.1) enables Earth scientists to calculate the age of most rocks and fossils they encounter in the rock record.

Three types of radioactive mechanism occur. The different types of radioactive decay are classified based on the radiation produced. The first radiation type is composed of positively charged and alpha (α or ^4He) particles (Fig. 10.13a) which typically affects heavy nuclei composed of atomic numbers greater than 83. An example of an α decay:

$$^{210}_{84}\text{Po} \rightarrow {}^{4}_{2}\text{ He} + {}^{206}_{82}\text{ Pb} \tag{10.1}$$

Alpha decay produces daughter nuclides composed of mass numbers that are smaller by four units and atomic numbers that are smaller by two units compared to those of the parent nuclide (Fig. 10.13a).

The second type of radioactive decay involves the emission of an electron from a nucleus—beta (β) decay. An example of a nuclide that undergoes β decay is iodine through

$$^{131}_{53}\text{I} \rightarrow {}^{0}_{-1}\text{ e} + {}^{131}_{54}\text{ Xe} \tag{10.2}$$

During β decay, emission of an electron does not affect the mass number of the nuclide but decreases the number of neutrons by 1 and increases the number of protons by 1 (Fig. 10.13b).

Gamma (γ) emission is the third type of radioactive decay. During γ emission, a nuclide is formed in an excited state, and it then decays to its ground state through emission of a γ ray. An asterisk (*) denotes the presence of a

Fig. 10.12 Images showing **a** disconformity displayed by sedimentary rocks in the western US (*credit* D.T. Nelson), **b** angular unconformity in Colorado (*credit* M.E. Ellis), and **c** nonconformity between the Cape Granite below and the overlying Table Mountain Group (*credit* E. Mervine). Arrows indicate the unconformities

Table 10.1 Common parent-daughter isotope pairs used in radiometric age determination

Parent	Daughter	Half-life	Target minerals/material	Effective dating range
^{14}C	^{14}N	5730 years	Organic compounds	≤50,000 years
^{235}U	^{207}Pb	713 million	Zircon, uraninite, baddeleyite	>10 million years
^{40}K	^{40}Ar	1.3 billion years	Mica, feldspar, hornblende	10,000–3 billion years
^{238}U	^{206}Pb	4.5 billion years	Zircon, uraninite, baddeleyite	>10 million years
^{235}Th	^{208}Pb	14 billion years	Thorite, monazite, zircon, loparite, REE[a]	>10 million years
^{87}Rb	^{86}Sr	48.8 billion years	Mica, feldspar, hornblende	>10 million years
^{147}Sm	^{143}Nd	106 billion years	Garnet, mica	>10 million years

[a] REE—rare-earth elements

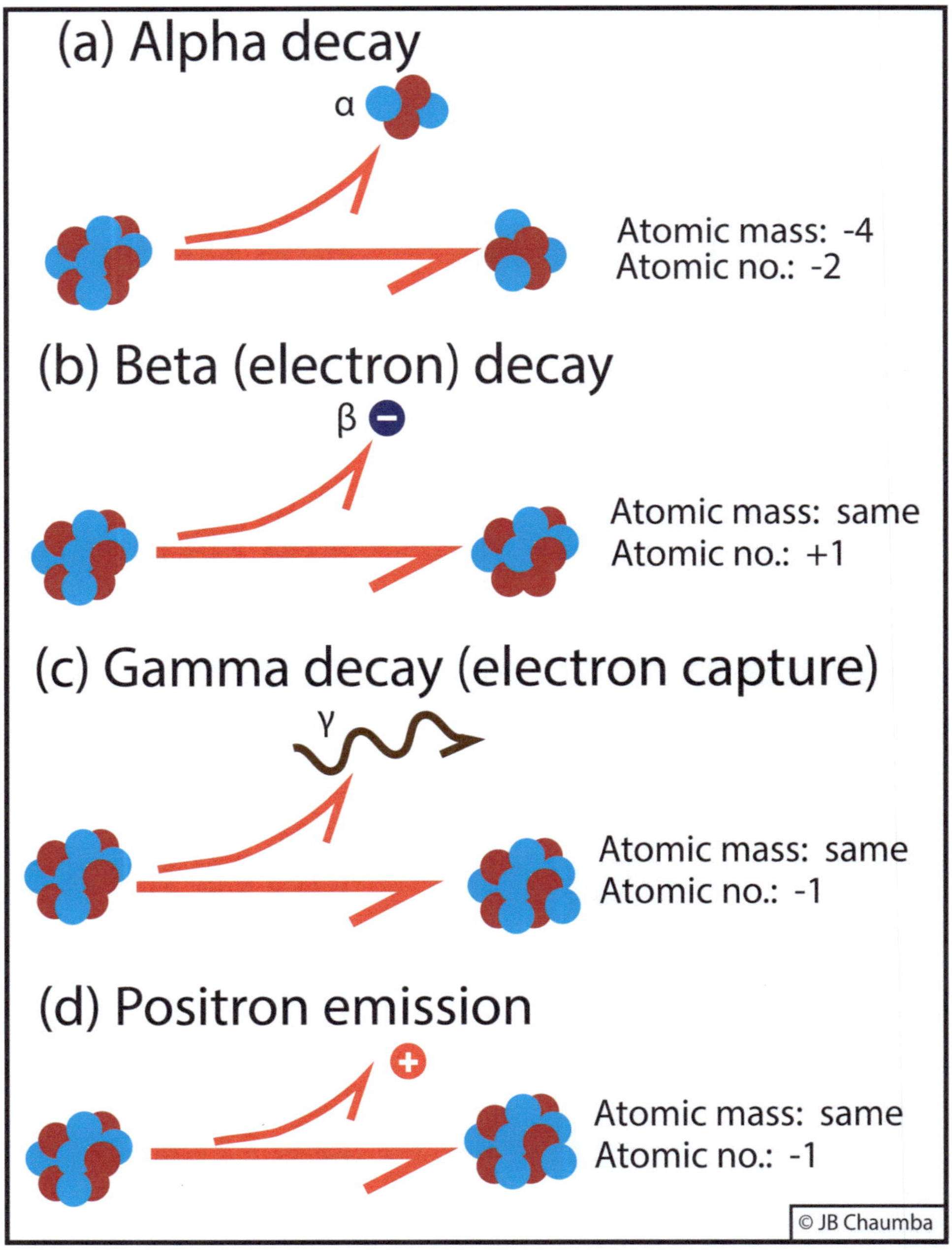

Fig. 10.13 Cartoons illustrating radiometric decay mechanisms of alpha decay where the mass number decreases by 4 and the atomic number decreases by 2 (**a**), beta decay where the mass number is unchanged and the atomic number increases by 1 (**b**), gamma decay where both the mass number and the atomic number are unchanged (**c**), and the positron emission where the mass number is unchanged but the atomic number decreases by 1 (**d**)

nucleus in an excited state. An example of radioactive decay involving γ emission is

$$^{60}_{27}\text{Co}^* \rightarrow ^{0}_{0}\gamma + ^{60}_{27}\text{Co} \qquad (10.3)$$

During γ emission, no change occurs either in the mass number or in the atomic number (Fig. 10.13c) unless the γ emission accompanies another radioactive decay mechanism. This $^{60}{}_{27}$Co radiation is useful in the treatment of cancer.

Yet another radiation, the positron emission (β^+ decay), can also occur during radioactive decay which involves the emission of a positron from the nucleus (Fig. 10.13d). An example positron decay is

$$^{15}_{8}\text{O} \rightarrow ^{0}_{+1}\text{e} + ^{15}_{7}\text{N} \qquad (10.4)$$

The positron emission nuclides, however, lie below the band of stability. Positron decay is the conversion of a proton into a neutron with the emission of a positron.

Naturally occurring radioactive isotopes of the heavy elements undergo successive radioactive decay. Radioactive species in any one chain form a **radioactive decay series**. Examples of radioactive decay series are the uranium series, the actinide series, and the thorium series. Each radioactive nuclide has a **half-life** ($T_{1/2}$), which is the time required for half of the atoms in a sample to decay (Fig. 10.14). The half-life is characteristic and constant for that element.

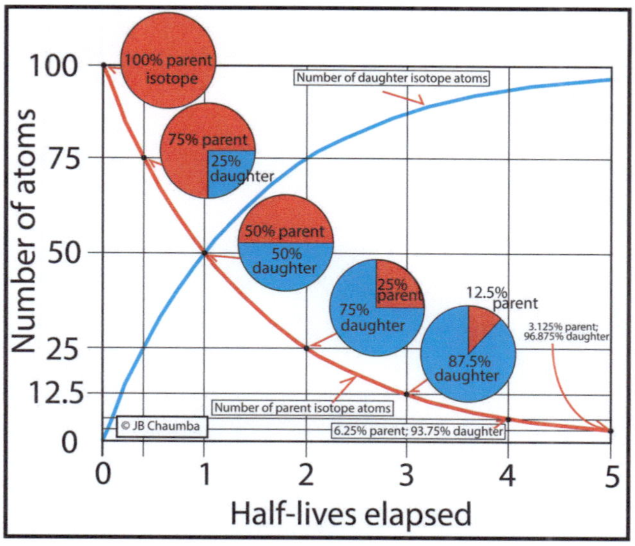

Fig. 10.14 Plot showing the change in the proportion of the atoms of parent isotope as it decays to form atoms of the daughter isotope from an initial 100 atoms of the parent isotope. After 1 half-life, 50 atoms of the parent isotope would have decayed to form 50 atoms of the daughter atom. As the number of atoms of the parent isotope decays (or decreases), the number of atoms of the daughter formed increases. Note that the rate of decrease of atoms of the parent decreases with time, while those of the daughter isotope increase with time

An isotope's half-life permits Earth scientists to determine how long a sample of an undesirable or dangerous isotope is retained before it decays to a low-enough radiation level that no longer poses a problem.

For any radioactive element, the intensity of radiation that it produces is directly proportional to the rate of decay of the substance and the amount of the material. The radioactive atom that undergoes decay is called the parent isotope which decays to form the daughter isotope. In the case of the $^{238}U \rightarrow {}^{206}Pb$ pair, ^{238}U is the parent isotope and the daughter isotope is the ^{206}Pb.

10.8 Deciphering Earth History Utilizing Relative and Radiometric Ages

Not all rocks can be dated using radiogenic isotopes. In areas where metamorphic, igneous, and sedimentary rocks crop out, one can use a combination of stratigraphic relations and radiometric ages to provide the geologic history of the area under study. In the case of the package of rocks shown in Fig. 10.15, all three rock types occur, and radiometric ages have been obtained for the crystalline rocks. A complete geologic history of the area can be attempted.

Metamorphic rocks, dated at 405 Ma, are the oldest rocks in the area. These rocks have been eroded to form a nonconformity above which beds of sedimentary rocks were deposited. A 290 Ma granitic pluton was then intruded into

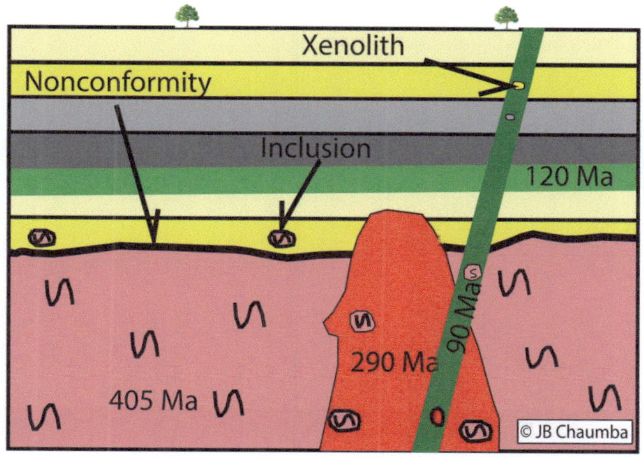

Fig. 10.15 Geologic map of an area in which crystalline rocks, sedimentary beds, and a nonconformity occur. Intrusion of a granitic pluton and a diabase dike, as well as emplacement of a sill, also affected the area

the rocks. However, the granitic intrusion does not affect beds above the second sandstone layer. A 120 Ma sill was then emplaced between the sandstone and shale beds. The last event was the intrusion of a subvertical diabase dike which intrudes all lithologies in the area 90 million years ago. A similar approach can be applied to any area where ages of some rocks are available which helps the Earth scientist decipher the geologic history of an area.

10.9 The Geologic Time Scale

The vast interval of time represented and recorded in rocks on Earth constitutes geologic time. The geologic time scale can be thought of as the "calendar" for events that occurred throughout Earth history. Geologists working on outcrops and drilling cores from all continents of our planet have helped compile a composite stratigraphic column known as the **geologic time scale**. Subdivisions of the geologic time scale are into specific intervals of time which represent different segments. In descending order of duration, subdivisions of geologic time are **eons**, **eras**, **periods**, and **epochs** (Fig. 10.16). One way of distinguishing and defining a segment of time is by the occurrence of major geologic events together with the appearance, and disappearance, of major life forms, beginning with the formation of crust of our planet. The appearance of ever-changing forms of life on Earth followed the formation of Earth's crust. Eons, which are the largest of the subdivisions, range from the Hadean through the Archean and the Proterozoic to the Phanerozoic (Fig. 10.16). The Phanerozoic eon, for example, is divided into three eras: the Paleozoic, Mesozoic, and Cenozoic Eras (Fig. 10.16). Epochs (Fig. 10.16) are the smallest subdivisions of the geologic time scale. Eons are the largest

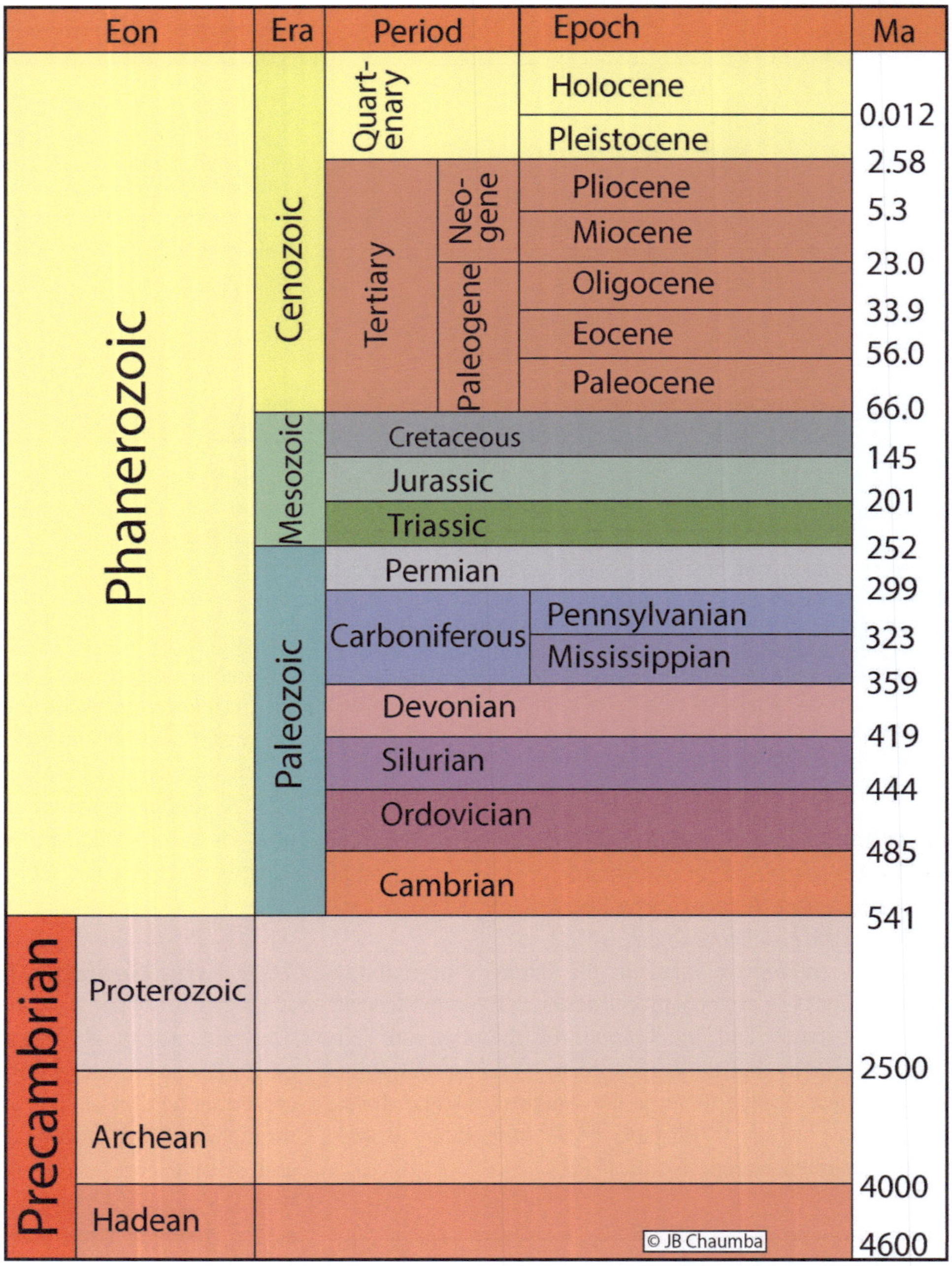

Fig. 10.16 Geologic time scale. Abbreviation: Ma—million years. Ages in Ma are based on version 5 of the geologic time scale of the Geological Society of America

subdivisions of the geologic time scale, followed by Eras (Fig. 10.16). Epochs are the smallest subdivisions of the geologic time scale (Fig. 10.16). The need to organize the immensity of geologic time, as well as correlating geologic events on a worldwide scale, resulted in the establishment of the geologic time scale.

The onset of geologic time in the rock record is generally agreed to have begun at the start of the Archean eon (4.0 billion {Ga, or billion years ago} to 2.5 Ga) and continues to the present day. The Hadean eon, an informal interval that extends from about 4.6 billion years ago (corresponding to Earth's initial formation) to 4.0 billion years ago, is included in modern geologic time scales (Fig. 10.16). Geologic time is, in effect, that segment of Earth history that is represented by and recorded in Earth's strata.

The Precambrian is composed of the Hadean, Archean, and Proterozoic eons, with the suffix *zoic* meaning life. Thus, Proterozoic means first life and Phanerozoic denotes visible life. However, the earliest life, in the form of Bacteria and Archea, is now known to have appeared on Earth during the Archean. Only the Phanerozoic eon is subdivided further, from the oldest to the youngest, into the Paleozoic (ancient life), Mesozoic (middle life), and Cenozoic (recent life) (Fig. 10.16). Eras can be further subdivided into Periods, some of which can also further be subdivided into Epochs as shown in Fig. 10.16.

The geologic time scale was established long before radiometric dating was discovered. As ages for critical rocks that coincide with, or constrain, boundaries of the geologic time became available, dates were, and continue, to be added and refined to precisely denote the boundaries on the geologic time scale as shown in Fig. 10.16. Consequently, shifting of specific boundaries on the geologic time scale can continue to occur as more and more ages become available.

Glossary

Alpha (α) decay radioactive decay that leads to loss of an alpha particle

Amber fossilized translucent yellowish resin that was produced by extinct coniferous trees during the Tertiary.

Angular unconformity a type of unconformity whereby older packages of sediments were tilted and later truncated by erosion, and younger packages of sediments were later deposited on the erosion surface.

Beta (β) decay breakdown of a neutron to form a proton, which remains in the nucleus, and an electron, which is emitted as a β particle.

Cast an imprint left in sediment by a dead animal whose body decays and the imprint then fills in with minerals from sediment which hardens to form a fossil.

Carbonization the process where carbon forms from organic matter.

Catastrophism the view that ascribes geological changes in the Earth's crust to be a consequence of sudden, unusual and violent events.

Coprolite trace fossils composed of fossilized feces.

Daughter nuclide stable or unstable nuclide that is formed by the radioactive decay of another nuclide.

Disconformity gap or break in the sedimentary record of parallel strata on either side of the break or unconformity.

Electron capture a combination of an electron and a proton to yield a neutron in the nucleus.

Eon a major unit of time equal to approximately a billion years that is subdivided into eras.

Epoch subdivisions of a Period that are subdivided into ages which correspond to series.

Era a major division of time that is subdivided into periods. Eons are subdivided into eras.

Fossil assemblage groups of fossils that occur together in strata.

Fossilization—the process by which an animal or plant becomes preserved in a petrified form.

Gamma (γ) emission decay of a nuclide accompanied by emission of a gamma ray.

Geologic history reconstruction of the geologic history of Earth.

Geologic time scale a chronological sequence of geologic events typically represented in the form of a chart showing names showing various rock layers and indicating the estimated duration of each geologic unit.

Half-life ($T_{1/2}$) the time that is required for half of the atom in a radioactive sample to decay into a daughter product.

Index fossil a fossil that is useful for age determination and correlation of strata in which it is preserved.

Impression trace fossil that forms when a foot, leaf, shell, for example, leaves an imprint in soft sediments which hardens to form a fossil in the shape of the original.

Mold an imprint that is left by a shell on the sediments that surrounded it.

Nonconformity a surface that separates overlying younger sedimentary strata from underlying eroded igneous or metamorphic rocks and represents a long gap in the rock record.

Paleontology the study of the history and evolution of life on Earth as based on fossils contained in rocks.

Paraunconformity a type of unconformity whereby overlying and underlying sedimentary are parallel with no obvious erosional break between them.

Parent nuclide unstable nuclide that naturally decays into a daughter nuclide.

Period a basic unit of geologic time which lasts tens of millions of years, generally the time it takes for rock system type to be formed.

Permineralization a process of fossilization whereby mineral deposits carried by water form internal casts of organisms by filling the spaces within organic tissue.

Pleistocene the time period between about 10,000 years before present and about 1,650,000 years before present. As a descriptive term applied to rocks or faults, it marks the period of rock formation or the time of most recent fault slip, respectively. Faults of Pleistocene age may still be active though their activity rates are commonly lower than younger faults.

Positron emission also called β^+ decay, is the conversion of a proton into a positron which is emitted, and a neutron which remains in the nucleus.

Quaternary the geologic time period comprising about the last 1.65 million years.

Radioactive decay natural decay of an unstable nuclide into another nuclide.

Radiocarbon dating accurate age determination of carbon-bearing materials 30,000–50,000 years old; using the carbon-14 dating method.

Radiometric dating age determination of rocks and other materials using radiogenic isotopes.

Range interval in time between when a particular fossil first appears and when it last appears in the rock record.

Stratigraphic formation the consequence of a geological system that consists of processes and setting of sediment production.

Trace fossils or ichnofossils, are geologic records of activities and behaviors of past life such as burrows, footprints, and nests.

Unconformity strata where older rock layers were deformed or partially eroded before deposition of younger rock layers.

Uniformitarianism the idea that changes of Earth occur in uniform ways and thus that the present is the key to the past.

11 Weathering, Soils, and Mass Wasting

Abstract

After defining weathering, physical, chemical, and biological weathering are described, and each is subdivided into various categories and examples given. Diagrams are used in illustration. Descriptions of wide varieties of landforms that are exposed and/or formed by weathering are provided, and some world-famous examples given. Soils, soil-forming factors, and soil profiles are also described using diagrams and photographs. Soil classification is illustrated by the soil classification diagram, whereas the twelve soil orders are explained using a table, soil maps of the US and the world. Mass wasting is defined, types of mass wasting, and factors that affect it are described and summarized using diagrams. Mechanisms that trigger mass wasting are summarized. Maps of the US and world showing landslide incidences are also provided. The chapter ends with a discussion of some strategies for mitigating mass wasting using diagrams.

Learning Goals

Studying this chapter will enable you to:

- Define weathering and to recognize weathered material and how rocks are weathered
- Summarize physical weathering and list some examples
- Describe chemical weathering and summarize the role of water in chemical weathering
- Differentiate between physical and chemical weathering
- List factors that influence rate and type of weathering of rocks
- Describe the formation of soil and the controls of soil formation
- Summarize activities that help reduce soil erosion
- Label and describe a typical soil profile and explain why soils are classified
- Recognize and describe human impact on soil and how rates of erosion can be reduced
- Define mass wasting and its triggering mechanisms
- Identify and summarize types of mass wasting
- List and describe ways of alleviating mass wasting

11.1 Introduction

If you have ever gone rock-climbing, like me you may not easily connect that huge mass of rock you are trying to conquer may end up being part of the sand you relax on when you visit the beach. No matter how huge any mass of rock appears, it is constantly being attacked by both chemical and mechanical or physical weathering. In this chapter we will learn about these processes and more. We will begin by defining weathering.

11.2 Weathering

Any rock type or other broadly comparable substances, such as materials made of concrete, are weathered and break down with the passage of time into smaller fragments. **Weathering** can be defined as the in-situ breaking down (**physical weathering** or **mechanical**), dissolving (**chemical weathering**), and/or **biological processes** affecting rocks and minerals at or near the Earth's surface due to exposure to the atmosphere. Sediments (or clasts) and/or dissolved ions are a product of weathering. Recall that silicates are the dominant minerals forming the crust of our planet (Chap. 3). Hence quartz, iron oxides,

Supplementary Information The online version contains supplementary material available at https://doi.org/10.1007/978-3-032-20231-4_11.

J. B. Chaumba, *Introductory Physical Geology*, Springer Textbooks in Earth Sciences, Geography and Environment,
https://doi.org/10.1007/978-3-032-20231-4_11

aluminum oxides, clay minerals, as well as dissolved ions just mentioned, comprise the most common products of weathering. Weathering, thus, causes the unweathered, or fresh, rock to break down into smaller fragments which often occur together with water containing dissolved ions which ultimately form the ingredients of sedimentary rocks we discussed in Chap. 4. Erosion of overburden, any rocks and sediments overlying any buried rock, results in the formation of cracks of joints due to pressure release as the once buried rocks expand in response to reduction in both pressure and temperature. The reduction in pressure and temperature causes the rock to split or crack, i.e., develop **joints**. Joints are a common feature at almost all outcrops.

11.2.1 Physical Weathering

Physical weathering occurs because of the changing temperature of rocks which physically alters rocks by causing them to break down into smaller fragments. The presence of water often assists physical weathering. Different types of physical weathering are recognized, caused by wind, rain, freezing, thawing, sand, and other natural forces that physically alter rocks. One type is **exfoliation**, which occurs due to cracks which develop parallel to the Earth's surface and intrusion's outermost surface due to the reduced pressure that is experienced during uplift accompanied by sheeting (Fig. 11.1).

Exfoliation, which is also referred to sheeting, is a process where concentric the outer portions of plutons develop layers that resemble an onion (Fig. 11.1). Sheeting is thought to be generated due to unloading, which is the reduction in pressure which occurs because of removal of the overburden (Fig. 11.1b). Expansion causes sheeting as the reduced pressure from erosion causes expansion which is accompanied by separation of the outermost onion-like layers from the main body of the pluton (Fig. 11.1). As exfoliation continues, blocks of rock to slide down the sides of the exposed plutons and accumulate on the surface to form piles of broken rectangular rock of varying sizes termed **scree** (Fig. 11.1b). Joints (see Chap. 9) typically form when the overburden is removed (Fig. 11.1b). Such layers which develop during unloading produce layers which resemble those exemplified by an onion when it is peeled off exposing its concentric layers.

Plutons and batholiths form beneath the Earth's surface from slow crystallization of magma in a magma chamber. Recall that batholiths and plutons are intrusive features (Chap. 3) which can ultimately become exposed through stripping on overlying material.

Freeze–thaw action, or frost wedging, another type of physical weathering, occurs when water seeps into cracks, which then freezes at night and during the winter but expands at night and during the summer (Fig. 11.2) which leads to breaking the rocks apart. Rainwater or snowmelt collects in cracks in rocks of mountainous regions such as the Himalayas, the Rockies, the Andes, the Appalachians, the Alps, and the Eastern Highlands and Drakensberg Escarpment in Southern Africa. During the night, temperatures drop resulting in the water freezing and expanding.

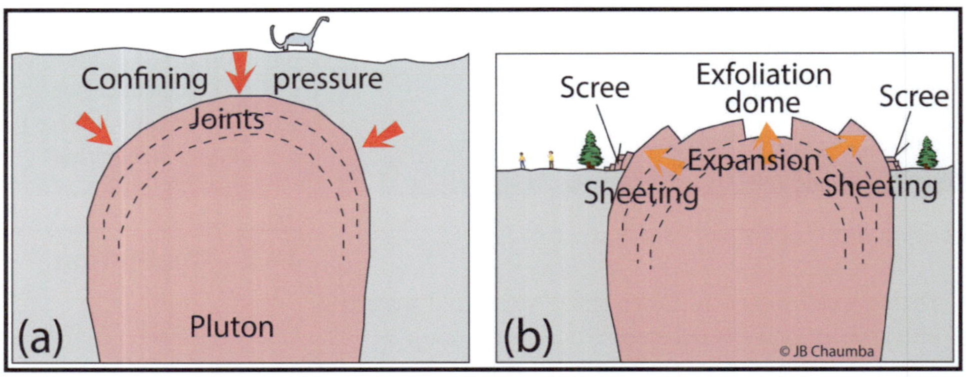

Fig. 11.1 **a** A pluton experiences confining pressure at depth with dashed lines indicating possible orientation of future joints. **b** However, with removal of overburden, unroofing, and sheeting accompanied by expansion in response to the reduction in pressure due to unloading leads to sheeting. Joints are indicated by the concentric dashed lines

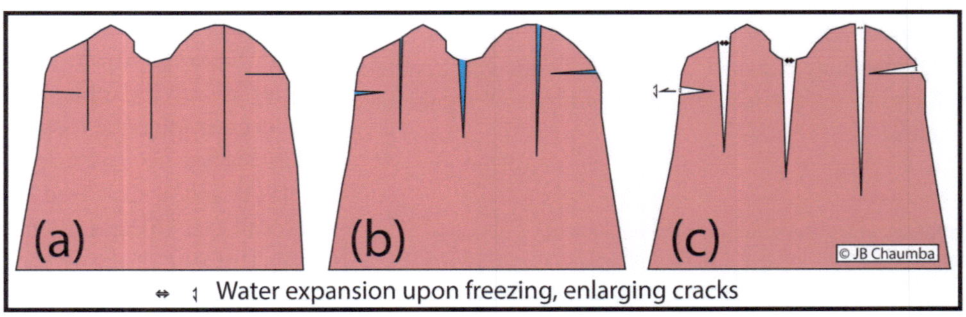

Fig. 11.2 Illustration of freeze–thaw action or frost wedging. Images of an outcrop with cracks (**a**). Water penetrates the cracks (**b**), and the water freezes (**c**) at night or during the winter which expands upon freezing leading to widening of the cracks

The expansion of the water in cracks leads to increases in volume of the ice in the cracks which then exerts more pressure on the cracks in the rock leading to widening and deepening of the cracks. This results in further breaking down of the rocks into smaller pieces. Eventually, the small pieces accumulate at the bottom of the mountain to produce scree. Freeze–thaw action, therefore, occurs either through repeated thawing and freezing of water particularly in mountainous regions and in the tundra. Frost wedging, thus, is common in mountains located in mid-latitude and polar regions. Expansion and contraction of layers of rocks at or near the Earth's surface that are baked by the Sun, particularly in tropical and some subtropical hot deserts, also induces free-thaw action.

Sea water is saline, and in coastal regions where there are rocky coasts, **salt crystal growth** occurs. Crystals of salt exert expansive forces upon enclosing rocks resulting in splitting up of rocks. Sprays from breaking sea waves enter cracks in rocky coastlines. Crystals of salt then grow from the salts dissolved in ocean water, exerting forces that enlarge the crevices in rocks. This process also takes place in rocks at coastal regions when saline groundwater accumulates in pores in rocks. No chemical reactions are considered here.

Animals, including human beings, can also cause physical weathering on rocks. **Burrowing** animals such as mountain goats and other animals, for example, dislodge some rocks, and these rocks, upon hitting other rocks, are themselves broken into even smaller fragments. Human beings contribute to physical weathering during construction of roads, bridges, and buildings. We also break down rocks into smaller fragments as we hike mountains and hills. Mineral exploration and mining operations also contribute to physical weathering. Additionally, field geologists often break rocks down into smaller pieces with a geologic hammer and other bigger hammers, contributing to mechanical weathering. Vegetation and organisms that live in rocks eventually die and decay, resulting in the production of acids. These acids from decaying organisms help promote chemical weathering, which is described next. Physical weathering typically takes place where there is very little soil and a general lack of plant growth, such as in mountain regions and arid regions. Among some of the agents of physical weathering are the effects of changing temperature on rocks, causing the rock to break apart. The process is sometimes assisted by water.

11.2.2 Chemical Weathering

Chemical weathering is caused by molecules in air and weakly acidic rainwater reacting with minerals in rocks to produce new minerals (such as clays) as well as soluble salts through breaking down of chemical bonds. Processes of chemicals in precipitation that cause changes to minerals in rocks contribute to this type of weathering. Carbon dioxide (CO_2) that is dissolved in water results in the formation of a weak acid called carbonic acid (H_2CO_3). Rainwater, thus, is also slightly acidic, and it also combines with carbon dioxide in the atmosphere to form carbonic acid. As rainwater becomes surface and groundwater, it reacts with the rocks it comes in contact leading to some elements to be removed from the rock while other elements are added. These reactions result in the alteration of the chemical composition of the original rock. Rates of chemical weathering occur more rapidly at higher temperatures; hence, warm and damp climates are associated with higher weathering rates. The early stages in the production of soils involve oxidation and hydrolysis.

Hydrolysis, oxidation, and solution are examples of the different types of chemical weathering. **Hydrolysis** involves the breakdown of rocks by reaction with acidic water leading to the production of clays and soluble salts. Acid rain and carbonation are examples of hydrolysis. Hydrolysis in granite is when feldspar is weathered to form clay minerals. The majority of feldspars as well as other silicates are transformed into clay minerals through chemical reactions with water leading to potassium, sodium, and calcium ions being carried away in solution. Weathering of iron and magnesium-bearing silicates such as amphiboles and olivines result in the residual clay minerals and iron oxides, whereas silica, calcium, and magnesium irons are transported away in solution. Quartz is largely affected.

Dissolution is the breakdown and removal of rock in solution by reaction with slightly acidic rainwater. Dissolution tends to easily weather rocks that are composed of carbonates such limestone and jointed marble. Carbon dioxide in the air can dissolve in rainwater resulting in the water forming the formation of a weak acid called carbonic acid. This acid eventually falls as rain on rocks and structures as acid rain. Carbonic acid tends to dissolve rocks such as limestone and dolomite. Acid rain weathers rocks, monuments, and buildings through solution and hydrolysis. With the passage of time, continued seeping of carbonic acid into cracks and cavities can lead to the formation of spectacular cave systems such as the Carlsbad Caverns in the western US, the Mammoth Caves in Kentucky, the Chinhoyi Caves in Zimbabwe, the Cango Caves in South Africa, the Caverna Pontes do Sumidouro in Bahia, Brazil, the Longmen Grottoes Caves in Henan Province, China, and the Cheddar caves, Somerset, UK, and elsewhere (Chap. 12).

Oxidation occurs when the breakdown of rocks interaction with oxygen and rainwater containing dissolved CO_2 results in the formation of iron-rich rocks which have a rusty-colored surface (this process is sometimes called carbonation).

11.2.3 Biological Weathering

Biological weathering is ascribed to living organisms that contribute to the weathering process which includes the action of growing plant roots (Fig. 11.3a, b) which push joints apart resulting in further breaking up of rocks. Bacteria and lichens, some of which can be seen on some rock outcrops, produce chemicals that help disintegrate rocks on which they live and obtain the nutrients they need in the process. Movements of plants and animals also cause biological weathering. In addition, some animals bore into rocks for protection either by secreting acids which dissolve the rock or by scraping away the grains of rocks, contributing to biological weathering. Human beings also cause biological weathering primarily through exposing soil to erosion from altering the landscape. Human activities such as mining, agriculture, construction, and foot traffic. Such activities often disrupt vegetation, leading to the breakdown of rocks. In addition, these human activities often expose new surfaces to agents of weathering like plants and even microscopic organisms occurring in water, soil, and air.

Weathering, thus, deals with how rocks and minerals, as well as earlier formed soils, are changed into progressively smaller fragments. The agents of weathering are water, plants, animals, ice, salts, acids, and changes in temperature. Weathering is an integral part of soil development. Depending on the soil-forming factors in an area, weathering may proceed rapidly (at rates of decades) or slowly (at rates of millions of years).

Although physical and chemical weathering were described separately, it should be emphasized that these two types of weathering occur together and affect rocks simultaneously. The two types of weathering are always at work, complementing one another, and greatly help in sculpturing the landscape and some of the landforms we see on the surface, and beneath it such as in caves. As rocks are broken up into progressively smaller pieces, increasing the surface area on which chemical weathering occurs.

Chemical weathering in turn aids physical weathering by weakening chemical bonds of strong minerals such as feldspars into relatively soft ones like clays which can then be easily broken up by physical weathering. Even some relatively hard cementing materials such as silica and hematite which occur in some sedimentary rocks are weakened by chemical weathering, making them susceptible to being broken down through physical weathering. Weathering rates obviously vary depending on the varying strength of bonds making up different rock types, a reflection of their chemical make-up. Consequently, rocks that are less resistant are weathered at faster rates in comparison to rocks that are more resistant to weathering. Differential weathering refers to the differences in degrees breaking down of rocks due to the differences in the chemical composition of rocks, reflecting the non-uniform nature of weathering on rocks. An example of differential weathering is exemplified in the red Rocks near Morrison, Colorado where layers of more resistant conglomerates protrude outwards with relatively indented soft sandstone layers occupying the layers which form the indentations (Fig. 11.4). This type of weathering is also referred to as sawtooth weathering.

Oxygen is an element that is reactive, and it, thus, can easily react with rocks through oxidation reactions. The formation of rust of rocks such as those shown in Fig. 11.4 is an example of this type of weathering called rust forma-

Fig. 11.3 Photographs of root wedging. **a** Looking Glass Falls, western North Carolina, USA, showing trees growing within cracks in the rocks, with their roots widening the cracks as the trees grow. **b** Numerous tree roots (some indicated by arrows) growing along joints in rocks exposed on creek bed, pushing apart rocks

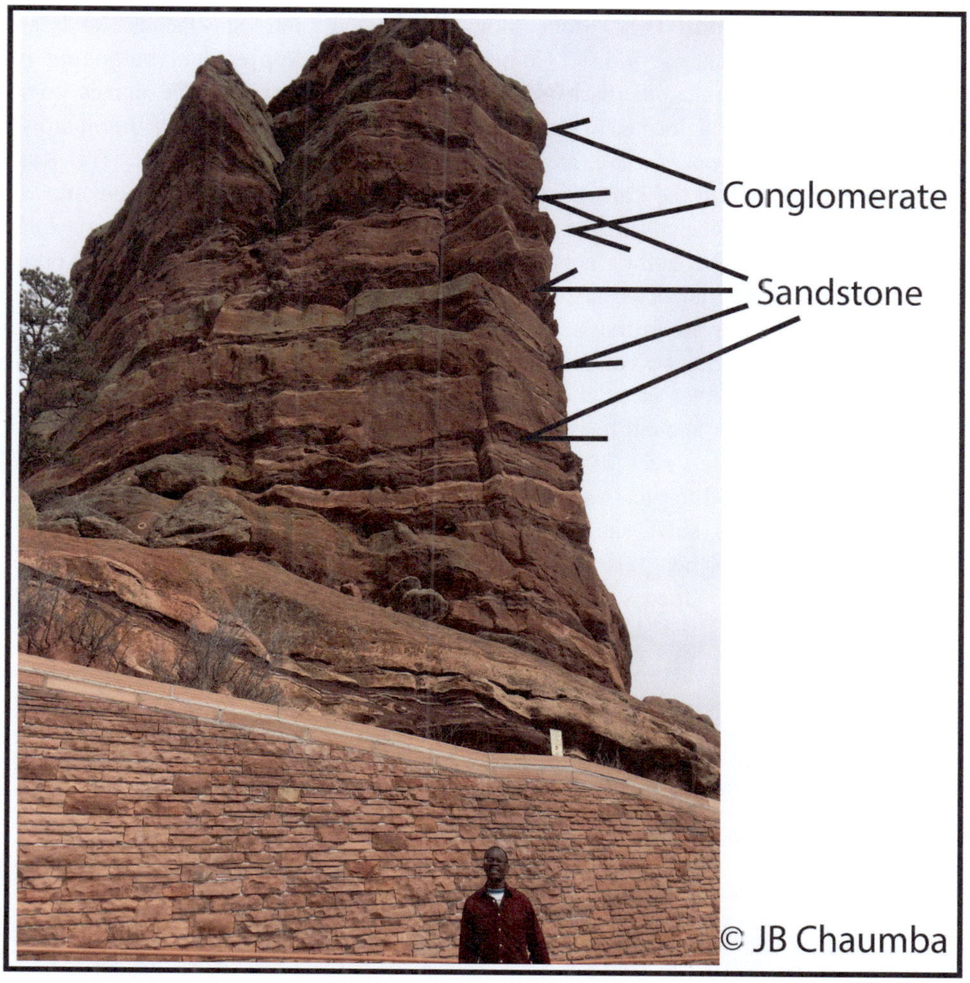

Fig. 11.4 Differential weathering, also called sawtooth weathering, is exemplified by alternating beds of more resistant sandstone forming the protruding layers where the relatively weak beds of shale form the intervening indented layers

tion. Rusting occurs when oxygen combines with iron to form iron oxide (rust). During rust formation, the color of the rock is changed. In addition, iron oxide is much more fragile than iron, hence the weathered rock becomes more susceptible to breaking into smaller fragments and contributing to chemical weathering.

In arid regions and on shorelines which consist of rocks, the rocks are weathered by salt crystal growth. Sea sprays from breakers penetrate any cracks in rocks. When the water evaporates, the salt is left behind and crystallizes to form salt crystals which increasingly get bigger with continued crystallization of more crystals of salt resulting in progressively larger salt crystals which break up rocks as they grow. As salt crystals grow in size due to the addition of more atoms to form salt, expansive forces are generated which exert increasing forces until they ultimately split enclosing rocks. A similar process also occurs in desert regions where salt crystals are increasingly enlarged within the pore spaces in rocks and, similarly, eventually split up rocks.

11.3 Some Landforms Exposed or Formed by Weathering

Intrusive igneous rocks form within the Earth's crust where very little of both oxygen and water occur. As the material above is weathering and carried away through erosion, the rocks that formed within the crust such as batholiths and plutons (see Chap. 3) would eventually be exposed at the Earth's surface (Fig. 11.1a, b) where there is enhanced weathering rates due to the abundance of water and exposure to the atmosphere. Additionally, the decrease in both temperature and pressure and the presence of organisms help enhance weathering. Even some metamorphic rocks (Chap. 5) that formed at relatively great depths also experience higher weathering rates as overlying materials are weathered and eroded away.

The process of exfoliation we described in Sect. 11.1 also plays a role in the formation of exfoliation domes such as those encountered in the Idaho batholith, Half Dome in California's Yosemite National Park, Stone Mountain located just east of Atlanta, Georgia, and the Chinamora batholith in Zimbabwe. Recall from our discussion in Chap. 3 that batholiths are the largest formed within the Earth's crust. Extensive periods of erosion are required to ultimately expose these batholiths at the Earth's surface.

An isolated dome-shaped hill or knob or ridge that abruptly rises from a surrounding generally level or gently sloping plain is termed an **inselberg** (which means island mountain in German) or **monadnock**. Inselbergs are largely comprised of rocks of granitic composition.

Inselberg is a term that is used not only in savannah regions, but it has been applied to isolated mountains in a variety of climatic regions such as arid, semi-arid, humid-subtropical, arctic humid-tropical as well as Mediterranean.

In Archean cratons such as the Kalahari craton in southern Africa and others elsewhere, weathering and erosion over millennia helped sculpture granitic batholiths into spectacular landscapes such as **bornhardts**, **castle kopjes**, and **balancing rocks**. Bornhardts (Fig. 11.5a, b) display large domes composed of massive granitic rocks. The sides of bornhardts range from almost vertical to gently inclined. The Kalahari craton in southern Africa, for example, contains classic examples of these landforms. Bornhardts are inferred to form as a consequence of exhumation of rocks originally formed at depth as release of pressure is usually accompanied by the formation of cracks resembling onion layers as well as some subvertical cracks. These cracks break down the rocks further leading to the formation of other landforms such as castle kopjes and tors. Bornhardts are common landforms in areas underlain by granitic rocks, and they are common throughout the Kalahari craton's areas underlain by granites. Bornhardts are large domes of massive rocks with faces which range from gently inclined slopes to near vertical slopes (Fig. 11.5).

Fig. 11.5 Denzva Mountain (**a**) and Murehwa Mountain (**b**) exposed in eastern and northeastern Zimbabwe, respectively, are classic examples of bornhardts. Denzva Mountain is shaped like a loaf. Bornhardts are thought to develop on resistant portions of batholiths, and they are surrounded by highly fractured rock although open fractures are few. The domical profile of these bornhardts is inferred to be a consequence of exfoliation

Inselbergs come in several different forms depending on the rock type being weathered. In addition, the rate of erosion as well as the orientation of the rock types if they are layered, leads to the formation of inselbergs that have the shape of loaf as exemplified by Uluru or Ayers Rock in Australia's Northern Territory (Fig. 11.6a, b). Uluru is thought to have been formed from a segment of more resistant sandstone layer that was vertically tilted during compressive deformation. The other adjoining folded beds were eroded away, and alluvium was later deposited over the remnants of the leveled surrounding beds resulting in Uluru sticking above the surrounding surface. Denzva Mountain, which we have just discussed above (Fig. 11.5a), is also another example of a loaf shaped inselberg.

Bornhardts, balancing rocks, and castle kopjes are scattered across the Archean cratons of southern Africa, North America, and Western Australia in particular. These landforms have also been described from almost all continents.

Castle kopjes, which are assemblages of boulders contiguous with pockets of soil and vegetation (Fig. 11.7a, b) form as a consequence of the disintegration of massive rocks into boulders that are dominated by rectangular joints. Occasionally, the jointed rocks balance precariously on top of other rocks as can be observed in areas of castle kopjes (Fig. 11.7). The balancing rocks in Epworth on the outskirts of Harare, Zimbabwe, are world class examples (Fig. 11.8a, b).

After rocks are weathered, and even as rocks weather, they are transported away via **erosion**. Weathering and erosion tend to occur concurrently, and they are responsible for carving some impressive landforms on planet as well as other planets and some moons within our solar system. Erosion can be defined as the process by which the Earth's surface gets worn down by water, wind, and glacial ice due to the removal of soil, rock, and/or dissolved materials. Landforms such as coastlines, mountain peaks, and valleys, for example, found on Earth and other planets were formed and/or shaped by weathering and erosion, among other processes. Throughout geologic history, weathering has been responsible for the formation of unique and impressive landforms on Earth. The Badlands (Fig. 11.9a) are layered and spired rock sedimentary formations with incised steep canyons that are exposed in South Dakota, USA. These rocks are composed of limestones, volcanic ash, sandstones, siltstones, mudstones, claystones, and shale deposited from different regions. In response to regional

Fig. 11.6 **a** Side view of Uluru, another example of an inselberg which is shaped like a loaf (credit: J. Chitambo). **b** Photograph of Uluru from an airplane window (credit D. L. Reid)

Fig. 11.7 **a**, **b** Photographs of two different views of the classic balancing rocks from Epworth in Harare, Zimbabwe, with an 11 year-old girl shown for scale (credit: J. Dzoro)

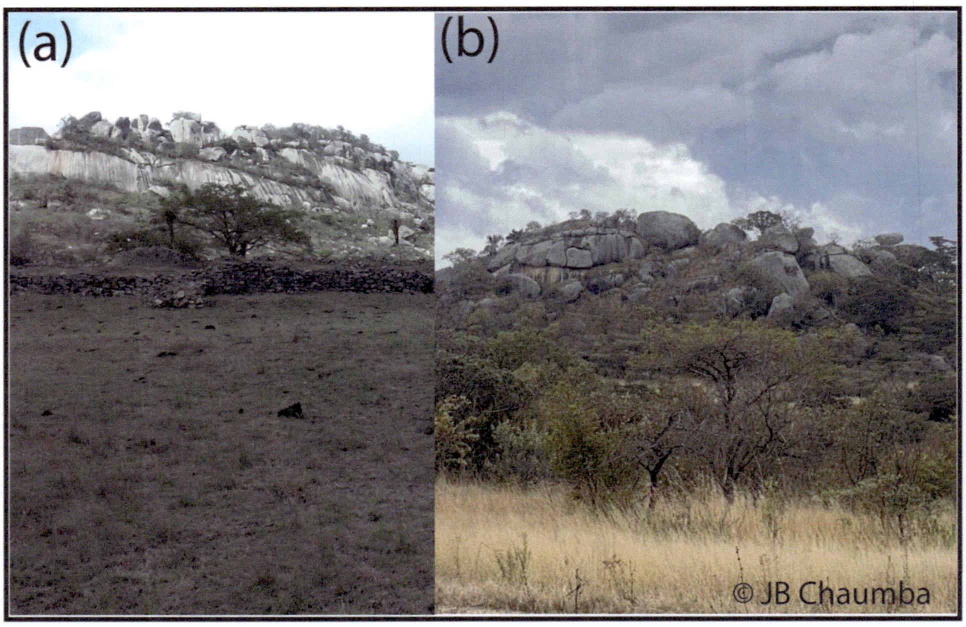

Fig. 11.8 **a** Photograph of part of the rocks at Great Zimbabwe, south of Masvingo, Zimbabwe, showing a bornhardt overlain by castle kopjes at its upper part. Weathering of the underlying bornhardt produces the overlying castle kopjes which are part of the terrain of the Zimbabwe Ruins, Masvingo Province, Zimbabwe. **b** Photograph of a castle kopje from south of Rusape, Zimbabwe, along the Harare-Mutare Highway

uplift that raised the land, the soils of the Badlands have been under extensive erosion by both water and wind since ~5 million years ago. The uplift is interpreted to have initiated the erosion that formed the Badlands.

Located in an area that straddles the borders of Arizona (its northeastern part) and Utah (the southeastern part) within the Navajo Nation reservation is Monument Valley. The area is ~1,100 square miles (~2849 km^2), and it is dominated by sandstones. Monument Valley forms part of the Colorado Plateau which was formed when sediments derived from the ancestral Rocky Mountains were deposited in the region concurrently with regional uplift which affected the Colorado Plateau. Bulging and cracking occurred because of uplifting of the plateau. Water and wind subsequently began to erode the area, widening and deepening the cracks which resulted in the formation of canyons. **Mesas** and **buttes** of red sandstone rise ~300 m high. Mesas can be defined as isolated, broad, flat-topped mountains with at least one steep side, whereas buttes are relatively smaller, flat-topped hills or mountains completely surrounded by steep slopes. These mesas and buttes (Fig. 11.10) that make Monument Valley scenic are interpreted to be the last remnants of beds of sandstone that previously covered the area straddling the

Fig. 11.9 Photographs of remnants of layered rock formations of the Badlands, Badlands National Park, South Dakota, USA (credit: D. T. Nelson)

northeastern Arizona and southeastern Utah border region. Colorado and New Mexico are the other states where mesas and buttes are exposed.

Where the bedrock undergoing weathering is composed of thick beds of sandstone traversed by vertical joints that have wide spaces between them, erosion is concentrated on the joints to form walls of sandstone called **fins**. At the bases of the fins localized erosion can develop, and if it continues, **natural arches** such as those shown in Fig. 11.11a, b can be developed.

Tafoni (singular is tafone), also referred to as "swiss-cheese rock" or cavernous or honeycomb weathering, are rounded and small smooth-edged openings exposed on rock surfaces. Tafoni resemble sponges, and they have a tendency of forming clusters (Fig. 10.12). Tafoni can form when sand from ancient dunes is cemented together by calcite, such as tafoni encountered in the Entrada and Navajo sandstones at Arches National Park in Utah. In these sandstones, moisture percolates through the porous rock, dissolving the calcite in the process and depositing calcite crystals at the surface when it evaporates. Weathering tends to be concentrated in pockets of rock with relatively hardened bands between those pockets leading to the formation of an intricate, lacy structure that often resembles a honeycomb. Salt spray or salty air spray also leads to the formation of tafoni, comprised of pits or holes in sandstones exposed by the

Fig. 11.10 Photograph of buttes exposed in Monument Valley straddling the Arizona and Utah in the southwestern USA

Fig. 11.11 Photograph of Delicate Arch in Grand County, Utah, USA (**a**), and another arch near Moab, Utah, USA (**b**) (credit: S. D. Clark)

Golden Gate Yacht Club. Here, wind carrying salty seawater splashes the seawater on the sandstones resulting in the formation of tafoni (Fig. 11.12). Salt air, specifically salt spray, causes this dramatic pitting in sandstone of the breakwater out by the Golden Gate Yacht Club toward Wave Organ.

Although numerous relatively large landforms that existed for millions of years on our planet are inferred to have been erased by continued weathering and erosion, and there is no more record of them, some landforms, however, that have recently been wiped out from the face of the Earth come to mind. Mukurob, which was also referred to as

Fig. 11.12 Tafoni exposed near Golden Gate Bridge, San Francisco Bay, California

Finger of God (Fig. 11.13a) in Namibia, was a structure that was composed largely of sandstone measured 12 m high, up to 4.5 m wide, and weighed 450 tons. Mukurob had a base of only 3 m long and 1.5 m wide which was much narrower than the mass of rock which it supported. Mukurob formed part of the Weissrand Plateau before 50,000 years of erosion slowly eroded the rest of the plateau leaving Mukurob sticking above the plain. Mukurob, which was a sandstone rock formation exposed near Asab in the desert of Namibia, collapsed on December 7, 1988. The collapse of Mukurob is ascribed to both weathering and erosion.

The Great Stone Face and the Profile, which was commonly referred to as the Old Man of the Mountain in the White Mountains of New Hampshire, USA (Fig. 11.13b) was composed of five granite cliff ledges on Cannon Mountain in Franconia, New Hampshire. This feature was so named due to its appearance resembling a jagged profile of a human face when viewed by a viewer facing south from the north. The Old Man of the Mountain was inferred to have been formed approximately 300–12,000 years ago and had a height of 12 m (40 ft). The Old Man of the Mountain collapsed on May 3, 2003, leaving a different view (Fig. 10.13c) to the mountainside on which it had been a part of hundreds of years. Mukurob and Old Man of the Mountain demonstrate that our planet is constantly changing, albeit slowly, over time, and it has been changing since the formation of our planet close to 5 billion years ago.

Evidence of continued weathering and erosion is scattered almost everywhere on the continents of our planet. Canyons, such as the Kali Gandaki Gorge in Nepal which is thought to be the world's largest canyon, are some of the great examples of the powerful forces of weathering and erosion. A few of some of the relatively large and most impressive canyons are shown in Fig. 11.14. The Grand Canyon, located in the southwestern USA (Fig. 11.14a), is a major feature in North America. The Gorges of Gorropu (Fig. 11.14b), exposed in Sardinia, Italy, is one of Europe's largest canyons. The Yarlung Zangbo Grand Canyon, China (Fig. 11.14c), forms one of Asia's largest Canyons that is still being scoured by the Yarlung Tsangpo River in the Tibet Autonomous Region, China. The Blyde River Canyon (Fig. 11.14d), located in Mpumalanga Province, South Africa, is a 26 km long Canyon carved by weathering and erosion.

Weathering and erosion ultimately result in the deposition of the material weathered, eroded, and transported, resulting in the formation of soils. We next discuss how soils form and then examine how soils are classified, as well as the different types of soils.

11.4 Soils

The Soil Science Society of America defines **soil** as: (i) the unconsolidated mineral or organic material on the immediate surface of the Earth that serves as a natural medium for the growth of land plants, and (ii) the unconsolidated mineral or organic matter on the surface of the Earth that has

Fig. 11.13 Some examples of landforms that have been wiped off the face in recent years through weathering and erosion. **a** Mukurob, Namibia (credit: Road Trip SA). **b** Old Man of the Mountain, New Hampshire, USA (credit: Boston Globe), before the "face and profile" fell off, and **c** remnants of Old Man of the Mountain, New Hampshire, USA, after the "face and profile" fell off (credit: Jim Cole)

been subjected to and shows effects of genetic and environmental factors of climate (which includes effects of water and temperature) and micro- and macro-organisms, conditioned by relief, acting on parent material over a long period of time. Soil, as a product, therefore, is different from the material from which it was derived in its physical, chemical, biological, and morphological properties and characteristics.

11.4.1 Formation of Soils

The formation and development of soil is a dynamic rather than static process. The development of a specific soil is influenced by five soil-forming factors which are: (i) parent material, (ii) topography, (iii) Climate, (iv) biota (organisms), and (v) time. If these five factors don't vary with time on the same landscape, then the soil does not change; it remains the same. If, however, topography, climate, living organisms, and time act on parent material, it is broken down through weathering into smaller particles which form soil. Should any one or more of the factors change, however, the soil will change. The formation and development of soil is dynamic, and it is a progressive process that is constantly changing. Soil development reflects weathering and the dynamic environment in which the soil was formed.

Minerals and rocks formed in place, or remnants of minerals and rocks that were transported by water, ice, wind, or even due to gravity can all constitute parent material. There are many types of parent material possessing different mineral contents. Mountain ranges have been

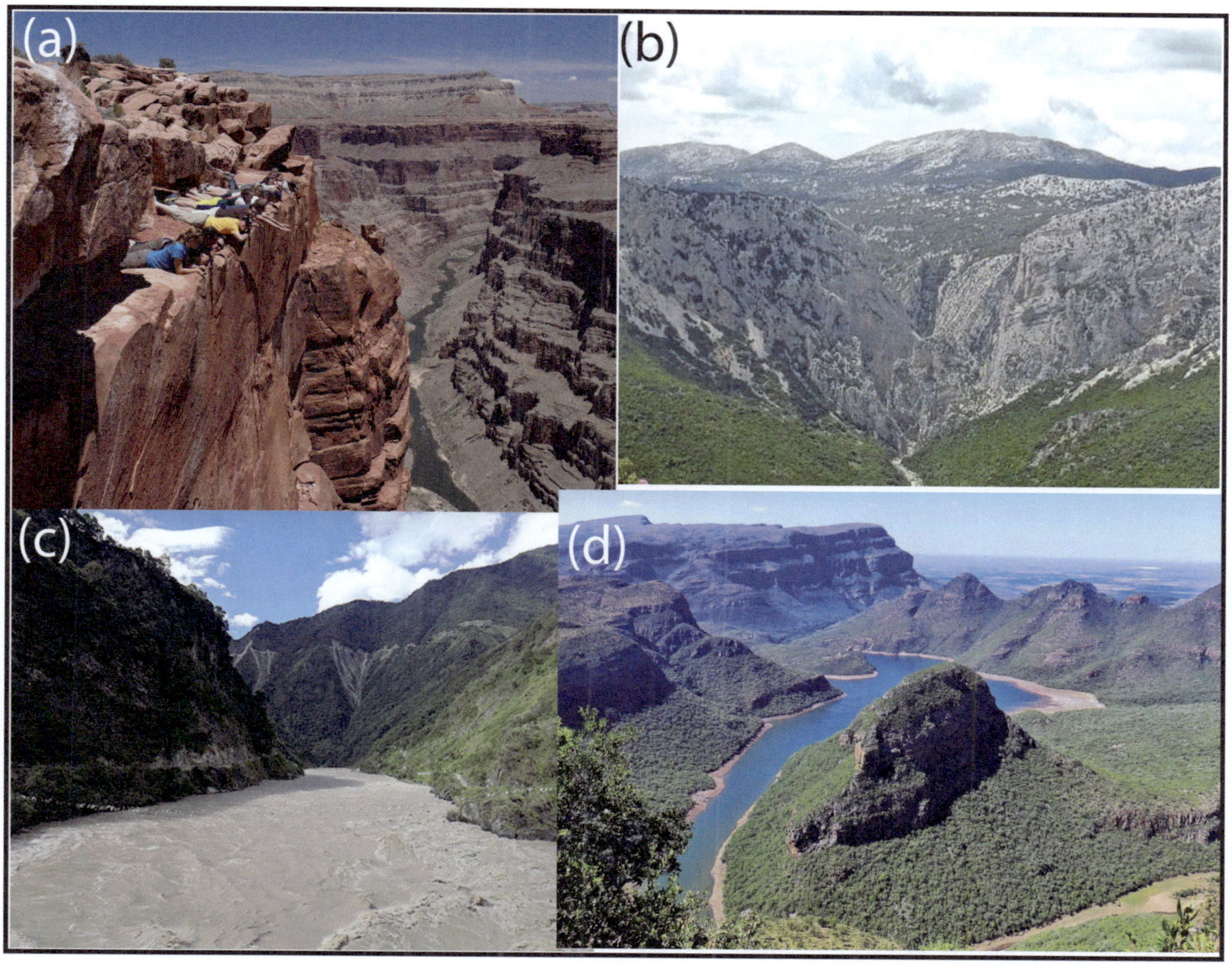

Fig. 11.14 Some examples of some canyons from the world: **a** Grand Canyon, western USA (credit: D. T. Nelson), **b** Gorges of Gorropu, Italy (credit: Sardinia Exploring), **c** Yarlung Zangbo Grand Canyon, China (credit: Top Asia Tour), and **d** Blyde River Canyon, Mpumalanga Province, South Africa (credit: TripAdvisor)

formed and wiped out through erosion and new ones formed again. In the geologic past, seas have flooded the continents depositing layers of varying thicknesses of sand, mud, and carbonates, and receded. Erupting volcanoes accompany mountain building episodes, and glaciers form during long periods of cooler climate, but they melt away during periods of warm climate.

Rainwater infiltrates and percolates downwards due to gravity, incorporating dissolved ions in the soil and transporting clay minerals downwards. The depth to which dissolved ions and clay are removed by downward percolating water is referred to as the **zone of leaching**. New minerals crystallize from the downward infiltrating water to form the **zone of accumulation**. Downward infiltrating clay also accumulates in the zone of accumulation. These processes of the formation of soil occur at different depths below the Earth's surface, and they result in the formation of specific and recognizable zones termed **soil horizons**. A soil horizon is a layer with physical, chemical, and biological characteristics which are different from those in the layers above and beneath it. Soil horizons form parallel to the surface of the soil, and they are denoted by physical features such as color and texture.

11.4.2 Soil Profiles and Soil Formation Factors

Soil horizons develop from the occurrence, at different depths, to form a vertical succession known as a **soil profile** (Fig. 11.15), of different processes that form soil. A soil profile is a vertical section of soil which permits one to examine the structure of soil. A soil profile is divided into soil horizons. Typical soil horizons are O, A, E, B, C, and R, which represent, respectively, organic layer, topsoil, eluviation layer, subsoil, parent rock, and bedrock, as shown in Fig. 11.15.

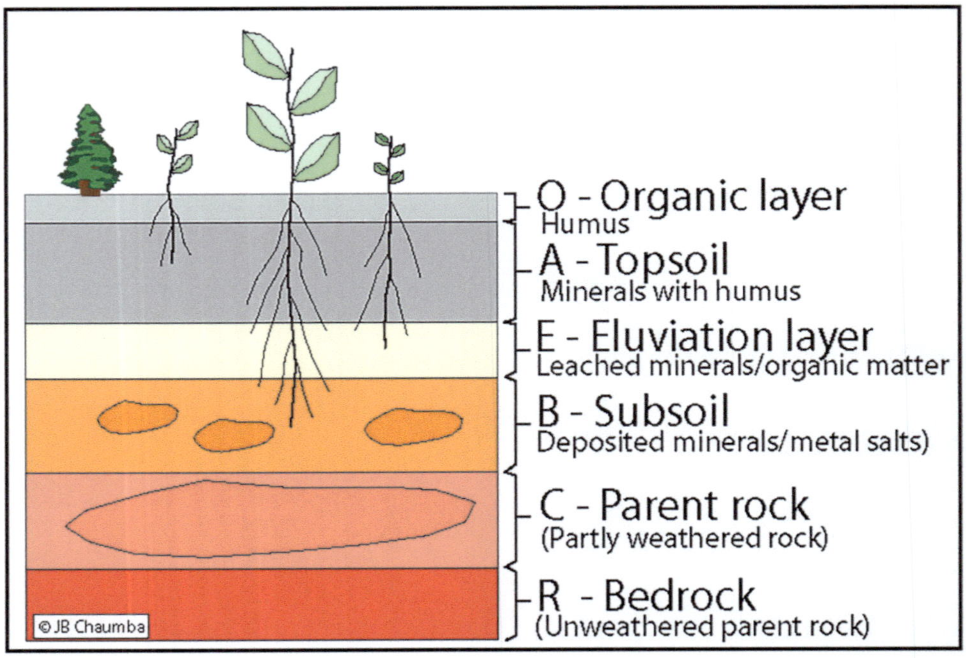

Fig. 11.15 Diagram illustrating a typical soil profile

The B horizon is also referred to as the **zone of accumulation** since leached materials from the A horizon precipitate in this horizon. The B horizon is reddish-brown in color due to iron minerals, and it is also rich in clay. This horizon is also called the subsoil. The combined O–A–B section is sometimes called the **solum**. Known as the "true soil," the solum is where most of the biological activity in soil is restricted. Horizons that are crucial for plant growth are the top two layers of the soil: the O and A horizons. The overlying sedimentary material which occurs above the bedrock is called the **regolith**.

These horizons are briefly described next, keeping Fig. 11.15 in mind. The O, short for organic or humus, is dominated by organic matter, in particular leaves that are decomposing. The O horizon may be missing in some profiles, and it is generally thin in some soils, but it can be thick in others. The A horizon, which is the topsoil, is largely composed of minerals derived from the parent material with some organic matter incorporated as well. Plants and other organisms live in this horizon. The E, or eluviated, horizon is largely leached of minerals, organic matter, and clay, leaving a concentration of resistant materials such as quartz, sand, and silt particles. The E horizon can be lacking in some soils, although it is found in soils in forests and older soils.

Horizon B, also referred to as the subsoil, is enriched in minerals that are moved down, or leached, from the overlying A and/or E horizons. Beneath the B horizon is the C horizon, the parent rock which represents the rock from which the soil was derived. Underlying the C horizon occurs the bedrock (R) which is the lowermost horizon. The R horizon is comprised of rocks, for example, gabbro or basalt, granite or rhyolite, sandstone or quartzite, and even limestone or dolomite which form the parent material for some soils provided the bedrock is close enough to the Earth's surface for weather to occur. If the R horizon is located below the C horizon, then it does not form soil.

Different types of soils are encountered. In some tropical regions such as Zimbabwe where temperatures are relatively high but moderate annual precipitation is received, soil profiles tend to be relatively deep (Fig. 11.16a). Mottles in some of these soils from Chinhoyi, Zimbabwe (Fig. 11.16a), are formed from fluctuating water table between the wet summer and dry winter seasons. Where soils form from weathering of gneisses, such as in Chiredzi District, Masvingo Province, Zimbabwe, also in tropical regions but low annual precipitation is received, the soil profile lacks the eluviation (E) zone (Fig. 11.16b). Due to the dry climate in the Chiredzi area located in the Limpopo lowveld, the O horizon tends to be very narrow (Fig. 11.16b).

Soils from the southeastern United States, where the climate is hot and humid, are characterized by deeply weathered soils due to high rates of weathering to produce thick soils which lack rock fragments (Fig. 11.17a). This is particularly the case in states such as Georgia and adjoining ones. Some soil profiles from the Dallas-Fort Worth, Texas region, for example, contain rock fragments as the limited annual precipitation ensures that rocks decompose much more slowly. Soil profiles in sedimentary rocks can reveal thick formations, such as within the Bend Arch–Fort Worth Basin Province in North Central Texas. A soil profile exposed on the west bank along Rock Creek south of the bridge along County Road 1015 in Johnson County, Texas, USA,

Fig. 11.16 **a** Soil profile derived from weathering of metasediments from Chinhoyi, Mashonaland West Province, Zimbabwe, with soil horizons indicated. The mottles are formed from fluctuating water table. Depth of profile is 1.07 m (3 ft 6 inch). **b** Soil profile derived from weathering of gneisses in Chiredzi District, Masvingo Province, Zimbabwe, with soil horizons indicated. Length of stick is 1 m. Note that most of the O horizon has been eroded or stripped away (credit: C. Bangira)

shows the O, A, E, and C horizons with rock fragments of varying sizes often encountered in all the horizons except O (Fig. 11.17b). This soil profile from the bank of Rock Creek, thus, is within a basin that was once a major producer of petroleum.

Parent material, topography, climate, biota/organisms, and time are the five factors that soil scientists attribute to formation of soil. Some soils form directly from the weathering of the underlying bedrock. The depth of such soils depends on the age and hardness of the bedrock; where soils form above relatively young igneous and metamorphic rocks, the soils tend to be shallow, and consequently crop production on such soils is limited. However, in soils that form on Archean rocks that have been extensively eroded, crop yields can be very lucrative. Till is predominant in the south-central, west-central, and southwestern parts of the state. As the last glacier was melting, these materials were deposited. Soils formed in glacial till tend to produce silty clay loams with many different rock sizes. Internal drainage tends to be poor. Soils that are developed in loess are silty loam textures and often lack rocks.

The rate at which parent materials weather and form soil properties is directly influenced by temperature and amount of precipitation. Temperature and precipitation influence the rate of chemical reactions as in general the higher the temperature, the faster the rate at which reactions occur. Additionally, fluctuations in temperature promote physical weathering. The amount of precipitation determines the rate of water movement in the soil. In equatorial regions, for example, where both temperature and precipitation are high, weathering rates are high which results in the formation of thick soils.

Evapotranspiration is the combination of water evaporated from the soil surface and water transpired by growing plants. As air temperatures increase, the rate of evapotranspiration increases. High evapotranspiration rates relative to precipitation imply that less water is available to move through the soil.

Biota include organisms such as bacteria that live in the soil, and vegetation that grows at the surface which both affect the soil formation process. Organisms can either speed up or slow down soil formation. Microorganisms, for example, can either facilitate chemical reactions and/or excrete organic substances helping improve infiltration of water in the soil. Organisms like gophers can slow down formation of soil through digging and mixing of soil materials. Gophers can also destroy earlier formed soil horizons.

Topographic features of aspect and slope affect formation of soil. Slope refers to the steepness from horizontal. Slope affects the amount of soil that is eroded or deposited. Flat soil, which doesn't lose or gain material, is the most developed. It is the soil-forming process that is slowed by the change in material.

The direction the slope faces relative to the Sun (compass direction) is termed the aspect. Aspect influences the amount of water that moves through the soil. In the northern hemisphere, the south side tends to have less water because there's more evaporation resulting in potentially less vegetation, the opposite of what occurs in the southern hemisphere. The south aspect's colder soil temperatures tend to increase chemical processes in the soil. Soils with a northern aspect tend to have tree vegetation, colder soil temperatures, and less evaporation. The result is that less soil aging with

Fig. 11.17 **a** Soil profile from Chatham County near Savannah, Georgia, with the O, A, and E horizons. **b** Soil profile exposed on the west bank of Rock Creek ~0.5 miles south of the bridge on County Road 1015 in Johnson County, Texas, USA, showing the O, A, E, and B horizons and parent rock (R)

a southern aspect compared to soil with a northern aspect, even with the warmer soil temperatures.

Vegetation and climate act on parent material and topography over time. The age of a soil is determined by development, not its age. The degree of aging of soil is dependent on the intensity of the other four soil-forming factors. Soil formation is slowed, for example, by steep slopes, high water table, high lime content in parent material, cold temperature, severe erosion, low humidity, low rainfall, and rock parent material as well as accumulations and mixing by animals or man, i.e., constant deposition.

11.4.3 Classification of Soils

The properties and nature of soils display wide variations from place to place, even within distances of a few meters. The same properties of soil, however, also exhibit similar characteristics, such as vegetation and climate, over large areas. The soil-forming factors we described above, such parent material, topography, vegetation (biota), climate, and time, produce soils that characterize the environments in which they were formed. Based on soil properties we have described, soils are mapped and classified. This has resulted in soil maps of regions, states, countries, and the world being produced.

Soil textures are classified by the fractions of sand, silt, and clay in the soil. Twelve major soil texture classifications are defined by the United States Department of Agriculture: sand, loamy sand, sandy loam, loam, silt loam, silt, sandy clay loam, clay loam, silty clay loam, sandy clay, silty clay, and clay. Quartz and clay particles of varying sizes predominate what constitutes soil. Quartz grains are important in ensuring that soil is porous. The importance of clay particles lies in retention of nutrients and water which are essential for the growth of plants. **Loam** contains sand, silt, and clay with abundant organic material. Roughly equal proportions of sand, silt, and/or clay characterize loams. A triangular diagram (Fig. 11.18) is typically used to easily visualize the proportions of the primary constituent particle size of sand, silt, and clay in any soil. Combinations of the most abundant particle sizes, such as sandy clay, silty clay (Fig. 11.18), are also deduced from the soil classification triangle. The upper part of a section of loam, which is generally the most fertile layer and contains the highest organic content, is the **topsoil**. Underlying the topsoil and often containing less organic material is the rocky **subsoil**. Hundreds of thousands of years are generally required for soil to form. Soil texture also affects properties of soil such as **porosity**, **infiltration**, **structure**, **water holding capacity**, and the chemistry of the soil.

The United States Department of Agriculture (USDA) uses a soil classification system called Soil Taxonomy (soil naming) which is used widely across the globe. The soil taxonomy classification system reflects the dominant soil-forming factors that are active during the formation of soil at any particular location. The Soil Taxonomy consists of a hierarchy of six levels. These levels, listed in order from most general to most specific, are: (1) Order, (2) Suborder, (3) Great Group, (4) Subgroup, (5) Family, and (6) Series.

Order—Twelve soil orders are recognized. Differences among orders reflect the dominant soil-forming processes

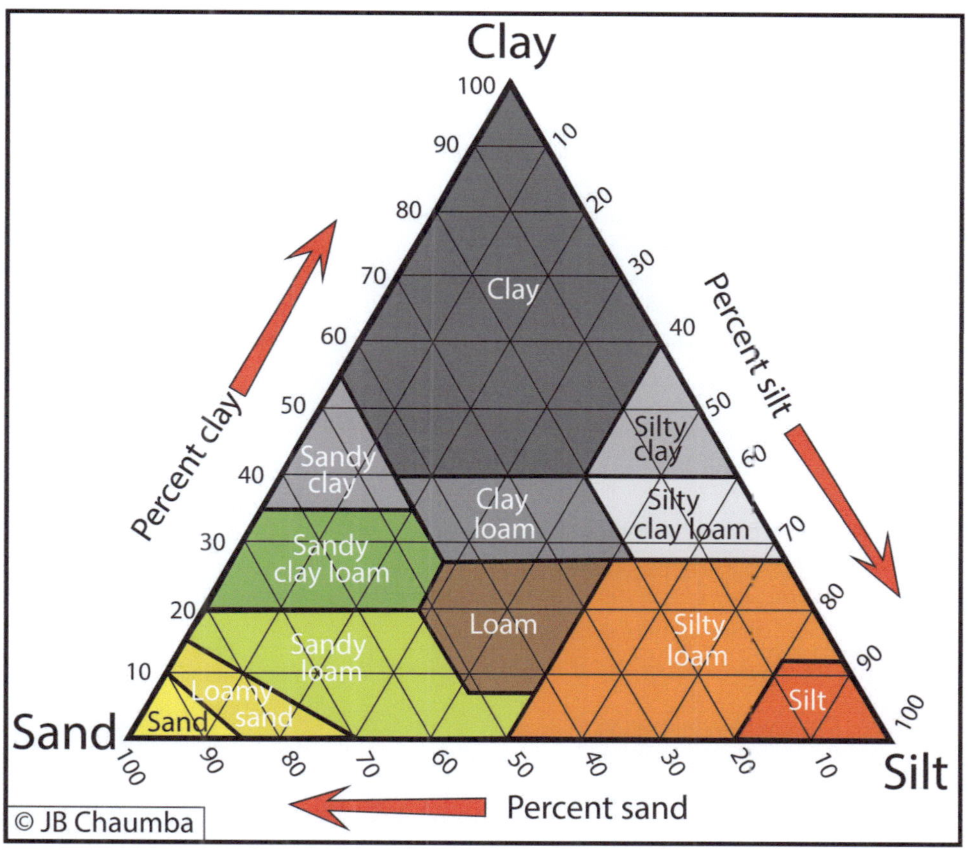

Fig. 11.18 Soil classification diagram

together with the degree of soil formation. A word ending in "sol," such as Alfisols, identifies each order.

Suborder—Subdivision for each order into suborders is made primarily on the basis of properties that are important to plant growth and/or influence soil formation.

Great Group—Based on the similarities in horizons present, temperature regimes, soil moisture, or other significant soil properties, each suborder is divided into great groups.

Subgroup—Each great group is characterized by a "typic" (typical) subgroup that is defined by the Great Group. The transitions to other orders, suborders, or great groups are due to properties that distinguish it from the great group from other Subgroups.

Family—Based on the physical and chemical properties along with other characteristics that affect management, Families within a subgroup are established.

Series—Soils within a family that have horizons similar in color, texture, structure, reaction, consistence, mineral and chemical composition, and arrangement in the profile comprise Series.

A summary of the 12 soil orders and their brief descriptions is shown in Table 11.1.

The Soil taxonomy system provides information that can be utilized to make inferences pertaining to the utility of a particular soil for various purposes such as waste management, construction sites, urban/residential construction, as well for farming to feed the world's continuously increasing human population. Some areas of our continents lack soil as they are either covered by glaciers and/or ice, or by shifting sands such as most parts of the Sahara Desert. Additionally, areas where bedrock is exposed at the Earth's surface lack soil.

Table 11.1 The 12 soil orders and their brief descriptions

Type	Description
Alfisols	Soils located in deciduous forest in semi-arid to humid regions
Andisols	Soil formed in volcanic material containing high proportions of glass and amorphous colloids
Aridisols	Soils located in arid and semi-arid climates, dominant in deserts and xeric shrublands
Entisols	Lack any soil profile development other than an A horizon
Gelisols	Soils of cold climates with permafrost within two meters of the soil surface
Histosols	Consist primarily of organic materials and shrink and swell
Inceptisols	Develop quickly through alteration of parent material; beginning of horizon development
Mollisols	Soft grassland soils in semi-arid to semi-humid areas
Oxisols	Acidic, coniferous tropical rain forest soils
Spodosols	Soft, grassland soils typical of coniferous or boreal forests
Ultisols	Extensively weathered soils, also called red clay soils, are typical in humid temperate or tropical regions
Vertisols	Extremely weathered, high content of expansive clay minerals soils that form deep cracks in dry seasons

Clayey soils that are characterized by wide and deep cracks for part of the year are called vertisols. Vertisols shrink upon drying, and they swell when they become wet or moist due to their high proportions of clay minerals. Shown in Fig. 11.19 is a soil profile through vertisols from Chisumbanje in southeastern Zimbabwe.

A representation of the types of soil for an area of interest, as well as the soil properties such as soil color, soil horizons, organic matter content, and soil reaction (pH), is called a soil map. Soil maps are available for every country such as the US (Fig. 11.20) and even for our planet (Fig. 11.21).

Soil maps have traditionally been compiled on the basis of field observations conducted during soil surveys at scales that would have been pre-determined. The generalized resulting maps at the scale of the maps are then published with the status of the soils as well as their distribution over the geographic area. A map legend groups and provides brief descriptions of the main soil types. An accompanying soil survey report provides details on the main types of soils, their distributions, properties, and capabilities. Typically, results of traditional soil surveys are presented as polygon maps. These maps are used for assessment of agricultural extension, agricultural potential (land evaluation), and planning land use and management.

11.5 Mass Wasting

Some of us get outside our houses to be greeted by relatively flat terrane, provided you live on the Interior Plains or the Plains of Nunavut in North America, the Veld in Southern Africa, the Great European Plain in Europe, the Plains of Australia and new Zealand, the Plains of Asia (Aaiha, Indo-Gangetic, Jiaolin, Liyang, Tihamah, and Ulu Baram), or the Plains of South America (Pampas, Plains of Argentina, Brazil, Venezuela) depending on where you live, you are largely greeted with relatively level expanses of land. Slope failure, or **mass wasting**, or even mass movement, is rare on the Plains, except where major rivers scour channels across the Plains. Mass wasting can be defined as the downhill movement of rock and/or soil due to the force of gravity. Forces that resist movement of material downslope include cementation, cohesion, the strength of the material making up the slope, internal friction between the grains as well as any external support on the slope. These forces which resist mass wasting have a bearing on the shear strength of a slope.

Fig. 11.19 Soil profile of vertisols from Chisumbanje, southeastern Zimbabwe. This soil was derived from weathering of mafic rocks such as basalts (credit: C. Bangira)

Shear strength is the maximum shear stress sustained by a soil without experiencing failure. The downslope movement of soil and rock due to the influence of gravity is called mass wasting. However, if you live in mountainous regions on any continent, you have probably heard of, or even experienced, mass wasting.

If you one those people who enjoy the beach, remember that in order to swim in coastal water or simply take a stroll on the beach, one has to first climb up and then down the pile of sand that forms ridges (sand dunes) particularly on United States southeastern beaches. If the dunes are mostly covered by grass, their slopes to be steeper than those without any vegetation on them. Before you get your swim or play with beach sand whenever to take a trip there, you typically must first go over the sand dune(s) (Fig. 11.22a). The dunes vary in size, depending on a variety of factors. Shown in Fig. 11.22a, b are two dunes: Dune 1 sustains a smaller angle with the horizontal compared to Dune 2. Dune 1 has less vegetation in comparison to the vegetation on Dune 2. Sand on Dune 1 has less vegetation to hold it to support a steeper angle than that sustained by Dune 2 which had relatively more vegetation.

The reason Dune 1 sustains a smaller angle with the horizontal than Dune 2 is that more sand particles easily roll down the flanks of the dune. Drier sand with no grass at all growing on it sustains an even smaller angle. The steepest angle maintained by a slope without collapsing is called the **angle of repose** (Fig. 11.23). The angle of repose varies depending on the material. For dry sand, the angle of repose ranges from 30° to 37° and for gravel and grains with irregular shapes, the angle of repose can be as high as 45°.

The reason why sand rolls down the slopes of sand dunes and mounds of sand is shown, respectively, in Figs. 11.22 and 11.23 is that the force of gravity acts on the sand particles which run downslope to suspend at an angle corresponding with the grain size, moisture content in the ground, and grass binding the sand together. The sand and gravel particles shown in Fig. 11.23 the sand and gravel particles are dry, thus eliminating the effect of moisture which affects beach sand. The angle of repose is the angle of maximum slope inclination that is sustained by free-standing dry sand at which the soil is barely stable. The angle of repose varies from horizontal (0°) to vertical (90°). For dry sand, the angle of repose is approximately 30°.

For any mass on an inclined surface, the normal force is oriented in a direction perpendicular to the inclined surface ($F_{perpendicular}$) and the parallel being parallel to the inclined surface ($F_{parallel}$) which can be added to give the weight ($F_{gravity}$) (Fig. 11.24). The resultant or net force on the object is shown in Fig. 11.24. Any force that is directed at an angle to the horizontal, for example, is resolved into horizontal and vertical components. In Fig. 11.24, the weight vector ($F_{gravity}$) is resolved into two components: one that

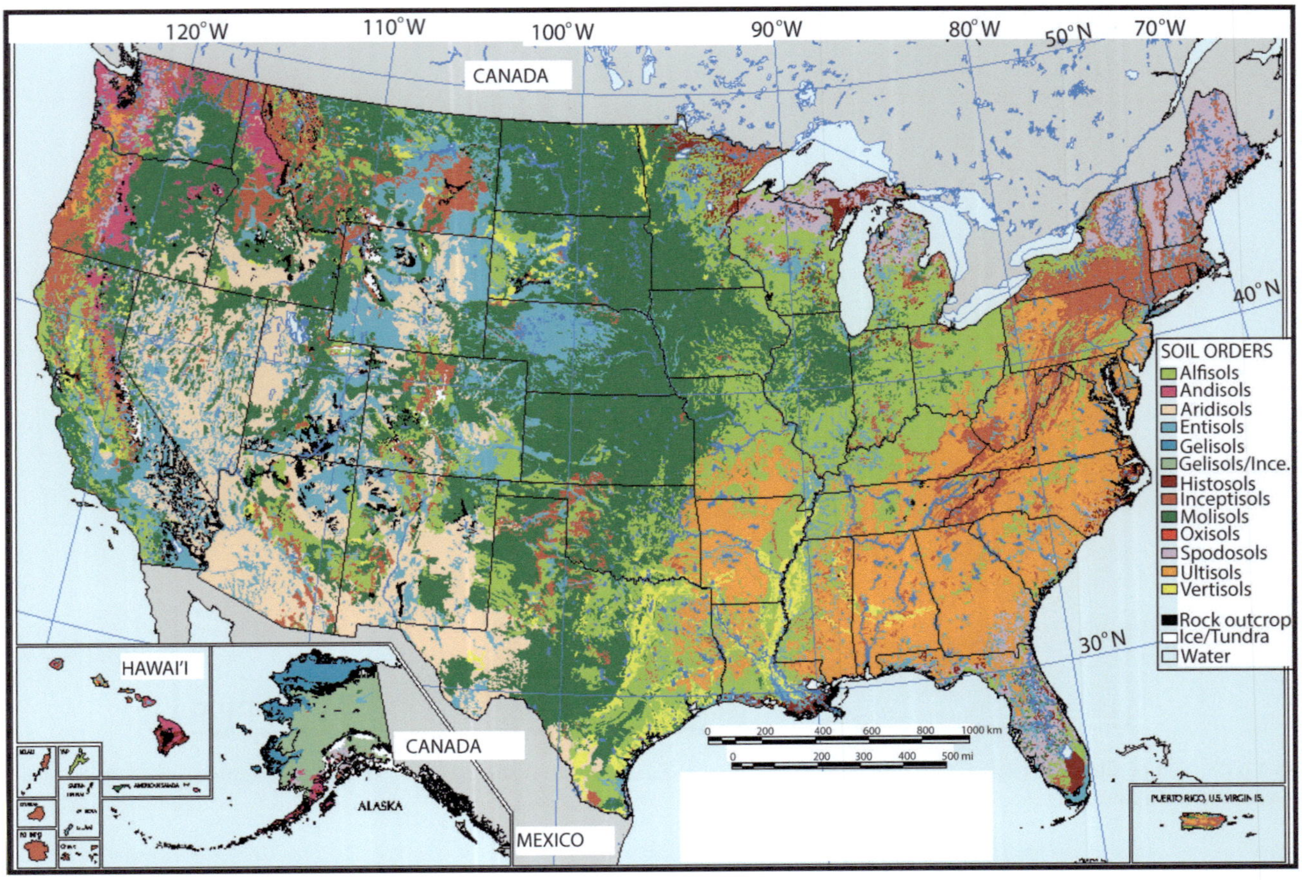

Fig. 11.20 Map of United States soil regions (after *USDA-NRCS*)

is directed parallel to the inclined surface ($F_{parallel}$) and the other one directed perpendicular to the inclined surface ($P_{erpendicular}$). Mass wasting occurs only if the gravitational force parallel to the slope ($F_{parallel}$) is greater than the force resisting motion ($F_{friction}$), that is, when ($F_{parallel}$) > ($F_{friction}$) (Fig. 11.24).

11.5.1 Triggering Mechanisms of Mass Wasting

In mountainous regions, movement related to landslides, mud and debris flows, and rockfalls occurs naturally and rather continuously. Triggering mechanisms of mass movements can be due to human activity such as land development for various purposes including construction of roads and houses, mining, quarrying, as well as other disturbances. These events can occur at any time of the year from almost any location along a slope. Any change in elevation is likely to trigger mass wasting, and such changes occur in any mountainous region. Other triggering mechanisms of mass wasting involve heavy rainfall over relatively short time periods such as during hurricanes or cyclones, or even light rains over prolonged time periods.

Natural triggers of mass wasting include high intensity rainfall during a relatively short time period such as during hurricanes (Fig. 11.25a) or less intense over a prolonged period such as weeks and months, earthquake, coastal or stream erosion, volcanic eruption rapid snowmelt. Hurricane Idai struck Mozambique, Zimbabwe, and Malawi from March 4 to 19, 2019 and triggered deadly landslides and extensive flooding. In total, 1,593 people died. In mountainous regions, the rainfall adds weight to the soil, and steepening slopes during construction of roads both help trigger mass wasting (Fig. 11.25b) resulting in the occurrence of mass wasting such landslides (Fig. 11.25c). Running water confined to river channels also widens riverbanks and widens the channel through undercutting the banks. Floods often lead to extensive destruction of infrastructure such as bridges (Fig. 11.25d) in populated areas.

11.5.2 Types of Mass Wasting

Landslides, slumps, rockfalls, rockslides, debris flows, mud flows, earthflows, and creeps are among the many geologic and soil hazards that impact us. Landslide is a general term for mass wasting that involves fast-moving geologic mate-

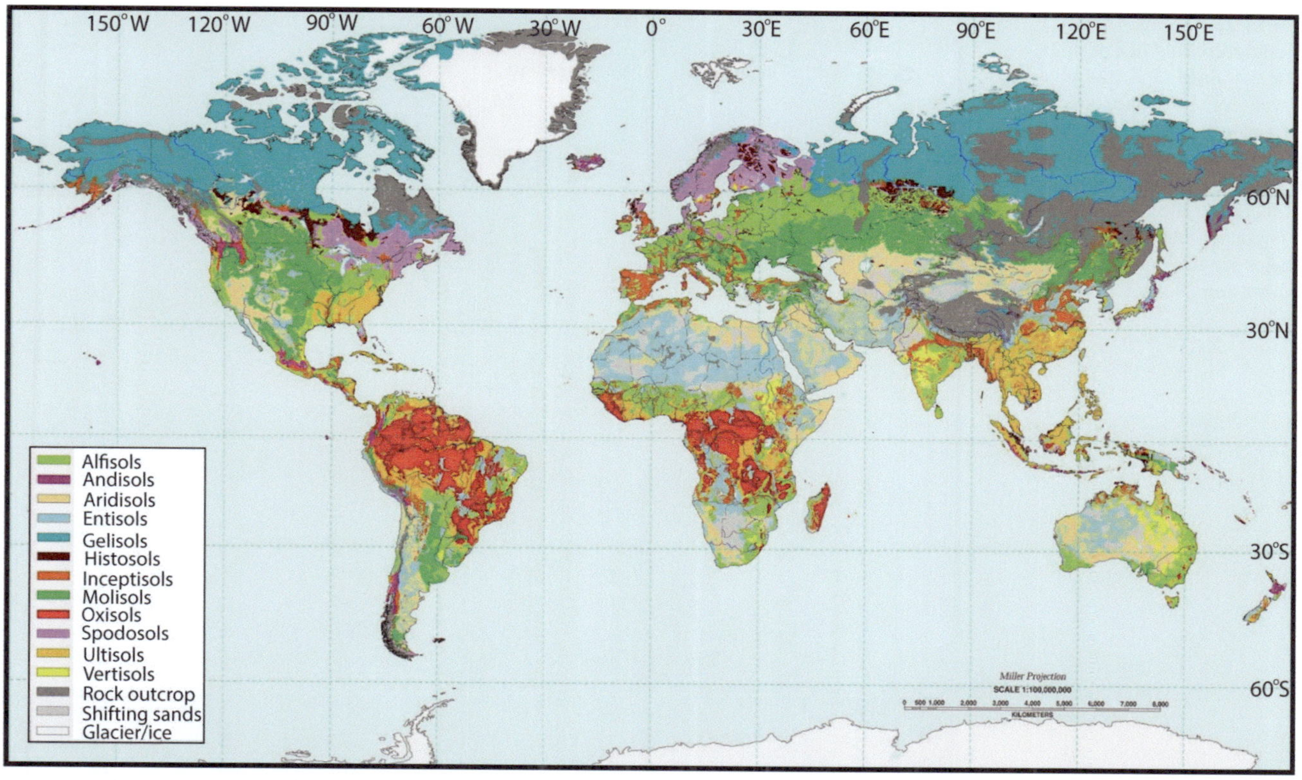

Fig. 11.21 Map of global soil regions, modified after Vivid Maps (credit: https://vividmaps.com/soil-map/)

Fig. 11.22 **a** Photograph of sand dune with the Atlantic Ocean behind the sand dune, Oak Island, North Carolina. **b** Photograph of a coastline showing Fernandina Beach, Florida looking north with the Atlantic Ocean to the far right and Dune 1, with less grass with the sand dune supporting a smaller angle with the horizontal in comparison with Dune 2 which has more grass than Dune 1

rial. Loose material along with overlying soils are what typically move during a mass wasting event. Moving blocks of bedrock are called rock topples, rockslides, or rock falls, depending on the dominant motion of the blocks. Mass movements of dominantly liquid material are called flows. Movement by mass wasting can be slow or rapid. Rapid movement can be dangerous, such as during debris flows. Areas with steep topography and rapid rainfall, such as the California coast, the Rocky Mountain Region, and the Pacific Northwest, are particularly susceptible to hazardous mass wasting events.

Landslides occur when there is downslope movement of slopes that are composed of rock, soil, and/or artificial fills. Landslides occur due to decreases in frictional forces

Fig. 11.23 Mounds of dry fine-grained sand (**a**), of coarse-grained sand, and of gravel (both in **b**) support varying sizes of the angle of repose (r) which increases with increasing grain size (r_0 and r_1; **b**). Note that the angle of repose r_1 is greater than r_0 in (**b**). All images are from a gravel plant in southeastern North Carolina, USA

binding the Earth mass in place and/or due to increases in the downslope components of gravitational forces that promote movement. Movement in landslides occurs by sliding, falling, and flowing along surfaces demarcated by differences in characteristics of rock or soil. Rates of movement for landslides vary widely from very slow (less than centimeters (inches) per year) to rapid, which is tens of meters (feet) per second. Areas where landslides have previously occurred or are active are recognized by their physical appearance and topography.

A **debris flow** (Fig. 11.26a) occurs when more than half of the solids involved are composed of boulders and rocks, i.e., the material moved coarser-grained than sand. A **mud flow** (Fig. 11.26b) is a mass of water and fine-grained Earth materials that flows downwards in a stream, canyon, ravine, or an arroyo. Movements in debris and mud flows involve both fast-moving water as well as large volumes of sediment and debris surging downslope with high force. A **slump** is a type of slide (movement as a mass) involving movement along a curved, concave upward surface of failure (Fig. 11.26c) within thick (>10 m) unconsolidated deposits. The downward motion in a slump occurs near the top to expose a scarp, whereas an outward motion occurs toward the bottom.

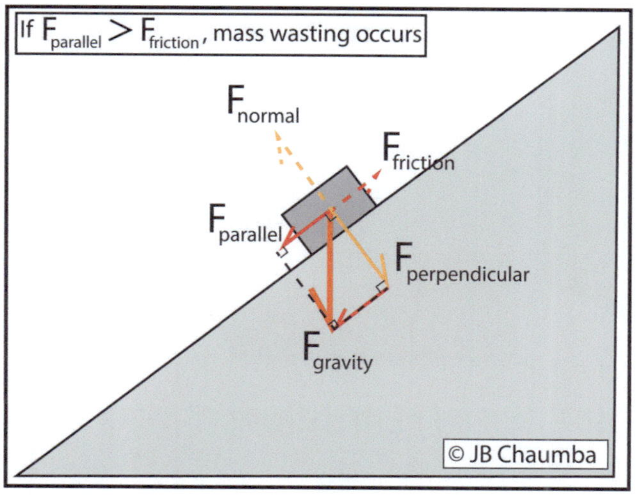

Fig. 11.24 Relationships between the forces of gravity > ($F_{gravity}$), friction > ($F_{friction}$), and downslope component of gravity > ($F_{parallel}$) acting on an inclined surface. Mass wasting occurs if ($F_{parallel}$) > ($F_{friction}$)

A **rockslide** is when sudden and rapid movement occurs along a roughly planar surface (Fig. 11.26d). A **rockfall** is when a newly detached rock falls at rapid velocities down very steep slopes or from cliffs (Fig. 11.26e). Rockfalls, which are the fastest type of mass movements, commonly occur in mountain regions. An **earthflow** (Fig. 11.26f) is the downslope movement of debris as a viscous fluid that typically develops at the lower end of a slope to produce a lobe-shaped landform. The flow of debris saturated with water over impermeable materials is called **solifluction**, whereas very slow, continuous, downslope movement of debris saturated with water is called **creep**.

Hurricane Mangkhut made landfall on September 15, 2018, at ~13:00 h local time in Barangay Ucab, in the northeast of the Philippines. The hurricane, locally referred to as Ompong. Triggered landslides (Fig. 11.27a, b) which resulted in 79 fatalities. The resulting debris flows affected the entire Benguet Province during the onslaught of Hurricane Mangkhut.

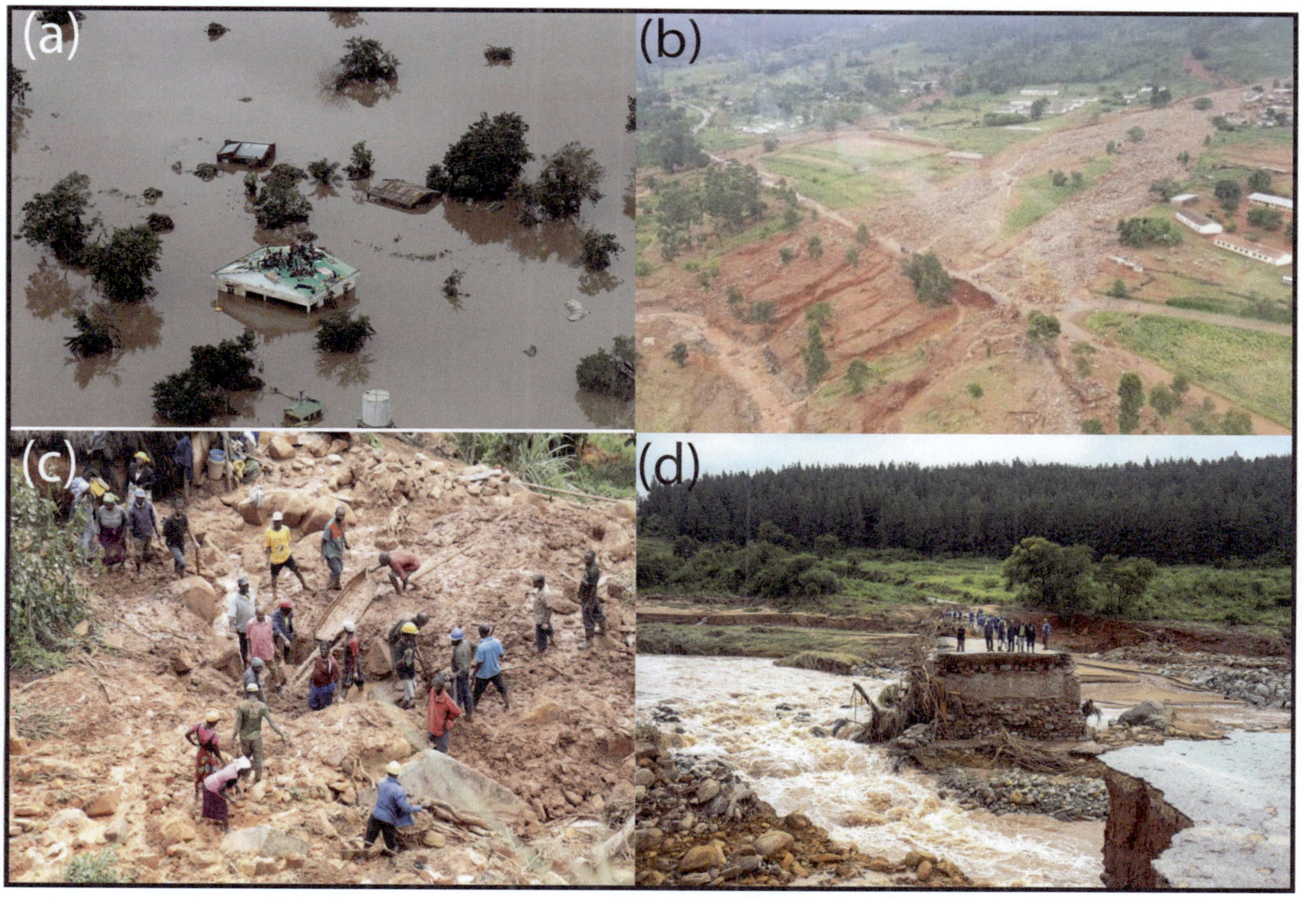

Fig. 11.25 Some examples of triggering mechanisms of mass wasting. **a** intense rainfall such as during Hurricane Idai in 2019 which caused extensive flooding. In mountainous regions, the rainfall adds weight to the soil, and steepening slopes during construction of roads both help trigger mass wasting (Max Bearak, Washington Post). Mass wasting such landslides as in (**b**, **c**) (credit: New Scientist) occurs. **b** Aftermath of Cyclone Idai in the village of Ngangu near Chimanimani, Zimbabwe (credit: Povo News). Running water in river channels also causes riverbanks to widen through widening of the channels as well as destruction of infrastructure such as bridges (**d**) (Max Bearak, Washington Post)

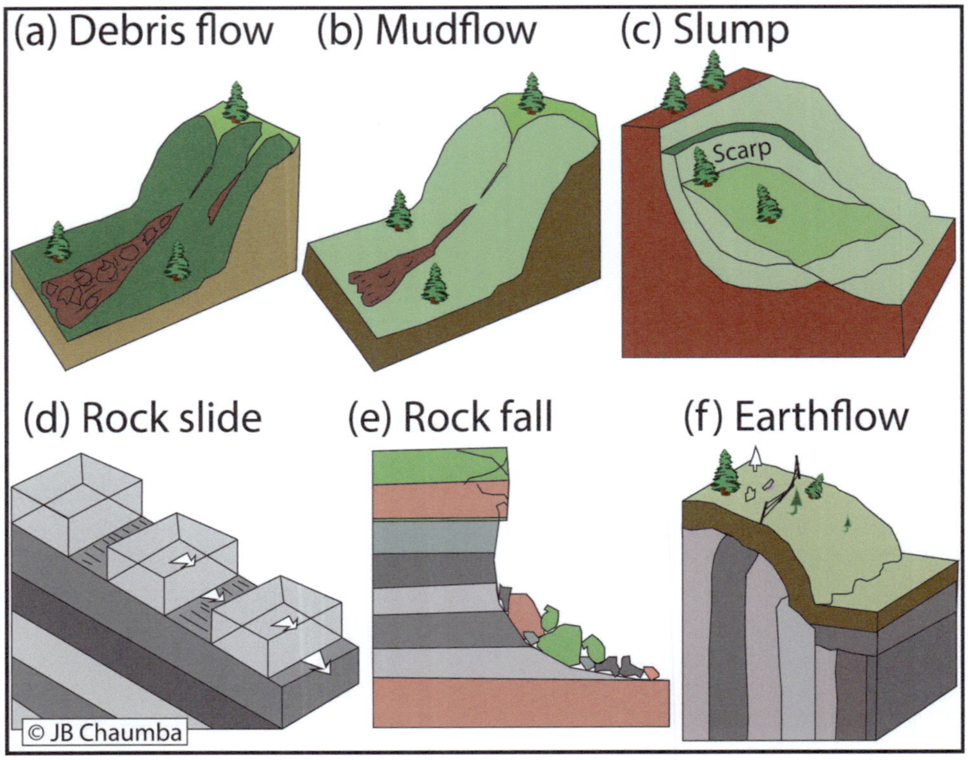

Fig. 11.26 Types of mass movements: **a** debris flow, **b** mudflow, **c** slump, **d** rockslide with arrows denoting direction of movement, **e**, rock fall, and **f** earthflow showing tilted trees and fence and trees with bent trunks

Fig. 11.27 **a**, **b** Debris flows in Barangay Ucab, Philippines, on September 15, 2018, at ~13:00 h local time resulted in the death of at least the death toll in the landslide in Barangay Ucab rose to 44, with a total fatality of 79 due to the landslides

Some examples of relatively recent landslides that affected different regions of our planet are shown in Fig. 11.28. A debris flow caused by the storm on July 20, 2021, which resulted in damage to property near the outlet of Black Hollow drainage basin following. The debris flow deposited sand, mud, cobbles, boulders, and woody debris on the property (Fig. 11.28a). Shown in Fig. 11.28b is a photograph of a slump. A mudflow (Fig. 11.28c) caused by heavy rains in mid-March 2017 led to extensive flooding shown in Lima, Peru. A rockslide is shown in Fig. 11.28d. The rockslide occurred in October 2009, and it led to the closure of Interstate-40 in the Pigeon River Gorge, North Carolina, USA.

A landslide that severed Highway 1 at Paul's Slide at Rat Creek near Big Sur, California, USA, on Jan. 28, 2021, is shown in Fig. 11.29 a and b. Paul's Slide is located at mile marker 22, and it is the same locality that previously slid in May 2017. The rains which triggered this landslide transported boulders, trees, mud, and ash downslope from areas that had been burned by the fire the previous fall. A section of Highway 1 near Big Sur, Calif., collapsed into the Pacific Ocean on the evening of Jan 28, 2021, following

Fig. 11.28 Photographs of some examples of mass wasting. **a** Debris flow. Damage to property near the outlet of Black Hollow drainage basin following the storm on July 20, 2021. The debris flow deposited sand, mud, cobbles, boulders, and woody debris on the property credit: Jason Kean, USGS. **b** Slump in Noatak National Preserve, Alaska, USA. Credit: National Park Service. **c** Residents tightly holding onto a rope as they walk to safety in Lima, Peru, during flooding in mid-March 2017, credit: Martin Mejia/Associated Press. **d** Rockslide of October 2009 that closed Interstate-40 in the Pigeon River Gorge, North Carolina (credit: North Carolina Geological Survey)

Fig. 11.29 Big Sur Highway 1, California, landslide. Aerial view (**a**) (credit: USGS), and view of the landslide damage to the road, Highway (**b**) (credit: Caltrans, Associated Press) on Jan. 30, 2021, of a part of Highway 1 near Big Sur, California, USA

heavy rains for a week. A landslide incised into the rocks where Highway 1 runs parallel to the coastline ~44 miles south of Monterey, California (Fig. 11.29).

During Hurricane Helene which made landfall in Florida, USA, on September 26, 2024, the United States Geological Survey reported that over 2000 landslides were linked to the hurricane, which affected the Southern Appalachians. Western North Carolina was seriously affected by the hurricane. The landslides were triggered by excessive rainfall which occurred in a very short period of time. Shown in Fig. 11.30a, b are photographs sent to the writer by one property owner whose driveway was affected. Figure 11.30c shows a landslide that affected the San Francisco area.

On Friday, May 24, 1968, a M 7.1 magnitude earthquake struck Inangahua, South Island of New Zealand which triggered landslides (Fig. 11.31a). The earthquake led to the fatalities of 3 individuals. The area was also affected by a landslide about a decade later (Fig. 11.31b). In 2006, a photograph taken from nearly the same position as in 1970 and 1971 (Fig. 11.31c) shows that the valley has recovered from the effects of the previous two landslides.

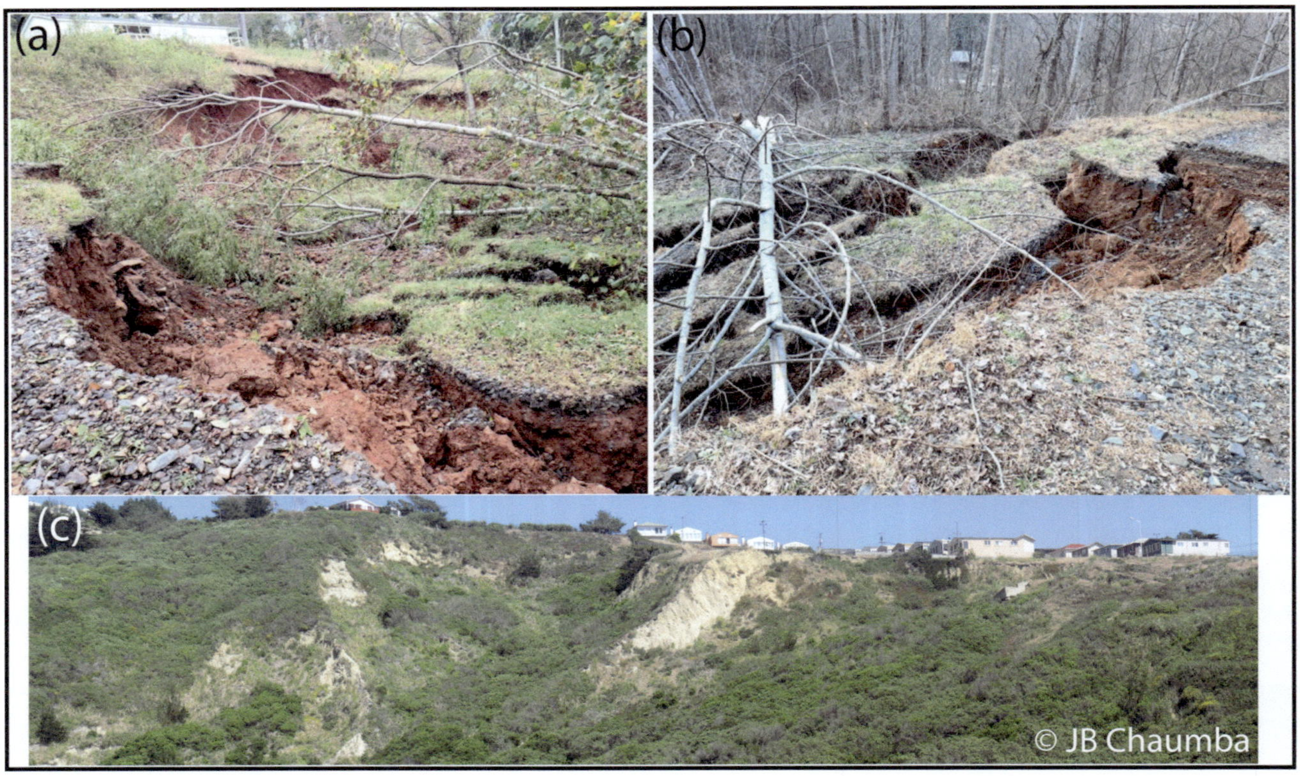

Fig. 11.30 Slumps. (**a**, **b**) are photographs of slumps from the Appalachians of western North Carolina triggered by hurricane Helene (credit: property owner), whereas (**c**) shows a slump from San Francisco, California, USA (credit: D. T. Nelson)

Fig. 11.31 **a** Photograph showing the aftermath of landslides after the 1968 Inangahua earthquake. The same area was affected by another landslide about a decade later (**b**), and the photograph shown in (**c**) shows how the slope has regenerated itself (credit: D. L. Reid)

Fig. 11.32 **a**, **b** Photographs of gabions (credit: Enviro-mesh.com)

Wire-mesh cages and structures that are constructed and then filled with rock fragments and used as retention walls for streambanks, precarious areas, and erosion control systems are called **gabions**. Gabions (Fig. 11.32), which are referred to as "rock cages," "wire cages" or "gabion baskets" can also be utilized as sound barriers. The porous nature of gabions, their flexibility, and their stability provided by their mass render them as cost-effective and durable solutions for civil landscaping and engineering applications. Gabions, composed primarily of wire mesh baskets typically in square or rectangular shapes, can be filled with various rocks, gravel, or even recycled concrete. Gabions prevent soil from collapsing on uneven terrain. Gabions stabilize sloping areas such as shorelines, streambanks, and other areas susceptible to erosion by holding soil in place as well as absorbing the pressure of water. Gabions can also form decorative garden walls, and they also find use militarily as both barricades and protective barriers, and even in fortification. Wire-mesh structure permits gabions to resist cracking and remain stable. The porous design of gabions allows water to flow through them, which maintains the stability of wire meshes and prevents buildup of water. Made with heavily galvanized steel wire, gabions are resistant to weathering and can maintain stability for decades. The design of gabions often promotes vegetation growth over time, thereby reducing carbon emissions. Gabions are often filled with locally sourced or inexpensive rock, making them cost-effective.

In general, mountainous areas which have steep slopes experience high incidences of slope failure resulting in mass movements such as landslides, whereas flat to gentle regions tend to lack mass movements, or if they occur, they are mild. In the United States, areas such as the Coastal Plain and the Midwest hardly experience any landslides (Fig. 11.33). Regions of the country with mountains experience high incidences of landslides (Fig. 11.33). The highest landslide incidences, therefore, occur within the Appalachians, the Colorado Rockies, and on the US West Coast (Fig. 11.33).

Even on an international scale, the picture remains the same with high incidences of mass movements occurring in orogens where slopes are generally steeper and much lower in regions that are relatively flat (Fig. 11.34).

Proper design during construction projects can eliminate the potential for increased mass wasting. Human activities such as undercutting the base of the slope during construction of buildings or roads, railroads, and airports increasing the risks of wasting. Additionally, additional weight to upper parts of the slopes through the construction of large structures, removing vegetation for recreation purposes such as sporting fields may enable saturation of the ground with water, which increases the risks of mass wasting.

11.5.3 Mitigating Mass Wasting

Engineering solutions to mitigating mass wasting include slope terraces, barriers and retention walls, reducing the steepness of cuts, revegetation soon after the cuts, and installation of drainage pipes. The use of rock bolts, cables, and screens, cutting back slopes to more gentle gradients control or eliminates rockfalls.

Mitigating mass wasting can be achieved through special stretchable fencing that is applied in areas susceptible to rockfalls. Diversion channels (Fig. 11.35a), benching or terracing slopes (Fig. 11.35b), screening or installation of rock bolts (Fig. 11.35c), retaining walls (Fig. 11.35d), construction of avalanche sheds (Fig. 11.35e), and revegetation of slopes (Fig. 11.35f) are useful practices in mitigating mass wasting. Retaining walls are useful in supporting the bases of slopes. To avoid water building up behind these

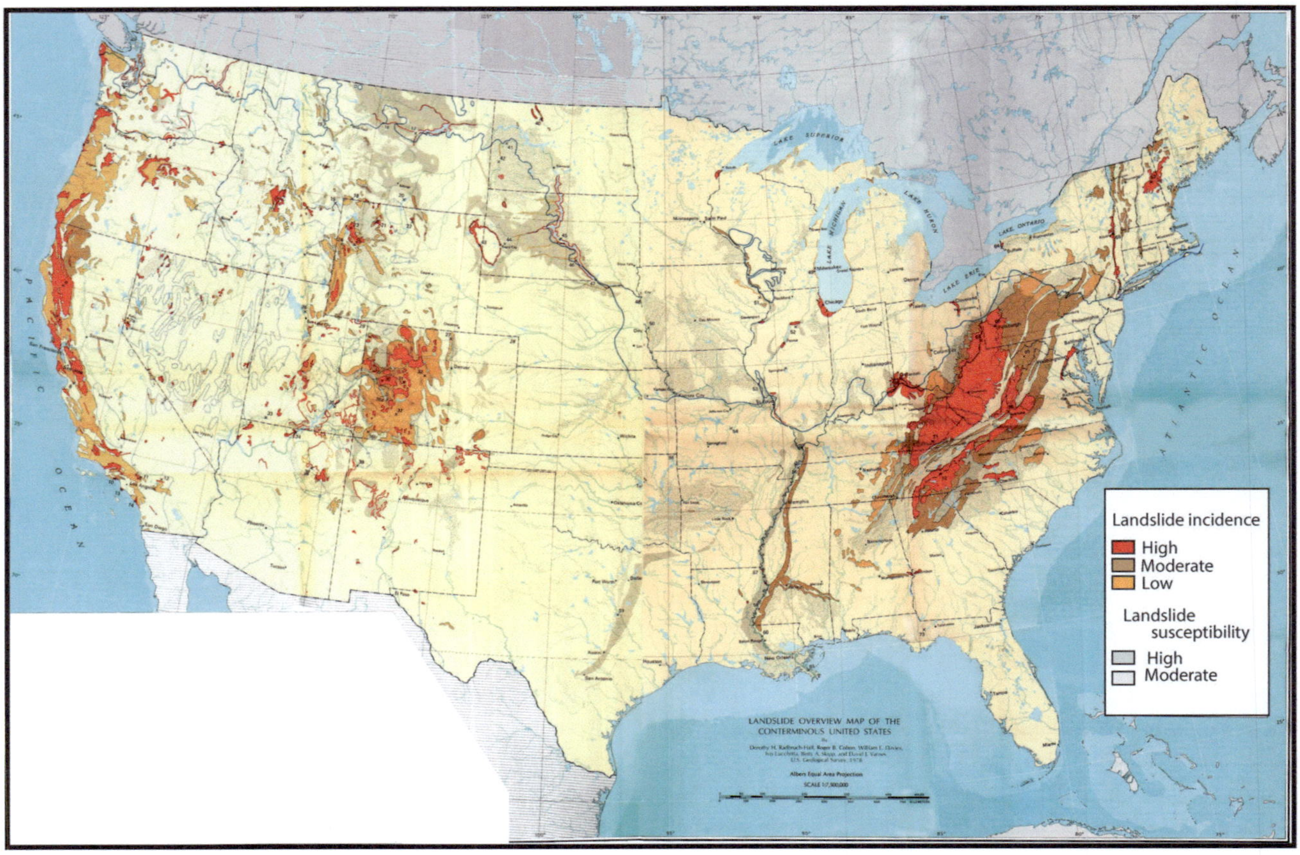

Fig. 11.33 Map of the United States showing landslide incidence. Note that the highest landslide incidents occur in the Appalachians, the Colorado Rockies, and on the West Coast (credit: USGS Professional Paper 1183)

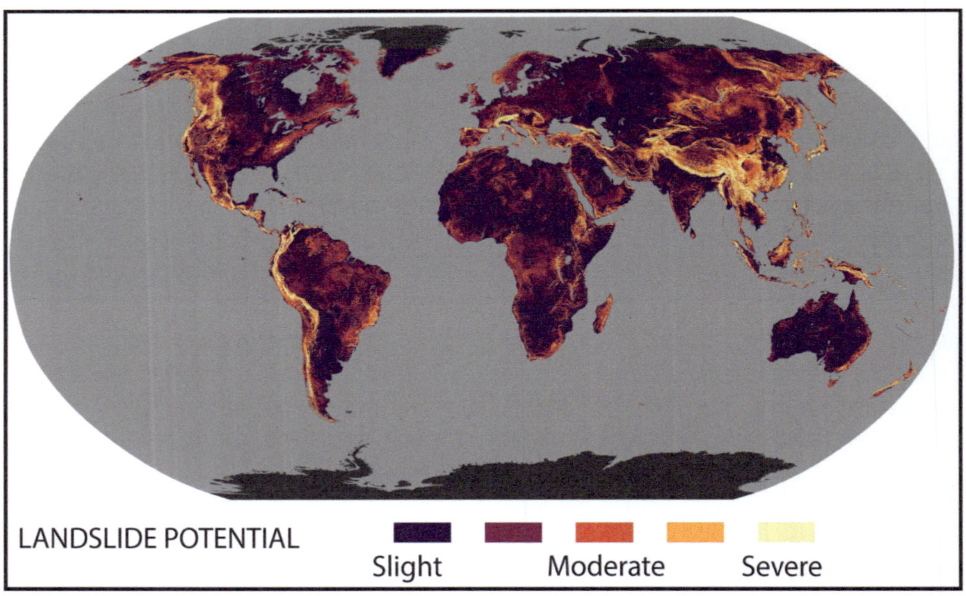

Fig. 11.34 Global map showing landslide potential 2000–2023. Notice that areas with the highest landslide potential are associated with west coasts of the Americas and with the Alpine-Himalayan orogen (credit: NASA Earth Observatory)

retaining walls and risk mass movement, drain holes and pipes are used (Fig. 11.35d). Cutting back slopes to produce gentler gradients helps minimize or sometimes even eliminate rockfalls. In addition, runout channels and check dams are useful in slowing down debris flows as well as diverting them around structures to be preserved. Landslides can be best diverted by construction of special tunnels over some highways in mountainous regions. It is recommended to practice revegetation on steepened slopes immediately after construction of the highways. Installation of drainage pipes

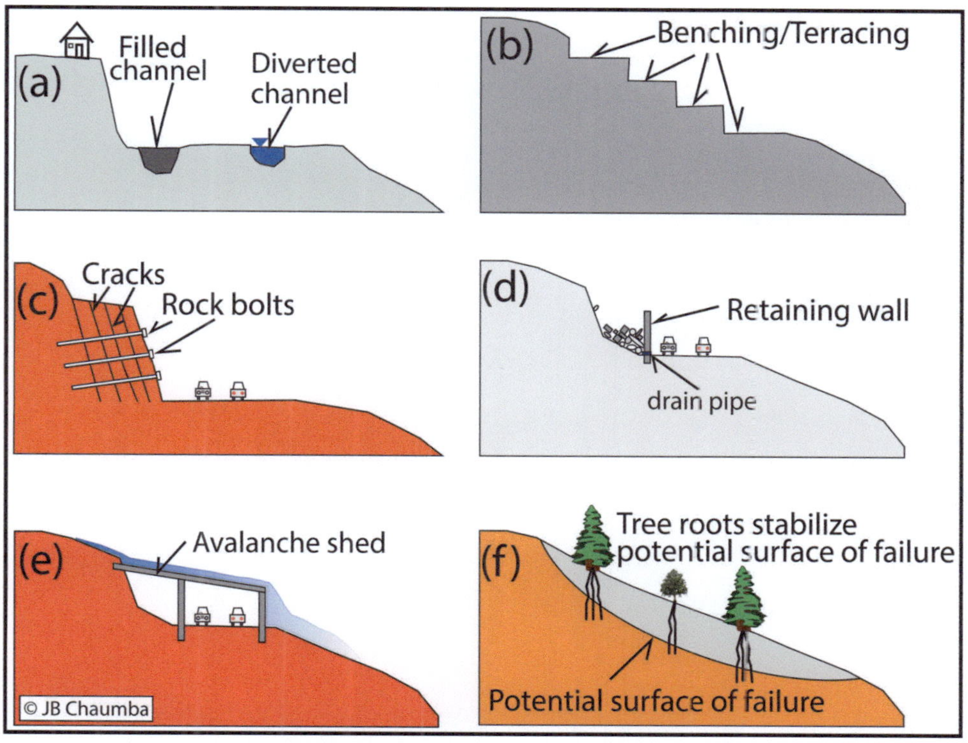

Fig. 11.35 Some strategies for mitigating mass wasting. **a** Channel diversion, **b** benching terracing, **c** rock bolting, **d** retaining wall, **e** avalanche shed, and slope revegetation (**f**)

also helps to mitigate mass wasting. Terracing the slope to reduce the steepness of the cuts, retaining walls and barriers through the use of rock bolts, cables, screens, as well as installation of avalanche sheds are structural solutions to the prevention of mass wasting.

Terracing is widely practiced widely, particularly in agricultural areas in mountainous regions. In the Eastern Highlands in Manicaland Province of Zimbabwe (Fig. 11.36a), in wine-producing mountains of Stellenbosch in the Cape Fold Belt, South Africa (Fig. 11.36b), benching or terracing helps reduce erosion of disturbed slopes for plantations. In Shaanxi Province, China, slope stabilization in the Pleistocene Loess Plateau (Fig. 11.36c) has greatly reduced extensive mass wasting that was occurring here prior to stabilization efforts. Loess Plateau with a total area of 4 km^2 is located in Gaoxigou village in northwestern China (Fig. 11.36c) is considered by some to be a model of Loess Plateau governance. The Chinese have made great progress in controlling soil erosion and advancing ecological conservation. The village of Gaoxigou used to be barren with extensive soil erosion, but its problems were reversed, and the region is considered by some to have currently achieved ecological prosperity. The Gaoxigou village in Mizhi County even warranted a visit by the Chinese President Xi Jinping in September 2021.

One way of stabilizing a slope of interest is the installation of perforated drainpipes on a slope of interest lowers the water table and stabilizes the slope by removing groundwater (Fig. 11.37a). The perforated side faces upwards to enable the water to seep into the pipe along which it is transported to the hillside and becomes surface water (Fig. 11.37a). The lowered water table ensures the slope is relatively more stable due to increased friction and removal of additional mass through draining the water out of the soil. Cracks in rocks also help drain water from the slope making up the hillside, thereby increasing the stability of the slope.

Slopes can also be stabilized by using the cut-and-fill method. The cut-and-fill involves the removal of material from the upper part of a hillside which is generally the steepest part of the hill. The removed material from the slope's upper part is then added to the base or lower part of the slope to make it more gentle by reducing the angle of the slope (Fig. 11.37b, c). This provides a more stable slope from the material added to support the bottom of the slope (Fig. 11.37b, c).

Reduction of any slope reduces the possibility of mass wasting. If the angle of a slope is reduced, this reduces the likelihood of slope failure. In addition to the cut-and-fill method of reducing the angle of a hillside (Fig. 11.37b), material can also be removed from the upper part of a slope and piled at the base of the slope (Fig. 11.37b, c), helping make the slope stable. Furthermore, the material piled at the bottom of a slope provides additional support for the slope.

Fig. 11.36 Slope stabilization of plateau of Pleistocene loess in Nyanga, Zimbabwe (**a**), Stellenbosch, South Africa (**b**), and slope stabilization in Pleistocene Loess Plateau, Shaanxi Province, China (**c**) (credit: D. L. Reid)

Fig. 11.37 Diagrams illustrating some strategies used to mitigate mass wasting. **a** Installation of perforated drainpipes on a slope of interest lowers the water table and stabilizes the slope by removing groundwater. **b**, **c** Removing material from the top of the slope reduces the angle, and piling the material at the base of the slope provides additional support

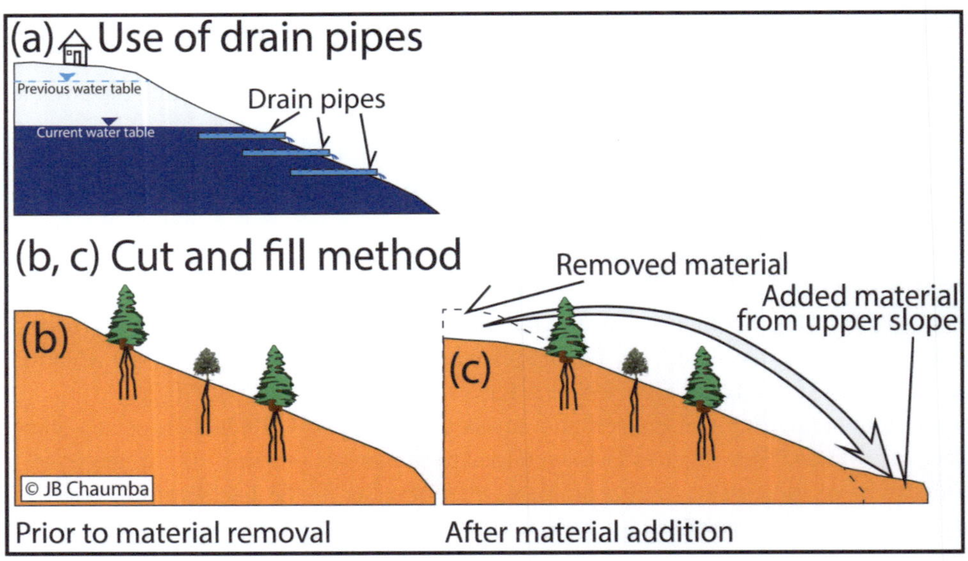

There typically must be a causal mechanism, a trigger, for any mass wasting event to occur. A trigger could be a change in any parameter that results in a mass wasting episode such as a volcanic eruption, an earthquake, intense precipitation over a relatively short period of time, spring snowmelt. Other triggering mechanisms could be activities caused by human beings such construction of roads or houses on relatively steep slopes. Houses add weight to a slope thereby helping trigger earthquakes. A number of mass wasting events such as those in the Eastern Highlands of Zimbabwe which occurred during and/or soon after Hurricane Idai occurred in 2017 led to increases in water content within the slopes. Addition of water to slopes could be the most common trigger of mass wasting since water contents can increase as a consequence of rapid snow or ice melting snow following an intense precipitation event such as a hurricane. For example, during El Niño years, intense precipitation events occur more often leading to the United States west coast receiving more precipitation than normal causing higher landslide incidences. Plastic and tarps sheeting are typically used for draining water off of landslide bodies for the prevention infiltration into the landslide. Drains are also useful in dewatering landslides.

Any oversteepened slope is more likely to trigger landslides. Erosion is accelerated by human activities such as formation of steep slopes through cultivation, or when humans modify the landscape through construction of houses, parking lots, businesses, and roads. In addition, natural processes including stream bank erosion, coastal erosion, and during storms or floods, result in oversteepening of slopes. Implementing proper designs during construction can greatly reduce the potential for mass wasting.

Glossary

Angle of repose The steepest angle of dip relative to the horizontal plane on which the material is piled without slumping, and where the material on the slope face is on the verge of sliding

Biological processes Weathering ascribed to the movements of plants and animals such as burrowing animals including plant growth. Humans also cause biological weathering

Bornhardt A dome-shaped rock formation composed of steep sides, several hundred meters wide, and at least 30 m in height

Burrowing A type of movement involving digging of holes or tunnels in the ground for refuge, shelter, or as a consequence of movement

Butte A small, isolated hill with a flat top and steep sides that are often vertical

Castle kopje A large rock formation that is free-standing and rises abruptly from a ridge crest or rounded hill

Chemical weathering The decomposition or break down of rocks as a result of chemical reactions that occur between the minerals making up the rocks and mineral solutions in the environment

Compaction The process of sediment particles being compressed together over time thereby reducing their volume

Creep The imperceptibly slow and steady downward movement of slope-forming soil or debris

Debris flow A fast-moving landslide composed of a combination of debris moving downslope

Dissolution A chemical weathering process whereby water combines with acid and removes mineral compounds from solid and transports them away

Earthflow A type of mass wasting that occurs when saturated soil and/or rock material moves downslope due to gravity

Eluviated (E) horizon A layer from which clay, minerals, and/or organic matter have been moved downwards by percolating fluids

Erosion The process by which the Earth's surface gets worn down by water, wind, and glacial ice due to the removal of soil, rock, and/or dissolved materials

Exfoliation Also referred to as onion weathering—is a type of mechanical weathering that occurs when large sheets of rock break off and separate from outcrop as a consequence of the continued reduced pressure during uplift and erosion

Family A group of soils within a subgroup that describes the chemical and physical properties affecting the soil response to both agricultural management and engineering applications

Fins Narrow, wall-like structures composed of hard sedimentary rock that form through erosion. They are an intermediate stage in the formation of other geological features, such as arches, windows, and hoodoos.

Frost wedging When water in cracks and crevices of rocks freezes as commonly occurs in high-altitude and high-latitude areas, it expands in volume up to 9% which causes the cracks to expand and deepen, leading the rock to break apart.

Freeze–thaw action Thawing occurs when water seeps into cracks in rock, freezes at night, and expands during the day, leading to breaking apart of the rock

Gabion A modular containment system enabling stone, rock, and other weathering resistant materials that are used as construction materials

Great group A category in the classification system of soil that is a subdivision of a suborder. Great groups are distinguished by similarities in soil horizons, moisture and temperature regimes, as well as other soil properties

Horizon Soil layer possessing properties that are different from those layers occurring above or beneath it

Humus Dark, organic material that forms in soil and produced from decaying plant and animal matter

Hydrolysis The process whereby the chemical composition of rocks is changed due to reaction with water

Joint An example of an extension fracture that is formed when movement of rock occurs in a direction perpendicular to the plane of fracturing

Leaching The removal of minerals and nutrients from a soil or a horizon as water passes through it

Landslide Various types of downslope movement of a mass of Earth, debris, or rock due to effect of gravity

Mass wasting The downslope movement of soil and rock due to the influence of gravity

Mesa derived from Spanish word *mesa* meaning table, a mesa is a broad and isolated mountain or flat-topped hill with steep sides that is isolated and broad.

Mud flow A mass of soil and water flowing downslope in a channel but are more water-saturated and contain more sand than debris flows

Mechanical weathering See physical weathering

Marine regression Occurs when sea level drops and exposes submerged areas of the seafloor

Marine regression Occurs when the sea covers previously exposed land

Mud cracks Polygonal cracks that are formed in mud or clay-bearing soils when they dry and contract indicating their formation in sedimentary basins which experience wetting and drying climatic conditions

Natural arches Also known as natural bridges or rock arches, are rock formations produced by the erosion of rock that have an opening or hole running through them

Offlap sequence Encountered when younger rock layers pinch out before they reach the full extent of the older beds underneath. Offlap is caused by a marine regression

Onlap sequence Encountered when younger rock layers develop a wedge and extend across an erosion surface in older rocks. Onlap is caused by a marine transgression

Order A group or category of soils containing both distinct characteristics and ecological significance

Organic matter Material containing carbon and hydrogen compounds that are derived from the decay of plants and animals

Organisms Living organisms such as plants, animals, bacteria, and fungi

Oxidation A chemical weathering process where oxygen reacts with minerals in a rock, particularly iron, causing it to break down and produce a rusty appearance

Parent material Materials that form soil formed such as bedrock or materials that is transported and deposited by water, glaciers, wind, and/or gravity

Peat Partly decomposed organic matter that accumulates in an environment that is always wet

Physical weathering The process whereby rocks break down into smaller fragments, or change shape, with no change to their chemical composition. Also referred to as mechanical weathering

Regolith The overlying sedimentary material which occurs above the bedrock

Rockfall Rock that has freely fallen from cliff face

Rockslide Rock that has been deposited by an avalanche

Series The lowest category of the national soil classification system, and soil series name is the common reference term that is used to name soil map units. The most homogenous class in the system of taxonomy is the soil series

Shear strength The maximum shear stress sustained by a soil without experiencing failure

Slump A mass wasting type of where loosely consolidated material or rock moves downslope along a concave upward surface

Soil A mixture of soil particles, sediment, air, moisture, and organic materials which occur at Earth's surface

Soil profile Portion of the soil that has been exposed vertically to reveal its layers/horizons

Soil structure arrangement of soil particles into peds, or clusters, of various shapes resembling columns, balls, plates, or blocks

Soil texture Relative proportions of particles of sand, silt, and clay

Solifluction The slow movement of saturated soil downslope indicating that no frozen ground is present in the moving layer

Subgroup A subdivision of a great group and it is defined by properties that are similar or transitional to other great groups, suborders, or orders

Suborder A division of a soil order on the basis of soil properties that are important for both plant growth and soil development

Subsoil Soil horizon enriched in minerals leached down or eluviated from the horizons above it, but it can be absent in some horizons. The subsoil is also termed the B horizon

Topsoil Soil horizon that forms at the land surface and it is dominated by weathered minerals derived from parent material with a minor organic matter added. It is also called the A horizon

Stratum A layer of bed

Till Poorly sorted and unstratified glacial sediment that is deposited by glaciers

Turbidites Sedimentary rock or deposit that was formed by turbidity currents

Turbidity current A gravity flow of sediment laden fluid that moves downslope in the ocean

Weathering The process whereby rocks and minerals break down into smaller fragments and/or at or near the Earth's surface

12 Hydrologic Cycle and Flowing Water

Abstract

The chapter begins with a discussion of the hydrologic cycle including definitions of its components such as evaporation, transpiration, evapotranspiration, condensation, precipitation, among others, with the aid of a diagram of the hydrologic cycle. The proportions of water in various reservoirs on our planet are summarized and illustrated in tabular form whereas the distribution of its water in those reservoirs is graphically represented by pie charts. Drainage, drainage basins, drainage patterns and are discussed and clarified by means of illustrations, similarly for ephemeral, intermittent, and perennial streams. River profile, graded stream, knickpoint, changes in a river channel from its source to mouth the relationship between bed load, flow velocity, channel gradient, channel width, channel depth, stream discharge, and the volume of stored alluvium are similarly described. Effluent and influent streams are also described and illustrated by means of diagrams. Stream terraces, incised meanders, erosion, transportation, and deposition by rivers, landforms associated with meandering streams, floodplains, natural levees, Yazoo tributary/stream, and bedrock and alluvial channels are similarly discussed. Bedrock and alluvial channels, floods, kart topography and living with floods and how to alleviate flooding are also discussed.

Learning Goals.

Studying this chapter will enable you to:

- Define the hydrologic cycle
- List the water reservoirs of our planet
- Sketch and summarize the water cycle and the processes involved
- Sketch the longitudinal profile of a river and label its important components
- Define and differentiate drainage basin from divide
- Differentiate between incised meanders and water gaps
- Define a graded stream and differentiate its upper, middle, and lower courses
- Sketch the common drainage patterns
- Recognize and define ephemeral, intermittent, and perennial streams
- Sketch and explain why different cross sections determine stream velocity
- Differentiate the types of loads carried and deposited by streams
- Describe how meanders form and where meandering streams erode and deposit their load
- List and identify stream erosional and depositional features
- Describe how and why stream characteristics vary from river's source to mouth
- Differentiate incised meanders from water gaps
- Explain how flash floods are caused and how to lessen the impacts of flash floods
- List causes of flash floods and some common flood control methods

Supplementary Information The online version contains supplementary material available at https://doi.org/10.1007/978-3-032-20231-4_12.

12.1 Introduction

You may have heard of, or seen in news, or even have read on the internet, about flooded rivers carrying away vehicles, infrastructure, or even people. You may also have seen on the internet snow-related accidents on roads during the winter depending on where you live. The Texas, USA floods of July 4, 2025, may come to mind for many when, in some areas, more than 15 cm (6 in.) of precipitation fell in south-

J. B. Chaumba, *Introductory Physical Geology*, Springer Textbooks in Earth Sciences, Geography and Environment,
https://doi.org/10.1007/978-3-032-20231-4_12

central Texas, a region in south-central Texas that forms the southeast part of the Edwards Plateau. This area is also referred to as the Texas Hill Country. Such floods and/or heavy snowstorms, however, don't occur all year round or even every year, but they occur. Sadly, at least 135 fatalities occurred from these floods, with 9 children and 18 adults at Camp Mystic on the banks of the Guadalope River near Hunt, Texas. The flood caused damage worth US$1.1 billion. What causes floods? Where will this water be stored when there is no flooding? What about some houses being destroyed or swallowed up by sinkholes? In this chapter we will discuss these events and more, including some of their causes and how to minimize and reduce the risks of you and your house falling victim to such events.

12.2 Hydrologic Cycle

All water on Earth can be stored in any one of these reservoirs: oceans, seas, lakes, ponds, glaciers, snowfields, groundwater, rivers, soil, and atmosphere. The oceans cover approximately 71% of our planet's surface area. The storage and circulation of water from the oceans and the ground to the atmosphere and back again through the hydrosphere, lithosphere, biosphere, and atmosphere is termed the **hydrologic cycle** (Fig. 12.1). This continuous circulation of water in the Earth-Atmosphere system is critical in generating currents which are responsible for the movement of warm water from the tropics to the poles thereby greatly assisting in regulating the Earth's temperature.

Several processes involved in the hydrologic cycle and the most important are briefly described here. The change of state in a substance from a liquid state (water) to a gaseous state (water vapor) is called **evaporation**. The bulk of evaporation of water on our planet occurs from the oceans, with additional evaporation occurring in continental storages of water such as lakes, dams, rivers, and canals, for example. For evaporation to occur, energy is required, and the Sun powers the bulk of this energy with smaller contributions from the atmosphere and our planet. When evaporation occurs, only freshwater evaporates, leaving behind salts dissolved in water. Ice sheets contain in excess of 99% of all the freshwater ice on Earth.

Transpiration is the evaporation of water from plants through small openings found on the underside of leaves that are connected to vascular plant tissues called stomata. **Evapotranspiration** is the combined effect of evaporation and transpiration. The process by which water vapor in the air is changed into liquid water and is responsible for the formation of clouds is called **condensation**. Condensation, therefore, is the opposite of evaporation. **Precipitation** occurs when tiny droplets, through the collision and coalescence process, combine and become too large for the rising air to support, and they fall to the Earth's surface. Precipitation occurs in various forms such as rain, snow, ice, sleet, and hail.

Table 12.1 Inventory of water on Earth and proportions

Reservoir	Volume ($\times 10^6$ km^3)	Percent of total
Oceans	1370	97.25
Glaciers and ice caps	29	2.05
Groundwater	9.5	0.68
Lakes	0.125	0.01
Soil moisture	0.065	0.005
Atmosphere	0.013	0.001
Streams and rivers	0.0017	0.0001
Biosphere	0.0006	0.00004

Precipitation is the primary way our planet receives freshwater which results in an annual average of ~980 mm (38½") falling on both the continents and the oceans.

Two types of flow can be distinguished: **stream flow** and **sheet flow**. Streamflow is confined to conduits and channels. Sheet flow is, in general, shallow (less than an inch deep), and the water moves slowly. Runoff whereby water flows over relatively flat land surface, such agricultural fields, lawn and parking lots, for example, is sheet flow. Sheet flow, therefore, is shallow stormwater that moves over relatively flat land.

When there is excessive precipitation after the ground cannot absorb any more water, i.e., when the ground is saturated, **runoff** occurs. When all the reservoirs of water on Earth are considered (Table 12.1), it can be noted that streams and rivers account for 0.0001% of the total water on our planet, i.e., the smallest percentage, accounting for even less than what the atmosphere contains (0.001%). During and after precipitation, there is runoff (Fig. 12.1; Table 12.1) as most of the water does not infiltrate into the ground and becomes runoff. Runoff ultimately enters creeks, streams, and rivers in which it flows downhill toward the lakes and ultimately into the oceans. As water in the rivers flows downslope, some of the water gets diverted for human use. Ultimately, water in rivers which do not disappear is emptied into oceans.

It should be noted that some of the runoff that flows into lakes that have no outlets to oceans only has evaporation as the means for the water to return to the atmosphere. Water contains dissolved substances, and water in inland lakes evaporates, leaving impurities or salts behind to form progressively saltier lake water, such as the Dead Sea in Israel or the Great Salt Lake in Utah. When all the water in a lake eventually evaporates, a white-coated **salt flat** is formed. Some of the water which is at the Earth's surface percolates into the ground to form water occurring below the ground is called **groundwater**. We pump groundwater

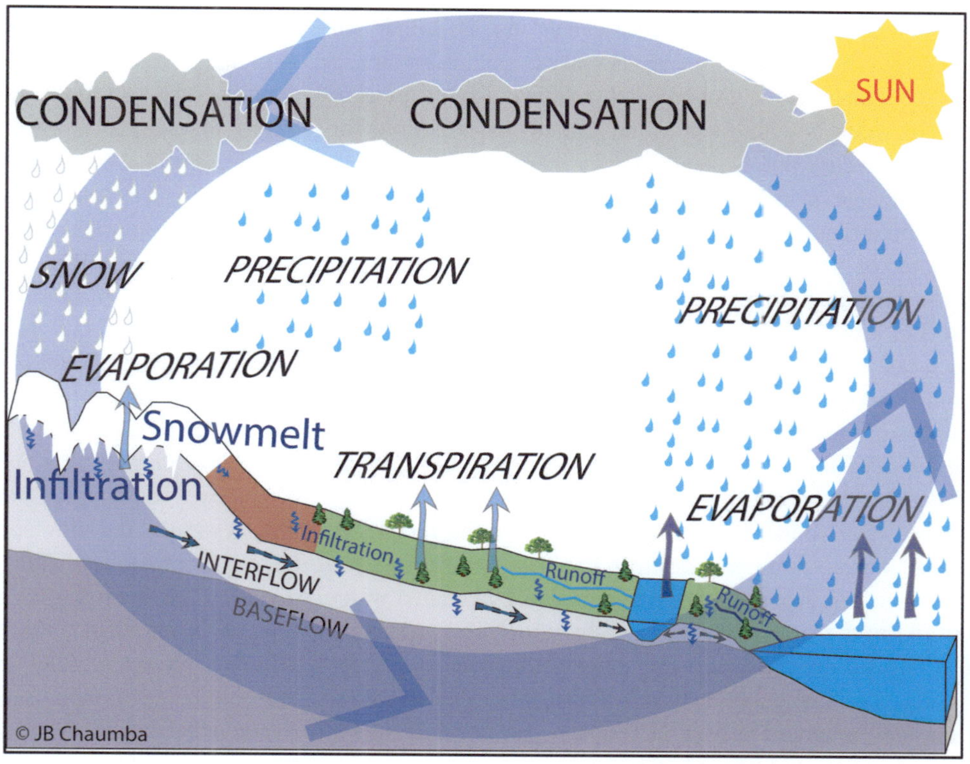

Fig. 12.1 The hydrologic cycle

for domestic, irrigation, and industrial uses, and some of it is drawn into plants again for use during transpiration.

When we are thirsty, we search for a glass or bottle of water to quench our thirst. The water in the bottle was obviously obtained from a freshwater source. Freshwater contains less than 500 parts per million (ppm) of dissolved salts. The reason why piped water is found in our homes, schools, offices, and all other public places is that it would have been treated and determined to be safe for drinking. Water in the oceans contains more dissolved salts than that which our bodies can tolerate hence we don't consume it if we don't want to be sick or even lose our lives. Ocean water, however, constitutes the lion's share of water on Earth (Fig. 12.1).

Our planet is unique in that it is the only known planet containing liquid water in our solar system. Although water covers 71% of our planet, only 3% of this water is freshwater. It is estimated that 2.5% of Earth's freshwater is not available for our use. This is due to the freshwater being locked up in glaciers and polar ice caps, in the atmosphere and soil. Some of our planet is located too deep below the Earth's surface to be extracted at affordable costs. Additionally, some of the freshwater is also highly polluted with pollution likely to increase as our population continues to increase due to new settlements being established in areas previously unsettled. Consequently, a mere 0.5% of our planet's water is freshwater that we can access!

The sources of freshwater on our planet are runoff, groundwater, and **snowmelt**.

Runoff occurs when water from precipitation collects in channels such as streams, rivers, in lakes, wetlands, and reservoirs. Runoff, therefore, is precipitation that neither infiltrates into the ground nor returns to the atmosphere through evaporation. Water which infiltrates into the ground through porous materials to various depths into the Earth is referred to as groundwater. Underground rock formations, sediments, pores, and fractures beneath the ground which contain water that can be pumped out for any use are called **aquifers**. Some of this groundwater, however, lies too deep beneath the Earth's surface to be economically extracted. It is estimated that snow with a thickness of 10 cm (4 inches) can contain approximately the same amount of water as 1 cm (1/3) inch of rain.

In the United States, it is estimated that water use is 59% for industry, 33% for agriculture, and 8% for domestic use. It is also estimated that in excess of 600 gallons per day of water per person in the U.S. is being diverted for irrigation at farms and use for livestock from natural aquatic sources. More than half the people in the U.S. are estimated to obtain their water from groundwater. Water withdrawal is when water is sourced from groundwater and/or surface-water source and transported to a location of use. Water consumption refers to water that has been withdrawn and is no longer available for reuse in the area from which it was withdrawn. It is estimated that in the United States approximately three-fourths of the freshwater that is withdrawn each year is sourced from rivers, lakes, and reservoirs; a quarter comes from groundwater aquifers. Eighty percent of water that is withdrawn in the United States is used for cooling electric power plants and for irrigation.

Discharge is the quantity, or amount of water, which flows in a river per given amount of time, such as per hour, minute, or second. The term streamflow also means the same thing as discharge. Water on our planet is continuously moving from one reservoir to another through processes such as evaporation, transpiration, or a combination of both evaporation and transpiration termed evapotranspiration, as discussed above, condensation, sublimation, precipitation, deposition, infiltration, runoff, melting, and baseflow. Water that evaporates from the oceans comprises the bulk of the water found in the atmosphere, with the remainder being sourced from lakes, streams, snow, vegetation, and other sources. Approximately 91% of this evaporated water returns to the oceans through precipitation. The remaining 9% falls on the continents and other landmasses as precipitation. The imbalance between the quantities of evaporation and precipitation over land and ocean is made up for through runoff and groundwater flow to the oceans.

The oceans dominate the water supply on Earth (Table 12.1; Fig. 12.2a). Approximately 97% of all the water on the Earth is stored in the oceans (Fig. 12.2a). Glaciers and icecaps, groundwater, lakes, soil, the atmosphere, and the biosphere account for 3% which constitute all the freshwater on our planet (Table 12.1; Fig. 12.2a). Although our planet contains an abundance of water, unfortunately only a small percentage is usable by humans (Fig. 12.2b). The other 99.7 percent is in the oceans, icecaps, soil, or as water vapor in the atmosphere. Some water can be strongly concentrated in, or even saturated with, salt to form saline groundwater and lakes—**brines** (Fig. 12.2b).

12.3 Drainage and Drainage Basin

Across the land surface and in the lithosphere, water flows naturally downhill. **Drainage** refers to the downslope movement of water through channels that are well defined. When precipitation occurs on land, it eventually finds its way into creeks, streams, and rivers where it flows as surface water where it begins, or its **source**, and it can converge to a single large channel or a network of channels where it empties into a stationary body of water such as a lake/sea, or its **mouth**. Lake Tana in Ethiopia and Lake Victoria, for example, are the sources of the Blue Nile and White Nile, respectively. The river water subsequently flows into water bodies like a lake and eventually flows into a sea or an ocean. Drainage occurs within a **drainage basin** (Fig. 12.3). A drainage basin can be defined as the land area in which all water that is drained eventually merges to form a common outlet. A drainage basin is composed of a succession of elevated features, such as hills or ridges, and it may consist of smaller basins that merge at river confluences, forming a hierarchical pattern. Any drop of rain that falls within drainage basin A, for example, contributes to the water Stream 1 if it does not evaporate. Similarly, any drop of rain that falls within drainage basin B contributes to the water Stream 2. The drainage basin or catchment for Stream 1 is drainage basin A. A drop of rain that falls right at the drainage divide between streams 1 and 2 will be split such that half of it contributes to drainage basin A, whereas the other half contributes to water in drainage basin B.

Drainage basins (Fig. 12.3) are separated from adjacent basins by a topographic boundary called **drainage divides** that separates the flow of water into different drainage basins. Continental divides are the highest order drainage divides. Continental divides are natural boundaries that separate a continent's watersheds by channeling water to different oceans or seas. In North America, for example, the continental divide (see insert in Fig. 12.3) separates rivers which drain westwards to the Pacific Ocean from those draining eastwards to the Atlantic Ocean. A succession of elevated features, such as ridges and hills, make up the drainage divide. A drainage basin is also referred to as a catchment area or a watershed. Drainage basins vary widely in size, with the total land area that drains water into one single river system ranging from small areas such as tributaries or creeks to massive regions that can encompass large portions of continents. Although the Amazon River is not the longest one in the world, it has the largest drainage basin in the world. The longest river on Earth, the Nile, is 6 650 km (4 130 mi) long (Fig. 12.4), and it is followed by the Amazon River which has a length of 6 400 km (3 976 mi). The Yangtze River, the third longest one is 6 300 km (3 917mi). The Mississippi River in North America, which is 6 275 km (3 902 mi) long, is the world's fourth-longest river. The Yenisei River in Asia has a length of 5 539 km (3 445 mi) round the top five longest rivers of the world.

One of the important features of the Nile River is that it passes through a total of 11 countries with a population that exceeds 258 million people, or roughly a fifth of the population of the African continent. Although several rivers act as tributaries to this mighty river, the Blue Nile and the White Nile are its main tributaries. Additionally, the Nile River passes through some of the driest and most populous regions of the world in Egypt and northern Sudan, often triggering animosity and friction between the governments of the countries within the drainage basin of this River. It is worth noting that the Nile River basin is very varied and ranges in climatic regions from very dry deserts in its lower reaches near the Mediterranean Sea to tropical rainforests in its upper parts. In the great Lakes region of central Africa.

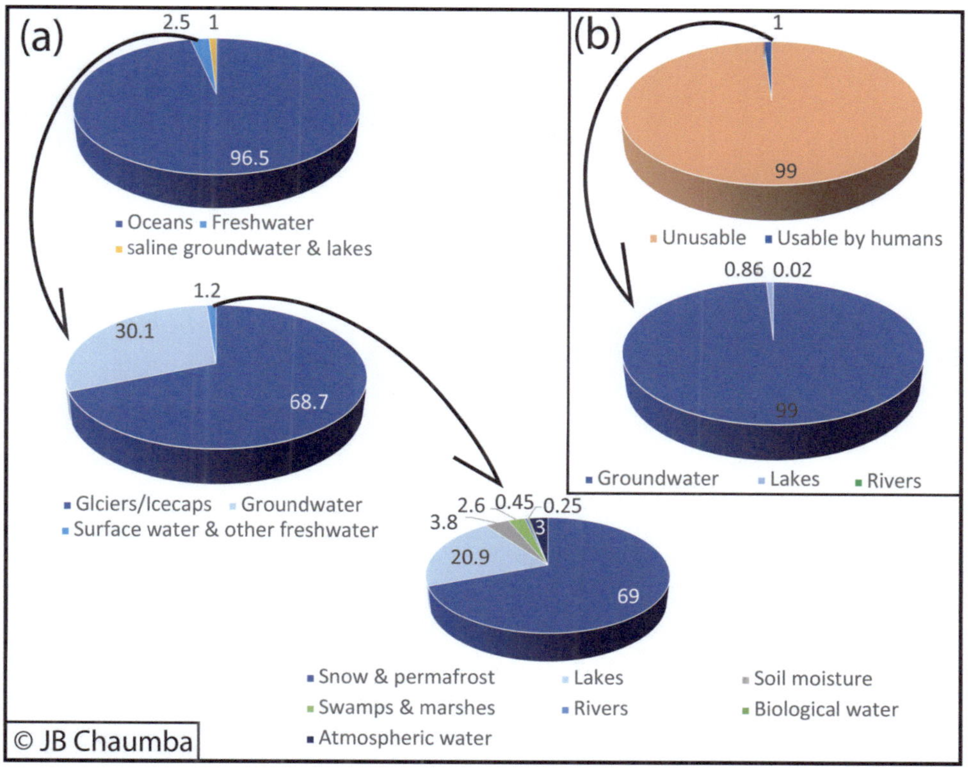

Fig. 12.2 Pie charts showing the distribution of water on Earth. **a** The bulk of the water is salty ocean water, and a big percentage of freshwater (68.7%) is locked up in glaciers and ice caps. Groundwater, representing 30.1% of freshwater, is our largest resource of freshwater. **b** Only 1% of water on Earth is usable, of which 99% occurs as groundwater. Groundwater, therefore, represents our most important source of freshwater. Consequently, the importance of preserving this resource cannot be overstated

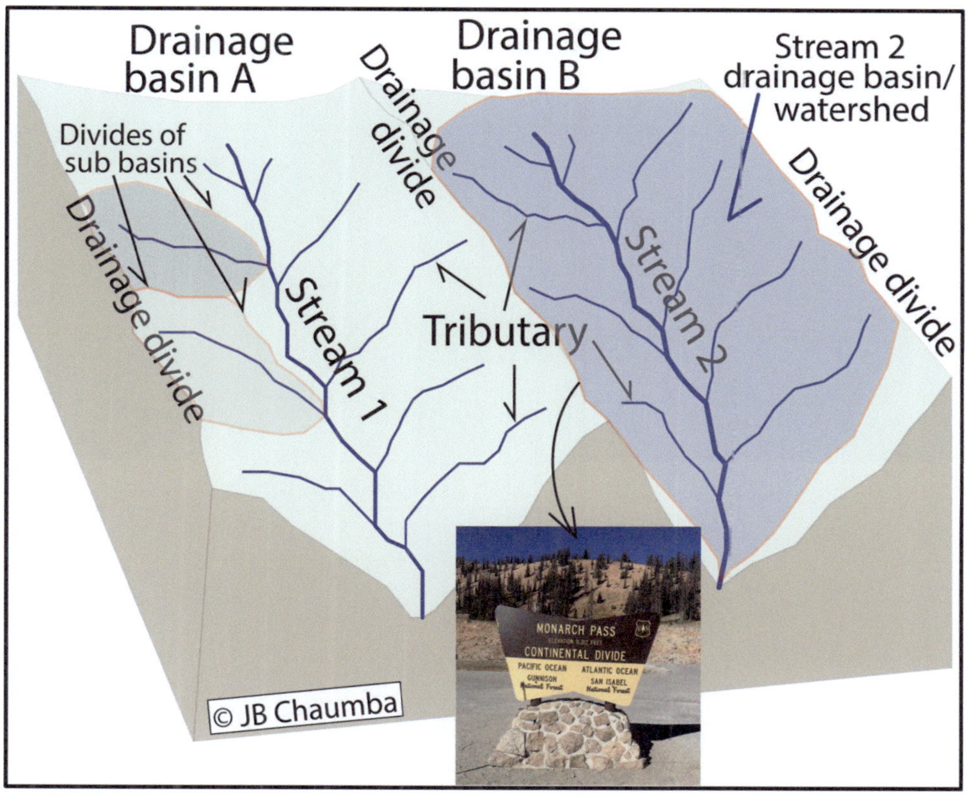

Fig. 12.3 Diagram illustrating drainage basins and sub-basins, drainage basin, drainage divides, tributaries, and streams. Insert shows a picture from the location of continental divide in North America at Monarch Pass, Colorado, along Colorado's Highway 50 between Gunnison and Poncha Springs

Although the Nile River is the world's longest river, it does not have the largest drainage basin in the world. The rivers with the largest drainage basin areas in world are the Amazon River, 7 million km^2 (2.7 million mi^2), the Congo River, 3.4 million km^2 (1.3 million mi^2), the Nile River 3.254 million km^2 (1.256 mi^2), and the Mississippi River, 3.22 million km^2 (1.245 million mi^2), respectively. The Rio de La Plata River, with a drainage basin area of

Fig. 12.4 Drainage basin for the Nile River, the longest river in the world

3.17 million km^2 (1,22 million mi^2), rounds off the list of the five largest drainage basins in the world. That honor belongs to the Amazon River which has the largest drainage basin in terms of both area and volume of water. This river flows through the Amazon rainforest, the world's largest tropical rainforest, and it therefore plays a crucial role in the global water cycle and climate based on its size. It is worth noting that the Ganges River in India, which exceeds 2,500 km (1553 mi) long, is the most populated river basin in the world as hundreds of millions of people and a variety of wildlife also rely on the waters of the Ganges River.

Almost every river has its own **brook**, or **tributary**, or **creek**, which is a smaller stream which empties its water into the stream (Fig. 12.3). Tributaries of any stream, such as the two tributaries of stream 1 shaded in Fig. 12.3, also have their own drainage divides and drainage basins called sub-basins.

12.4 Drainage Patterns

Factors such as the structure of the bedrock, the slope of the land, and the climate of the area influence the drainage patterns of streams which can be produced within a drainage basin. The pattern that streams, rivers, and lakes produce in a particular drainage basin is called the **drainage pattern**. The pattern that is formed by erosion of a stream over time reveals some characteristics of the rock types and geologic structures in a landscape region that is drained by streams. Consequently, different drainage patterns can be produced as illustrated in Fig. 12.5.

Dendritic drainage pattern formed where the trunk and its tributaries resemble the branches of a tree (Fig. 12.5), which resulted in the name dendritic. In the dendritic drainage pattern, the river channel follows the regional slope. Dendritic pattern is developed where there is almost no structural control of the direction that rivers flow. **Radial drainage patterns** develop where rivers flow radially away in different directions from a dome-like structure or central peak such as a volcanic peak (Fig. 12.5b). Where the underlying rocks are highly jointed by two sets of rocks intersecting more or less at right angles to each other, the **rectangular drainage pattern** is developed (Fig. 12.5c). The **trellis drainage pattern** is formed where tributaries join a river at approximately 90° to produce a trellis pattern. In regions where alternating parallel hard and soft rocks are developed, the trellis drainage pattern is formed (Fig. 12.5d).

Centripetal drainage pattern forms where streams flow inwards toward a central depression, hence all streams converge at an area such as a lake or dam (Fig. 12.5e). Centripetal drainage pattern is considered to be the opposite of the radial drainage pattern. A **deranged drainage pattern** is characterized by an irregular, broken drainage system, and it is formed where a coherent pattern to the rivers and lakes is lacking (Fig. 12.5f). This pattern is formed in areas that have experienced a lot of geological disruption. Rivers would not have had adequate time to adapt to the bedrock but tend to mimic the irregularities of deposits from glaciation.

A stream that has flowing water only during or for a short duration after precipitation events in a typical year is termed an **ephemeral stream** (Figs. 12.5a, 12.6a). In many states, this term refers to streambeds that are located above the water table year round and streams where groundwater is not a source of water for the stream. An **intermittent stream** (Figs. 12.5b, 6b) is one that has flowing water only during certain times of the year, when groundwater feeds the stream. A supplemental source of water in an intermittent stream for stream flow is runoff from precipitation. Intermittent streams typically do not have any flowing water during dry periods such as winter months in tropical regions of late summer and early fall in temperate regions. A stream containing flowing water throughout the entire year is termed a perennial stream. For a **perennial stream**, the water table is located higher than the bed of the stream for most of the year (Figs. 12.5c, 12.6c). The most important source of water for a perennial stream is groundwater, as can be observed in Figs. 12.5c and 12.6c. A supplemental source of water for perennial stream flow is runoff from precipitation. Most of the world's largest and longest rivers such as the Zambezi River in Southern Africa, the Indus River in Asia, the Murray River in Australia, the St. Lawrence River, and the Mackenzie River in North America, for example, are perennial.

Shown in Fig. 12.7 are photographs of an ephemeral (a, b), intermittent stream (c, d), and perennial stream (e, f).

Streams encountered in mountainous regions (Fig. 12.8) are characterized by steep gradients with waters transporting rocks and other debris with very coarse-grained with angular edges and corners due to the relatively short transport distances. The rocks are sources from falling rocks and other mass wasting activities due to steep mountainous slopes. Additionally, mountain streams tend to have swift flowing but variable volumes of water.

When a stream sometimes reaches a state of equilibrium where its slope and channel characteristics are well adjusted to transport sediment supplied to it, it would have achieved a balanced relationship between erosion and deposition throughout its course. Such a stream is referred to as a **graded stream**. A graded stream neither erodes nor deposits large amounts of sediment over time. The profile of a graded stream is steeper at its headwaters and grad-

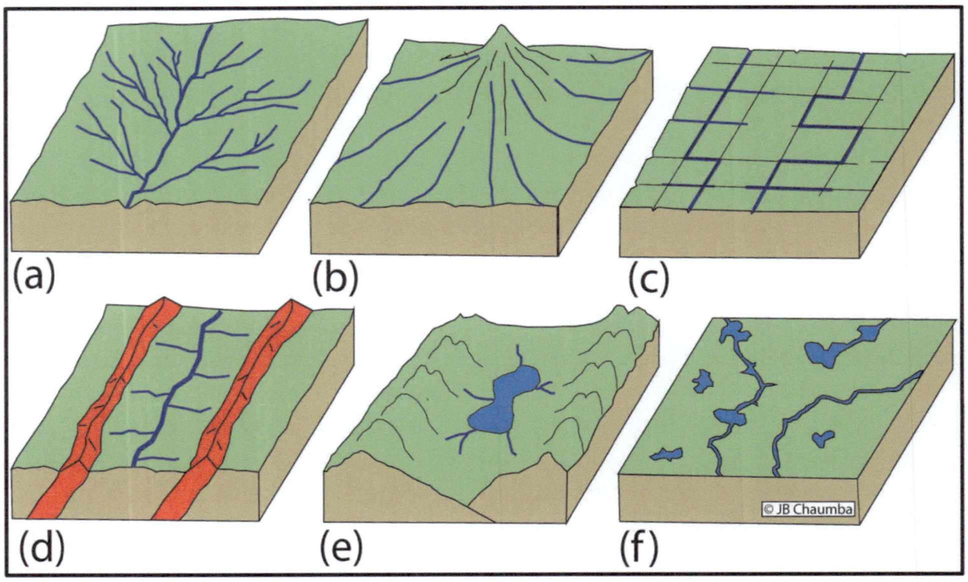

Fig. 12.5 Illustrations of some common drainage patterns. **a** dendritic, **b** radial **c** rectangular, **d** trellis, **e** centripetal, and **f** deranged pattern

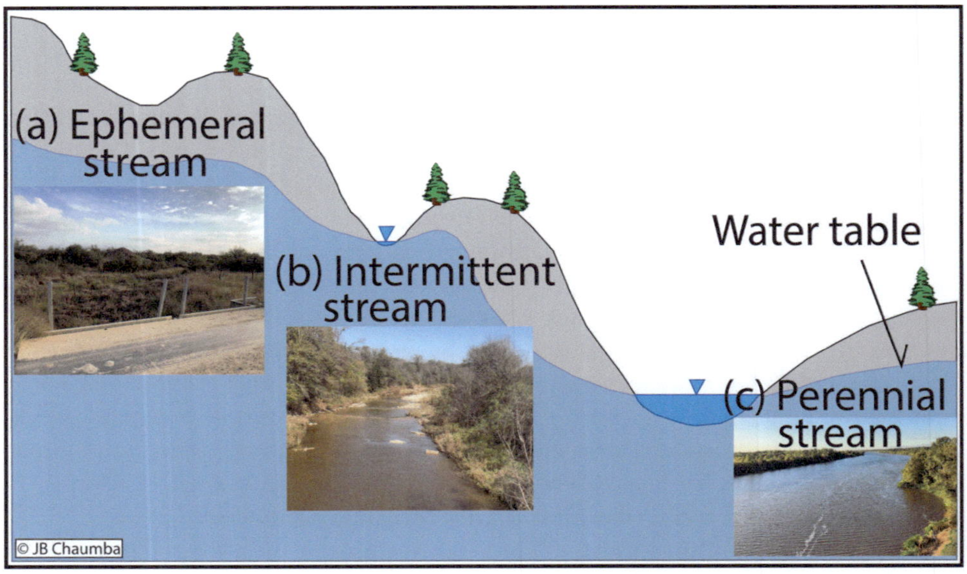

Fig. 12.6 **a** Ephemeral stream, Tierra Blanca Creek located southwest of Catarina, Texas, USA. **b** Rock Creek, an intermittent stream photographed in Johnson County, Texas, USA, **c** Red River, a perennial stream, photographed near where US route 371 crosses this river between the towns of Armistead and Coushatta in Louisiana, USA

ually decreases downstream to produce a concave profile (Fig. 12.9). If, for example, a change in conditions occurs, such as an increase in the supply of sediments or a change in base level, a graded stream adjusts its profile through erosion or deposition until equilibrium is re-established.

Erosion occurs which is concentrated in steep parts of the stream that experience high velocity which is accompanied by an increase in both competence and capacity, whereas deposition occurs in gentle segments of the stream with decreased velocity and capacity (Fig. 12.9a). Material removed from zones of erosion, which are steep, is deposited in the zones of deposition with gentle slopes (b), resulting in the formation of a graded stream profile shown in (12.9c).

12.5 Profile of a River

Based on gradient, the longitudinal profile of a hypothetical river which can be subdivided into three courses (Fig. 12.10a). The longitudinal profile of a river begins at its source and ends where it empties into a standing water body such as a lake or ocean. The upper course of a river is encountered in the region of high relief. A river channel is formed when water trickles from saturated ground, merges, and then flows downslope following a steep gradient from both gravity and potential energy. Rapids and waterfalls may develop in the upper course of the river where its cross section is characterized by V-shaped valleys. **Erosion**, largely vertical by hydraulic action, occurs in the upper

Fig. 12.7 A **a**, **b** photographs of an ephemeral stream, Tierra Blanca Creek located southwest of Catarina, Texas, USA, **c**, **b** Rock Creek, an intermittent stream, in Johnson County, Texas, USA, and **e**, **f** Mississippi River, a perennial stream, near where Interstate 20 crosses this river between Vicksburg, Mississippi **e** and Delta, Louisiana **f**, USA

course of the river. **Transportation** involves mostly large boulders which are moved by traction. **Deposition** is largely of large boulders that are deposited in the river's upper course.

The **graded profile** of a stream is concave upward, and it is steeper near its source or where it begins (Fig. 12.10b), U-shaped in its middle reaches (Fig. 12.10c), and gets progressively flatter near where it empties into a lake or ocean, that is, its mouth (Fig. 12.10d). In the upper reaches of a stream, its discharge is smaller due to the smaller size of the channel. As the stream merges with other streams progressively further downstream, the discharge, cross-sectional area, and hydraulic radius of the stream all increase. As the distance from the stream's source increases downstream, the stream grows in size as its waters flow more freely. In the upper reaches, a small stream must be steeper in order to transport its sediments. Higher gravitational energy on the steeper slope is needed for the stream to overcome the force of friction in the shallow stream. Sediments accumulate if the slope is too gentle and the velocity is too low to transport sediments supplied by weathering and erosion. Consequently, the gradient increases resulting in faster flowing water which increases both erosion and transport leading to a reduction in the gradient.

Fig. 12.8 Mountain streams in Golden, Colorado, USA **a**, and Beartooth Mountains, Montana, USA **a** Very coarse-grained rocks and other debris are common in mountain streams

Discharge can be defined as the volume of water that passes through a given point on the river. To calculate discharge, the formula: discharge = velocity x cross-sectional area is used. The units of discharge are cubic meters per second (m/s) or feet per second (ft/s). As tributaries join the river and add more water to it, its discharge increases along its course. Velocity is the other parameter that also increases along a river's course. Although the gradient in the upper reaches of a river is steeper, its lower reach has the greatest velocity as velocity is dependent on the volume of water that is in contact with the riverbed and its banks. The narrow V-shaped channel in the upper reaches of a river results in more friction which reduces the velocity of the water. In the lower reaches of a river, however, a smaller proportion of the water is in contact with the riverbed and banks resulting in less friction with the riverbanks leading to higher water velocity. Additional water from a river's tributaries increases water velocity due to higher discharge. A steep region along a river profile which can vary in form from a single waterfall to a high-gradient region extending for many kilometers (miles) is called a **knickpoint** (Fig. 12.10a).

A knickpoint (Fig. 12.10a) is defined as a steep region along a river profile and can vary in form from a single waterfall to a high-gradient region extending for many kilometers. The origin and mobility of knickpoints are being debated. Stable knickpoints are thought to be formed where rivers flow over resistant rocks or active faults. Stable knickpoints may also be formed where the supply of coarse sediments is ample to inhibit transport of sediments or erosion. Such conditions exist as tributary junctions or at the toes of massive landslides. Such knickpoints occur at a fixed in position along the river channel ascribed to local controls. Some knickpoints are mobile, and these are initiated by some event downstream and then propagate upstream. A signal of erosion is communicated throughout trunk and tributary streams of the river network. A variety of processes can initiate a downstream incision that generates a signal of relative fall in base level which includes a fall in sea level, fault rupture, stream straightening, and drainage capture.

Rivers tend to have three parts: upper, middle, and lower courses (Figs. 12.10a, 12.11). The upper course of a river is where we find its source (often in mountainous regions of the river basin), where rivers flow down steep slopes as fast flowing streams (Fig. Fig. 12.10a, b). In the upper course of rivers, erosion and transport processes are the dominant processes. Erosion of the riverbed leads to the development of gorges, canyons, and V-shaped valleys (Fig. 12.10a, b). The water has high oxygen contents. Waterfalls, rapids, and potholes are the most common features in the upper course of the river.

The velocity of water slows down in the middle course of the river (Fig. 12.10a, c) in comparison to that in the upper course. In this middle course, the river channel widens, discharge increases, and the steepness of the gradient decreases. Transport and sedimentation processes dominate over erosion. Consequently, flat and wide valleys called alluvial plains or floodplains are developed (Fig. 12.10a, c). The river channel may be anastomosing and meandering. On outer courses of meanders, erosion occurs on the outer, concave banks, whereas sediments are deposited on the inner, convex banks. The result is widening of the valley as the river meanders.

In the lower course of a river (Fig. 12.10a, d), both the gradient and the velocity of water decrease. Sedimentation and transport of very fine material are the dominant processes of rivers in the lower courses, as they no longer

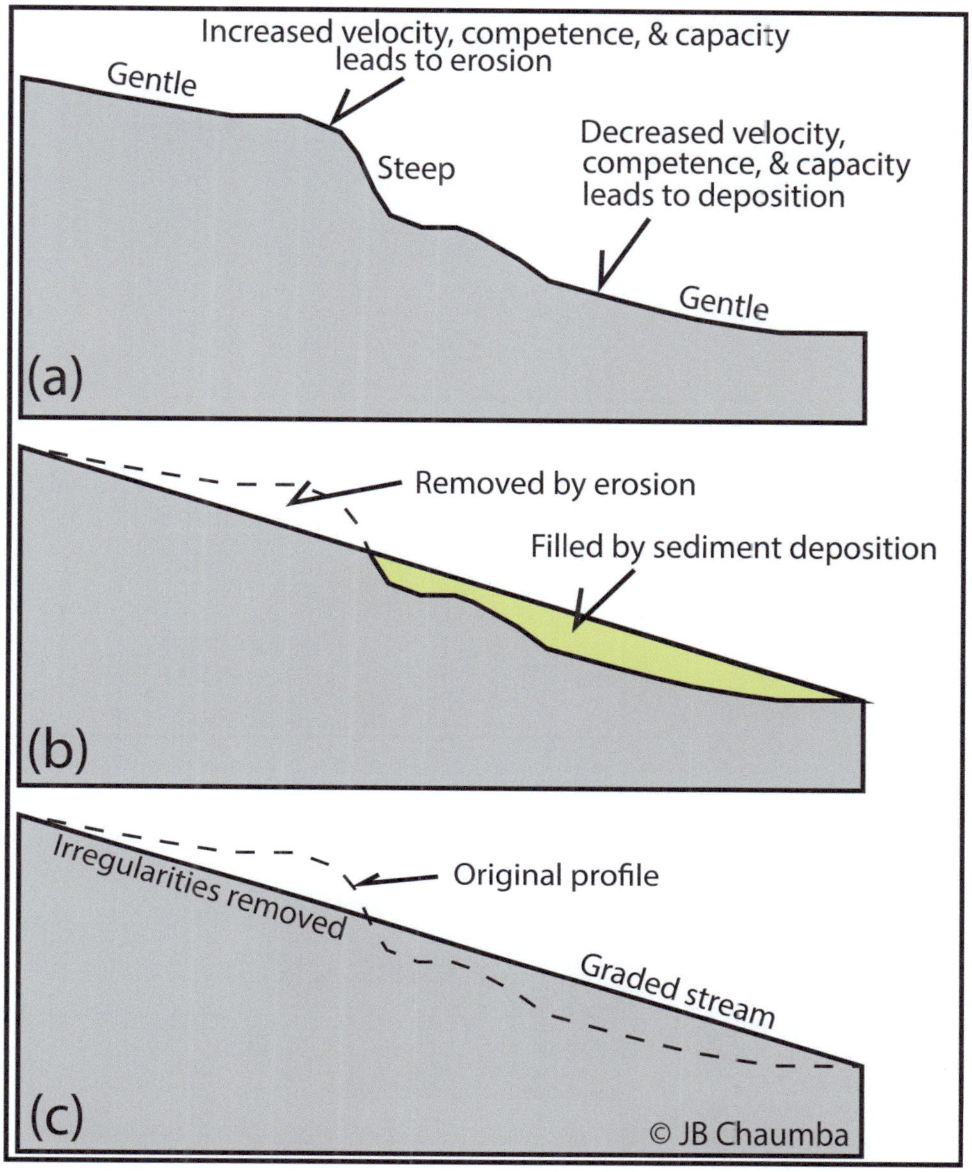

Fig. 12.9 Illustration of the formation of a graded stream from an initial profile of a stream containing changes in the gradient of the stream through erosion in steep areas of high velocity and deposition in areas of low velocity which have gentle slopes (**a**) Material removed from steep zones of erosion is deposited in zones of deposition with gentle slopes (**b**) to form a graded stream profile (**c**)

have high erosion capacity. The discharge of the river tends to spread over the floodplain during periods of high water. Transported sediments are sourced from the large plains, together with nutrients that fertilize the river valley. The valleys in the lower courses of rivers are fertile such that they are typically used to grow crops. The crops are irrigated by taking advantage of the availability of water as well as the fertile soils. Some settlements occur in such valleys, although they are at risk from flooding and high water flows. Additionally, groundwater aquifers are periodically recharged, and sometimes tributaries join the main channel in the lower river courses.

We can also briefly describe changes that occur from the source of a river to its mouth. Over the course of any river from its source to its mouth, several changes occur (Fig. 12.12). The bed load, flow velocity, and channel gradient all decrease from the source of a river to its mouth (Fig. 12.12). Other factors such as stream discharge, channel width, channel depth, and the volume of stored alluvium, however, all increase from the source to the mouth of a river as shown in Fig. 12.12.

Material that is deposited by rivers is called **alluvium**. The stored alluvium volume of a river (Fig. 12.10) is its total amount of sediment that is presently deposited and accumulated both within the floodplain and the riverbed. Alluvium tends to be extensively developed in the lower parts of the course of a river's course (Fig. 12.12) where it forms deltas and floodplains. Alluvium can also be deposited at any point where a river overflows its banks or where the river's velocity is checked, such as where it empties into a reservoir such a lake. Alluvium is comprised of gravel, sand, clay, silt, sand, as well as organic matter. Alluvium

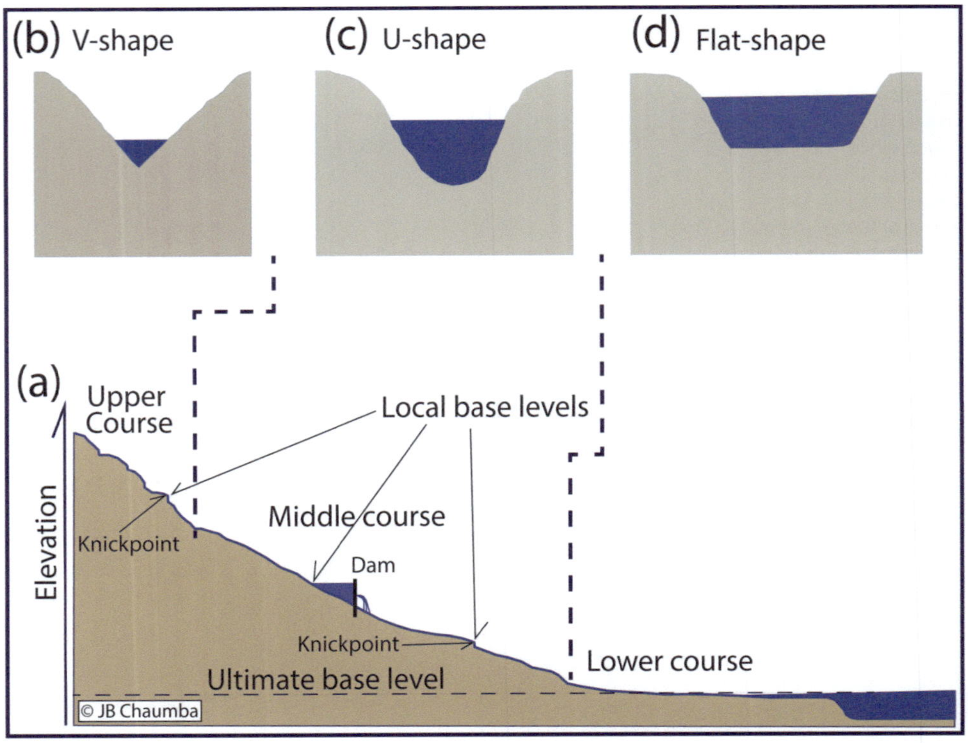

Fig. 12.10 **a** Longitudinal profile of a hypothetical river showing the upper course, middle course, and lower course. Typical cross sections of the upper course **b**, middle course **c**, and lower course **d** are also shown. Local base levels are represented by knickpoint, and a dam as shown

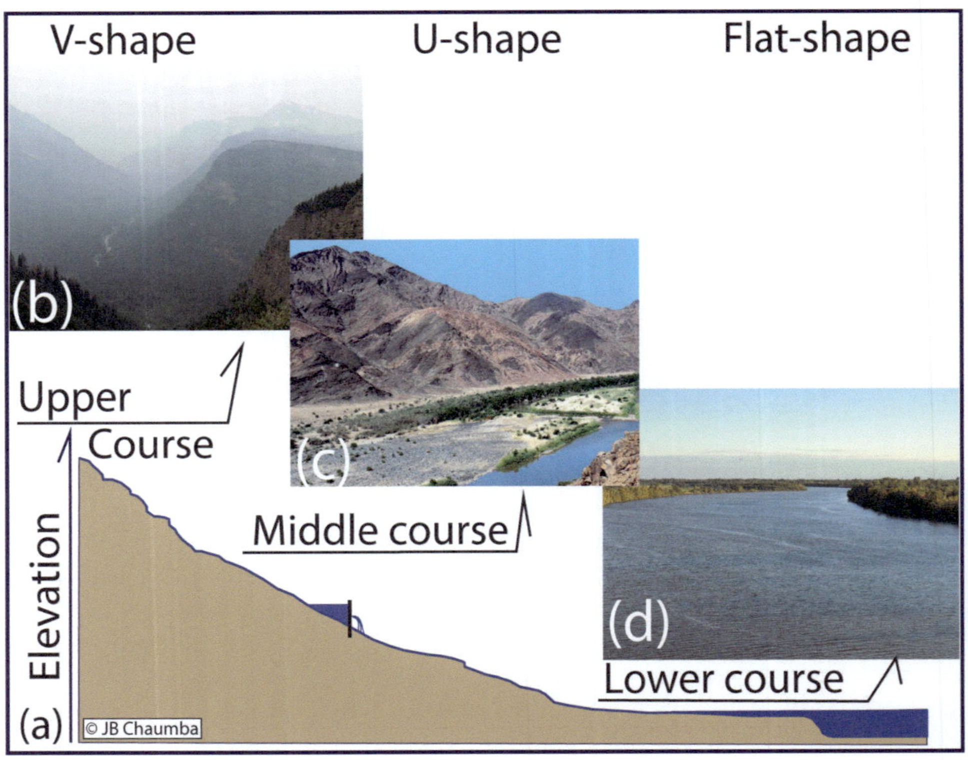

Fig. 12.11 **a** Diagram representing longitudinal profile of a hypothetical river showing the upper **b**, middle **c**, and lower **d** courses of a river. The dam and rapids represent knickpoints. **b** shows a representative photograph of the upper reaches of a river in Glacier National Park, Montana, USA. Photograph **c** is view of part of the Orange River north of Nous Wells, Namaqualand, Southern Africa representing the river's middle course, and shown in **d** is the Red River from US Route 371 between Armistead and Coushatta in Louisiana, USA representing the river's lower course

yields fertile soil. Deltas of the Nile, Mississippi, Huang, and Brahmaputra rivers are good examples. Although the Amazon River also has a delta, its soil, particularly in the hilly uplands, is largely infertile, although some areas of high delta soil fertility exist.

12.6 Effluent and Influent Streams

The location of the water table in comparison to the water level in streams is of great help in the determination of the stream's contribution of water to ground-

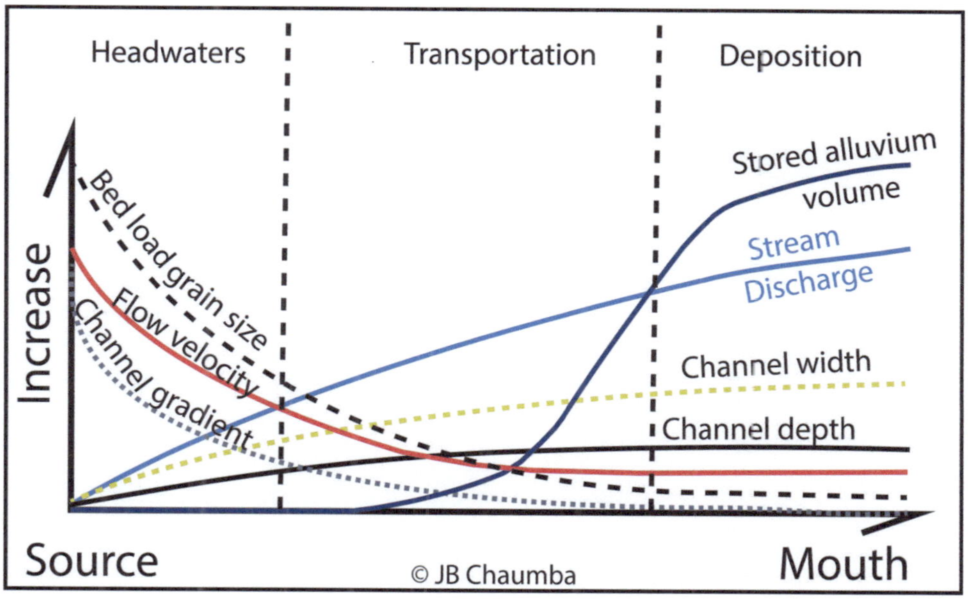

Fig. 12.12 Changes in a river channel from its source to its mouth. Bed load, flow velocity, and channel gradient all decrease from source to mouth, where stream discharge, channel width, channel depth, and the volume of stored alluvium all increase from the source to the mouth of a river

water (Chap. 13) storage. If the streambed's location is below the groundwater table, as occurs during periods of low streamflow, the groundwater contributes to the flow in the stream. Based on how streams interact with groundwater, we can categorize them into **effluent** streams and **influent** streams. Streams which gain water from groundwater are called effluent streams (Fig. 12.13a). Examples of effluent streams are perennial streams such as those located in the United States Coastal Plain, the Volga, Danube, Rhine, Thames Rivers in Europe, the Amazon and Parana Rivers in South America, the Yangtze and Ganga Rivers in Asia, or the Nile, Congo, Niger, Zambezi, and Limpopo Rivers in Africa. Effluent streams (Fig. 12.13a) receive some or all their water from the aquifer. The "gained water" is discharged into the stream from a groundwater reservoir. For a stream to "gain" water, the elevation of the water table must be higher than the level of water in the stream (Fig. 12.13a).

Influent streams, also referred to as "losing" streams (Fig. 12.13b), are streams that add water to an aquifer as they flow downstream. The water percolates into the ground where it recharges the local groundwater. As can be observed in Fig. 12.13b, influent streams are located above the water table (Fig. 12.13c). For a stream to "lose" water, the elevation of the water table must be lower than the stream bed (Fig. 12.13c). The losing stream can either be connected to the groundwater by a saturated zone (Fig. 12.13b) or by an unsaturated zone (Fig. 12.13c). The Nile River in Africa and the Colorado River in the USA are examples of influent rivers that are considered influent as they lose a significant amount of water through seepage and evaporation. These influent rivers contribute less water to the water body they flow into in comparison to the amount of water they carry upstream, hence they "lose" water along their courses. Other examples of influent rivers such as the Rio Grande River in North America, the Indus River in Asia, the Murray River in Australia, the Atuel River in Argentina, and the Desaguadero River in Bolivia are dependent on the region.

When there is excessive precipitation, the ground becomes saturated, and runoff occurs. Runoff results in the formation of rivers and lakes. Although some evaporation into the atmosphere occurs from runoff, the bulk of water in rivers and lakes ultimately returns to the oceans. Runoff accumulates only in interior basins called sinks, in lakes with no outlets called endorheic basins. When water accumulates in these endorheic basins, the water in lakes eventually returns to the atmosphere through evaporation, leaving behind impurities and/or salts, resulting in the formation of saline waters such as those in the Great Salt Lake in Utah, USA, or the Dead Sea in Israel, Jordan, and the West Bank areas under the control of Palestine. An endorheic basin may be a permanent lake, a dry lake, or a point where surface water is lost underground.

12.7 Stream Terraces

Consider a river or a stream that incises into an area of uniform rock (Fig. 12.14a). Terraces of the same elevation that are formed on opposite sides of either a stream or a river are called paired terraces. Paired terraces occur when a river or stream incises evenly on both sides resulting in terraces on one side of the river or stream of corresponding height with those on the other side, as shown in Fig. 12.14b and c. An unpaired terrace can be formed

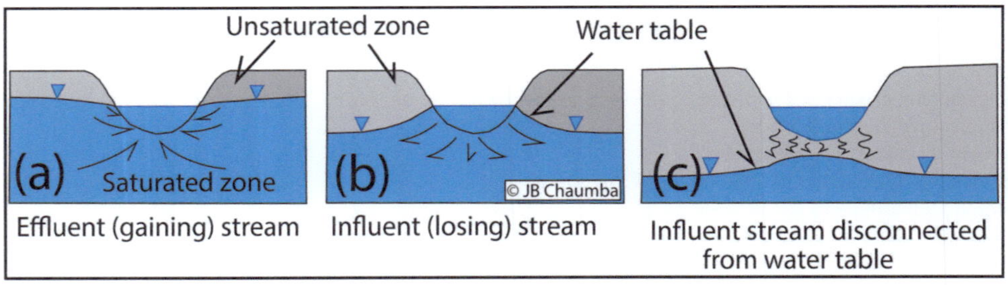

Fig. 12.13 Diagrams illustrating effluent (gaining) stream (**a**), influent (losing) stream connected to the groundwater (**b**), and influent (losing) stream that is not connected to the groundwater (**c**)

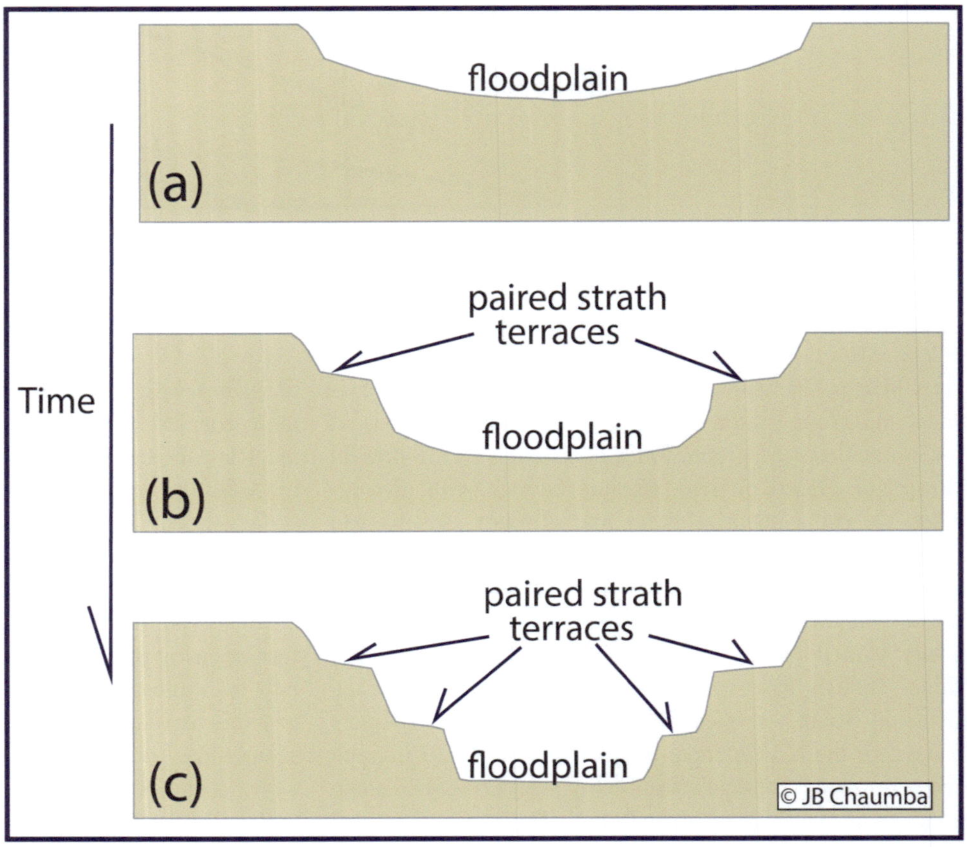

Fig. 12.14 Diagram showing the formation of paired strath terraces (**b** and **c**) with time from an initial stage where no terrace existed (**a**)

when either a stream or river erodes into rocks on one side that are relatively more resistant to erosion in comparison to the other. This results in a single terrace produced on the single side of the river or stream comprised of the more resistant rock but lacks a terrace on the corresponding side of the stream comprised of the less resistant rock. The origin of paired terraces is ascribed to rejuvenation of a river or a stream. Most rivers (e.g., the Mississippi River and the River Thames in southeast England) have terraces in the lower parts of their courses near where they empty their water into the ocean. The Amazon River is characterized by wide terraces that can rise tens of meters above the water. The Yangtze River in Asia has different terrace configurations across its upstream region having two sets of terraces, whereas the subsidence depression only has one set. Its uplifted upper gorge has two separate sets of terraces that were generated after the post-penultimate glacial period of valley filling (aggradation) and two distinct sets of thick valley-filling sequences when the Yangtze River was in a transport-limited state.

If the slope of an already established meandering stream is increased suddenly due to the base level of the stream being reduced caused by either tectonic uplift of the region or a global fall in sea level, the stream will resume downward erosion. As the stream continues to erode downwards, its established meandering pattern remains as a deep valley referred to as an **incised meander**. An incised meander, also referred to as an *entrenched meander*, therefore, is a meandering river valley that has cut down, or incised, its bed into the bedrock, due to lowered base level or uplift. Incised meanders typically can't erode laterally hence they lack floodplains. The stream is restricted to deep, meandering canyons and typically occupies the whole width of the floor of the canyon (Fig. 12.15). If the base level is lowered, downcutting by a river occurs to

Fig. 12.15 The Colorado River **a** in western USA, and the Fish River canyon **b** in southern Africa, are excellent examples of incised meanders

form incised meanders. Uplift of the land leads to downcutting which also leads to the formation of incised meanders. The Colorado River (Fig. 12.15) is an excellent example of an incised meander. During downcutting by a river, lateral erosion is restricted. Other excellent examples of incised meanders are the Fish River canyon in southern Africa (Fig. 12.15b) and the Kentucky River Palisades in the Ozark Plateau of central Kentucky.

For an understanding of the formation of incised meanders, consider igneous activity occurring at Time 1 with intrusion of dikes (Fig. 12.16a). The dykes, which are more resistant to weathering and erosion than the surrounding terrain, stick higher than the surrounding terrain which was lowered by weathering and erosion at Time 2. At Time 2, the river incised into the dikes at a rate equal to that of uplift during a period of uplift to result in the formation of wind gaps (Fig. 12.16b). The incised meanders are the deep and meandering canyons which the river produced by downcutting into the bedrock (Fig. 12.15b).

12.8 Erosion, Transportation and Deposition by Rivers

Erosion can be defined as the wearing away of sediment, incorporation and transportation of the sediment by a mobile agent such as water, ice, or wind. Rivers do work through erosion, transportation, and deposition. In a river, erosion is the process of wearing away of sediment along its banks and bed. V-shaped valleys are formed by vertical erosion and are encountered in the upper course of a river characterized by steep slopes. Fast-flowing water dominates the upper courses of a river.

Transportation of the eroded material downstream is by running water. The sediment carried by the river also breaks up into smaller fragments. When the energy of flowing water drops, the load gets deposited in slower water flow to create new landforms such as deltas, floodplains, and point bars. Deposition of the eroded and transported material occurs along its banks.

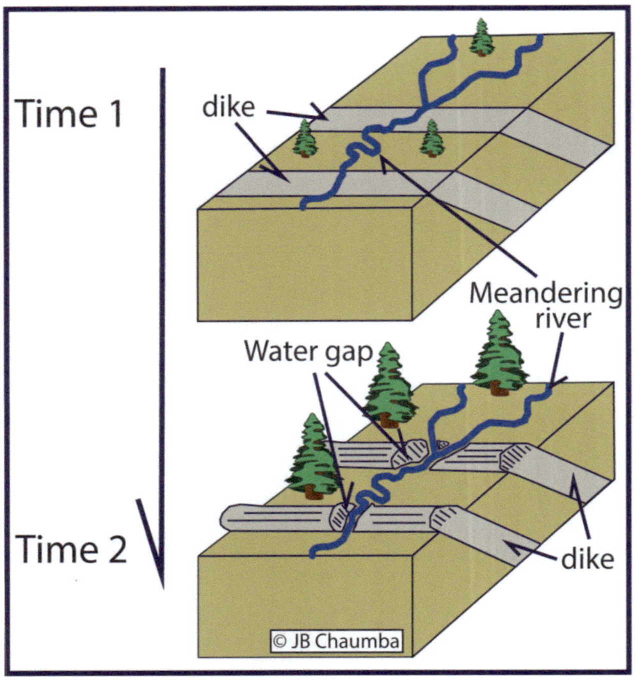

Fig. 12.16 Illustration of the formation of incised meanders. At Time 1, a river is flowing at the land surface (**a**). With the passage of time, the land gets lowered so that the more resistant to weathering dikes stick higher above the surrounding surface at Time 1 resulting in the formation of incised meanders at Time 2 (**b**)

Erosion by rivers occurs via abrasion, hydraulic action, and attrition. Flowing water always contains dissolved substances and particles of clay, silt, sand, and even gravel, depending on stream velocity and the lithology of the drainage basin. **Abrasion** occurs when sediments carried by a river scrape against its riverbed and banks, which wears the river down. Solution occurs when some soluble mineral salts dissolve in river water, contributing to a river's **solution load** (Fig. 12.17). The sheer force of water hitting against the riverbanks breaks off pieces of rock. **Attrition** is when particles carried by flowing water in a river collide with each other, breaking particles into smaller pieces in the process. **Dissolved** or **suspended load** is comprised of fine-grained particles of clay and silt that are transported within the water column (Fig. 12.17). If the water is still or when it enters a reservoir, the particles settle out.

Traction is when sediments are transported along the bed of a river as large, heavy objects that are dragged or rolled along the riverbed. **Bed load** is comprised of very coarse particles that are too heavy to be suspended in the water column and can be transported via sliding, rolling, and saltation. **Saltation** is when sand and pebbles that are relatively large but smaller than particles transported by traction hop or jump downstream along the river bottom due to the force of the flow of water (Fig. 12.17). Saltation plays a huge role in transporting particles that are too large to be entirely suspended in the water column. During floods, flowing water does the greatest amount of its work in the form of erosion, transportation, and deposition of the sediments during floods. When flow velocity decreases, flowing water deposits the load it carries (Fig. 12.17).

Meandering rivers derive their name from the Maeander River, or the Büyük Menderes River, a river in southwestern Turkey famous for its winding course. A meander forms when water in a stream channel erodes the sediments of an outer bend of a streambank and deposits the sediment on the subsequent inner bend downstream to produce a meandering pattern (Fig. 12.18). This depositional feature that is formed on the inside of meander bends in meandering rivers is called a **point bar**, as shown in Fig. 12.18. A point bars, therefore, is a usually crescent-shaped deposit formed by lateral accretion, or the gradual addition of sediment, to the inside bank of a meander with a gentle slope (Fig. 12.18a) resulting in an increase in both size and curvature of the meanders as shown in Fig. 12.18a-d. Deltas are huge deposits of sediment that are encountered where a river meets a larger body of water like a lake or ocean.

As water flows around the bend of a meander bend, the flow on the inside of the bend slows down relative to the flow on the outside of the bend due to the momentum of the water being directed toward the outer bank by centrifugal acceleration. A line that joins the lowest points along the length of a stream bed or valley in its downward slope to define the deepest part of the channel is called a **thalweg** (Fig. 12.18). A thalweg also corresponds to the fastest water along a river. The thalweg meanders even within sections of the river channel that are straight.

The slower flow results in the reduction of the river's ability to transport sediment, resulting in excess sediment being deposited and piling up over time to form point bars. Point bars are generally crescent-shaped and composed of sand and gravel. Point bars are typically formed in the shallowest parts of the river. A cut bank is an erosional surface on the outside of a meander bend (Fig. 12.18b). **Oxbow lakes** and **cut banks** are other features that are formed because of the interaction of meandering river processes. Oxbow lakes, also called cut offs or abandoned meanders, are formed when the bends of two meanders intersect, resulting in bypassing the curve of the river (Fig. 12.18d).

Examples of cut banks and point bars are shown in Fig. 12.19. Shown in Fig. 12.19a is a point bar along East Buffalo Creek in Cleburne, TX west of the intersection of TX 171 and Ramsey Avenue. On the opposite side of the point bar in the same figure is the cut bank. Similarly, a point bar and a cut bank are also shown on the meander shown in Fig. 12.19b. This image of the meander is from

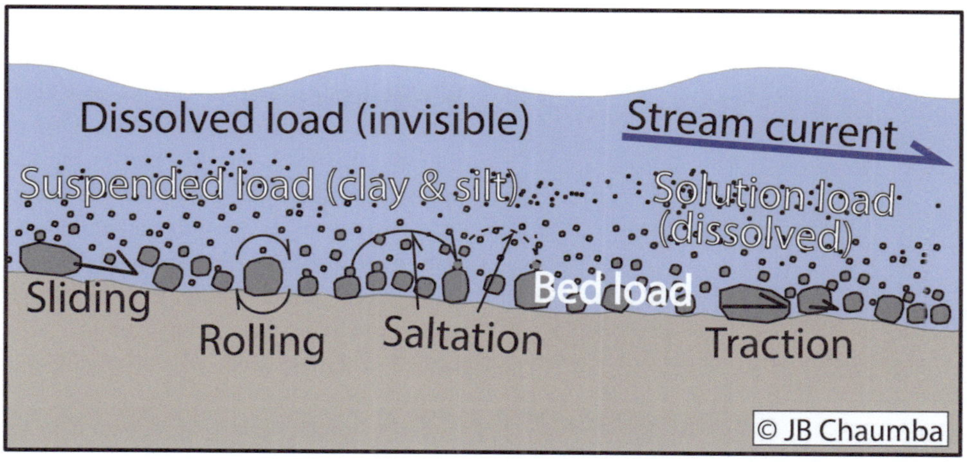

Fig. 12.17 Diagram showing various methods by which sediment is transported by flowing water in a river. Large particles, or bed load, are sediment that is transported by sliding and rolling on the riverbed

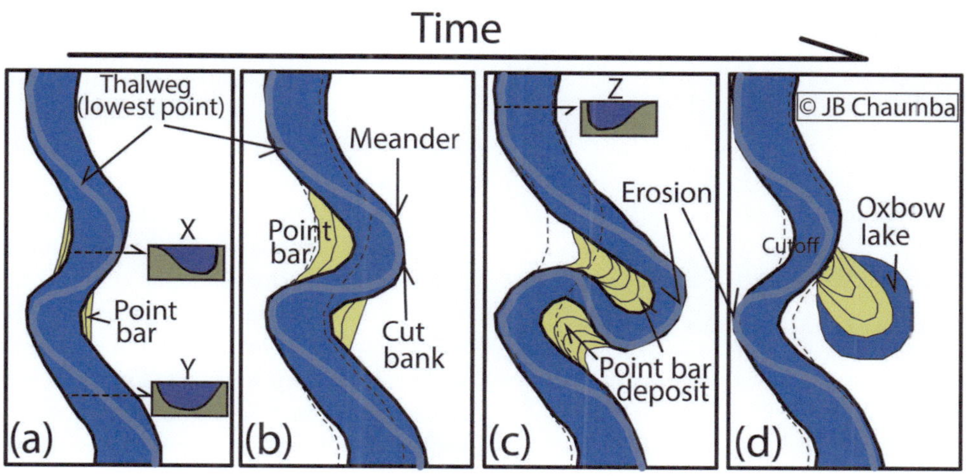

Fig. 12.18 Stages X, Y, and Z are cross sections of the stream and different indicated locations along the stream

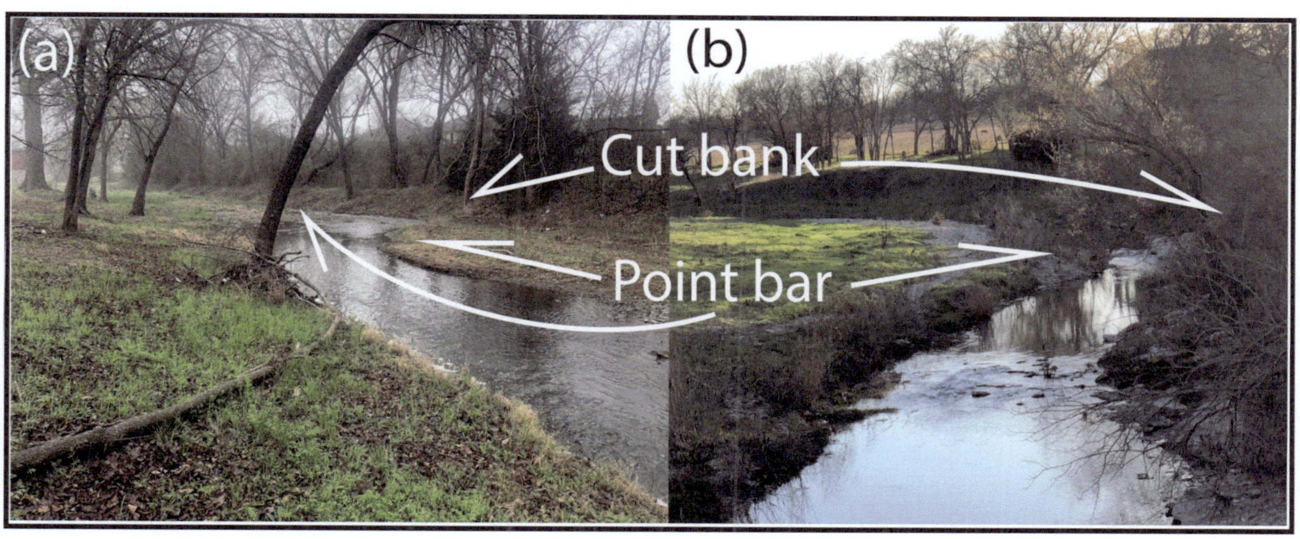

Fig. 12.19 Examples of cut banks and point bars. **a** Point bar (left side) and cut bank (right side) of creek along East Buffalo Creek in Cleburne, TX near the intersection of TX 171 and Ramsey Ave. Shown in **b** is image of the meander is from just the south side of the bridge along County Route 1015 in northern Johnson County, Texas. The point bar (left) and cut bank (right side) are located on opposite sides of the creek

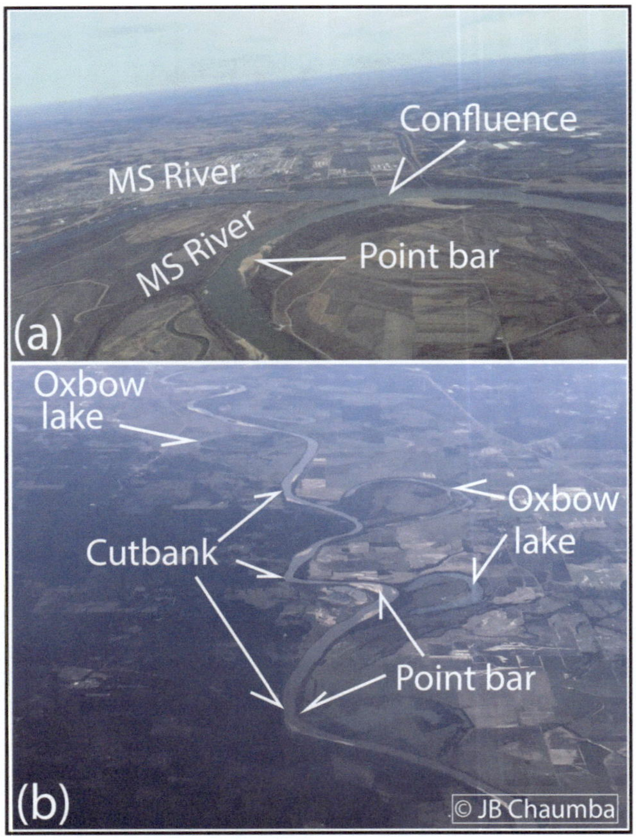

Fig. 12.20 Photographs through airplane window showing the confluence of the Mississippi and Missouri Rivers along the border of the States of Illinois and Missouri (**a**) and erosional and deposition features in the floodplain of the Red River northwest of Bradley, Arkansas, USA (**b**)

just the south side of the bridge along County Route 1015 in northern Johnson County, Texas. The point bar and cut bank are located on opposite sides of the creek, as shown in Fig. 12.19b.

Shown in Fig. 12.20 are photographs taken through airplane windows of the confluence of the Mississippi and Missouri Rivers (a) and cut banks, point bars, and oxbow lakes on the Red River northwest of Bradley, Arkansas (b). Cut banks are also labeled in this figure.

A low-lying area that is prone to flooding from a river, lake, or ocean is called a **floodplain** (Fig. 12.21. Flat land that is adjacent to the channel is constructed by deposition during flood events. **Natural levees**, which are ridges of sediment that form on the floodplains of rivers, are a common feature of river landscapes nearer the mouth of a river. Natural levees are formed by the movement of water which pushes sediments to the banks of the river, to form an elevated area that connects the channel to the floodplain. Natural levees are composed of fine-grained sand, silt, and other Earth materials, and they vary in height from a few centimeters to ~3 m. The width of natural levees ranges in width from 200 to 4000 m.

Yazoo tributary or Yazoo stream (Fig. 12.21) is tributary or stream that occurs within the floodplain of a larger river for considerable distance and runs parallel to the larger river before it eventually joins it. Yazoo tributaries are named after the Yazoo River that lies to the east of the Mississippi River and runs parallel to it for 280 km (170 mi) before it converges with the major river. Both natural and man-made levees prevent the Yazoo River from joining the major river further upstream. A backswamp (Fig. 12.21) is any low-lying and poorly drained area of a floodplain formed when silt and clay eventually accumulate after a flood. Backswamps form in natural levees of a stream. Backswamps are lower in elevation in comparison to the floodplain.

12.9 Bedrock and Alluvial Channels

Different channel patterns are formed from differences in bank full gradient, sediment supply, bank material, and discharge. Bedrock channels are scoured into solid rock. The bed and banks of bedrock channels are made of solid rock (Fig. 12.22a). Bedrock channels are typically encountered in mountainous regions where the stream gradient is steeper, and erosion is rapid (Fig. 12.22a). The flow of water is also fast in mountainous regions. Bedrock channels are typically found in mouths of river systems where streams have steep slopes. Bedrock channels have resistance to change as they have relatively slow, linear adjustments and are dominated by processes of erosion and transport.

Alluvial channels that are made up of loose sediments are called alluvial channels (Fig. 12.22b). The loose sedimentary material in which alluvial channels are encountered is known as alluvium. The beds and banks of alluvial channels are composed of material that was transported by the river itself. Banks of alluvial channels are constantly subjected to erosion, or transportation, and deposition by fast flowing water. Where the flow velocity of water is low, sediments settle out and deposition occurs.

Four types of alluvial channels are distinguished: straight and sinuous channels, meandering streams, braided streams, and anastomosing channels. Bedrock segments are composed of solid rock and are resistant to change with a limited supply of sediments, with relatively slow, linear adjustments, and predominantly erosion and transport. Alluvial channels, in contrast, are composed of loose sediments as they are formed in alluvium, and they are dominated by the production of organic matter. Mineral sediments are stored in alluvial channels. Bedrock channels are V-shaped and are narrow and deeper in comparison to alluvial channels. Bedrock channels are more resistant to erosion, whereas alluvial channels are less resistant to erosion as they are composed of alluvium. Since alluvial channels

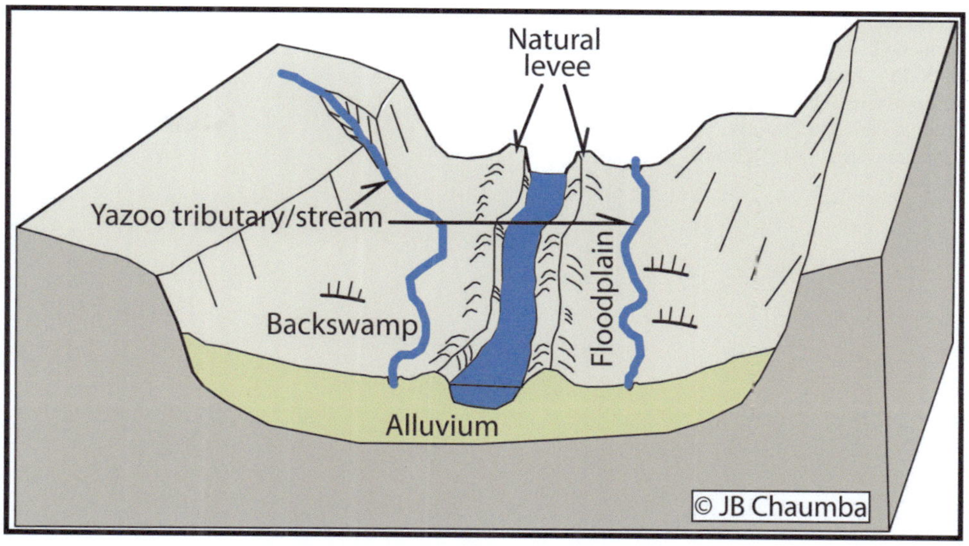

Fig. 12.21 Diagram illustrating a meandering river with natural levees, floodplain, backswamp, alluvium, and Yazoo tributary

are comprised of loose sediments, they are free to adjust because of the loose sediment, whereas bedrock channels do not easily adjust as they are scoured into solid rock.

12.9.1 Floods

Human beings don't always have it easy on our planet, but we experience a wide range of natural disasters such as droughts, floods, earthquakes, tornadoes, tsunamis, and volcanic eruptions, among others. Of the natural disasters we experience on our planet, floods are the most widely experienced. Flooding occurs when water overflows on land that is dry under normal circumstances.

Floods occur on every continent and their impacts often last long and are often accompanied by deaths and destruction. For example,

Floods occurred in the east central United States, where a total of 44 deaths occurred in July and August 2022. Although these floods affected several states, Kentucky was particularly hit the hardest by these deaths with 38 direct and 2 indirect deaths. Additionally, the death toll also included 2 indirect deaths in Missouri with Texas and Utah each recording a fatality each for a final tally of 44 deaths. At least two of the fatalities occurred during cleanup efforts. The deaths began on July 24, 2022, and lasted for a week, and it involved flash floods which affected several areas of the country from the southwestern United States to some of the eastern coastal states of the country. More than a hundred people were rescued from the ensuing floods. Pike County in Kentucky was one of the areas severely affected by the flooding. Floodwater washed out some bridges and destroyed power lines, with some residents sent scrambling to the rooftops of homes due to water gushing into their homes.

Floods can be caused by several factors, such as: i. heavy rain occurring during a short period of time or over a long period, ii. failure of water control structures such as dams or levees, iii. snowmelt occurring in the spring in colder regions often causes rivers to overflow, iv. size and shape of drainage basin may cause flooding as larger basins receive and quickly and often quickly discharge it, v. during heavy precipitation events such as tropical storms and hurricanes often lead to storm surges where water accumulates along the coast and floods coastal regions, vi. where there is inadequate drainage to channel water to the ocean, flooding occurs. Finally, the risk of flooding is increased by climate change which leads to frequent and more intense precipitation events and melting of glaciers increasing the risk of flooding.

Heavy rainfall from a low-pressure system called Subtropical Depression Issa triggered the April 2022 floods in KwaZulu-Natal, South Africa (Fig. 12.23a). Extreme flooding and landslides were triggered across the province during the April 11–12, 2022 extreme precipitation that affected country's east coast. At least 443 perished, and over 40,000 people were displaced. The floods were one of the deadliest and most destructive disasters in the country's recent history. The floods also damaged or destroyed thousands of homes and schools, and severely damaged infrastructure. The storm brought warm, moist air from the subtropics to the coast which caused heavy rain over several days. In the province, Virginia Airport recorded 12 inches of rain, whereas coastal areas received 18 inches. The precipitation was caused by clockwise flow of the low-pressure system that brought the warm and moist air to the coast. Compounding the disaster is that informal settlements are often built in flood-prone areas, and many urban areas in province and country lack proper drainage systems.

In Porto Alegre, Brazil, residents of two neighborhoods had to leave their homes in May 2024 due to the risk of new

Fig. 12.22 A **a** Bedrock channel, Glacier National Park credit: P.A. Bouker. **b** Braided stream in Great Sand Dunes National Park and Preserve, Colorado (*credit* D.T. Nelson)

floods. Shown in Fig. 12.23b is an aerial view of the flooded Rio Grande do Sul Art Museum (MARGS) in the city center of Porto Alegre, Rio Grande do Sul state, Brazil, on May 4, 2024. The storms which triggered the floods left at least 85 dead and 339 injured in the state, 134 missing, and 336 municipalities are in a state of public calamity. The floods were ascribed to the El Niño weather phenomenon which caused record rainfall and flooding from April 27 through mid-May 2024 in the southern Brazilian state of Rio Grande do Sul.

The summer 2020 was dubbed China's Summer of Floods (Fig. 12.23c). Aerial photograph showing partially submerged caravans and campers in flood waters at the camping site of De Hatenboer in Roermond, Netherlands, on July 15, 2021 Precipitation started in early June 2020 with heavy rains caused by the regional rainy season led to floods severely affecting large areas of southern China including the Yangtze basin and its tributaries. Rains and floods extended to central and eastern China during July 2024. Shown in Fig. 12.23d is an aerial photograph of

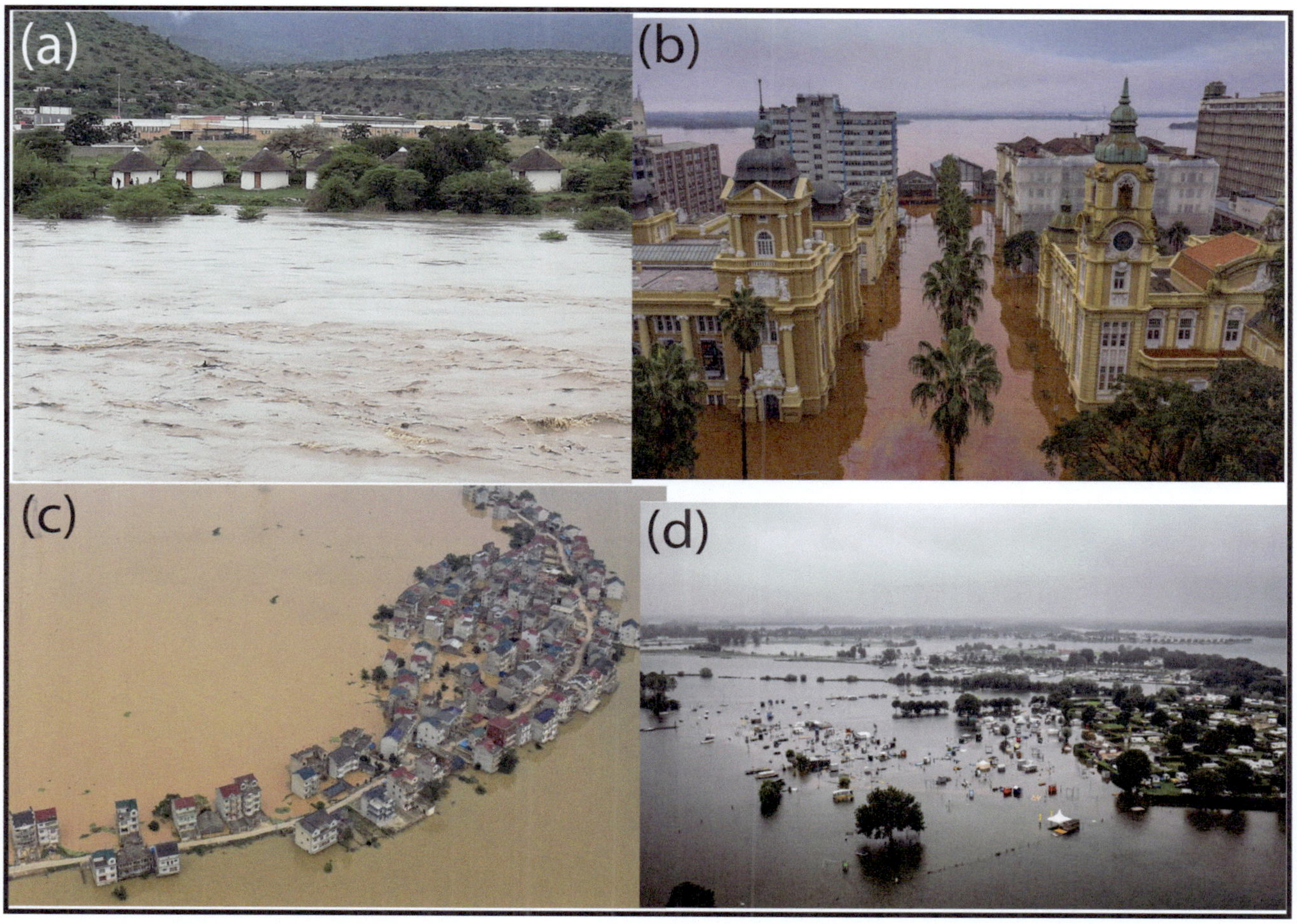

Fig. 12.23 Photographs of flooding from across some of the continents. **a** Photograph of floods in KwaZulu-Natal, South Africa in April 2022 (*credit KZN COGTA*), **b** Photograph released by the Rio Grande do Sul State Culture Secretary (SEDAC) showing an aerial view of the Rio Grande do Sul Art Museum (MARGS) that was flooded in the Porto Alegre, Rio Grande do Sul state, Brazil, city center on May 4, 2024 (*credit* SEDAC/AFP), **c** an aerial view showing submerged streets and inundated buildings after a dam was breached due to flooding in Jiujiang, China, on July 13, 2020, (*credit* AFP/Getty), and **d** aerial photograph showing partially submerged caravans in flood waters at the camping site of De Hatenboer in Roermond, Netherlands, on July 15, 2021 (*credit* Rob Engelaar, AFP/Getty Images)

caravans that are partially submerged and campers in flood waters at one De Hatenboer camping site in Roermond, Netherlands, on July 15, 2021.

Apart from climate change that is caused by human activities. Land subsidence ascribed to excessive extraction of groundwater, the weight of buildings as a result of urban growth, and poorly designed and constructed cities are thought to have contributed to flooding. Additionally, mining activities and urban transportation systems are also thought to have amplified the floods.

Finally, braided streams are discussed. A stream or river characterized by a network of channels that branch and rejoin to form a pattern that resembles a braid is termed a **braided stream** or braided river, as shown in Figs. 12.22b and 12.24b. A braided stream has several interconnected channels that are separated by small islands called braid bars. Bars and wide and shallow channels (Fig. 12.24 and b) are both constantly changing. Braided streams are encountered in broad valleys with well-developed floodplains. The bars and channels are formed, modified, destroyed, and reformed constantly. Braided streams are developed in floodplains where the gradient of the stream is low, often less than 0.5%. Additionally, braided streams form in regions with soils possessing high organic content where the water table is high for the bulk of the year.

Here, we only discuss floods and some mechanisms that can help alleviate damage from floods. When a water control structure breaks, it can cause flooding. Other causes of floods include ice or debris jams. A debris jam is a collection of debris, such as leaves, branches, or logs that can pile up and block the flow of water in a river.

Debris jams can cause problems such as flooding from increased water levels upstream resulting in flooding in nearby areas. Bridge damage can occur when debris jams increase the risk of bridge failure. Debris jams often result in lost revenue, damage to infrastructure, as well as shortages of supplies, which underscores the economic impact of debris jams. Although debris jams can occur naturally, they

Fig. 12.24 Braided channels developed in the Zambezi River by the Victoria Falls **a** (credit: African Safaris) and **b** Mutirikwi River, Zimbabwe (*credit* Herald, Zimbabwe)

can also be caused by human activities such as when a river or stream overflows due to a jam. Storm surges from hurricanes often move large amounts of water inland. Finally, when snow melts rapidly, it tends to release large amounts of water that results in flooded rivers.

Although flooding can develop slowly or quickly, it can last for days or weeks. If the ground is already saturated with water when precipitation occurs. Flooding can be triggered with just a few centimeters (inches) of rainfall. Consequences of floods are the triggering of landslides that can kill lots of people as what happened in the western mountains of North Carolina during Hurricane Helene (Tuesday, September 24, 2024, to Sunday, September 29, 2024). Floods also result in disruption of transportation, power outages, and extensive damage to property, particularly when hurricanes make landfall in densely populated coastal areas. Failure to heed evacuation orders often leads to residents being caught up by flood waters results in death or injury, as almost always happens whenever hurricanes make landfall in inhabited areas coastal areas. Hurricane Helene, which made landfall as a Category 4 hurricane on September 26, 2024, was one of the worst top five hurricanes in modern history in terms of fatalities to strike the United States. At least 234 people across 7 states lost their lives due to Hurricane Helene, which devastated Florida, Georgia, South Carolina, North Carolina, and Tennessee. One additional fatality occurred in Indiana due to a tree falling on a house during the hurricane. North Carolina, particularly the Asheville (Figs. 12.25a–d) and adjacent areas, bore the worst brunt of the hurricane in terms of fatalities with at least 103 deaths. Buncombe County in North Carolina, whose county seat is Asheville, was the worst affected with at least 43 fatalities.

According to the North Carolina's Division of Public Health, the bulk of the 95 deaths from western North Carolina were attributed to drownings (32) and landslides (20), blunt force injuries (18), wind or tree trauma (6), motor vehicle drownings (4), motor vehicle crashes (3), unknown circumstances (6) and other (5). One individual was reported to have died from an environmental exposure related cause.

We are all familiar with some beautiful features on our planet. When rivers flow over different layers of rocks consisting of varying levels of resistance to erosion, the softer rock formation erodes faster than the harder rock layer. This results in the harder rock layer hanging over

Fig. 12.25 Photographs of flooded houses **a–c** with flooding almost to roof level **a**, **c** in the mountains of western North Carolina triggered by Hurricane Helene in September 2024. **d** shows three cars crushed a huge fallen tree which thankfully spared the house (*credit* Michael Freas)

the river channel. With the passage of time, the hard rock can collapse, forming a waterfall. Waterfalls (Fig. 12.26) are among some of the most impressive features of our planet. These range from very small waterfalls that you all may have seen in creeks near where you live to intermediate falls such as Bushkill Falls in eastern Pennsylvania (Fig. 12.26a) to the world's most impressive waterfalls in the Victoria Falls (Fig. 12.26b) located along the Zambezi River. The Victoria Falls are at the border between Zimbabwe and Zambia in central southern Africa. Indeed, the Victoria Falls are one of the premier tourist attractions of our planet, as they are one of the seven natural wonders of the world.

Rapids are areas where the streambed is rough due to the presence of rocks and boulders, and the channel walls are narrower to constrict the water. Examples of rapids from the Yellowstone River, USA, and from Namibia are shown on Fig. 12.27a and b, respectively. At rapids, the water flows more turbulently, relatively faster, and over a steeper riverbed than where rapids are absent. Rapids generally form in or near mountainous areas (Fig. 12.27).

12.9.2 Flood Alleviation

A **flood** is when overflowing of water from a stream onto land that is normally dry occurs. A **flash flood** is any swift flooding that occurs within 6 h of heavy precipitation, or other causes such as dam failure, although flooding often occurs within three hours. Intense precipitation associated with thunderstorms, failure of a dam, breaching of natural or man-made levees are some of the leading causes of flash floods. Flash Floods can occur anywhere, and they can affect any area. Flooding that affected the Guadalupe River in the Texas Hill Country, USA (Fig. 12.28), in early July 2025 was a flash flood. In early August 2025, flash floods in northern India caused the deaths of at least 4 people, trapped more under debris, and swept away several homes and shops in northern India. Flash floods tend to affect inhabitants of the Himalayas mountains spanning across India, Nepal, Bhutan, Pakistan, and China (Tibet). In addition to helping trigger flash floods, these mountains also influence the river systems of these countries. The Himalayas mountains also act as a natural border between India, Nepal, and Bhutan.

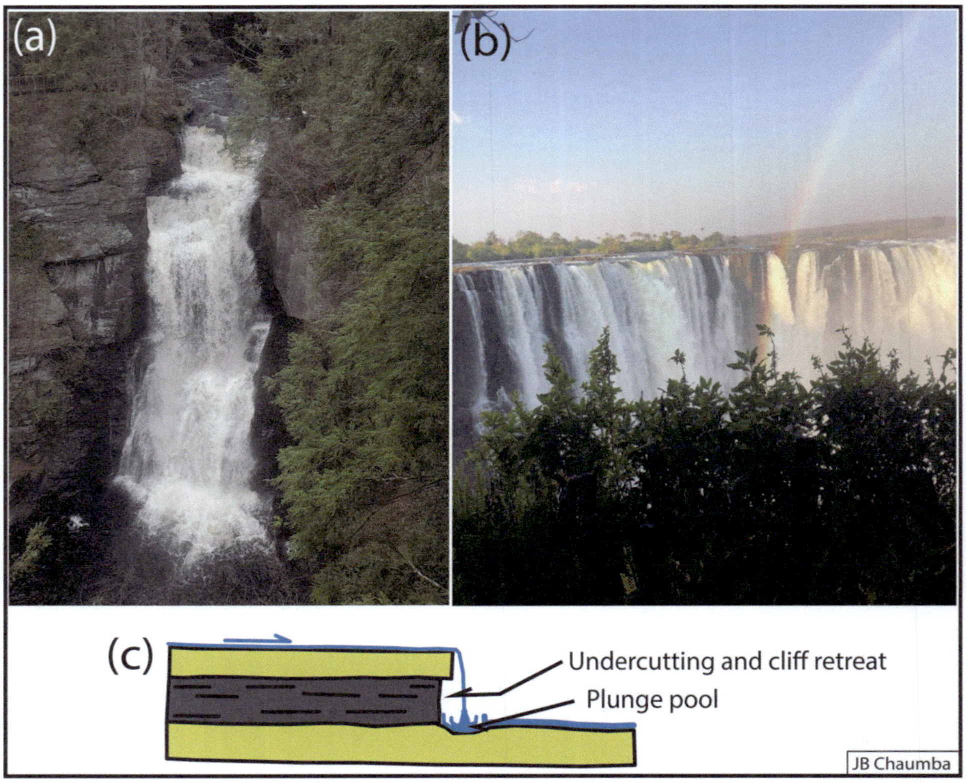

Fig. 12.26 Photograph of: **a** the Bush kill Falls on the Little Bush Kill River in eastern Pennsylvania, USA (*credit* S.D. Clark), and **b** the world-famous Victoria Falls on the Zambezi River (*credit* G.M. Chaumba). **c** diagram illustrating how erosion of shale undercuts the overlying sandstone over time resulting in upstream migration of the falls

Fig. 12.27 Rapids on riverbed in **a** Yellowstone Canyon, Wyoming, USA (*credit* D.T. Nelson) and Namibia (*credit* D.L. Reid). Note the turbulent flow of the water at the rapids

The intensity of the rainfall, the location and distribution of the rainfall, the land use and topography, vegetation types and growth/density, soil type, and soil water-content all determine just how quickly the flash flooding may occur, and influence where it may occur. Built-up areas are prone to the occurrence of floods in a short time span as the area to be occupied by rain is reduced due to buildings. Precipitation from the same event can result in faster flooding over urban areas relative to the neighboring countryside. Impermeable surfaces in built-up areas prevent infiltration of some of the rainfall into the ground, which quickly is channeled as runoff to low-lying areas.

Flash floods occur rapidly, catching people off-guard, with most engulfed while traveling. Those caught inside their offices or homes get trapped. Flooding often results in death as well as extensive damage to buildings, property, and infrastructure. Power is also lost, and it can take days, weeks, or sometimes even months before it is restored. Effects of flash flooding can be alleviated by heeding any evacuation warnings issued and promptly evacuating. Quickly moving to higher ground should be prioritized. One should avoid driving or walking into or through flood waters. Walking in or near flood waters is the second leading cause of flood-related deaths. Additional flooding may occur after a flash flood; hence, one should avoid areas that are ravaged by floods. By so doing, one avoids hampering any rescue and other emergency operations underway or about to be embarked upon. Before there is even a flash flood, you can protect your home from flash floods by installing a sump pump, sealing basement walls and floors, and installing barriers. Ensure that the walls are waterproofed. It is also recommended to have a backup

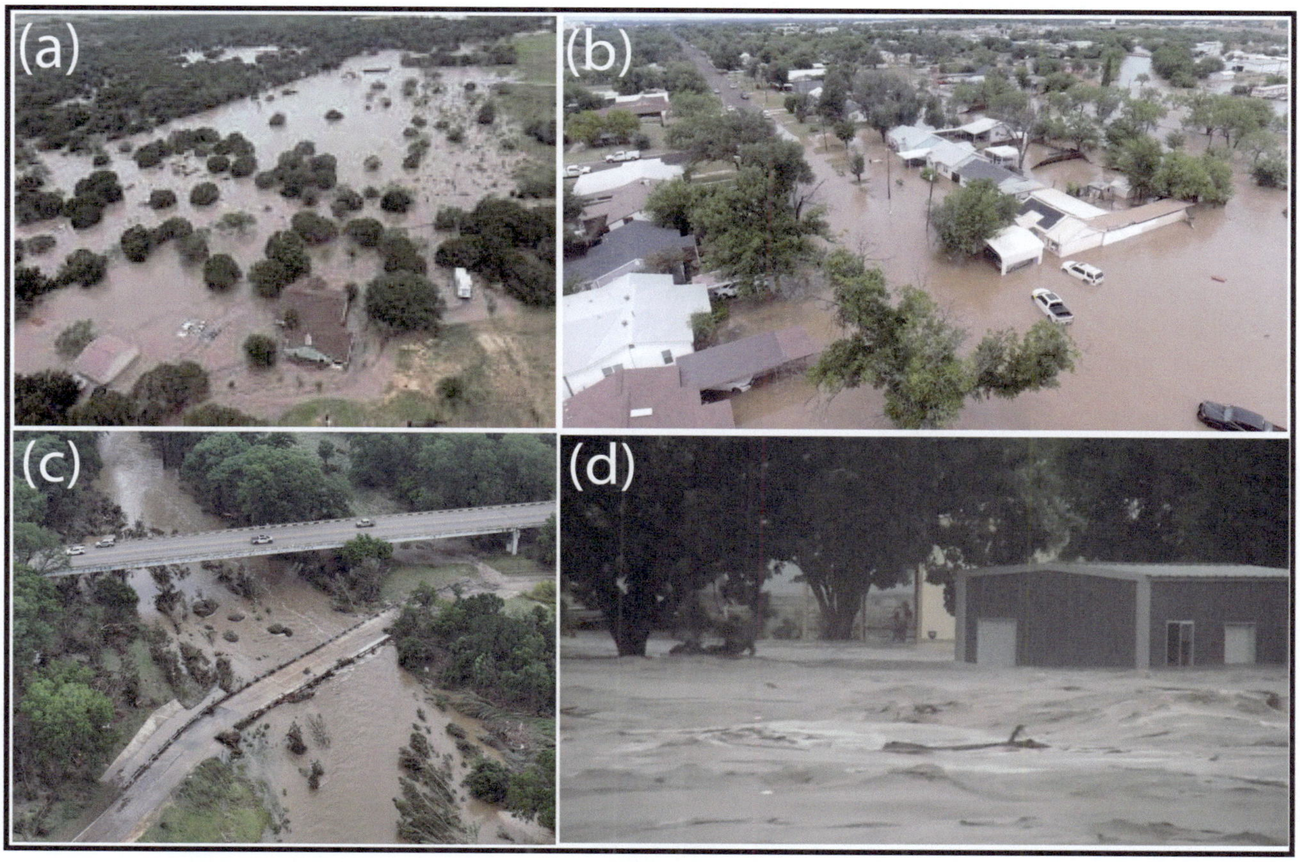

Fig. 12.28 Images from the Guadalupe River in the Texas Hill Country from a flash flood in July 2025. **a** Flooding of the Guadalope River, *credit* J.D. Capelouto, SEMAFOR, **b** View of part of flooded Guadalupe River, credit Patrick Keely, **c**, view from drone showing downed trees caused by flooding in Comfort, Texas, on July 5, 2025, *credit* Marco Bello, Reuters, **d** and severe flooding along the Guadalupe River, Kerr county, Texas, on July 4, 2025, *credit* AP

source of power. Elevating utilities, as well as sealing cracks and gaps, is also encouraged.

Eliminating floods appears to be an impossible task. Despite our best efforts, we cannot really *control* floods, but we can live with floods by lessening their impact on the human race by reducing both the loss of lives from floods and the damage to property and infrastructure. To lessen the impacts of floods, human beings can either invoke non-structural approaches or construct structures to "tame" rivers. Construction of structures such as channels, flood control dams, and artificial levees may work for some period, but there are disadvantages such as high costs as well as giving residence in harm's way wrong impressions of a problem having been solved. Flood control dams are specifically built to impound flood water during or soon after floods so that the water can then be slowly released, such that no flooding occurs. The water from the dam can also be stored and used later for various purposes such as irrigation. As water is released from flood control dams, it can be used for the generation of hydroelectricity. Additionally, dams are useful for recreation.

The disadvantage of flood control dams is that they inundate land which can be used for agricultural, transportation, forests as well as loss of historical sites and scenic views of river valleys. The volume of sediments trapped by a dam, with the passage of time, can greatly reduce the volume of water storage in that dam with time. Dams can also fail and cause even worse damage to property as well as higher fatality numbers had the dam not been built at all, defeating the original purpose of flood control.

Increasing the height of a riverbank by hauling huge quantities of Earth to raise the banks of the river creates artificial levees which have steeper slopes compared to natural levees. Construction of artificial levees increases the volume of water held by the river channel which reduces chances of water overflowing its banks to flood nearby homes and other structures. When water that is constrained in a river by artificial levees threatens to flood cities, it can be diverted from the river to intentionally flooded areas called floodways, that is, low risk or underpopulated rural areas through outlets in the artificial levees.

12.9.3 Preserving Kart Topography and Living with Floods

Both surface and subsurface water quality should be monitored and well managed to preserve karst terrain. Adhering to strict land use planning such as avoiding development in sensitive areas help preserve karst topography. In addition, careful construction guidelines in karst regions that maintain natural vegetation buffers where sinkholes and caves occur should be implemented. Karst terrain should be regularly managed in order to prevent damage and pollution underground system by diverting runoff away. Foot traffic can be properly managed and channeled in areas that limit human disturbances. The general population, in particular local communities near karst terrains, should be educated about the importance of karst terrains, their ecosystems, and how their actions impact the quality of both surface and groundwater. such fragile ecosystems.

To preserve karst topography, construction activities in karst regions and designation of protected areas should be prohibited (Fig. 12.29a). A sinkhole that formed in Pasco County, Florida, USA, in 2012 swallowed the back of a home along Shoal Drive in Hudson. The sinkhole which affected the house shown in Fig. 12.29a measured 40 feet wide and had a depth of 20 feet. Breaching of dam walls (Fig. 12.29b) can also lead to flooding which kills people and destroys homes. When near Pickstone Mine in Chegutu hectares of crops were destroyed after three dams near Pickstone Mine in Chegutu burst their dam walls on Wednesday, January 27, 2021. Additionally, livestock were also swept away by the floods. The dams store water that is used for household use, irrigation purposes, drinking water for livestock, as well as for mining purposes.

Drainage systems to prevent surface water from rapidly entering sinkholes, potentially causing erosion and instability, should be designed. Natural vegetation cover around karst features to act as vegetation buffers and filters to

Fig. 12.29 **a** Picture of a house that was constructed on karst topography in Pasco County, Florida, USA (*credit* ZUMA Press Inc/Smithsonian Magazine). **b** Photograph of breached Masterpiece Dam wall in Zimbabwe (*credit* Herald, Zimbabwe)

stabilize soil should be maintained. Prevention of pollutants from entering the groundwater system through sinkholes and cracks should also be strongly promoted through careful management of waste disposal. Regulations on protection of caves through avoidance of introduction of non-native species, vandalism, and disturbance of cave formations should be implemented.

Minimizing failure of dam failures requires implementing routine inspections, proper maintenance, and promptly addressing any potential problems. Furthermore, comprehensive emergency action plans need to be developed in order to minimize damage to the dam in the event of failure occurring. The condition of any dam needs constant monitoring and adhering to safety standards. Local communities need to be educated on potential risks as well as evacuation procedures. Engineering assessments need to be conducted periodically so as to evaluate the structural integrity of a dam. Qualified personnel should routinely conduct visual inspections for signs of wear and tear, cracks, leaks, or erosion. If there are any problems they need to be repaired promptly.

A brief discussion about **geothermal energy** is provided here to end this chapter. *Geothermal* is a word derived from the Greek words "geo," meaning *Earth*, and "thermal," meaning *heat*. Geothermal energy taps heat from the Earth's core that is generated by the decay of radioactive elements. Geothermal energy, therefore, is heat energy that originates from the core of planet Earth. It is considered a renewable form of energy due to the heat from the core of planet Earth being continuously produced within the planet. This energy is being presently harnessed in electricity generation such as at the Geysers located in Sonoma County, California, USA. Additionally, geothermal energy is also being utilized directly to heat buildings and even for cooling them in some situations. To tap into this type of energy, wells are drilled into underground reservoirs to intersect hot water or steam. It is anticipated that as oil reserves decline, geothermal energy will likely become an increasingly more attractive source of energy.

Glossary

Alluvium term for loose material such as gravel, sand, silt, and clay deposited by running water

Aquifer a permeable medium that can contain and transmit groundwater at rates high enough to be useable

Brines water containing high concentrations of salts

Brook small stream empties its water into a stream

Centripetal drainage pattern drainage system where streams flow toward a central depression to converge at a single point. This drainage pattern is opposite to a radial drainage pattern

Condensation process whereby water vapor changes to liquid due to either the air being cooled to its dew point or it becomes too saturated with water vapor such that it cannot hold any more water

Creek a stream or minor tributary of a river

Dendritic drainage pattern river system whereby tributaries join the river at acute angles to form a branching pattern resembling roots or branches of a tree. It develops in areas with gentle slopes and uniform rock

Deposition process of adding rocks, sediment, or soil to a surface and it occurs when particles settle down from a fluid such as ice, water, wind, or from the ocean

Deranged drainage pattern a drainage system lacking a clear pattern to the rivers and it develops from disruption of a pre-existing drainage pattern

Drainage removal of surplus water

Discharge volume, of water passing through a given point in a certain amount of time, and has units of gallons per second, or cubic meters per second

Drainage basin the land area in which all water that is drained eventually merges to form a common outlet

Drainage divide a boundary line separating any two adjacent drainage basins

Drainage pattern pattern produced by rivers, streams, and lakes in a particular drainage basin

Effluent stream a stream that receives water directly from groundwater and it is also called a gaining stream

Ephemeral stream stream that flows only during and/or shortly after precipitation events as it is primarily fed by runoff

Erosion gradual process occurring when ice, water, or wind breaks down and transports away rock, soil, or dissolved material from one location to another

Evaporation the process whereby water changes from liquid water to water vapor

Evapotranspiration combined processes of both evaporation and transpiration through which water moves from the Earth's surface to the atmosphere

Flash flood sudden flooding of low-lying areas typically caused by excessive or heavy rainfall within a short period of time of less than 6 h and can occur even if no rain has fallen. Meltwater from snow or ice, tropical storms, or hurricanes also cause flash floods.

Flood inundation of a land that is normally dry occurs due to rising water in an existing river, stream, or drainage ditch

Geothermal energy heat energy produced comes from the Earth's core, and it is a renewable resource since the heat is continuously produced within this planet

Graded profile longitudinal profile of a stream in equilibrium with a smooth, concave curve where the slope gradually decreases downstream which permits it to effectively transport its load without much erosion or deposition along its course. It represents a balanced state between a stream's energy and the material it carries

Graded stream a stream that has reached equilibrium with neither erosion nor deposition of large amounts of sediment occurring over time

Groundwater water occurring underground in soil, pores, or cracks in rock

Headwaters where a river originates or begins, usually in where there is a lake, spring, marsh or melting snow

Hydrologic cycle the continuous circulation of water between Earth's land, surface water, groundwater, and atmosphere. Energy from the Sun powers the cycle

Incised meander curved river valley with steep slopes on both sides that has cut into bedrock

Influent stream stream that is located above the water table and providing water to a larger body of water or a treatment system, and it is also called a losing stream

Intermittent stream stream that contains flowing water only during some portions of the year

Knickpoint part of a river or channel, such as a waterfall or lake, where there is a sharp change in slope of channel bed

Mouth where a river ends by emptying into a stationary body of water such as an ocean or lake

Perennial stream stream containing flowing water throughout the entire year

Precipitation process whereby water falls from the atmosphere to the Earth's surface in various forms such as rain, snow, hail, or sleet

Radial drainage pattern drainage system where streams flow outward from a central high point such as the summit of a volcanic peak to resemble the spokes of a wheel

Rapid an area of a river where water flows turbulently and relatively faster and over a steeper riverbed

Rectangular drainage pattern drainage system whereby streams repeatedly bend at right angles to form a pattern resembling a grid. It forms in faulted areas with fractures intersecting at right degree angles

Runoff water flowing downslope at the Earth's surface, and it occurs when there is more water on the ground than the land can absorb or infiltrate into the ground

Salt flat flat, salt-encrusted portion of land that forms when water evaporates from a former body of water in dry regions. Also called a salt pan, it is whitish in color due to the presence of salts

Snow melt water that is produced from melting of snow

Sheet flow shallow and slow-moving stormwater that moves slowly over relatively flat land

Stream flow flow confined to channels and conduits

Yazoo tributary stream or river flowing on the flood plain of a larger river that runs subparallel to the larger river before it eventually joins it

Transpiration process whereby plants release water vapor from the ground into the atmosphere through their leaves

Transportation movement of rocks, sediments, and dissolved material across the Earth's surface

Trellis drainage pattern drainage pattern whereby tributaries intersect at approximately right angles

Tributary small river or stream that flows into a larger river or a stationary body of water

13 Groundwater and Karst Topography

Abstract

The chapter begins by defining groundwater, and defines and illustrates several terms such as water table, porosity, and permeability, saturated and unsaturated zone, capillary fringe, aquifer, unconfined and confined aquifers, and groundwater movement, among others. Definition of, and distinction between, cone of depression and recharge are made, as well as distinguishing between hot springs and geysers. Definitions of travertine terraces, mud pots, fumaroles, and geysers are provided using examples and photographs. Landforms associated with karst topography are discussed and distinguished with the use of diagrams, photographs, and examples are provided. Definition of potentiometric surface is provided. The causes of both groundwater contamination and saltwater intrusion are discussed, as well as some ways to minimize them. Finally, mining of groundwater, land subsidence, and ways of minimizing them through a case study in the latter example, are discussed.

Learning Goals

Studying this chapter will enable you to:

- Define groundwater
- Differentiate between porosity and permeability
- Explain the movement of groundwater
- Define an aquifer and its importance
- Distinguish between an unconfined and confined aquifer
- Describe an artesian aquifer
- Define springs, wells, and artesian wells and explain how they form
- List features associated with karst topography and how they form
- Discuss some common environmental problems affecting groundwater
- Define groundwater contamination and list measures that reduce it

Supplementary Information The online version contains supplementary material available at https://doi.org/10.1007/978-3-032-20231-4_13.

13.1 Introduction

Have you ever wondered what happens to rain drops that fall on flat vegetated ground during precipitation events? Unlike the raindrops that fall on ditches which are promptly transported away by storm drains to creeks and rivers and ultimately to lakes and the oceans, raindrops that fall on generally flat ground and seep into the ground eventually add to the water occurring below the ground. This chapter deals with this invisible water below the Earth's surface.

Most of us are very familiar with water in creeks, rivers, and reservoirs as we can see the water. You may have dug a pit in our backyard or in sand at the beach deep enough to try and intersect water in the ground or beach sand. If you did, chances are that you encountered an increase in moisture in the soil or sand, called soil moisture, until you intersected water in the soil, particularly at the beach. This water occurring beneath the ground surface occupying spaces between particles of rock, soil, or sand is called **groundwater**. You are probably wondering how the water ended up in the ground in the first place. During precipitation, water that doesn't collect in creeks, rivers, lakes, the ocean, and even in ditches eventually seeps into the ground through **infiltration** to become soil moisture and groundwater. Groundwater, thus, is water occurring below the Earth's surface occurring in spaces between rocks, soil, sand, in cracks in rock formations (Chapters 3, 4, and 5), and in cavities in rocks composed of calcium carbonate (limestone, see Chapter 4). Groundwater is stored in, and moves slowly, through geologic formations of soil, sand,

J. B. Chaumba, *Introductory Physical Geology*, Springer Textbooks in Earth Sciences, Geography and Environment,
https://doi.org/10.1007/978-3-032-20231-4_13

and rocks from which it can be extracted that are called **aquifers**.

Groundwater is water from precipitation, snow, or melting ice that infiltrates into the soil where it is stored in the tiny pore spaces occurring between soil and rock and particles that comprise soil. Groundwater represents a significant source of our planet. In the Unites States, groundwater accounts for almost 95 percent of the nation's freshwater resources. Groundwater may return to the Earth's surface as springs, or it can be pumped out from wells. Approximately 50 percent of United States municipal, domestic, and agricultural water supply is groundwater.

13.2 Water Table, Porosity, and Permeability

The region encountered above the water table where pore spaces contain both air and water is called the **unsaturated zone** (Fig. 13.1) whereas the region where all the pore spaces in the soil and rock are entirely filled with water is called the **saturated zone** (Fig. 13.1). The saturated zone is also referred to as the zone of saturation or phreatic zone. The unsaturated zone is also referred to as the vadose zone (Figs. 13.1 and 13.2). The zone between the saturated and unsaturated zones is called the epiphreatic zone. The unsaturated zone is also called the zone of aeration. Within the unsaturated zone, therefore, pore spaces are not entirely saturated with water. The topmost surface of the saturated zone where the soil and/or rocks with openings in them are permanently filled with water is called the **water table** (Fig. 13.1). The water table, thus, delineates the underlying saturated zone from the unsaturated zone above. In the unsaturated zone, any water in this zone is not readily available for extraction by wells due to capillary forces as it occurs as a thin film surrounding soil particles and rocks, as shown in Fig. 13.1. The saturated zone is what is considered to be true groundwater. The zone in which all pores are filled, or saturated, with water is the saturated zone as it occurs below the water table (Figs. 13.1 and 13.2). In the saturated zone, water can be extracted through wells drilled into this zone.

Capillary action is a property of surface tension that acts to draw water upward. **Capillary fringe** (Fig. 13.2) denotes the zone occurring at the contact of the unsaturated and saturated zones where groundwater is held in the soil by capillary action. Additionally, the capillary fringe is the zone where most groundwater recharge passes. The importance of capillary fringe is that it influences soil moisture and the availability of plant water. Most groundwater recharge (Figs. 13.1 and 13.2) passes through the capillary fringe.

Porosity is the percentage of open (or void) spaces in the ground, that is, a measure of its ability of a material to hold fluids. **Permeability** is a measure of how easily fluids flow through rocks and soil. The movement of water below the earth's surface is called groundwater flow. Porosity and permeability are related. The shape and sorting of grains in a rock or sediment are important in its porosity. Well-sorted grain or rock particles have high porosity relative to poorly sorted ones. Soils and rocks with high porosity and permeability make very good aquifers. Soil and rocks can be very porous yet have low permeability if its pore spaces are not connected as water doesn't easily move through them. Clay, for example, is characterized by high porosity values but

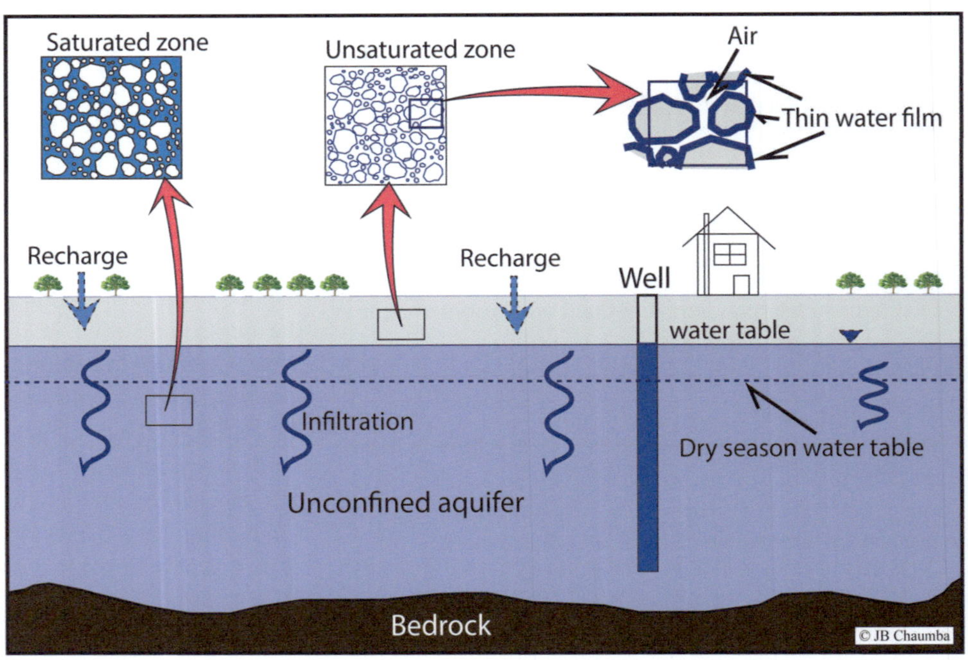

Fig. 13.1 Diagram illustrating an unconfined aquifer, saturated and unsaturated zones, water table, and a thin film of water that coats soil particles in the unsaturated zone. Recharge on an unconfined occurs from anywhere on the surface and the water infiltrates or percolates downwards

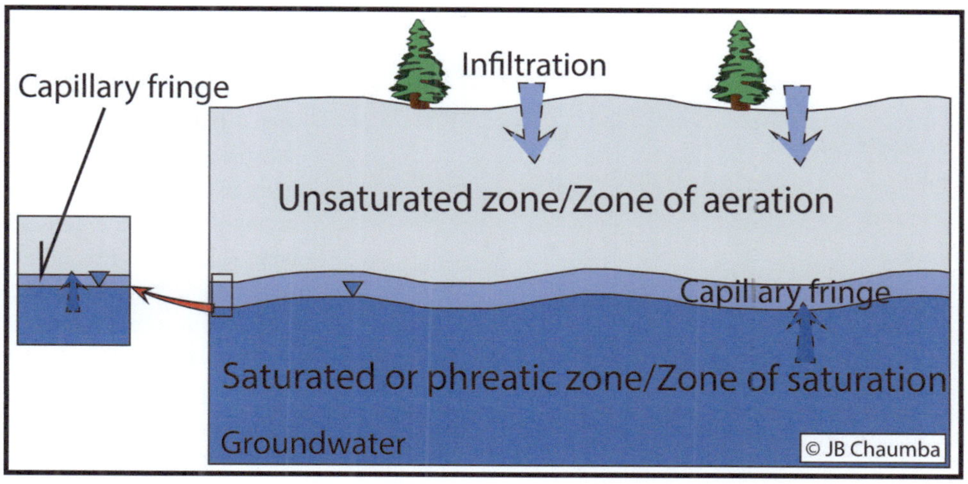

Fig. 13.2 Diagram illustrating the zone of aeration, zone of saturation, the water table, and the capillary fringe

low permeability values as its pores are small and are not interconnected. Permeability and porosity values for various rock types are shown in Table 13.1. It can be observed in Table 13.1 that unconsolidated rocks have high values of both porosity and permeability, whereas basalt at depth of > 1 km, andesite, and thermometamorphic rocks have low porosity and permeability values.

Porosity, therefore, measures the proportion of open spaces in a material, whereas permeability measures the ease with which fluids can pass through the material. Porosity and permeability are both dependent on the size, number, and interconnectedness of openings in a material. Clay, for example, is characterized by high porosity but its permeability is low due to its small particles that create several pore spaces that are very poorly connected. Both porosity and permeability are important as porosity improves soil aeration, availability of nutrients, as well as drainage and retention of water. Permeability, meanwhile, is important for engineering structures constructed on, in, or under the soil as water that passes through soil often affects its stresses and strength.

When the water table coincides with the earth's surface, the depth of the water table is zero at the land surface. The water table is usually shallower near permanent bodies of surface water such as lakes, some streams, and wetlands. The water table is not flat but has peaks and valleys that generally mimic the topography of the land above it. Groundwater is not static but flows to **discharge zones**. Groundwater discharge zones are areas where the water table coincides with the land surface. Visible groundwater discharge zones are springs. Where groundwater seeps into wetlands or contributes to stream flows are examples of groundwater discharge zones that are not obvious.

The process through which groundwater levels can be replenished with surface water or recycled water is called groundwater recharge. Groundwater recharge can either occur naturally or artificially, and it is a key strategy for improving water availability in California.

Recharge of groundwater can occur through natural, artificial, or in-lieu processes. Natural recharge occurs when precipitation and surface water infiltrate the ground through rivers, lakes, open spaces, and wetlands. When water is diverted to agricultural lands, injection wells, engineered infiltration ponds, or spreading basins, artificial recharge occurs. The importance of groundwater recharge is that it replenishes aquifers that are utilized for water supply purposes. Groundwater recharge also plays a role in improving the distribution of seasonal flows. Additionally, recharge of groundwater contributes to higher base flows. Baseflow is the proportion of a stream's flow that is contributed by groundwater and is sustained between precipitation events.

13.3 Aquifer, Unconfined and Confined Aquifers, and Groundwater Movement

An **aquifer** can be defined as a saturated rock formation or sediment that is capable of storing and releasing water in useful quantities to humanity. The largest proportion of most unconsolidated aquifers are relatively young sediments that occur at or close to the Earth's surface. Surface topography and climate both determine how groundwater flows within aquifers. Where topography is simple, groundwater flow in unconsolidated superficial sediments mimics drainage courses. The shape of the water table in relation to drainage, however, is dependent on climate, the balance between precipitation as well as the rate of evapotranspiration. Evapotranspiration exceeds precipitation in arid and semi-arid areas, hence many drainages in these dry climatic regions typically have ephemeral surface flow. Additionally,

Table 13.1 Permeability and porosity ranges for various rock types*

Lithology	Permeability (m^2)		Porosity (%)
	Min	Max	Range
Unconsolidated rocks			
Gravel	10^{-10}	10^{-7}	25–40
Clean sand	10^{-13}	10^{-9}	5–50
Silty sand	10^{-14}	10^{-10}	
Silty, loess	10^{-16}	10^{-12}	35–50
Unweathered clay	10^{-20}	10^{-15}	40–80
Consolidated rocks			
Shale	10^{-20}	10^{-16}	0–10
Unfractured igneous and met rocks	10^{-20}	10^{-17}	0–5
Sandstone	10^{-17}	10^{-13}	5–35
Limestone and dolomite	10^{-16}	10^{-13}	0–20
Fractured igneous and met rocks	10^{-15}	10^{-13}	0–10
Permeable basalt	10^{-14}	10^{-9}	0–25
Karst limestone	10^{-13}	10^{-9}	5–50
Fractured basalt			5–50
Basalt near surface	10^{-14}	10^{-12}	
Basalt at 1 km depth	10^{-18}	10^{-10}	
Andesite	10^{-20}	10^{-18}	0.2–0.3
Thermometamorphic rocks	10^{-18}	10^{-14}	2–17
Capi Flegrei trachy-phonolite			
Tuff, surface	10^{-16}	10^{-15}	48–52
Tuff, depth	10^{-17}	10^{-15}	19–52
Chaotic tuff/tuffites < 1 km depth	10^{-18}	10^{-14}	6–40
Chaotic tuff/tuffites > 1 km depth	10–18	10–14	5–36
Tuffites (HT altered)	10–16	10–16	.05–0.07
Lava	10–18	10–14	7–25
Montserrat andesite			
Pyroclastic flow deposit	10–18	10–13	
Lava	10–17	10–13	
Lahar deposit	10–14	10–13	

* After Alia Jasim, Brioch Hemmings, Klaus Mayer and Bettina Scheu, 2018, Groundwater flow and volcanic unrest, Advances in Volcanology, DOI: 10.1007/11157_2018_33.

recharge to an aquifer in desert regions is largely through the stream bed or the channel banks.

An unconfined aquifer, also called a water table aquifer, is an aquifer with a water table that is at atmospheric pressure and can therefore rise and fall and broadly as it mimics the surface topography. The upper boundary of an unconfined aquifer is the ground surface. The water table in an unconfined aquifer denotes the free water surface. The level of the water table at any point of interest in an unconfined aquifer indicates the static level of a well that penetrates the aquifer. The water table in an unconfined aquifer is not fixed as it changes constantly by adjusting its surface such that a balance between the recharge and outflow from the subsurface storage is achieved. In an unconfined well, fluctuations in the water table vary according to the seasons in any year. With excessive pumping out of water, the water table is lowered. Precipitation leads to groundwater recharge accompanied by rising of the water table if recharge exceeds withdrawal of groundwater.

The water table in an unconfined aquifer roughly mimics the topography of the surface; the water table is slightly higher beneath ridges and slightly lower in valleys. Where the water table intersects the land surface, groundwater becomes surface water through springs or seepage.

Lenses or localized patches of impermeable strata may be present inside an unconfined aquifer such that it retains a localized water table aquifer called **perched aquifers** (Fig. 12.3). Perched aquifers have perched water tables which occur above the regional water table, as shown in Fig. 12.3). A perched water table may often be confused with a regional water table during groundwater exploration.

A confined aquifer is one that is completely surrounded by impermeable layers which subject the water within it to be under pressure. If a well is drilled into a confined aquifer, the water rises above the top of the aquifer without pumping if it is confined between the confining layers and is protected from surface contamination. A confined aquifer is an aquifer that has a confining layer acting as an upper bound and another confining layer lower bound. Groundwater in a confined aquifer is under pressure that is greater than atmospheric pressure as it is confined by the impermeable layers bounding it both above and below it. If a well is drilled into a confined aquifer, the water flows naturally to the surface without pumping, to form an artesian well. Water in a confined aquifer can only be recharged in specific areas where the confining layer is missing (Fig. 13.3) or very thin, which allows water to infiltrate into it from the surface.

An **aquitard**, or confining layer (Fig. 13.3) is an almost impermeable layer that limits or restricts groundwater movement or allows it to flow at extremely slow rates that are negligible. An aquitard can store contamination and groundwater but slowly transmits it from one aquifer to another. It is composed of layers of clay or impermeable rock. Both porosity and permeability determine how much water an aquifer can contain and transmit. Aquitards act as barriers to groundwater flow due to their low permeability, yet they still hold a lot of water due their high porosity. For oil and gas to pass through a rock in the oil and gas field, the rock must be both porous and permeable.

A hole that is drilled into the ground to access groundwater from an aquifer is called a well. Wells drilled into unconfined aquifers are water table wells (Fig. 13.3). Wells

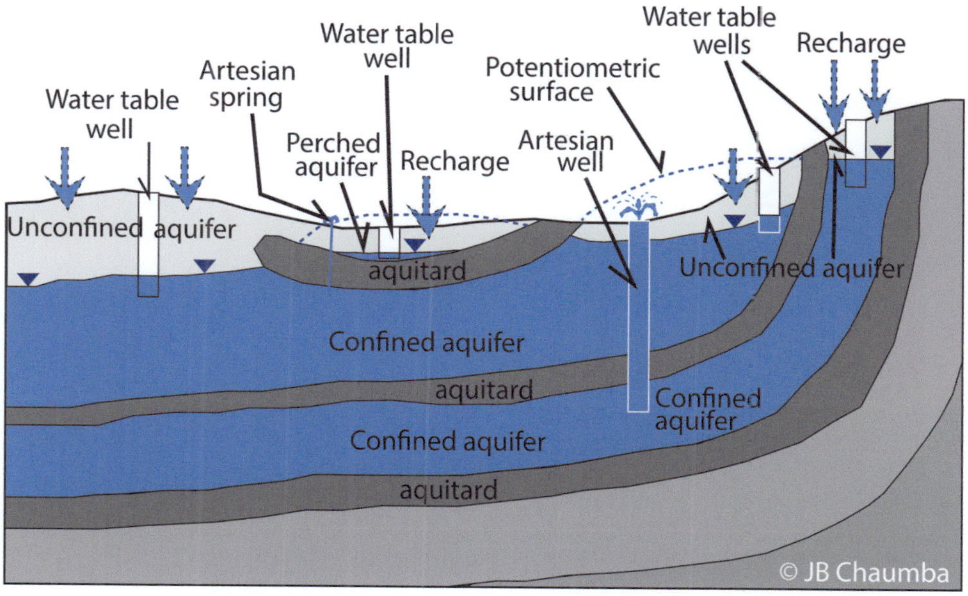

Fig. 13.3 A diagram illustrating unconfined, confined, and perched aquifers. Also illustrated are artesian wells, water table wells, and recharge zones. A confined aquifer is only recharged in a region that lacks a confining layer as the upper boundary. Also shown are recharge zones and a potentiometric surface

that are drilled to access water from confined aquifers such that water comes out without pumping are called **artesian wells** or free flowing wells. An **artesian spring** (Fig. 13.3) is a spring that forms when a fault of gap in an impermeable rock releases pressure that allows water to flow through to the Earth's surface. A forceful fountain of heated groundwater that is formed when the pressure is high enough produces a **geyser**. A geyser is generally caused by volcanic activity which causes buildup of pressure in a chamber beneath the ground.

Withdrawal or pumping of groundwater produces a **cone of depression** (Fig. 13.4). A cone of depression can be defined as a cone-shaped depression (Fig. 13.4) formed in a water table and it is caused by pumping groundwater from a well. Increased withdrawal of groundwater makes the cone of depression increase in the depth of the cone of depression from A to B as shown in Fig. 13.4. Continued pumping of groundwater from the aquifer causes the area of influence of the well to affect a greater and deeper area (to B) as shown in Fig. 13.4.

When snow melts or during and after precipitation events, some of the water accumulates in rivers and streams in addition to water that infiltrates into the ground. The water that infiltrates into the ground adds to water in aquifers through the process known as groundwater **recharge** (Fig. 13.3). The **recharge area** is the region on the surface where water infiltrates into the ground (Fig. 13.3). Groundwater flow is the movement of groundwater and represents a significant source of water for discharge from rivers. Groundwater moves from areas of high pressure to low pressure regions. The flow of groundwater of a proportion of streamflow that infiltrated into the ground, entered the phreatic zone, and gets discharged into a stream channel or springs, and seepage water. When groundwater flows through cracks, groundwater seepage occurs. Seepage can be caused by melting snow, heavy rain, melting snow, high water tables, as well as upslope rainwater that becomes groundwater.

Groundwater can move at rates of up to 1 ft/day (30 cm/day). Velocities of groundwater flow, however, can be as low as 1 ft/year (30 cm/year) or 1 ft/decade (10 cm/decade). Compare these low velocities of groundwater flow to those of streamflow which is characterized by rates of ft/s (cm/s). A velocity of 1 ft/s (30 cm/s) equals about 16 miles/day (26 km/day). The importance of groundwater flow is that it contributes to the water in streams and rivers. Furthermore, groundwater flow affects the habitats of both animals and plants. Human beings have been using groundwater for thousands of years for both drinking and irrigation.

13.4 Hot Springs and Geysers

Natural sources of hot water on the Earth's surface are **hot springs** and **geysers**. Distinguishing between these two hydrothermal deposits is important. Hot springs are pools of water that are heated by magma within the Earth's crust. Hot springs are formed when groundwater percolates through porous rock into fractures deep underground, where heat from a nearby magma chamber superheats the water.

Hot springs that are acidic enough to dissolve the surrounding rock are referred to as **mud pots**. Hot springs that flow through limestone and deposit calcite form **travertine terraces**. Formations of chalky white limestone that develop around hot springs from the deposition of

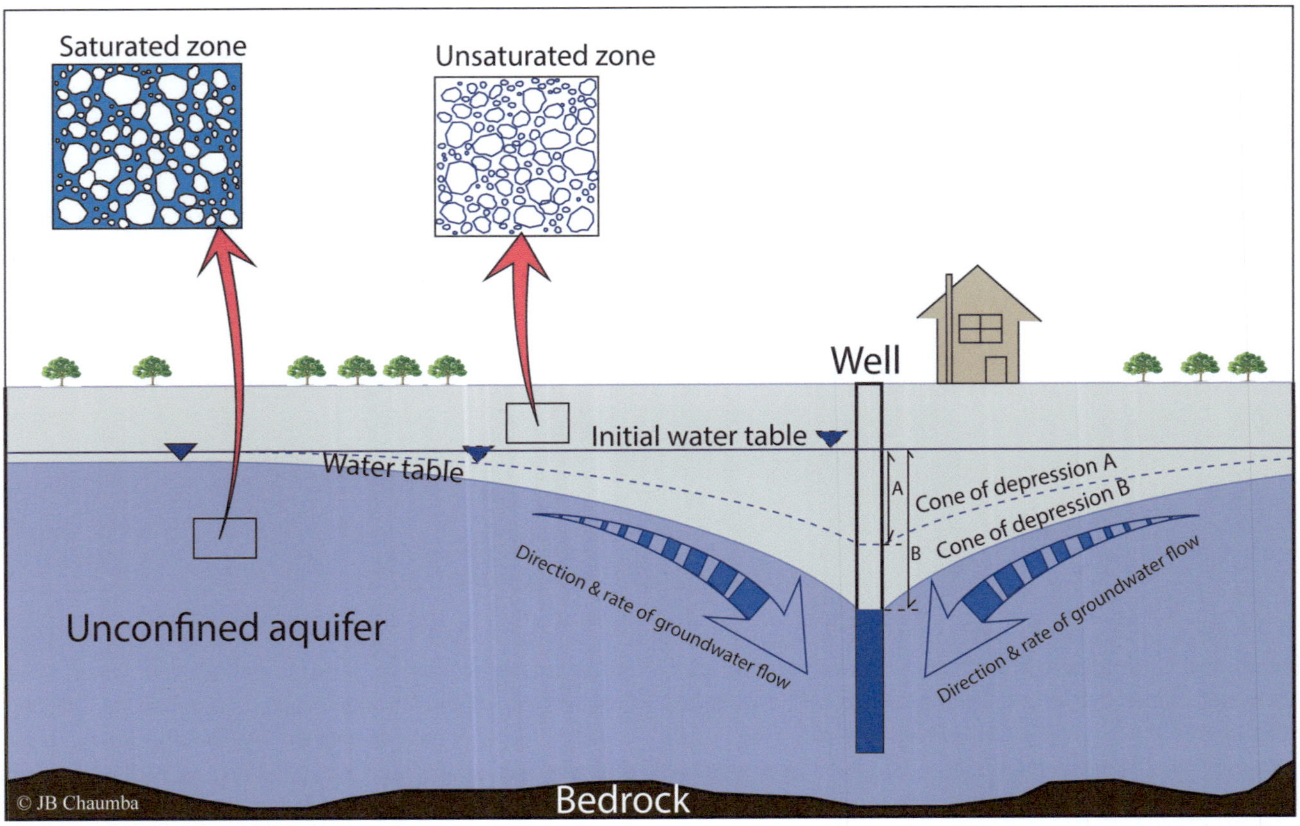

Fig. 13.4 A diagram illustrating how pumping groundwater from an unconfined aquifer produces lowering of the water table which forms a cone of depression. Distances A and B represent depths of the cone of depression from increased withdrawal of groundwater

calcium carbonate from geothermal waters are called travertine terraces. Travertine terraces form the rising of hot springs through limestone from which calcium carbonate is dissolved. When the heated water rises to the surface, a drop in pressure and temperature occurs which causes the escape of carbon dioxide. Calcium carbonate, which is less soluble at higher pH, causing precipitation of calcium carbonate out of the water resulting in the formation of travertine. Over time, the deposition of travertine builds up leading to the formation of terraces.

Travertine terraces change constantly due to the rapid rate of deposition. Blocking of one flow causes the water to move elsewhere. Mammoth Hot Springs in Yellowstone National Park are great examples of travertine terraces. Travertine deposits can be used as a chronometer in the determination of the age of these features. The Huanglong Travertine Terraces in southern China, which are known as the *dragon* based on their winding shape, are another example of travertine terraces.

Fumaroles, or steam vents, are hot formations that don't have water in their system but instead release hot steam constantly. **Geysers** are hot springs that periodically erupts by shooting jets of both water and steam into the air (Fig. 13.5). Geysers erupt periodically ejecting jets of heated water and steam into the atmosphere (Fig. 13.5). Geysers form when superheated water accumulates in underground pockets where pressure build-up occurs. Upon releasing of the pressure, the water flashes into steam, accompanied by expansion and ejection of a column to varying heights of both steam and water into the atmosphere (Fig. 13.5). Worldwide, geysers act as major tourist attractions. Examples of geysers include the Kitagata geyser in southwestern Uganda (Fig. 13.5a), the Chibwatata Hot Springs in Binga, Matabeleland North Province and the Nyanyadzi hot springs in southern Manicaland Province both in Zimbabwe (Fig. 13.5b and c, respectively), and the Old Faithful Geyser, Yellowstone National Park, USA (Fig. 13.5d). Some people believe that geysers and hot springs are said to have natural healing powers.

A mud pot (Fig. 13.6a) forms when surface water collects in an impermeable and shallow depression that lacks direct linkages to a flow of water underground. Heated water occurring beneath the depression forces steam to rise through the Earth's surface and heats water that accumulates at the surface. Recall our discussion from Chap. 7 of Yellowstone being a hot spot. The hot springs at Yellowstone emerge at temperatures of ~ 92 °C (~198°F) since it is an active geothermal area. The mean annual temperature

Fig. 13.5 Photographs of geysers. **a** Kitagata geyser in western Uganda. *Credit* Edna Namara, Global Press Journal, Uganda Bureau, **b** the Chibwatata Hot Springs and the Nyanyadzi hot springs (**c**), both Zimbabwe, and the Old Faithful Geyser, Yellowstone National Park, USA (**d**). Photograph (**d**). *Credit* D. T. Nelson

at Yellowstone, however, is 2.2 °C (36°F) which is slightly higher than the freezing point of water. The boiling point of water at the mean altitude of Yellowstone hot spot's steam vents, however, has a high value of 135 °C (275°F). This high temperature ensures that mud boils to form the mud pots that visitors view at Yellowstone (Fig. 13.6a). Devil's Thumb (Fig. 13.6b) is a hot spring shaped like a cone, and it is part of Mammoth Hot Springs on the Main Terrace Group.

Another interesting geologic feature exposed at Yellowstone which rises above the landscape near Mammoth Hot Springs Hotel is Liberty Cap (Fig. 13.6c). Liberty Cap was formed by a hot spring, and it is a travertine cone in Yellowstone National Park. Liberty Cap is a hot spring that has a single source of layers of travertine, a crystalline limestone, which was deposited by water during a period spanning at least hundreds of years. As long as the spring was active, the Liberty Cap cone increased in size. However, when the spring either found a new underground channel, or the orifice was sealed by travertine, the cone eventually became dormant. Liberty Cap, thus, is now a remnant mound of travertine.

A constant cycle of rising hot water, cooling, of Yellowstone National Park followed by falling formed the rings that are comprised of different temperatures around the center of spring resulted in the formation of the Grand Prismatic Spring at Yellowstone National Park (Fig. 13.6d). The Grand Prismatic Spring is the largest hot spring in the United States and the third largest in the world. The hot spring's colors are ascribed to different species of thermophile, or heat-loving, bacteria that thrive in the water. The temperature of the water in the center of the spring can be as high as 87 °C (189 °Ft). Most of life cannot survive at this high temperature. The origin of the rings of color around the spring are ascribed to different species of thermophile bacteria whereby the colors were produced

Fig. 13.6 Some of the geologic features at Yellowstone National Park associated with heated groundwater: **a** mud pot, **b** Devil's Thumb and travertine terraces, **c** Liberty Cap, and **d** Grand Prismatic Spring. *Credit* D. T. Nelson

from energy-harnessing pigments of bacteria. Mats with more chlorophyll, for example, are greener in color whereas those with more carotenoids are more orangey brown in color. The water in the center of the spring appears deep blue in color since there's very little life there which causes blue wavelengths to scatter.

13.5 Karst Topography

Groundwater, just like rainwater from where the bulk of it is derived, is weakly acidic, eroding through dissolution with the solid rock occurring in the phreatic zone being transported away. Carbonic acid is effective at dissolving limestone. Groundwater moves along small cracks over hundreds of years, dissolving and transporting away the limestone. Cracks are enlarged over time resulting in the formation of **caves**. A cave is hollow space in the ground formed by natural processes, and it is large enough for human beings to enter. Caves have rooms or passageways to explore. The agent of erosion through which the limestone is dissolved is chemical weathering. As groundwater flows through cracks and crevices in limestone, they are deepened and enlarged to create karst topography such as caves, sinkholes, disappearing streams, pillars, stalactites and stalagmites, and several others (Fig. 13.7). Karst topography is formed over extended periods of time.

Rainwater picks up carbon dioxide from the atmosphere, forming carbonic acid which is especially effective at dissolving limestone. Unlike surface erosion by flowing water, groundwater erosion is primarily caused by chemical reactions, dissolving minerals in the rock. As groundwater flows through cracks in limestone, it dissolves the rock and forms caves.

Karst topography, which is landscape characterized by sinkholes, caves, and underground drainage systems, is generated from extensive erosion caused by acidic ground-

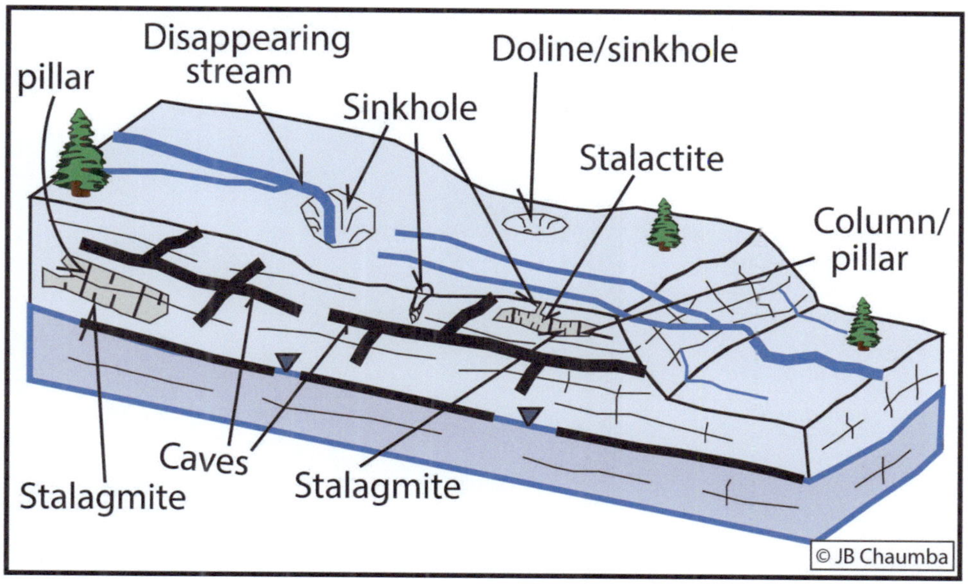

Fig. 13.7 Diagram illustrating karst topography and associated features such as disappearing streams, sinkholes, caves, stalactites and stalagmites, pillars, and dolines

water. As dissolved minerals precipitate from groundwater that drips within caves, they form stalagmites which rise from the floor as well as stalactites which hang from the ceiling. All continents have terrain composed of limestone. Europe, however, is the continent that contains the highest percentage of limestone and karst topography, accounting for 21.8% of its area. As the biggest continent in terms of size, Asia has the largest absolute area of karst terrain at 8.35 million km^2 (3.22 mi^2).

Karst topography is widespread in North America with karst topography occurring in almost every state, province, and territory. In the United States, the most extensive karst area, which is made up of limestones of Mississippian age, is located in the Interior Low Plateaus. A wide range of various types of karst topography, including arctic karst, forested karst, and alpine karst occurs in Canada. Additionally, Mississippian karst that occurs in Kentucky, Tennessee, and Indiana has many long cave systems, together with extensive areas composed of doline karst. Limestone exposed around the Ozark Dome in Missouri and northern Arkansas contains extensive karst, which include enormous caves as well as areas composed of doline karst and fluviokarst. There are a variety of karst landscape whereby surface stream channels and valleys can still be visible even though the bulk of the drainage occurs underground called fluviokarst. Landforms such as active stream valleys, blind valleys, dry valleys, and deranged streams. Approximately 10% of the Earth's surface is estimated to be composed of karst. Karst aquifers are very significant to mankind as a source of drinking water to a quarter of our planet's population.

In rocks that can be attached by carbonic acid, unique features such as **stalagmites**, **stalactites**, and **columns** (also called pillars) are also formed (Fig. 13.7). When water drops fall to the floor of a cave before evaporating completely, a portion of dissolved calcium bicarbonate is carried with them. The water droplets then evaporate and leave behind deposits of calcium carbonate on the cave floor. Continued deposited of the calcium on the cave floor results in cave formations that grow upwards cave towards the ceiling from the cave floor to produce stalagmites. Stalagmites are often located just below a stalactite, and when the two finally grow together a column is formed. Stalagmites, stalactites, and columns, thus, are formed in caves through water dripping from the roofs of caves which then deposits some of its dissolved load of calcium carbonate through the release into the air of carbon dioxide. Photographs of both stalagmites and stalactites are shown on Fig. 13.8a whereas photographs of pillars are shown on Fig. 13.8b. Stalactites are deposits shaped like icicles that form when water dissolves overlying limestone and then re-deposits the calcium carbonate along the ceilings of caves. Stalactites, thus, form along ceilings and hang downwards (Fig. 13.8a).

A **doline** is a bowl-shaped depression in the ground in karst landscapes. In North America, a doline is also referred to as a sinkhole. Dolines are the most common landform in karst areas, ranging in size from small to medium, with the depth and diameter both varying from meters to tens of meters. A doline forms processes of **dissolution** where limestone is dissolved from the surface downwards to produce hollows. The overlying rock can **collapse** into a cave and caves of limestone form below the insoluble material during **subsidence**. Dolines act as funnels after they are formed, which permits surface water to enter the underlying karstic bedrock aquifer directly. A doline may also appear

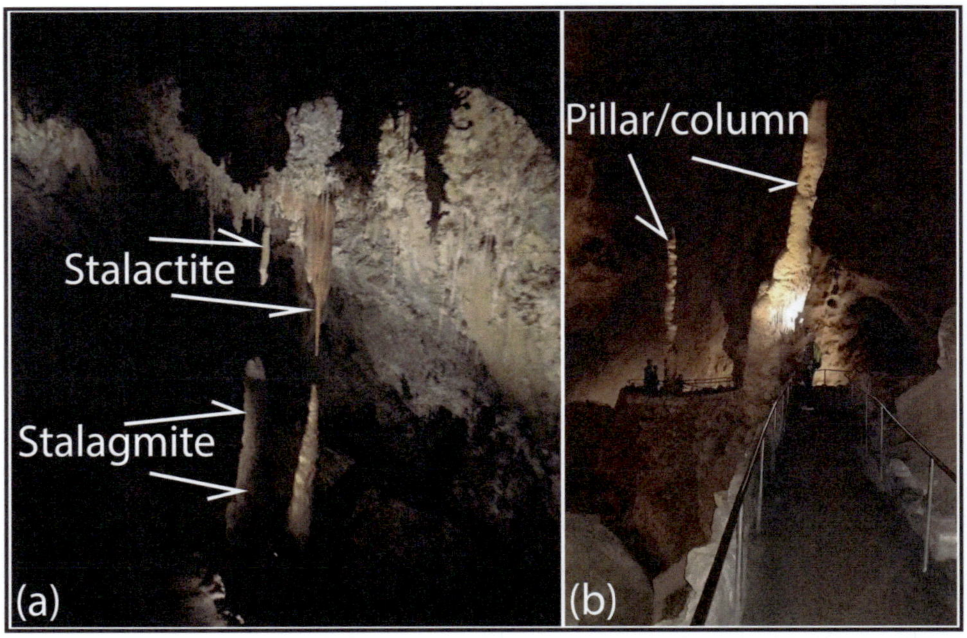

Fig. 13.8 **a** Photograph from a cave showing stalactites growing from the ceiling or roof of the cave downwards and stalagmites growing from the floor upwards, Mt. Timpanogos cave, American Fork, UT. *Credit* D. T. Nelson. **b** Photograph of a column that extends from the floor to the roof of the cave, Carlsbad Caverns National Park, New Mexico. *Credit* S. D. Clark

in clusters or as isolated features to generate a land surface that is pock-marked.

It is estimated that approximately 50% of the world's population is defendant on groundwater for drinking water, which underscores the significance of groundwater as a resource. Although this resource occurs below the ground and "invisible," groundwater constitutes up 99% of our planet's liquid freshwater. The proportion of all water that is used for irrigation that is sourced from groundwater is approximately 43%, testifying to the importance of groundwater in feeding the world population. Groundwater, therefore, is a critical resource. Replacement of some groundwater reservoirs can occur in periods estimated to range from hundreds to thousands of years.

On a global scale, we are using resources such as groundwater at rates that far exceed their renewal times, which has prompted some advocates for careful utilization of groundwater to have it treated as a nonrenewable resource. According to available data, approximately 50% of the world's population relies on groundwater for drinking water, meaning roughly half of the global population depends on this resource. Groundwater is the primary source for nearly half of the total amount of water used for domestic purposes globally. In addition, approximately 43% of all water used for irrigation is sourced from groundwater. Groundwater accounts for ~ 99% of Earth's liquid freshwater.

An imaginary plane that represents the level to which water would rise in a well if it were not trapped in a confined aquifer is referred to as a **potentiometric surface** or a piezometric surface. A potentiometric surface is an imaginary surface that represents the level to which water would freely rise in a well if it were not trapped in a confined aquifer. The potentiometric surface determines the highest level of high-rise buildings that can be constructed and supplied by water from an existing water tank (Fig. 13.9).

The bigger the difference between the potentiometric surface and level of the tap in a house, the higher the water pressure in the house. Thus, all the houses on Fig. 13.9a have water that comes out at high pressure as the houses are at a maximum of two levels and very close to the water tank. Any house whose floor is higher than the level of the potentiometric surface (Fig. 13.9b) won't have any water as the pressure of water at that high level will be too low to supply water to the house.

13.6 Groundwater Contamination

When harmful substances, or pollutants, enter the groundwater, **groundwater contamination** occurs. These pollutants are limitless but can range from motor oil to chemicals from agriculture to untreated waste. Certain diseases can occur from drinking contaminated groundwater. In the United States, the outflow (or effluent) from cesspools and septic tanks is among the top contaminants of groundwater. A quarter of all homes in the United States, particularly those in the country, rely on septic systems to dispose of their human wastes. In addition to septic systems, groundwater can also become contaminated if the casings or caps for wells are not properly installed. In addition to septic systems, arsenic, radon, and hydrogen sulfide are leached out of the soil as groundwater percolates into aquifers.

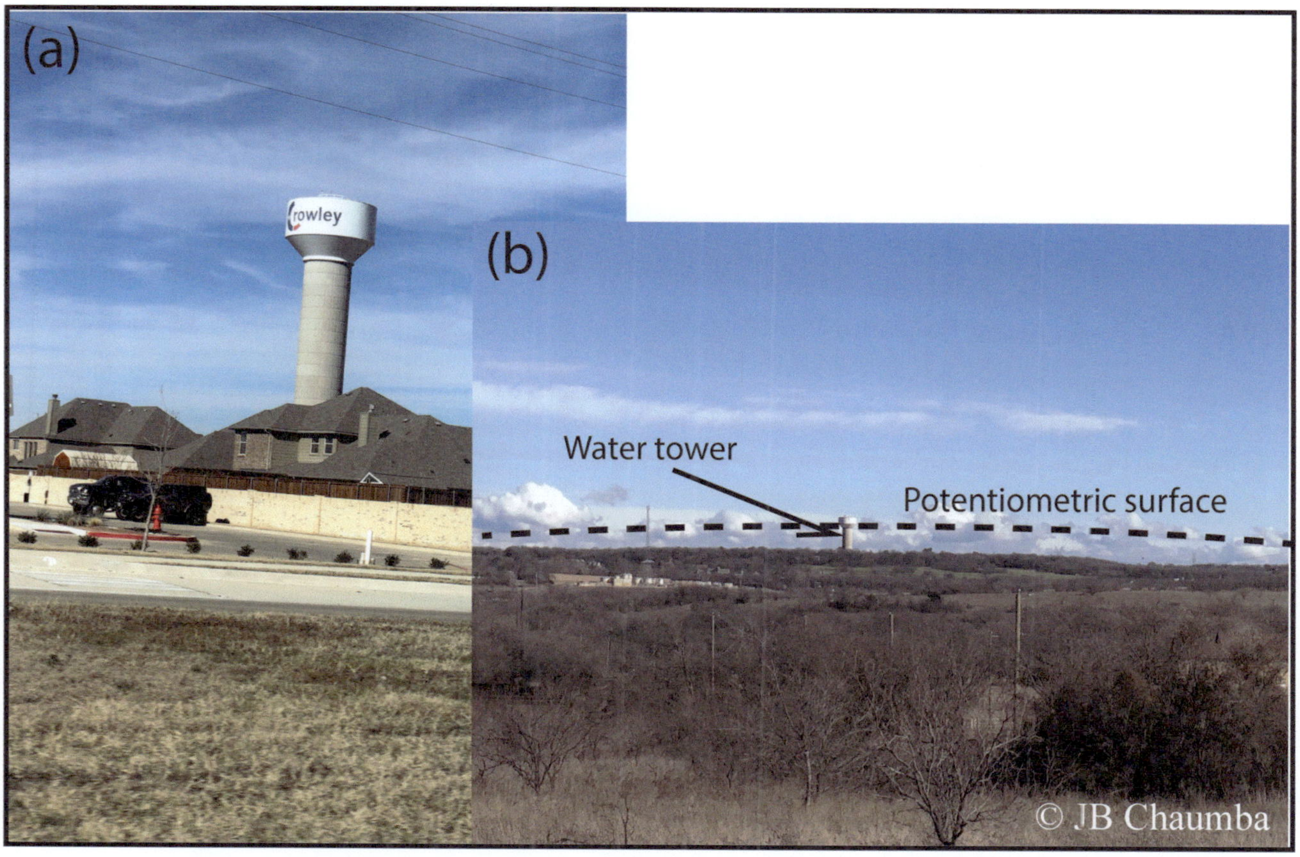

Fig. 13.9 **a** Photograph a water tank in Crowley, Texas. In order to be supplied with water from the water tank, the houses must be below the potentiometric surface of the water tank

These elements and compounds occur in some aquifers in California. Diseases such as dysentery, hepatitis A, cholera, diarrhea, typhoid as well as polio are linked to contaminated groundwater and poor sanitation. Additionally, gastrointestinal infections, vomiting, and nausea are some of the health effects emanating from drinking contaminated water.

Groundwater contamination can be ascribed to both human activities and natural resources. Natural sources of groundwater contamination involve elements and substances such as arsenic, iron, manganese, chlorides, and radionuclides that dissolve in groundwater. Decaying organic matter can also move through and pollute groundwater as particles. Human activities can cause contamination of groundwater via septic systems, spills, releasing of industrial waste, contaminated stormwater or irrigation water, leaking underground storage tanks, and several other factors. Activities like irrigation, high-volume pumping, and artificial recharge lead to mixing of waters which can cause aquifer rocks and sediment to react and release contaminants. For residents of coastal regions, seawater can seep into freshwater aquifers caused by either human activities or natural processes and contaminate it through a process called **seawater intrusion** (Fig. 13.10).

Seawater intrusion is caused by decreases in groundwater levels or rises in seawater levels can cause seawater intrusion. The height of the freshwater in the aquifer is lowered to produce a cone of depression. Saltwater is known to rise 12 m (40 ft) for every 0.3 m (1 ft) of freshwater depression, and it leads to the formation of a cone of ascension (Fig. 13.10). Seawater intrusion compromises water quality both at the pumping well sites as well as affecting portions of the aquifer that are not experiencing groundwater withdrawal and other nearby well sites.

13.7 Impact of Groundwater Extraction and Mining from Karst Topography: Example of Hideaway Hills Neighborhood Near Rapid City, South Dakota

Pollutants that affect groundwater include leakage from both underground storage tanks and containers on land. Gasoline, pesticides, diesel, solvents, and heavy metals can leak and lead to contamination of groundwater. In addition, radioactive waste from nuclear power plants, mining of uranium, as

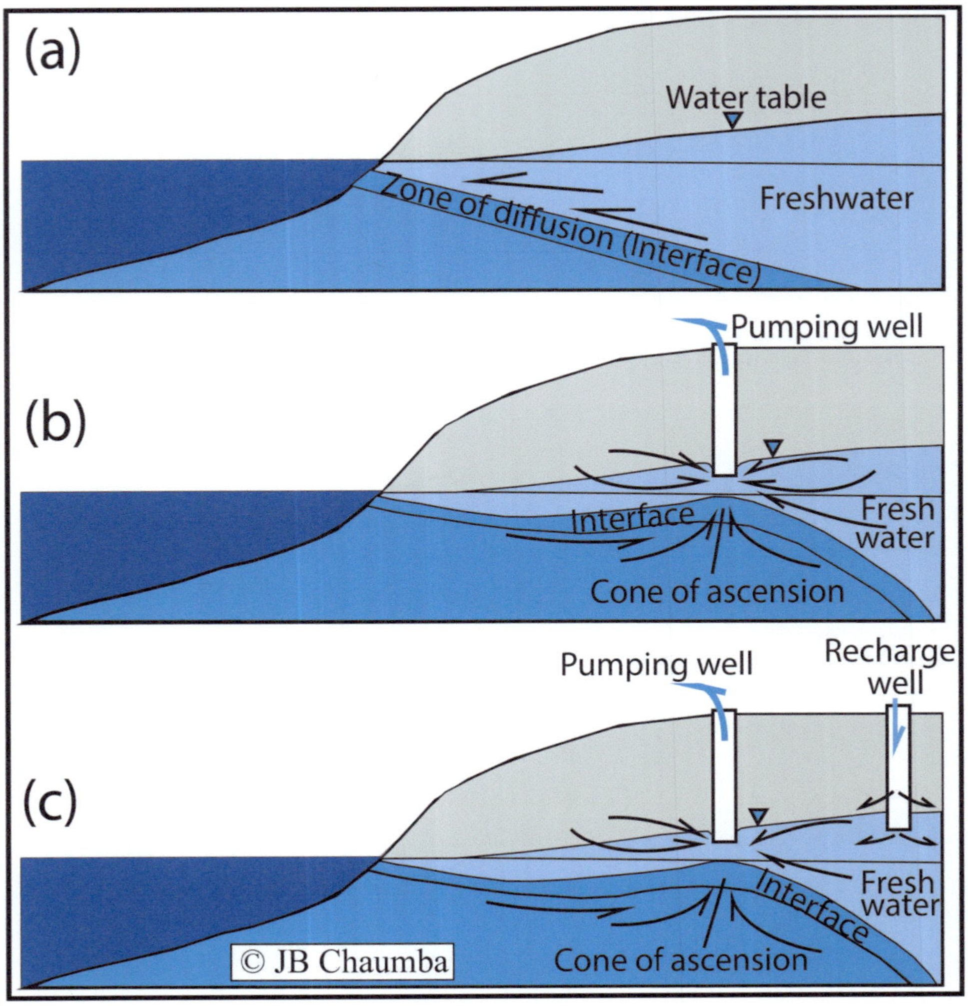

Fig. 13.10 Diagrams illustrating the interaction between fresh and sea water in coastal aquifers from a scenario where there is no withdrawal of groundwater (**a**) to one where there is pumping accompanied by formation of a cone of ascension (**b**, **c**). In **c**, recharging groundwater is conducted through a recharge well to minimize the effect of seawater intrusion

well as waste from livestock and other sources can contaminate groundwater. Iron, manganese, arsenic, chlorides, and radionuclides which occur naturally can also contaminate groundwater. Excessive withdrawal of groundwater at rates which cannot easily be replenished by precipitation, called **mining of groundwater**, is a serious threat particularly in semi-arid and arid regions. Mining groundwater is associated with other environmental problems such as lowering the regional and compaction of the ground which causes cracks in buildings due to **land subsidence**. Furthermore, the quality of water for drinking, irrigation, and other human needs is affected by groundwater. Frequent testing of groundwater that is pumped and used in households for contaminants is recommended. Groundwater quality has a bearing on ecological systems such as caves and sinkholes. Aquatic life is impacted by groundwater pollution. Excessive withdrawal of groundwater at rates higher than replenishment rates can have serious consequences such as lowering of the water table (Figs. 13.1 and 13.3) which can leave once prolific wells dry. Land subsidence is another consequence of excessive pumping of groundwater where pore pressure between grains is significantly reduced in unconsolidated sediments and some sedimentary rocks. Withdrawal of groundwater for irrigation, population growth, and industry is increasing with increase in world population, which is also increasing the risk of land subsidence. Land subsidence has a devastating effect on infrastructure such as tilted buildings and telecommunication lines, broken roads, and rail roads. Several buildings in Mexico City, Mexico's historic center, the Leaning Tower of Pisa in Italy, and the Millennium Tower in San Francisco, California, USA, are examples of buildings that were severely impacted and tilted due to ground subsidence.

In areas where oil and gas extraction is booming business, groundwater contamination occurs from hydraulic fracturing, or fracking, of rocks to enhance recovery. Several

Fig. 13.11 Oil and gas rigs targeting the Haynesville shale near Coushatta, Louisiana, USA (**a**) and near Catarina, Texas targeting the Eagle Ford shale in South Texas, USA (**b**), in 2023

rigs exploiting oil and gas-bearing formations in sedimentary formations (Fig. 13.11a and b) promptly maximize their profits through operating 24 h per day and seven days per week through two 12-h shifts per day. Pollution of groundwater as well as contribution to global warming from the greenhouse gas methane that is releasing from this industry is a huge concern to opponents of fracking. Proponents of the fracking, however, cite job creation, reduction of energy costs, and reducing dependence on oil and gas from other sources or even on the few countries with abundant oil and gas reserves.

Hideaway Hills, South Dakota, is a subdivision located a few miles northwest of Rapid City that was developed from 2002 to 2004. The land was previously state land prior to the establishment of this neighborhood. Where this subdivision now stands, gypsum was mined for use at a nearby cement plant which is also owned by the state. After mining operations ended, the state of South Dakota sold the land on which 94 subdivision homes were constructed, but held on to the subsurface, and it did not disclose it had compromised, through mining, the soil's natural ability to hold up the surface. The cause of sinkholes at the subdivision (Fig. 13.12) is an abandoned underground gypsum mine located beneath the northeastern part of the subdivision that was not properly sealed. Another open pit mine, 12 m (40 ft) in depth, is located in another corner of the subdivision.

In parts of the neighborhood, the land slightly sunk over time since the subdivision was constructed and a sinkhole subsequently formed beneath a back porch. Conditions escalated after a large sinkhole developed in April 2020, which prompted residents to the discovery of a large and improperly sealed mine beneath the northeastern part of the subdivision, and a 40 ft (12 m-deep) deep open pit mine in

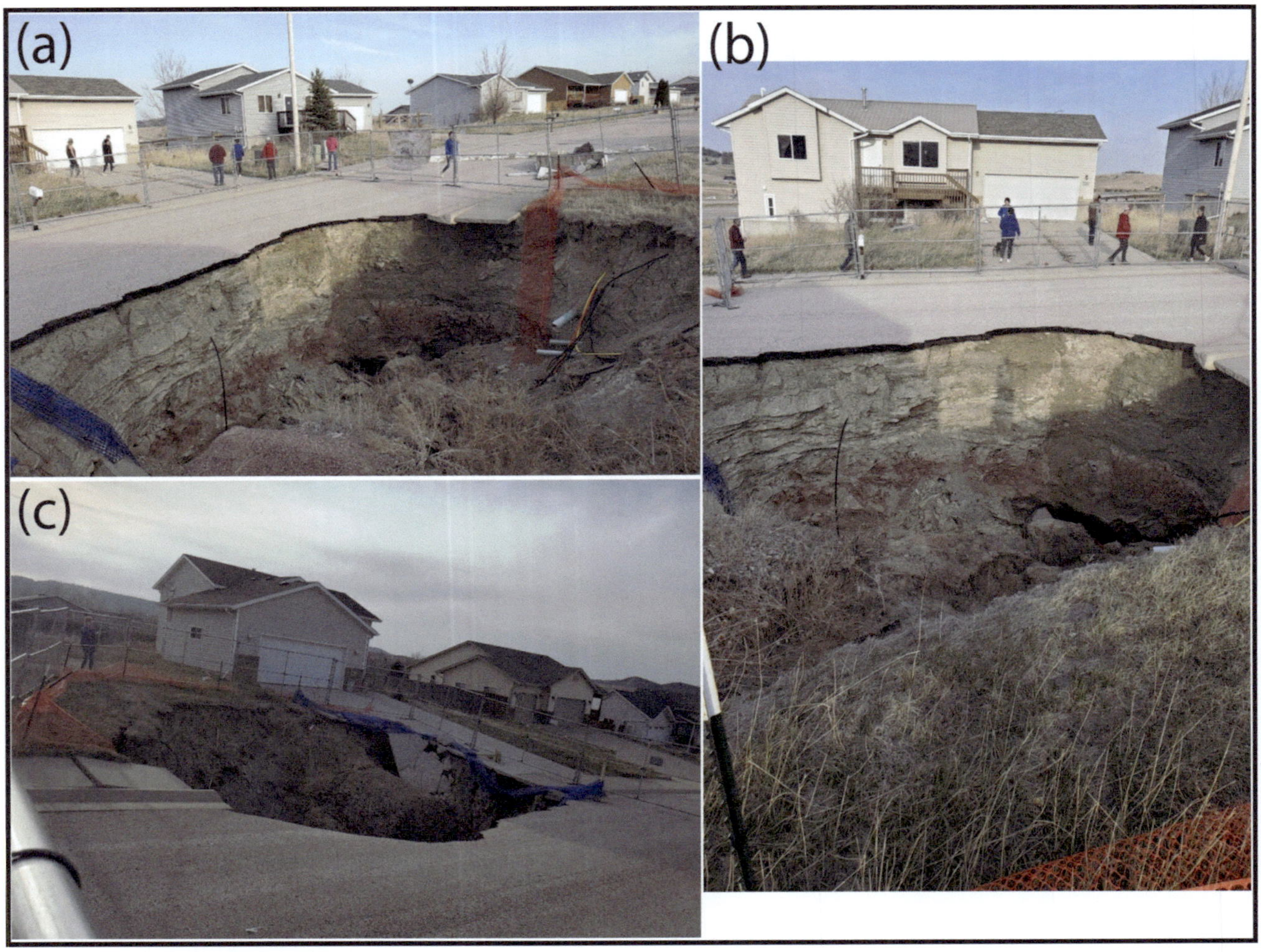

Fig. 13.12 Photographs of sinkholes from Hideaway Hills, South Dakota subdivision that is located a few miles northwest of Rapid City. Gypsum was mined for use at a nearby cement plant which is owned by the State of South Dakota. *Credit* Tonya Junker—AP

another section of the neighborhood. Sinkholes are often formed in areas that are underlain by rocks such as limestone that dissolve and create caves. Sinkholes can also form in areas where underground mining operations remove the underlying earth materials, leaving relatively narrow overlying material above to fall into hole left by mining. This latter scenario is what likely resulted in the generation of sinkholes in the Hideaway Hills, South Dakota subdivision. Approximately 150 residents in this neighborhood sued the state for US$45 million in an attempt to cover the value of their homes as well as legal costs.

We end this chapter with a brief discussion of areas in the United States formed by karst features (Fig. 13.13). The Appalachian region, some sections of the Ozarks, and almost the entire State of Florida account for the major areas that are composed of extensive karst topography. Although some states are more likely to develop sinkholes, soluble rock and potential for the formation of karst topography is generally low in all states in the Appalachians and the Ozarks. As discussed earlier in this chapter, karst topography is characterized by sinkholes, disappearing streams, caves, as well as underground streams. The Appalachian region with karst sizable regions of topography are Kentucky, Pennsylvania, Virginia, West Virginia, Tennessee, North Carolina, Georgia, and Alabama. The mammoth caves in Kentucky, USA, for example, contain 580 km of passages that are connected and they represent the world's largest cave system. Almost all of Florida is composed of karst topography. In the Ozarks, karst topography occurs in parts of Arkansas, Missouri, as well as Oklahoma. Additionally, a greater proportion of Texas is comprised of karst topography. In the south-central region of Texas, carbonate rocks which are highly faulted and fractured occur.

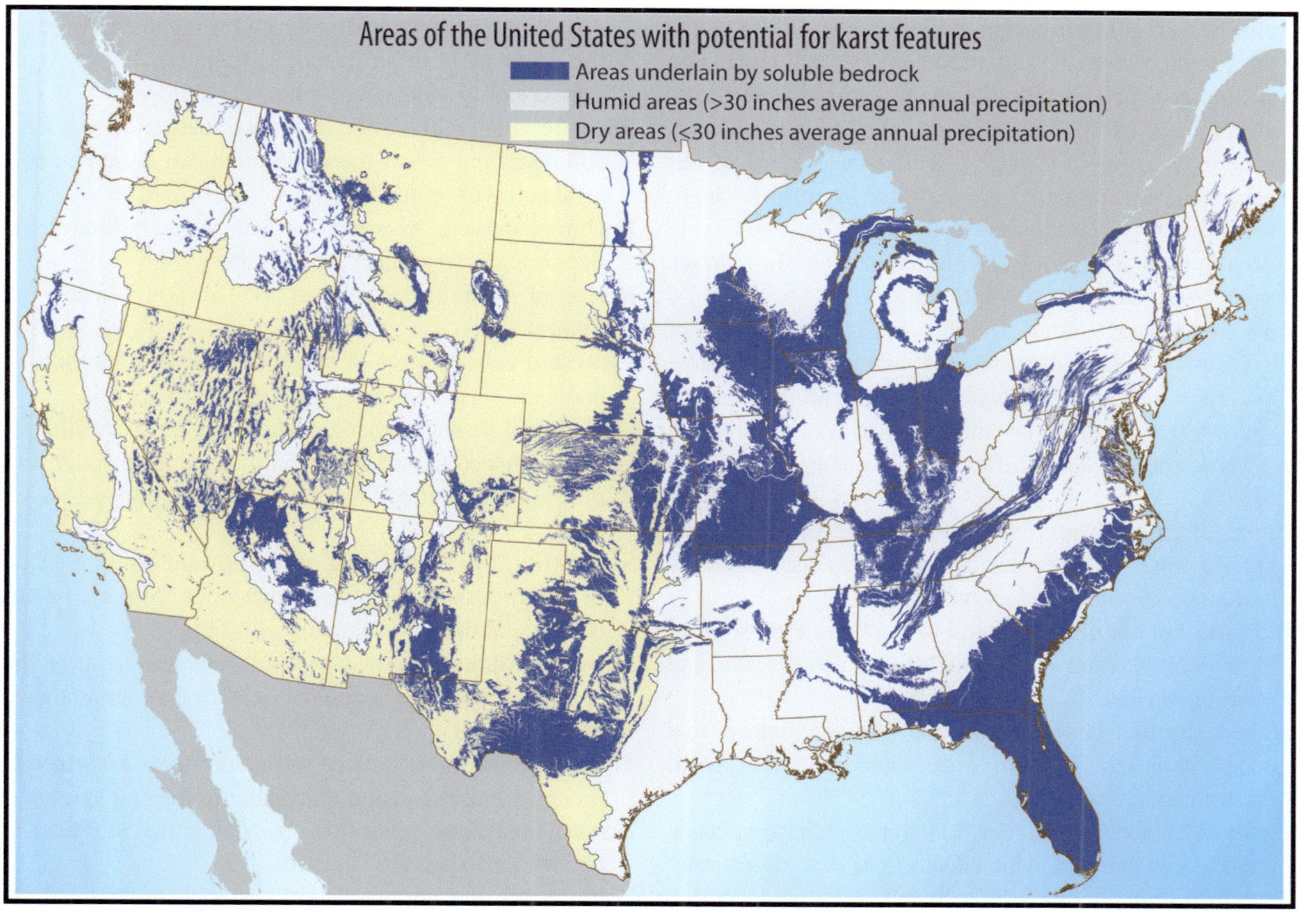

Fig. 13.13 Map showing areas of the United States with potential for karst features (after USGS)

Glossary

Aquifer a rock formation or sediment that stores and releases groundwater and rates high enough to be usable

Artesian spring a spring formed from water under pressure flowing to the Earth's surface through a fault of gap in an impermeable rock

Artesian well a well that is drilled into a confined or artesian aquifer and allows water, under pressure from overlying impermeable rock, to flow to the surface

Aquitard an underground rock formation that restricts groundwater flow

Cave an underground cavity that is large enough for a human to enter formed by the dissolution of limestone, dolomite, or gypsum by groundwater

Column a vertical column of rock a cave formed when a stalactite and a stalagmite join to create a single structure extending from the floor to the ceiling in a cave

Cone of depression an inverted conical-shaped lowering of the water table occurring around a pumped well when water is pumped out. When the groundwater is pumped from a well, the water level decreases accompanied by a gradient producing a depression resembling a cone

Capillary fringe a narrow subsurface zone in which groundwater seeps up above the water table by capillary action to fill up pores

Creek a stream or minor tributary of a river

Discharge zone areas where the water table coincides with the land surface

Discharge area where groundwater flows out of an aquifer naturally as springs or contributes to stream flow

Dissolution a process whereby soluble rock like limestone dissolves gradually in the presence of water leading to breaking down of the rock through chemical weathering to form karst topography

Doline an enclosed small to medium sized closed depression occurring in karst landscapes. Also called a sinkhole, a doline is the most common landform in karst regions

Fumaroles openings in the Earth's surface that emit volcanic steam and volcanic gases

Geyser a spring with an intermittent discharge of water fountain that ejected turbulently into the atmosphere and accompanied by steam

Groundwater water which occurs underground in the pore spaces between soil particles and/or fractures in rocks

Groundwater contamination pollution of water beneath the ground surface through introduction of pollutants into an aquifer which degrades its quality and renders it water unsafe for drinking or other uses

Got springs springs of water reaching the surface of our planet due to heating of groundwater by magma chambers within the Earth's crust

Infiltration the process by which water on the surface water percolates or enters into the soil

Land subsidence gradual settling, or sudden sinking, of the Earth's surface, due to the extraction of underground materials such as groundwater, minerals, oil, and natural gas due to pumping or mining activities, resulting in lowering of the land surface and/or compaction of the soil

Mining of groundwater extraction of the non-renewable ground water as the water cannot be replaced naturally at the rate at which it is extracted

Mud pots hot springs composed of bubbling mud sourced from a combination of volcanic activity, acidic water, and microorganisms

Perched aquifer body of water of limited extend that occurs above the regional water table and is a type of unconfined aquifer

Permeability measure of the interconnectedness between open spaces in a rock which determines how easily water can flow through a rock or sediment

Porosity percentage of voids, or open spaces, in a rock or sediment

Potentiometric surface an imaginary surface that represents the level to which water would rise in a cased well drilled into a confined aquifer and it is the equivalent of the water table in an unconfined aquifer. It depicts the pressure of groundwater within that aquifer

Phreatic zone rocks, soil, or a combination of rocks and soil occurring below the water table that is permanently saturated with groundwater

Recharge process by which surface water infiltrates into the ground to replenish groundwater

Recharge area the region on the surface where water infiltrates into the ground

Saturated zone portion of an aquifer that contains both air and water between the spaces of the rocks

Stalactite a formation hanging from the roof or ceiling of a cave shaped like an icicle that formed from water dripping through the ceiling and depositing minerals in the process

Stalagmite a formation that grows upwards from the floor of a cave produced by water that drips from the roof or ceiling and deposits minerals as it falls

Seawater intrusion process whereby seawater moves into coastal freshwater aquifers leading to groundwater contamination

Travertine terraces terraced formations that are formed by precipitation of calcium carbonate by mineral springs

Unsaturated zone portion of an aquifer that contains air between the spaces of the rocks

Water table the boundary between the unsaturated zone and the saturated zone beneath where water-saturated ground meets unsaturated ground

Deserts, Glaciers, and Associated Landforms

14

Abstract

The definition of desertification is provided at the beginning of the chapter, followed by a discussion of its causes. The distribution and types of deserts on Earth is then discussed with the aid of several maps. Erosion and deposition in deserts and associated landforms such as sand dunes and loess are also described with the aid of diagrams and photographs. Unique landforms produced by erosion, transportation, and deposition processes occurring in desert environments are discussed and illustrated. The consequences of desertification, such as loss of agricultural land and dust storms, are described. Some sways that help in avoiding and alleviating desertification are summarized based on the 15 Sustainable Development Goals recommended by the United Nations. Glaciation, types of glaciers and associated landforms are also discussed, as are landforms formed by glacial erosion and deposition. The chapter also discusses the Laurentide, Fennoscandian, and Antarctic ice sheets as well as Milankovitch cycles. Also discussed in concluding the chapter is recent evidence which disputes that long-term climatic predictions on which Milanković's cycles is based upon may be more inaccurate than realized in the scientific community.

Learning Goals

Studying this chapter will enable you to:

- Define desertification and its causes
- Discuss wind erosion and transport
- Recognize deposits associated with wind deposits
- Discuss how dunes form and migrate
- Identify the different types of dunes
- Distribution and characteristics of desert landforms
- Discuss the origin and formation of glaciers
- Define glacial budget and explain how glaciers move
- Describe how glaciers erode and transport
- Differentiate continental glaciers from valley glaciers
- Describe how glaciers erode, transport, and deposit sediment
- Recognize glacial deposits and identify them
- Moraines, glacial drift, glacial till, and drumlins
- Recognize landforms made of stratified drift
- Define and recognize glacial lakes
- Ice ages and their causes: Milankovich theory

Supplementary Information The online version contains supplementary material available at https://doi.org/10.1007/978-3-032-20231-4_14.

14.1 Introduction

This chapter deals with extreme realms of hot and cold regions. We start off hot, by discussing dry regions parched with heat with having little or no rain also called arid regions in the first part of the chapter. In the second part of the chapter, we cool things down and discuss cold regions and the causes of their distribution and origin.

A desert can be defined as a dry region with limited availability of water for plants and other organisms and it is therefore characterized by little to no vegetation. Deserts receive less than 250 mm (10 inches) of rain, snow, mist, or fog on average annually. In a desert, the amount of water that evaporates is typically much greater than the amount of rain received. Consequently, sparse or no vegetation is encountered in deserts (Fig. 14.1a and b). Even trees are usually absent in deserts, as shown in Fig. 14.1a. Vegetation, if present, such as the trees visible in the swamp, Sossus Vlei in the Namib Desert in Namibia shown in Fig. 14.1b, utilizes water fed to the swamp by the Tsauchab River.

Deserts are native habitats for most cacti (Fig. 14.2). Cacti store the water in their stem, or pads, as their waxy coating prevents loss of stored water. Cacti thrive where

J. B. Chaumba, *Introductory Physical Geology*, Springer Textbooks in Earth Sciences, Geography and Environment.
https://doi.org/10.1007/978-3-032-20231-4_14

Fig. 14.1 **a** Photograph of stunted grass east of Sun Valley, Arizona (credit: S. D. Clark). **b** Photograph of Roter Kamm impact crater from summit of Aurus Mountains Namibia (credit: D. L. Reid)

Fig. 14.2 Cacti thrive in desert environments. **a** Cactus from the Namib Desert, Namibia (credit: D. L. Reid), **b** Cactus from South Texas near the town of Encinal, Texas

there is plenty of light, high temperatures, low moisture, and good drainage. These plants adapted to deserts and arid grasslands.

Deserts, however, play a major role in their ecosystem which supports unique animals and plants that adapt to severe conditions. Dry or seasonal streams, such as Dry Creek in South Texas (Fig. 14.3a and b), are a characteristic of deserts unless the streams receive water from sources outside of their catchment areas. Are deserts fixed in space and time? Let us begin our study on deserts defining **desertification**. The process of land degradation that occurs in dry, semi-arid, and even dry sub-humid regions due to a wide range of climatic and human-induced factors is called desertification. Desertification is now considered to be one of the major environmental problems presently affecting our planet. Desertification is the process by which fertile land progressively becomes desert through desertification. Climate change and human activities are both responsible for desertification. Climate change, in particular, acts to multiply the factors responsible for desertification.

Precipitation in deserts is only 25.4 cm (10 in.) or less per year. Extreme diurnal temperature changes are characteristic of deserts as day temperatures average at least 38 °C (100°F) whereas night temperatures decrease by at least 4–10 °C (40–50°F). Humidity in deserts is low, hence after sunset the air rapidly cools. Since deserts are arid, they have very little water that is available for plants and other organisms. Hot deserts are distributed in the tropics and on western sides of continents with the greatest expanse of hot deserts located in North Africa through the Middle East

Fig. 14.3 **a** Photograph of Dry Creek signage where Texas 774 crosses this creek, and the creek with very little water flowing through it during the rainy season in July 2024. **b** Photograph of Dry Creek from the bridge on Texas 774 showing the low volume of water in its channel

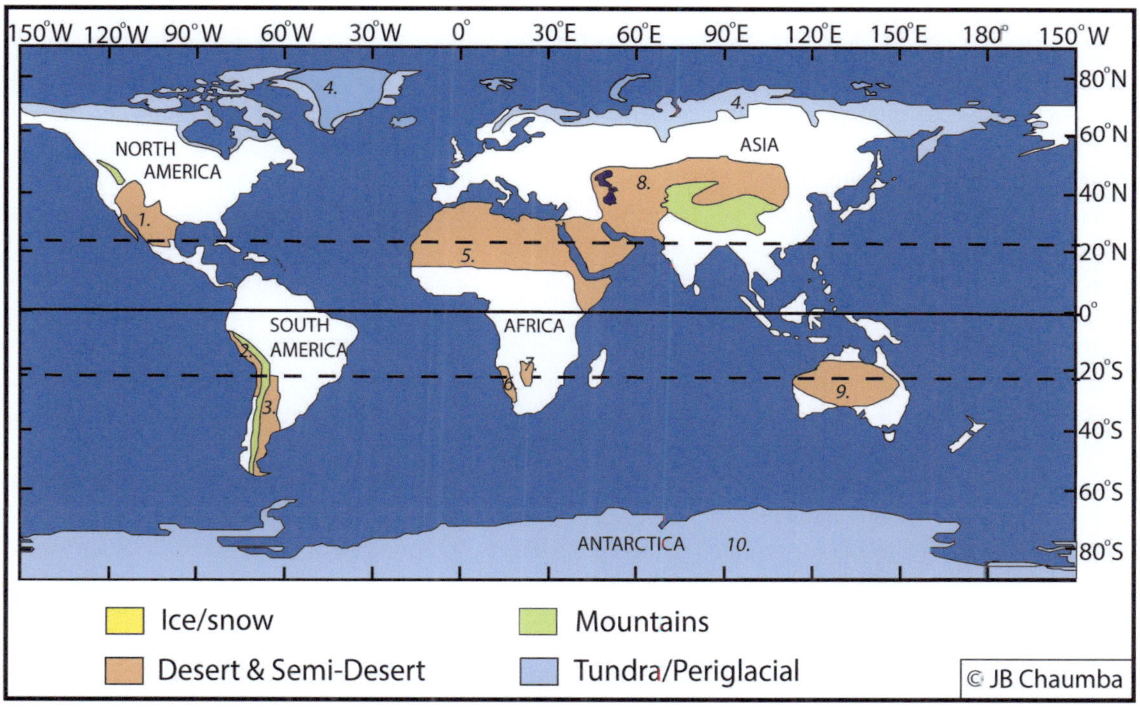

Fig. 14.4 Map of the world showing the distribution of deserts on Earth. Deserts indicated are: 1. Great Basin, Mohave, Sonoran, Chihuahuan, and Colorado Plateau, 2. Acatama, 3. Patagonian, 4. Arctic, 5. Sahara, 6. Namib, 7. Kalahari, 8. Syrian, Arabian, Karakum, Kyzylkum, Gobi, Taklamakan and Thar, 9. Great Sandy, Little Sandy, Tanami, Tirari, Sturt Stony, Gibson, Great Victorian, Strzelicki, and Simpson, and 10. Antarctic

and extending into western Asia (Fig. 14.4). There are a variety of plants and animals found in deserts that adapt to survive in the extreme prevailing conditions. Some plants, for example, store water in their stems, whereas others develop large root systems that gather water. Additionally, there are some animals that have adapted to dealing with the extreme high heat and scarce water in deserts.

14.2 Causes of Desertification

Poor agricultural practices, bad livestock practices, and human activities all cause desertification. Factors such as failure to practice crop rotation, use of both fertilizer and pesticides fertilizers, among other weed or pest control chemicals are **poor agricultural practices,** which cause desertification. Poor management or irresponsible management of both surface water and groundwater (see Chap. 12), vegetation, and natural resources are an example of over-exploitation of natural resources which also causes desertification. Desertification is land degradation in drylands. Drylands are arid, semi-arid, and dry sub-humid areas. Several factors such as human activities and climatic variations cause desertification. Overcultivation of marginal land, application of chemical fertilizers and pesticides, and crop rotation are also poor agricultural practices that contribute to desertification.

Overgrazing often leads to severe erosion of the soil as well as failure to prevent the regeneration of vegetation. Rates of soil erosion are, therefore, accelerated by **bad livestock practices** such as overgrazing and unchecked cutting down of trees for domestic use as the vegetation which binds the soil together through roots of vegetation. Through **overgrazing** by livestock, the topsoil is eroded away faster from hooves of animal hooves as well as reduction or removal of grass which help bind the topsoil to the ground. Removal of vegetation that binds the soil makes the affected area susceptible to soil erosion by wind and water.

In addition, regeneration of plants is prevented rendering the soil unusable and susceptible to erosion. The risk of fires and droughts, for example, is greatly increased by **deforestation**, among other factors. The stripping away of trees and bushes for fuelwood and timber, or to clear land for cultivation, causes desertification. Failure to implement **crop rotation**, and the use of chemical fertilizers and pesticides, as well as over-cultivating marginal land also cause desertification. Finally, **inappropriate irrigation** causes accumulation of salts in the soil which can rapidly transform productive land into salt flats, contributing to desertification. Desertification often results in the loss of land utilized for agriculture, livestock, and heat. Desertification often forces people to migrate to new locations that are still not badly affected.

Poor irrigation management often leads to a buildup of salts and chemicals in the soil and water, causing the soil to be less fertile and hinder crop growth, a process referred to salinization. Leaks can occur in irrigation systems caused by pests, yard work, or shifts in the soil. Inappropriate irrigation is caused by irrigation systems that are poorly designed often resulting in uneven water distribution resulting in over- or under-irrigation. In addition, applying too much water results in waterlogging and aeration deficit, which harm plants. New plantings can dry out quickly between waterings due to underwatering such that trees and shrubs that depend on lawn irrigation decline. Failure to practice correct irrigation practices often leads to degradation of lands under irrigation.

Global warming can result in climate fluctuations which are a result of human **greenhouse gas emissions** which greatly contribute to desertification.

14.3 Distribution of Deserts on Earth

Deserts don't occur randomly on our planet, but there are factors that are responsible for their formation which play a role in determining their distribution. However, desertification has a role to play in increasing the sizes of the deserts. If one examines a map showing the distribution of deserts around the world, it is noticeable that they are located in a belt centered on approximately at the 30° North and 30° South lines of latitude (Fig. 14.4). This distribution of deserts will be discussed later in this chapter and here we only mention that the global circulation patterns are responsible for rising of hot and humid air masses at the equator and their descending as dry air masses at the 30° North and South latitudes. It is the descending of the dry air masses that create desert environments at the 30° North and South latitudes.

Other factors can contribute to desert formation. Deserts that occur in the heart of continents are called **interior deserts**. Interior deserts are sometimes called inland deserts. Interior deserts form due to lack of moisture-filled winds that reach them; hence some deserts tend to form in regions that are far from oceans. By the time air masses from coastal areas reach the interior, they have lost all their moisture. Deserts typically occur in regions located far from oceans as they lack access to air from the sea that is moisture laden. Any winds that blow over these interior regions would have lost the bulk of their moisture by the time they travel far inland. Consequently, very little precipitation is received by interior regions which experience arid conditions that characterize deserts. The further from the ocean a region is located, the less the moisture that is available to form rain.

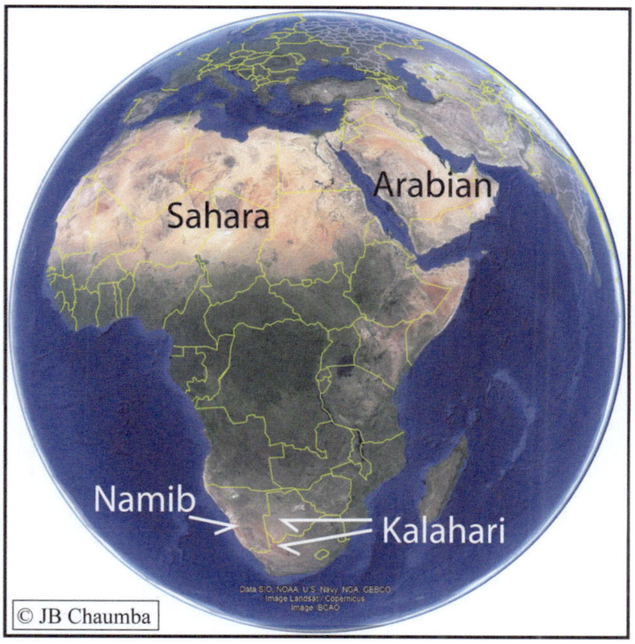

Fig. 14.5 Google Earth map showing the distribution and extend of the Sahara Desert in North Africa, the Namib Desert, and the Kalahari Desert in southern Africa

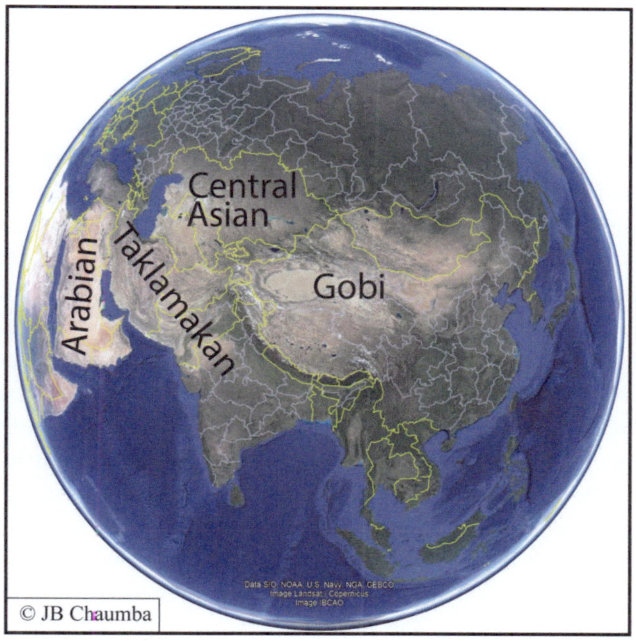

Fig. 14.6 Google Earth map showing the distribution of deserts in Asia. The location of the Gobi, Central Asian, and Taklamakan Deserts are shown

Hot and dry deserts or arid deserts, semi-arid deserts, coastal deserts, and cold deserts are among the common types of deserts. The Sahara Desert (Fig. 14.5) is an example of an arid desert. Year-round warm and dry temperatures characterize hot and dry deserts. The Mojave Desert in the southwest of the United States is another example of an arid desert.

Semi-arid deserts are slightly cooler than hot and dry deserts. The summers in these deserts are moderately long and dry. Similar to hot deserts, the winters in semi-arid deserts are often accompanied by low concentrations of rainfall. Summer temperatures usually average between 21–27 °C and (70–81 °F). Temperatures in semi-arid deserts rarely exceed 38 °C (100 °F). In semi-arid deserts, the dry and long summers are followed by winters which are often accompanied by some rain. Semi-arid deserts are encountered in North America (Fig. 14.7), Asia (Fig. 14.6), Europe, and Greenland (Fig. 14.4). Semi-arid deserts receive precipitation of 250–500 mm (9.8–19.7 in) per year, which is more than that received in arid deserts. Vegetation in semi-arid deserts is sparse, and includes cacti (Fig. 14.2) and grasses that have adapted to dry conditions. Semi-arid deserts occur in various parts of the world such as North America, Australia, Europe, and Asia.

The colloquial name for the vast and unpopulated mostly arid areas that comprise Australia's interior and remote coasts is termed the Australian Outback. The Outback encompasses the Red Centre of the Northern Territory. Although the Kalahari Desert in Southern Africa (Fig. 14.5), the sagebrush steppe of the Great Basin in the United States (Fig. 14.7), the central plains of Spain, portions of the Australian Outback, as well as the Kazakh semi-desert in Northern Asia all receive slightly more rainfall than a typical desert, they all still experience dry conditions and they are characterized by scarce vegetation.

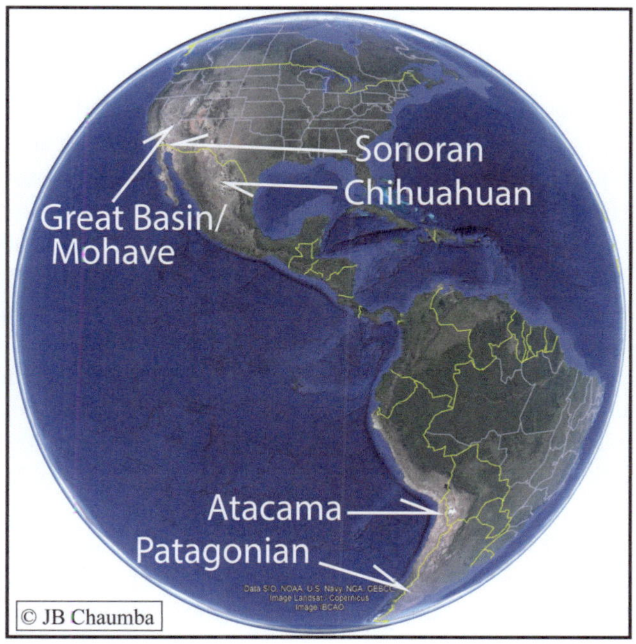

Fig. 14.7 Google Earth map showing the distribution of deserts in both North America (Great Basin, Mohave, Sonoran, and Chihuahuan Deserts) and Acatama and Patagonian Deserts in South America

The Atacama Desert is largely located in Chile, South America, but is also extends into parts of Argentina, Bolivia, and Peru (Fig. 14.7). It is the driest desert in the world. Located between two high mountain chains of the Andes and the Chilean Coast Range, moisture advection is prevented from reaching the Acatama Desert both from the Pacific and the Atlantic Oceans, making it a rain shadow from two-sides. Also located in South Africa, in the Patagonia region of southern Argentina and parts of Chile, is the Patagonian Desert (Fig. 14.7). The Patagonian, another rain shadow desert, was formed from the rain-shadow effect produced by Andes Mountains to the west creating an arid area to the eastern leeward side of these mountains (Fig. 14.7). Although continentality is not particularly pronounced in the Patagonia desert, the rain shadow effect still led to the formation of the Patagonian desert.

Coastal deserts are a bit more humid than other types of deserts. Although heavy fog blows in from the coast, rainfall is still rare. The Atacama Desert of Chile in western South America (Figs. 14.4 and 14.7) is an example of a coastal desert. Although small portions of this desert extend into western Bolivia and southern Peru, most of the Atacama Desert, however, occurs within Chile. Cold deserts are still dry but have extremely low temperatures in comparison to the other types of deserts. The Antarctic Desert is an example of a cold desert.

A desert that is dominantly comprised of parallel, large, long, and linear sand dunes that form a sea of dunes is called an erg desert. An example of an erg desert is the Simpson Desert in Australia (Fig. 14.8). The Simpson Desert encompasses portions of South Australia, Northern Territory, and Queensland in Australia. In the Simpson desert, sparse vegetation akin to spinifex grasses is encountered. In erg deserts, spinifex vegetation helps to stabilize the sand dunes.

Air above some ocean currents is cooled by **cold ocean currents**, hence contributing to the formation of deserts because cold ocean currents significantly reduce the air's ability to hold moisture. Less precipitation, therefore, falls on nearby land which leads to arid conditions and formation of desert. The air becomes too cool such that it is incapable of carrying appreciable amounts of water vapor prior to reaching the continent. Cold ocean currents create deserts due to air cooling down as it passes over a cold ocean current. Furthermore, cold air holds less water vapor in comparison to warm air. Instead of producing precipitation, cooled air often leads to the formation of **fog** along coastlines. Fog is composed of tiny water droplets that are too small to fall as rain. In coastal regions where cold currents flow close to the shoreline, coastal deserts are formed. The amount of water the air can hold is reduced by cold ocean currents. Consequently, water tends to fall as precipitation over the ocean before it reaches land. Land adjacent to the ocean, therefore, becomes arid and forms a desert. Cold ocean currents are capable of diverting air filled with rain away from coastlines, helping create a desert. Some examples of deserts that are partially formed by cold ocean currents are the Namib Desert in Namibia from the cold Benguela current and the Atacama Desert in in Chile from the cold Humboldt current. Cold ocean currents can form a layer of fog that drifts onto land. Fog can be formed by cold ocean currents drifting on land. Cold ocean currents can contribute to desert formation. The seasonal tilting of the Earth in relation to the sun can contribute to desert formation.

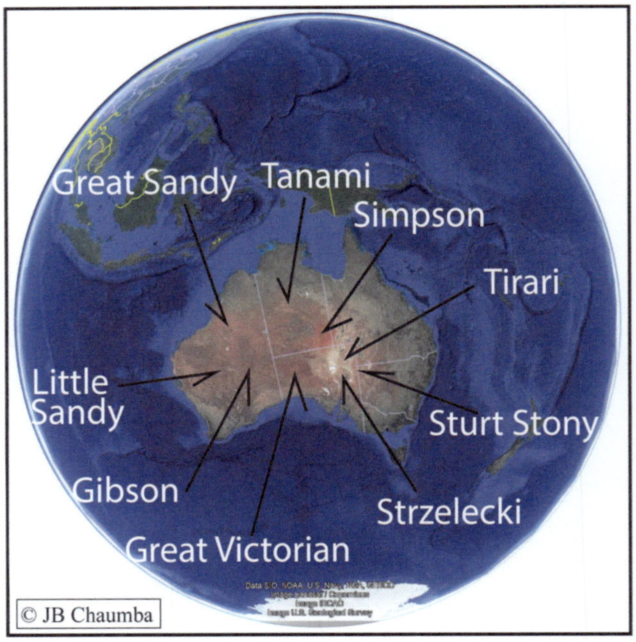

Fig. 14.8 Google Earth map showing the distribution of numerous deserts in Australia. The bulk of the interior of this landmass is comprised of various deserts

14.4 Erosion and Deposition in Deserts

In deserts, the major agent of erosion is wind. Due to little vegetation in deserts, wind erosion is responsible for creating several landforms when the eroded and transported sands are deposited. Typical desert landforms such as **sand dunes** (Fig. 14.9a) and **loess** are produced by deposition of landforms from wind erosion. Some sand dunes, however, can form at the beach where winds pile up beach sand which one must climb over before having a relaxed time at the beach (Fig. 14.9b). A sand dune is a ridge or mound of sand that is produced by wind depositing sand on a beach or downwind of an obstacle that is sheltered. Sand dunes are found in deserts, on beaches, and in some semiarid, or **steppe**, regions. Sand dunes are created when strong wind deposits sand on top of each other until a mound is formed

(Fig. 14.9). Once the first mound is formed, more sand piles up on the windward side until the edge of the dune collapses under its own weight.

A beach is where most sand dunes often form in bands that are parallel to the beach. Relatively taller dunes are encountered with increasing distance away from the beach (Fig. 14.9b).

As can be observed on Fig. 14.9b, dunes are often home to a wide range of plants and animals. Winds transport and deposit sand grains form sand dunes of varying shapes and sizes. Dunes typically have a crescent or "U" shape. Ocean grasses, wildflowers, and even shore birds are common where dunes develop. Dunes are not stationary landforms, but they are constantly changing structures. The size of dunes can even get reduced as a consequence of lack of supply of sand or when the sea level rises.

14.4.1 Types of Sand Dunes

Areas where dunes are produced can be simple landforms where isolated dunes of a basic type form, or compound where larger dunes with smaller dunes of the same type on top, or even complex dunes characterized by a combination of various types. The shape of sand dunes is dependent on the type of sand, both the wind direction and its strength and the environment. If direction from which wind blows from changes, the configuration of the sand dunes also changes. The availability and amount of sand also govern the shape of the sand dunes. Consequently, there are a variety of dunes that can be formed which include barchan, barchanoid, transverse, parabolic, linear, parabolic, and star.

One of the most common types of dunes found in sandy deserts worldwide is a barchan dune (Fig. 14.10a). A **barchan** dune has a crescent shape, and it is formed where the wind predominantly blows from one direction (Fig. 14.10a). Two arms or horns or tips that point downwind make a barchan dune one of the most common types of dunes found in sandy deserts on our planet. The prevailing wind for this diagram is from left to right based on the sand dune where the windward is to the left and the steeper leeward slope is on the right (Fig. 14.10a).

A **barchanoid** dune (Fig. 14.10b) is a crescent-shaped dune ridge that is composed of an amalgamation of many barchans or crescent-shaped dunes (Fig. 14.10b). Barchanoid dunes are also referred to as crescentic ridges or barchan chains. They form when barchan dunes join together as the sand supply increases. Barchan dunes have a tendence to align along a plane perpendicular to the wind direction. Should this line become linear, the result is a **transverse dune** (Fig. 14.10c). Transverse dunes march downwind as ridges. Dunes migrate downwind as particles of sand avalanches on downwind leeward slip face, releasing sand through periodic avalanches at a time.

A U-shaped dune with elongated arms that trail upwind with a convex nose that is trailed by elongated arms at the downwind end is called a **parabolic dune** (Fig. 14.10d). It is often created due to strong winds that erode a fraction of sand with vegetation. The arms of a parabolic dune are held place by the vegetation as the dune migrates forward. The crest of a parabolic dune points upwind. An elongated and large sand dune that is generated parallel to the direction of the wind is called a **longitudinal dune** (Fig. 14.10e). Longitudinal dunes form in areas with strong winds from two slightly different directions (Fig. 14.10e) and they are characterized by straight symmetrical rides. Sand supply in areas where longitudinal dunes are formed is limited. Longitudinal dunes are also referred to as linear dunes. **Star dunes** are depositional landforms produced in an area where wind blows from different directions within a given year (Fig. 14.10f). Other types of dunes such as dome and blowout can also be produced. In an effort to protect the environment where dunes form, visitors are recommended to stick to trails and boardwalks. In addition, uprooting vegetation should be avoided. Sticking to designated areas also helps prevent soil erosion and helps protect the environment in which dunes prevail.

Shown on Fig. 14.11 are photographs of sand dunes from various areas. Barchanoid dunes from the Namib Desert are shown on Fig. 14.11a. The slip face is the steeper one in the downwind direction (Fig. 14.11a). The barchanoid dunes are characterized by steep slip faces occurring on the downwind side whereas facing the wind are the gentle slopes to the left (Fig. 14.11b). Figure 14.11b and c are photographs of further barchanoids from dunes from Sand Dunes National Park in Southern Colorado, USA. The slip faces of the dunes, similarly, are also the steeper ones in the downwind direction (Figs. 14.11b, c). Massive dunes are shown on Fig. 14.11d. The dunes display cross bedding on a large scale, and they were formed by winds which reversed their directions. These huge dunes are also from the Great Sand Dunes National Park and Preserve in southern Colorado, USA.

14.5 Desertification and Its Causes

According to Worldmeter, the world population in December 2025 was 8.3 billion people. To feed this increasing population requires, as is the case today, increasingly clearing natural vegetation to expand the acreage for cultivation of crops. This increase in acreage for cultivation encroaches on land on the fringes of deserts thereby causing **desertification**. The balanced ecosystems at the margins of desert get disrupted, destroying the delicate buffer zone between deserts and relatively more humid regions.

Fig. 14.9 **a** Sand dunes developed at Sand Dunes National Park, Colorado, USA (credit: D. T. Nelson). **b** Sand dune developed by the beach at Amelia Island, Florida, USA, view looking south with the Atlantic Ocean to the left

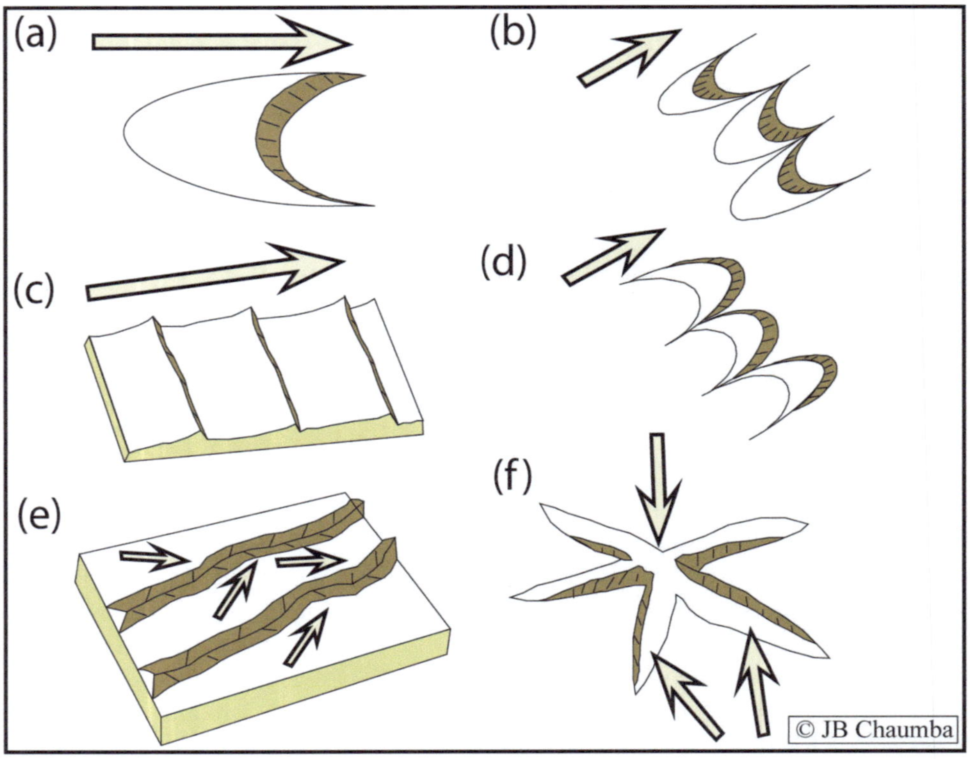

Fig. 14.10 Diagram illustrating the different types of common sand dunes: **a** barchan, **b** barchanoid, **c** transverse, **d** parabolic, **e** longitudinal, and **f** star dunes. Arrows indicate prevailing wind direction. The brown shading with lines indicates the steeper slope and slip face of the dune located in the downwind direction. The unshaded side represents the gentler slope which indicates the upwind direction of the sand dune

Desertification can be defined as degradation of land in arid, semi-arid, and dry sub-humid areas, which are collectively known as **drylands**, resulting from several factors, including human activities and climatic variations. Drylands currently cover approximately 46% of the global land area and they are home to about 3 billion people. Hotspots of desertification, as identified by a decline in vegetation productivity between the 1980s and 2000s, extended to approximately 9% of drylands. In 2015, desertification is inferred to have affected approximately 500 million people, with the highest numbers of affected people located in South and East Asia, the region surrounding the Sahara Desert in North Africa and the Middle East which encompasses the Arabian Peninsula (Fig. 14.4). Desertification also affected other dryland regions.

Both human activities and climate change are the major causes of desertification. Human activities which result in desertification are poor agricultural practices, overgrazing, and deforestation. Failure to practice crop rotation and the use of both pesticides and chemical fertilizers leads to desertification. In addition, inappropriate irrigation also contributes to deforestation through waterlogging, soil salinization, and disruption of natural water cycles as the ability of land to support vegetation is compromised. In addition, soil degradation and reduction in the ability of land to support vegetation are caused by poor agricultural practices. Exposing plants to excessive grazing for long periods of time with no sufficient recovery periods is overgrazing. Erosion of the topsoil and compaction of soil through hooves of animals is a consequence of overgrazing.

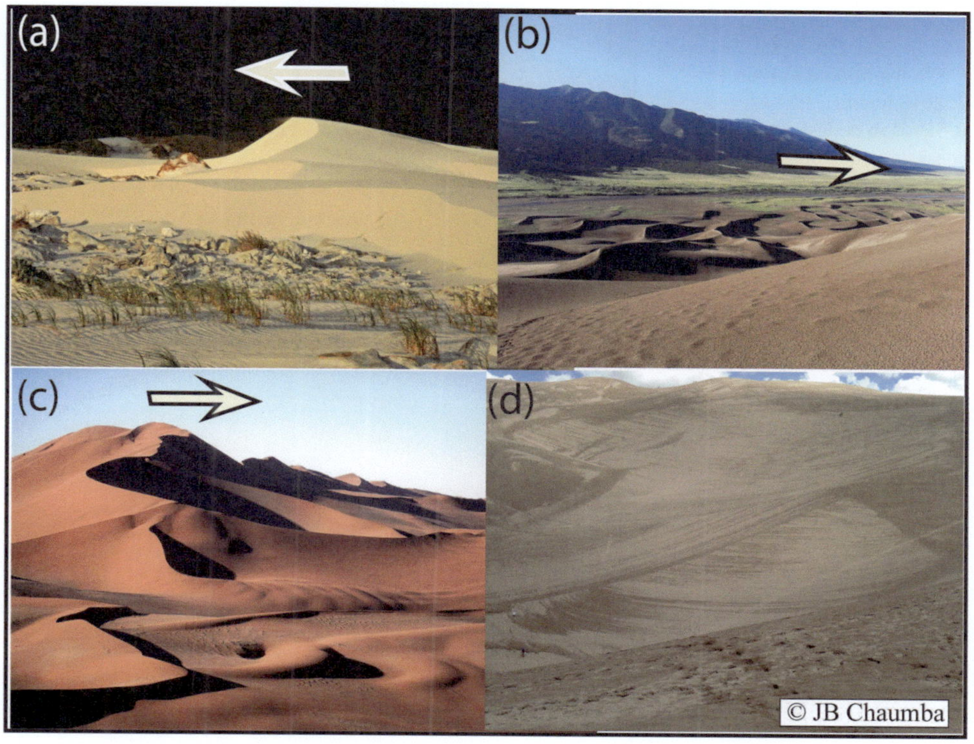

Fig. 14.11 Examples of dunes. Arrows indicate wind direction. **a** Barchanoid dunes from the Namib Desert (credit: D. L. Reid), **b**, and **c** show barchanoids from dunes from Sand Dunes National Park, Southern Colorado, USA (credit: D. T. Nelson). Shown on **d** are huge cross beds from Great Sand Dunes National Park and Preserve, Colorado, USA (credit: A. Gesualdo)

Deforestation is clearing land for cultivation as well as cutting trees and plants particularly for firewood and timber also contributes significantly to desertification. **Climate change** also contributes to deforestation through global warming caused by human-caused greenhouse gas emissions and fluctuations in climate. Desertification also leads to salinization and degradation of vegetation cover. Desertification often forces people to move to new land, as the land can no longer support vegetation.

14.6 Consequences of Desertification

The consequences of desertification are many and they vary widely. Desertification often results in soil erosion, land degradation, and a loss of biodiversity. Loss of forest cover due to desertification leads to shortages of wood resources. Additionally, desertification often results in food insecurity is caused by desertification as a consequence of reduced yields of crop failure. Consequently, huge economic costs are experienced by the countries experiencing desertification. Importation of food and other commodities affected by reduced harvests is more costly than food produced in the country. People living in areas undergoing desertification can experience poor health due to malnutrition and lack of access to clean water supply. People in regions undergoing desertification are more vulnerable to zoonotic diseases such as COVID-19. Desertification has a tendency to result in exacerbated poverty. An increase in the concentration of gases in the atmosphere which play a role in climate change has been linked to desertification. Desertification forces people living in the affected regions to migrate to other areas not affected by desertification. This presents some geopolitical challenges such as civil unrest.

Drylands cover approximately half of the earth's land surface that is free of ice. The consequences of desertification are worsened by the observation that large proportions of drylands are located in some of the poorest countries in the world. Insecurity of foods occurs due to reduced yields or crop failure. There is loss of vegetation cover and therefore of food for humans and livestock. It is estimated by the United Nations Convention to Combat Desertification that the lives of 250 million people are already affected by desertification and 135 million are maybe be displaced by desertification by 2045. Reserves in drinking water will decrease as a consequence of loss of aquifers.

Desertification often leads to both water and food borne diseases, respiratory diseases, malnutrition, as well as spreading of infectious diseases. Crop failure or reduction in crop yields which result in food shortages are other consequences of desertification. Thus, increases in poverty in impacted populations is another consequence of desertification. Desertification also results in worsening of living conditions for many species leading to loss of biodiversity.

Desertification can lead to forced migration of people as movement to more hospitable areas occurs. Land degradation is often a consequence of desertification due to soil erosion and the land being sterile resulting in food insecurity caused by the greenhouse effect. With scarcity of food

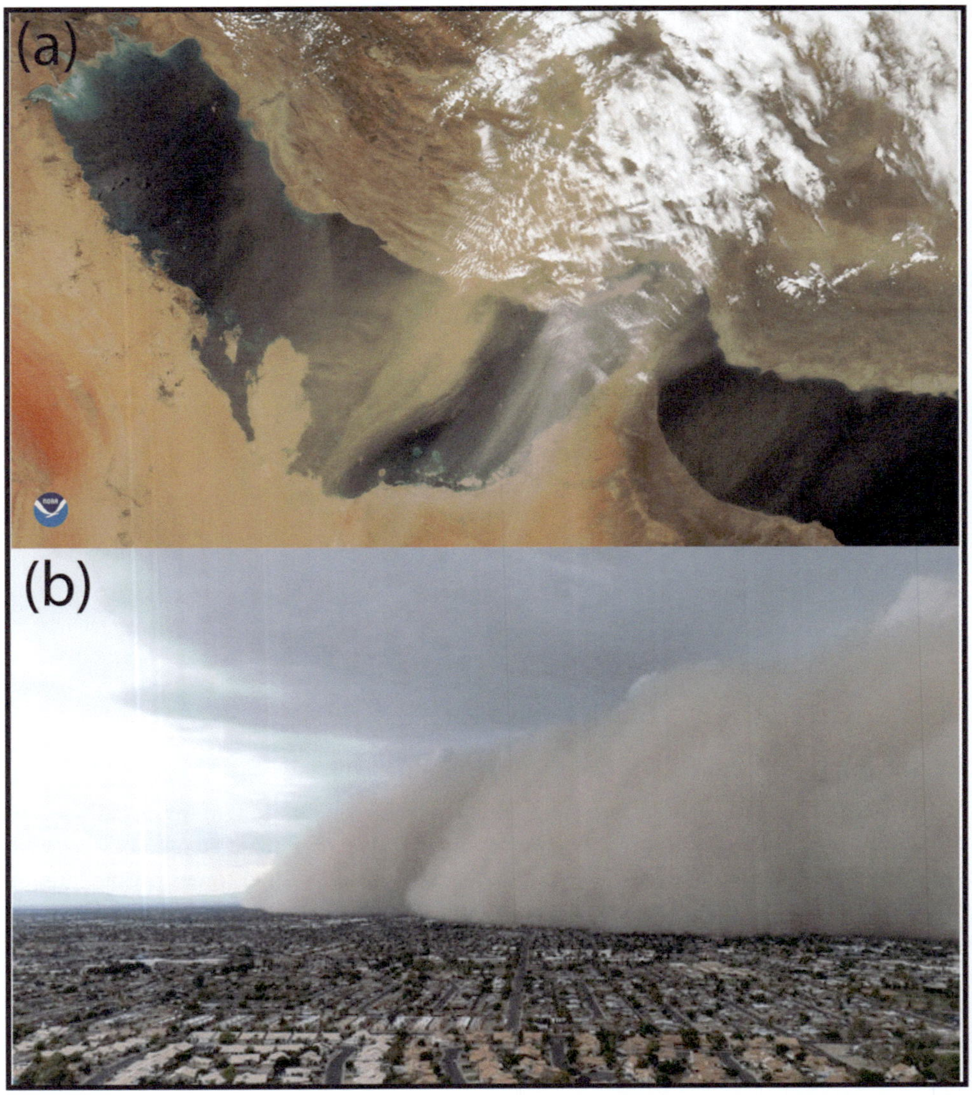

Fig. 14.12 **a** Photograph of a large dust storm over the Persian Gulf in the Middle East on May 13, 2018 (credit: NOAA/NESDIS), **b** A dust storm approaching downtown Phoenix on August 25, 2025 (credit: NOAA)

and resources caused by desertification, conflicts caused by increased competition for scarce resources occur. Loss of forest cover with desertification is accompanied by shortage of wood resources.

Downstream flooding reduced water quality, and sedimentation in rivers and lakes are caused and/or worsened by desertification. A wall of dust and debris of up to kilometers (miles) long and several meters (feet) high that is blown into an area by strong winds from thunderstorms is called a **dust storm** (Fig. 14.12a and b). Dust storms are caused by desertification. These dust storms are capable of transporting both soils and nutrients over great distances. Extremely strong winds, such as those produced by thunderstorms, cause dust storms. In dry regions, strong winds pull dust up into the air from the ground, resulting in the formation of dust storms. The geography of an area, as well as its scarce plant life, increases the likelihood of producing dust storms. An intense dust storm formed by strong winds either from weather fronts or the downdrafts of thunderstorms is called a haboob (Fig. 14.12b).

Dust storms, therefore, are commonly encountered in flat regions with very few plants and trees. Flat regions barren of vegetation permit winds to build up momentum which cause the winds to become stronger and propel more dust into the atmosphere. Dust storms happen in many places around the world. The bulk of the world's dust storms affect North Africa and the Middle East (Fig. 14.12a). The United States is also affected by dust storms. Dust storms in the United States tend to occur in the Southwest (Fig. 14.12b). Dust storms are a health hazard when inhaled due to the huge amount of dust they contain. In addition, they are disruptive on the ground, causing ground transportation and flight delays for airlines due to their associated very limited visibility. One recent haboob affected Phoenix, Arizona on August 25, 2025.

The Aralkum desert in central Asia represents an example of desertification. This desert formed on the seabed that was previously occupied by the Aral Sea. The Aralkum desert is currently the youngest desert in the world. The Aral Sea is located between Kazakhstan and Uzbekistan in Central Asia in the heart of Central Asia to the east of the Caspian Sea. The Aral Sea, which was once the fourth-largest saltwater lake in the world, covering 63,000 square kilometers (24,324 sq. miles), but has now shrunk to one tenth of its former size due to desertification. From approximately 1960, however, the Aral Sea began to shrink systematically and drastically in size mainly due to diversion of its main water sources for irrigation largely from the Amu Darya and Syr Darya rivers. Consequently, the Aral Sea has since lost approximately half of its area and three-quarters of its volume. Shrinking of the Aral Sea continued into the twenty-first century, which split it into several separate small lakes.

Agricultural productivity and incomes are reduced by desertification. Loss of biodiversity in some dryland regions has also occurred due to desertification. Spreading of invasive plants that affected many dryland areas has resulted in losses in ecosystem services. Lowering of the water table has already occurred due to overextraction of groundwater has led to depletion of groundwater resources. Unsustainable management of land, especially in land where droughts are frequent, has contributed to higher incidences of dust storms. In 2005, dust storms were linked with global cardiopulmonary mortality of about 402,000 people. Damage and disruption to solar and wind energy harvesting infrastructures as well as transportation are ascribed to higher intensities of both sandstorms and movements of sand dunes.

Desertification also leads to declining vegetation. Additionally, biodiversity also decreases due to desertification. Climate change is also exacerbated by desertification. An increase in global annual dust emissions of 25% is believed to have occurred on Earth since the late nineteenth century. Furthermore, dust storms are becoming more frequent as well as increasing in intensity, particularly in the Middle East. The United Nations estimates that 15.5% of land is being degraded, and drylands are increasing on our planet. Soil enrichment, reforestation, water management, soil buttressing, and Farmer Managed Natural Regeneration (FMNR) can all help in reducing desertification.

14.7 How to Avoid Desertification

The United Nations (UN) adopted 15 Sustainable Development Goals (SDGs). The SDGs aim to "protect, restore and promote the sustainable use of terrestrial ecosystems" to sustainably manage forests and stop and reverse land degradation. Additionally, the SDGs also have a goal to "combat desertification and stop biodiversity loss." At the local level, the solution to curb desertification is sustainable management of natural resources, particularly the conservation of fertile soils and water resources. Accordingly, the following are some ways that help in avoiding desertification: i. promotion of coordinated land-use planning, as well as the management of water resources, livestock, and agricultural activities, ii. preserving vegetation cover, which plays a key role in protecting the soil from wind and water erosion, by building barriers and stabilizing dunes, iii. promotion of climate change education to raise awareness, particularly through showing the consequences of desertification and ways of preventing it, iii. focusing both on organic farming and sustainable practices, such as cover crops or rotational crops, which prevent soil erosion and drought, iv. committing to reforestation with the goal of regenerating vegetation cover, reactivating moisture circulation, as well as generation of biodiversity, and v. encouraging rotational grazing which limits pressure to a particular area whilst others regenerate through co-existence with crops that permit more efficient nutrient cycling. The continent most affected by desertification is Africa. The southern margin of Sahara Desert in North Africa is one of the most obvious natural borders on the landmass where desertification is occurring.

Some parts of the United States are also undergoing desertification. Based on climate and soil type, 17 states that are all located in the western United States can be categorized as arid, semi-arid, or dry subhumid. This characterizes these regions as susceptible to desertification. New Mexico, Arizona, the Navajo Indian reservation in New Mexico, and Texas are the most severely affected areas in the United States.

14.8 Erosion, Transportation, and Deposition in Deserts

In deserts, wind is an active agent of both erosion and deposition as wind transports, erodes, and deposits sediment. We discussed above how desert landforms such as dunes and loess are formed by wind. However, water remains the main agent of erosion even in deserts. Although wind still plays an important role in erosion and deposition in deserts, wind still plays a secondary role in deserts in comparison to flowing water. Wind is an effective agent of erosion where vegetation cover is sparse, and where there is a lack of soil moisture but with ample supply of unconsolidated sediments. Fine-grained grains of sand in deserts are preferentially transported away by wind leaving behind coarser-grained material through **deflation**. Deflation, thus, is the process whereby wind blows away small and loose particles

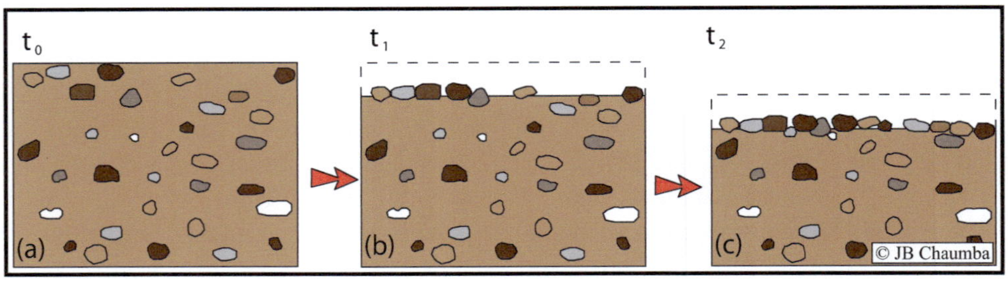

Fig. 14.13 Diagram illustrating formation possible formation of a desert pavement. **a** A mixture of particles of a variety of grain sizes making up desert soil at time t_0. **b** As finer-grained material if preferentially carried away by wind, an increase in the proportion of coarse-grained material that remains occurs at time t_1. **c** Continued removal of finer-grained material leaves behind a surface covered by coarse-grained material forming a desert pavement at time t_2

from the ground leaving behind the coarse-grained ones. A **deflation hollow**, or blowout, is a relatively small and shallow depression with diameters of up to several kilometers that is produced where wind has removed the loose sediment. The removal of sediment reveals a layer of coarse-grained particles that are too heavy for the wind to transport. Wind deflation occurs through the lifting and transportation away of lighter particles from a dry surface, leaving behind a surface with a high concentration of coarse grained sand and rocks. Deflation hollows, therefore, are formed as a consequence of the removal by wind of particles. The removed particles are transported to other regions where sand dunes are formed in a desert or on a beach.

A **desert pavement** is created from the coarse-grained material that is left behind (Fig. 14.13) after the fine-grained particles are transportated away. A desert pavement can be defined as a surface layer composed of closely packed pebbles and fine- to coarse-grained rock particles that cover large areas of drylands. A desert pavement is produced by the combined action of both wind and water over a long period of time.

A desert pavement, therefore, is a surface composed of angular and interlocking fragments of pebbles, boulders, or gravel that are formed in arid areas. Gradual removal of sand, dust, and other fine-grained materials by wind and intermittent rain leaves behind dominantly the larger particles. The Cuchillo geomorphic surface in the state of New Mexico, USA, is an example of an area where a desert pavement can be observed.

Desert winds transport sediment through three processes of **suspension**, **saltation**, and surface **creep**. The process whereby very fine particles like dust and silt sediment particles get lifted by wind into the atmosphere is called suspension (Fig. 14.14). The fine grained size of the particles permits them to stay suspended in the air for long periods of time, enabling them to travel long distances before they settle back on the ground. The process by which wind moves sand and other particles in a desert through a series of short bounces or hops is termed saltation (Fig. 14.14). Saltation represents the primary mechanism by which sediment transport by wind occurs in deserts. Particle sizes that are transported via saltation are typically in the 0.1–0.5 mm range. Through saltation, particles that are 1 cm (0.4 in) or less above the ground are usually moved. Particles transported by the saltation method are half to one-third the speed of wind. Upon hitting the ground, particles transported via saltation also dislodge other particles which cause them to be also moved through saltation. Saltation is the primary method of sediment transport in deserts. It involves the movement of sand particles by the wind through a series of short leaps or bounces. When wind blows over the desert surface and its speed exceeds a certain threshold, it lifts the sand particles and carries them forward. Wind speed fluctuates, which causes particles to fall back to the ground and rebound, resulting in a series of jumps which characterize saltation. The height of the hops of the particles generally ranges from a few millimeters (inches) to a few centimeters (inches. In addition to being responsible for the movement of the bulk of sand-sized particles in deserts, saltation also contributes to the formation of sand dunes.

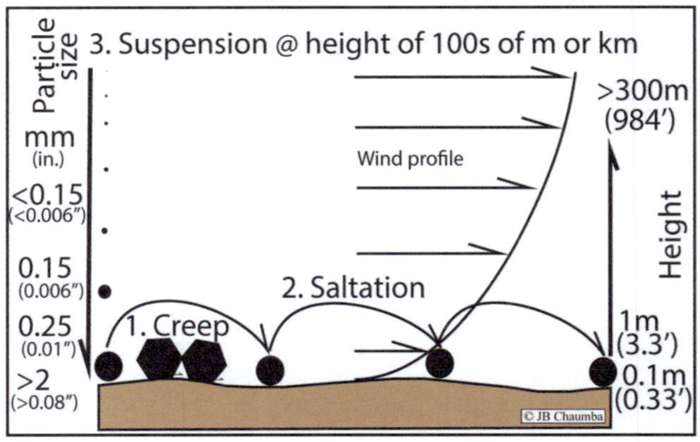

Fig. 14.14 Diagram illustrating the processes of suspension, saltation, and creep in desert regions. Particles sizes as well as height of the transported particles are also indicated

Landforms that are formed when wind deposits sand are called sand dunes. Desert winds also transport sediments through suspension. Through suspension, wind carries fine-grained particles of silt and clay as suspended load in the air for long distances. The last method through which wind carried material is through creep (Fig. 14.14). Creep is when larger particles along the ground slide or roll on the desert surface under the influence of wind and the momentum of saltating particles. The momentum of saltating particles together with wind both influence creep. Since saltating particles impact the larger grains, their energy is transferred, which causes the larger particles to be in motion. However, surface creep tends to move smaller volumes of sediment relative to saltation.

14.9 Wind Erosional Landforms

Strong winds lead to the formation of various landforms on Earth through eolian erosion. Eolian erosion refers to wind activity and the various landforms that are produced through this wind activity. **Abrasion**, or sandblasting, is the process where sand grains transported by wind cause erosion of rocks through grinding. The wearing and tearing of sand particles during transportation of sand particles is called **attrition**.

Winds blow faster in deserts compared to regions with vegetation due to the impact of the flying grains which loosens, or chips off, very fine-grained pieces of the rock. Abrasion is the process of erosion caused by particles suspended in wind that impact on solid objects at time t_0 (Fig. 14.15a). Windblown particles of sand that are transported at high speed by desert winds are very effective at sandblasting away rocks by abrasion (Fig. 14.15a). As grains of sand and dust are hurled at a rock or outcrop by wind through abrasion, **ventifacts** are formed at time t_2 (Fig. 14.15c). Ventifacts are wind-eroded rocks that are derived from fine-grained and hard rocks such as quartzite, chert, or obsidian. Ventifacts form in arid environments that are characterized by sparse vegetation, a steady supply of sand as well as strong winds. Wind-blown sand and dust eroding rocks lead to the formation of ventifacts. Dust and particles of sand grains hit rocks and in the process chip off pieces of the rock. With time (t_1, Fig. 14.15), flat sides form (Fig. 14.15b). Flat sides, called **facets**, which meet at sharp angles are formed in hard rocks such as basalt (Fig. 14.15b). In relatively softer rocks, such as sandstone, erosion occurs which leads to the formation of grooves, flutes, and pits.

Wind direction plays an important role in the shapes of ventifacts. Rocks being eroded often tip over, which exposes new surfaces to the wind. With time, rocks wear away and become smaller and more elongated. Ventifacts occur in arid environments that are characterized by strong winds but with little vegetation. Deserts, sandy coastlines, and paraglacial and periglacial environments produce these conditions that favor the development of ventifacts. Most ventifacts are comprised of hard and fine-grained rocks such as obsidian, chert, or quartzite. Ventifacts are considered to be natural records of environmental conditions over time, which display the strength and paleo-patterns and direction of historic winds.

Erosion can be defined as the processes that facilitate loosening, removing, and transporting away of weathered and/or unweathered solid material through downhill or downwind displacement through wind, flowing water, waves, ice (glaciers), gravity, and groundwater. Erosion of weathered solid material leads to the exposure of unweathered surfaces to weathering, ensuring continuation of weathering. Human beings can cause and/or increase the rate of weathering through human land use. Burrowing animals and decaying organisms, for example, also cause weathering.

In desert environments, wind erosion is called eolian erosion. In desert areas where vegetation is sparse or even absent, differences in atmospheric pressure often cause movement of air that erodes the surface material if velocities are high enough for particles to be moved. Eolian erosion, thus, is more pronounced in deserts as well as in areas where precipitation is insufficient for vegetation to grow and root systems to develop. Wind is incapable of transporting

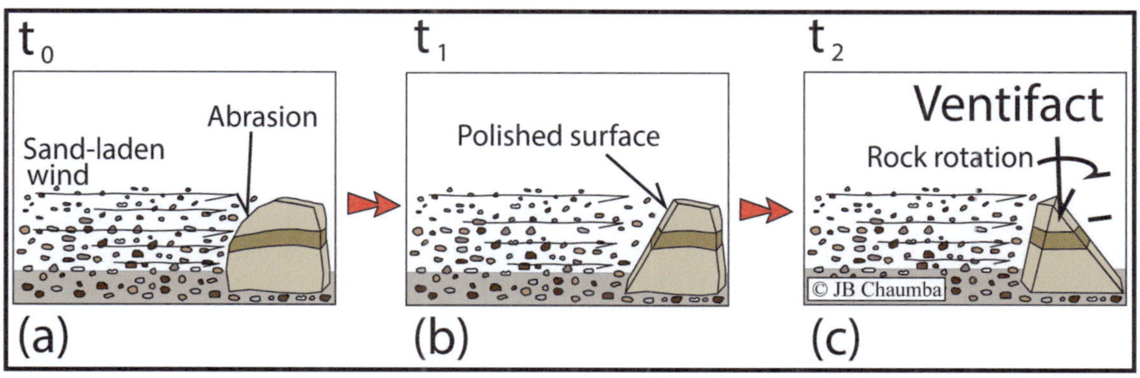

Fig. 14.15 Diagrams illustrating the processes of abrasion, suspension, and polishing in desert regions

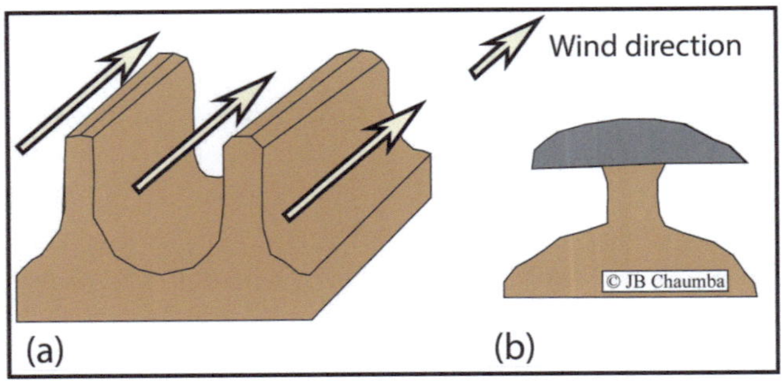

Fig. 14.16 Diagram illustration yardangs (**a**) and a mushroom rock (**b**)

large particles such as flowing water, but it can easily pick up soil, sand, and dust particles and transports them away. Wind is responsible for soil erosion through deflation and/or wind abrasion. Farmers can plant wind breaks in an effort to reduce wind erosion.

Erosional features of wind are widespread. **Inselbergs**, or monadnocks, are isolated hills, ridges, knobs, outcrops, or small mountains that abruptly rise from gentle slopes. **Pediplains** are the high relief structures that are formed by being reduced to low featureless plains by wind erosion in deserts. **Yardangs** (Fig. 14.16a) are ridges of rocks that are formed by wind, and they are aligned parallel to the prevailing wind direction.

A **rock pedestal**, or mushroom rock, or gour (Fig. 14.16b), is a rock that resembles a mushroom. Rock pedestals are primarily caused by weathering and wind erosion. A **zeugen** is an area of rocks encountered in both arid and semi-arid regions that is shaped like a table. A zeugen is formed when a relatively more resistant rock erodes slower in comparison to the soft rocks around it. **Wind bridges** and **wind windows** are formed due to powerful wind which continuously abrades stone lattices, creating holes in the process. These holes are sometimes widened to such that they reach each other at the end, to create an effect similar to that of a window to form wind windows. Wind bridges, on the other hand, are produced when the holes become wide enough to make an arch-like structure. **Demoiselles** are pillars of rocks rising above softer rocks due to differentiated erosion of both hard and soft rocks.

The first effect of wind erosion is the winnowing of light particles. Wind erosion, being selective, transports only the finest particles such as organic matter, clay, and loam for many kilometers. Dunes are other spectacular forms. Dunes are mounds of approximately sterile sand particles which move as the wind takes them, burying both oases and cities. Degradation of sedimentation crusts on the surfaces of stripped soils or the weathering of rocks at their base is another effect. Sheets of sand that travel close to the ground can damage crops like millets and cotton seedlings in semi-arid zones. The soil's capacity to hold nutrients and water is reduced by wind erosion, which makes the soil and environment drier.

14.10 Alleviating Wind Erosion

If soil cohesion is increased, the soil's capacity to resist wind erosion is improved. This results in decreased soil erosion. Pieces of soil that have detached from the main body of soil, yet not compressed or disturbed in any way, are called soil clods. Cloddy soil tends to occur when soil is too soft when seeds are planted, which causes the seeds to be planted too deep. An increase in the roughness of soil is caused by cropping techniques that leave behind large clods.

The formation of controlled ridges often leads to the soil's capacity to resist wind erosion. Leaving behind crop residues is a proven, yet inexpensive, way to reduce soil erosion due to wind. An increase in soil cohesion leads to an improvement in a soil's capacity to resist wind erosion. An effective way to reducing soil weathering due to wind is to increase vegetation cover. Installation of windbreaks leads to reduction in wind speed near the ground. Both soil erosion and evaporation are reduced by construction of windbreaks. An increase in plant cover leads to an increase in soil cohesion leading to the formation of fertile soils. Fertile soils, in turn, form in environments that contain more moisture and are also greener.

14.11 Glaciation and Glaciers

A **glacier** can be defined as a large and perennial accumulation of ice, snow, sediment, rock, often with liquid water originating on land and moving down slope due to the influence of its own weight as well as gravity. The process by which land is covered by glaciers is referred to as **glaciation**. An extensive, thick, and dense mass of glacial ice covering at least 50,000 square kilometers (~19,300 square miles) of land is called an **ice sheet**. There are only two ice sheets on our planet (Fig. 14.17). One of these two ice sheets, the Antarctic ice sheet, spans across the continent of Antarctica (Fig. 14.17a). The Antarctic ice sheet is the largest in the world, and approximately 98% of the Antarctic continent is covered by ice. Areas free of ice on this continent are located on nunataks (high mountains that poke through the ice sheets) such as James Ross Island and Alexander Island on the Antarctic Peninsula and the McMurdo Dry Valleys in East Antarctica. The other ice sheet, the Greenland ice sheet, covers most of the largest island in the world in Greenland (Fig. 14.17b). As we will discuss later in this chapter, ice sheets are not static, but they are moving constantly and spreading out under their own

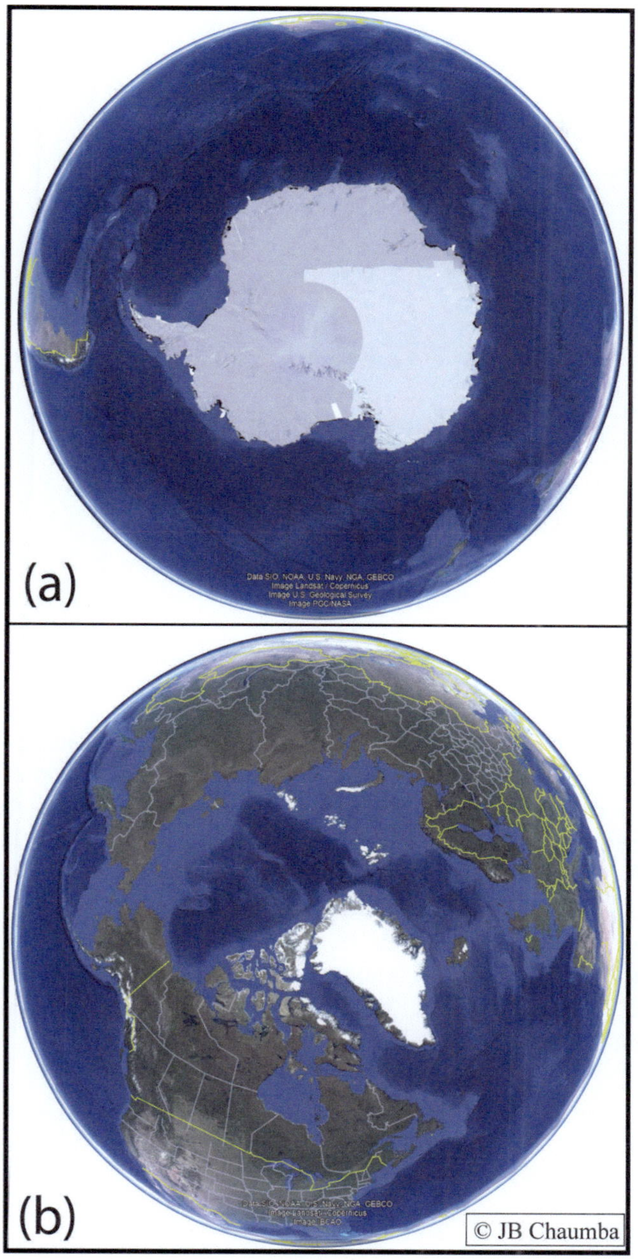

Fig. 14.17 Google Earth images of Earth showing the distribution and extent of ice sheets. **a** the distribution and extend of the Antarctic ice sheet and **b** the distribution and extend of the Greenland ice sheet

weight forming dome shapes that slope downwards toward the edges.

Ice sheets form in the winter due to snowfalls that accumulate to form layers and do not melt entirely in the summer. Over long periods of time such as thousands of years, these layers of snow build up to form thick ice. Ice sheets are not static, but they slowly flow downhill. The interior of the ice moves thousands of meters (feet) thick whereas the ice around the edges moves tens of meters (feet) thick. Ice sheets are sensitive to climate changes, and they contribute to sea level rise upon losing mass through melting of ice. The Greenland ice sheet, for example, has been losing ice every year since 1998 due to climate change. Ice sheets once covered much of the Northern Hemisphere during a series of Pleistocene Ice Ages. The Pleistocene epoch is a period of geologic time period that began about 2.6 million years ago and ended 11,700 years ago. The last ice age, as the Pleistocene Ice Age is also known, began 2.5 million years ago and it is still ongoing. During the last ice age, glaciers covered huge portions of our planet. The Pleistocene Ice Age is also called the Pleistocene era, or simply the Pleistocene.

Snow traps both dust particles and gases. Ice sheets contain records of our planet's past climate in dust particles and gases that are trapped by ice when it accumulates. Ice cores are recovered from ice sheets for studying past climates.

Pleistocene and ice sheet studies focus on investigating large continental ice sheets that existed during the Pleistocene epoch, a period spanning roughly 2.6 million to 11,700 years ago. This epoch is characterized by repeated glacial cycles where massive ice sheets advanced and retreated across the Earth's surface. The ice sheets impacted global climate as well as cause changes in sea levels. Ice cores from the ice sheets are investigated, as well cores of marine sediments, glacial deposits, among other evidence with the goal of understanding the mechanisms of the cycles, what causes them, and their impacts on ecosystems.

14.12 Types of Glaciers

There are two primary types of glaciers which are **continental glaciers** and **alpine glaciers**. Continental glaciers (Fig. 14.17) are large ice sheets that cover large areas, and they are primarily encountered in polar regions like Antarctica and Greenland. Four types of alpine glaciers are recognized: valley, cirque, piedmont, and ice field. Alpine glaciers, also known as **valley glaciers**, occur in mountainous regions (Fig. 14.18). Alpine glaciers are formed at high altitudes in mountainous regions. Here, temperatures are favorable for the formation of alpine glaciers. Four types of alpine glaciers can be distinguished which are valley, piedmont, cirque, and ice field glaciers. Valley glaciers originate from cirques and flow down the valleys of mountains. A valley glacier fills the pre-existing stream valleys with ice as the glacier pushes outward from its zone of accumulation. A cirque glacier that grows beyond the limits of the depression in which it forms spill out to form valley glaciers. **Piedmont glaciers** are glaciers which spread out at the base of a mountain range to produce a broad lobe where the terrain is flatter. A piedmont glacier forms when valley glaciers flow out of canyons and onto the lowland adjacent to a mountain front. The lengths of valley glaciers range from a few kilometers to over 100 km (60 miles).

Fig. 14.18 Photographs of a valley glacier, Franz Josef Glacier in New Zealand in 2006 (credit: D. L. Reid) and **b** Sawyer glacier, a tidewater glacier in Tracy Arm Fjord, Alaska, USA (credit: P. A. Bouker)

Valley glaciers often occur as single glaciers or they are comprised of multiple glaciers that branch out. A piedmont glacier is produced when two or more valley glaciers join. Valley glaciers flow downslope, and in the process erode the walls of a valley. Valley glaciers, however, are limited by topography and are limited to any preexisting valleys. Valleys comprised of steep sides supply the glacier with debris to transport downslope. A tidewater glacier (Fig. 18.18b) terminates in the ocean where calving may occur to produce icebergs. Valley glaciers form in high mountain environments, and they are common in the Andes, the Alps, the Himalayas, as well as in New Zealand. Glaciers that form in bowl-shaped depressions called *cirques* (Fig. 14.18) are called **cirque glaciers**. Cirque glaciers are formed in bowl-shaped depressions on mountain sides which are often the source of these glacier types. Alpine glaciers are widespread and are encountered in high mountain ranges in the Alps, the Rocky Mountains, the Himalayas Mountains, and the Andes Mountains. Except for Australia, alpine glaciers occur on every continent. Cirques are bowl-shaped depressions that are carved by glaciers into mountains as well as sidewalls of valley at high elevations. Weathering and erosion loosen bedrock at the head of any preexisting stream valley during accumulation of ice. With continued erosion, a depression is formed in which the cirque glacier grows in size. An **ice field** is confined to large basins or plateaus with tall mountain peaks called **nunataks** that protrude above the ice. Valley glaciers are fed by ice fields.

Ice caps are smaller when compared to ice sheets. Ice caps are often found in polar and subpolar regions. A **rock glacier** is a mass of debris that is mixed with ice and is encountered at high altitudes. The Franz Josef Glacier in New Zealand (Fig. 14.18b) is an example of a valley glacier. The Tracy Arm of South Sawyer Glacier in Alaska is a tidewater glacier that extends deep underwater to produce a blue iceberg (Fig. 14.18b). Valley glaciers often begin in cirques at the top of valleys. Valley glaciers can also originate in plateaus of ice caps.

14.13 Landforms Formed by Glacial Erosion and Deposition

Several landforms are created through erosion and deposition by glaciers. Glaciers are capable of crushing, abrading, as well as scouring the landscape through expansion which results in the formation of landforms such as **U-shaped valleys** (Fig. 14.19). U-shaped valleys are steep-sided and flat-bottomed valleys produced by glaciers modifying originally V-shaped stream-cut valleys. **Hanging valleys** (Fig. 14.19), also created by glacial erosion, are valleys that are carved into the upper part of a broad glacier valley by a former tributary glacier. Bowl-shaped depressions, called **cirques** or **corries** (Fig. 14.19), are carved by glaciers into mountains and valley sidewalls at high elevations. They are also formed by glacial erosion. When two glaciers erode alongside each other, a sharp, steep and narrow knife-like

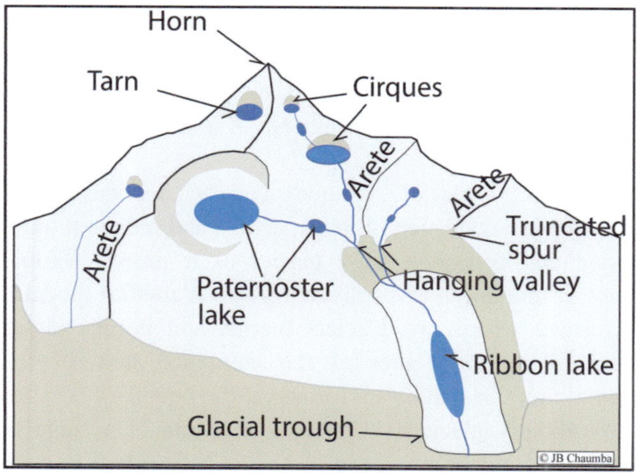

Fig. 14.19 Erosional landforms that are formed by alpine glaciers after the ice has melted away

ridge of rock called an **arête** is formed between the two glacial valleys (Fig. 14.19). A **horn**, or a **pyramidal peak**, is a pointed mountain peak formed where at least three arêtes meet (Fig. 14.19). **Paternoster lakes** (Fig. 14.19) are a series of lakes that form in low depressions of a U-shaped valley. The small basins which later get filled with water to form the lakes were scooped out by the glacier as it retreated. Paternoster lakes, thus, are erosional landforms that were produced by scouring of a valley bed that contains rocks composed of varying resistance.

A long and narrow lake that is carved out by a glacier in a U-shaped valley is called a **ribbon lake** (Fig. 14.19). Ribbon lakes are formed through glaciation. These lakes are formed when a glacier erodes the floor of a valley floor more deeply in comparison to softer areas of rock. Ribbon lakes can also be formed when a valley fills up behind a moraine dam.

Additionally, when a tributary glacier joins a main glacier, a ribbon lake may also be formed.

After melting of the ice, the valley bottom can be filled up by rainwater to form a ribbon lake. Misfit streams and rivers can also meander through the flat valley floor. Buttermere in the Lake District in England, Glencoe in Scotland, and Newtondale in Yorkshire are examples of ribbon lakes (Fig. 14.20).

Fig. 14.20 Photographs of glacial landforms. **a** Matterhorn, Switzerland is a horn or pyramidal peak (credit: E. J. Chopamba), **b** Cirque, Mount Blue Sky, Colorado, USA (credit: M. E. Ellis), **c** Tarn, Mount Blue Sky, Colorado, USA (credit: M. E. Ellis), and **d** Paternoster lakes, Mount Blue Sky, Colorado, USA (credit: M. E. Ellis)

14.14 Glacial Depositional Landforms

Glaciers are capable of picking up chunks of rocks and transporting them over long distances, eventually dropping them often far from the outcrops or bedrocks from which they were plucked. Moraines, erratics, and drumlins are among the most common glacial depositional landforms. Moraines, which form pronounced ridges of unsorted glacial till along the terminus and edges of a glacier, are the most prevalent glacial depositional landforms. A mixture of unsorted sediments, such as clay, sand, gravel, and boulders, that a glacier deposits as it moves is called **glacial till**. **Drumlins** are elongated hill deposits of glacial till that are shaped like teardrops. Drumlins occur in clusters called drumlin fields. A drumlin (Fig. 14.21) is a long streamlined hill made out of till. Drumlins are not round mounds like kames but are asymmetrical. Drumlins are shaped like a carrot and are streamlined asymmetrical hills that are made up of glacial tilt. They seem to originate when glaciers advance over the previously deposited drift and reshape the material to form drumlins. There is a steep side and a gentler side. The steep side of the hill faces the direction from which the ice advanced. The gentler slope faces the direction the ice moved. Drumlins have a height between 15 and 60 m and about are 0.4 to 0.8 km in length and they often occur in clustered referred to as drumlin fields. **Kames** (Fig. 14.21) are small, rounded, and steep-sided hills composed mostly of stratified drift that are formed by the deposition of meltwater sediments. Kames are created when glacial meltwater washes the sediment into the openings and depressions in the stagnant wasting area and the terminus of a glacier. As the ice melts, the stratified drift that is left over looks like a mound or hill. They are also formed when glacial meltwater washes the sediment into openings and depressions in the stagnant wasting terminus of a glacier. When the ice eventually melts away, the stratified drift is left behind as mounds or hills. These hills are symmetrical and round like perfect hemispheres. Kames are shaped like an apple.

Different types of moraines can be documented. A **moraine** is a collection of rocks, debris, and sediment transported and deposited by a glacier as it moves. Various types of moraines can be distinguished. A **medial moraine** is formed where two glaciers merge, and it runs downwards along the center of the combined flow of ice. Medial moraines, thus, are formed at the boundary between two tributary glaciers. A **lateral moraine** is a ridge of till that is deposited along the sides of valley glaciers. Medial and lateral moraines mark the edges of a body of ice. A **terminal moraine** is a ridge composed of till that was deposited at the glacier's furthest extent. **Ground moraines** (Fig. 14.21) are sediments from underneath a glacier that contribute to fertile transported soils. An **esker** (Fig. 14.21) is a long, narrow and winding ridge of sand and gravel that is deposited by meltwater flowing beneath under a glacier by glacial meltwater. An **outwash plain** is a relatively flat area that was formed by meltwater depositing sediments at the edge of a glacier.

A **kettle lake** or **kettle pond** (Figs. 14.21 and 14.22) is a bowl-shaped deep depression that is formed when a block of ice left behind by a glacier melts. A block of stagnant ice, called a **serac**, can detach from a glacier in the late stages of melting and becomes completely buried. Though the glacier has long melted away, the buried chunks of ice eventually melt to form bowl-shaped depressions called kettles. Eventually, the wholly or partially ice block buried in the sediment slowly melts. As the ice melts, it leaves behind a pit, or kettle, that often later gets filled with water to create a

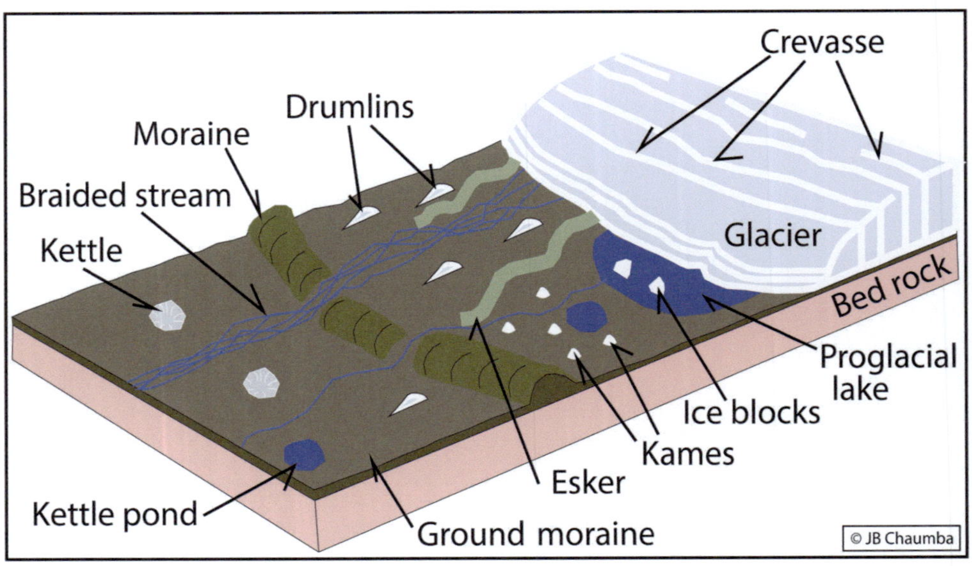

Fig. 14.21 Diagram illustrating depositional landforms formed by glaciers

Fig. 14.22 Photographs of a Kettle Lake, North Dakota. View of full Kettle Lake (**a**), and close-up of part of Kettle Lake (**b**) (credit: J. J. Donovan). Glacial striations from Devil's Postpile National Monument in California, USA formed on bedrock by rocks and sediments transported by the glacier (**c**) (credit: US National Park Service). A fjord from Norway is shown in (**d**) (credit: D Valdemaras via Pexels)

kettle hole lake. A kettle can also form when a dammed lake of ice suddenly drains. A kettle is also referred to as a kettle hole, or pothole. A kettle that fills up with water from melting ice generally does not exceed 10 m (33 ft) deep in depth. The bulk of lakes encountered in Michigan, USA, and adjoining states and Canada are kettles.

Material eroded and transported by glaciers finally gets deposited when glacial motion ceases, and melting occurs. The rocks, sediment, and debris deposited by melting glaciers form a variety of landforms. Medial and lateral moraines are composed of rock and debris transported by glaciers. **Glacial erratics**, or simply erratics, are large boulders that were dropped by a glacier. Erratics differ in composition from the surrounding bedrock. A lake that is formed at the edge of a glacier as it retreats is called a **proglacial lake**. It is often dammed by glacial ice (Fig. 12.22) or a moraine as meltwater gets trapped against the glacier. A proglacial lake is a body of water that is located directly in front of the edge of glacier due to melting ice as the glacier retreats. Depression of the surrounding land caused by the weight of the glacier helps produce the proglacial lake.

Glacier types are largely defined by their location and the topography they occupy. As glaciers move over land, they incorporate rocks and sediment into the ice. These inclusions make the bottom of the glacier, the glacier sole, into a coarse sandpaper-like material that scratches the bedrock during glacial movement. As the glacier moves over rock and sediment, it leaves scratches or grooves in a rock surface called **glacial striations** (Fig. 14.22d) that are created by a glacier's movement. Glacial striations reveal the direction of glacial flowage. Glacial striations are also called striae. A **fjord** is another landform that is formed by glacial erosion. It is a long, narrow, and deep inlet of water that was formed by carving out of a valley by glaciers and later filled with seawater. Fjords form due to glaciers carving U-shaped valleys into the bedrock (Fig. 14.22d). Upon melting of glaciers, seawater floods the valleys. Norway is famous for its fjords, containing more than a thousand of them along the country's coastline.

At least three bowl-shaped valleys or hollows on the side of a mountain were formed by glaciers or ice erosion, referred to as **corries**. Corries, which are also called **cirques**, grow in hollows on all sides of a mountain, where a horn or pyramidal peak is formed. A horseshoe-shaped valley formed from glacial erosion is called a corrie. The other name for a corrie is a cirque. As corries erode the mountain behind them, the remnant rock is eventually weathered to produce a sharp point. Corries are north-facing, away from the sun which stops the ice from melting. As snow and ice build-up, the underlying rock is eroded. The formation of corries happened a long time ago, during the last ice age. The Matterhorn in Switzerland is a classic example of a horn. Another example of a horn is Mount Snowdon located in Snowdonia National Park in Wales.

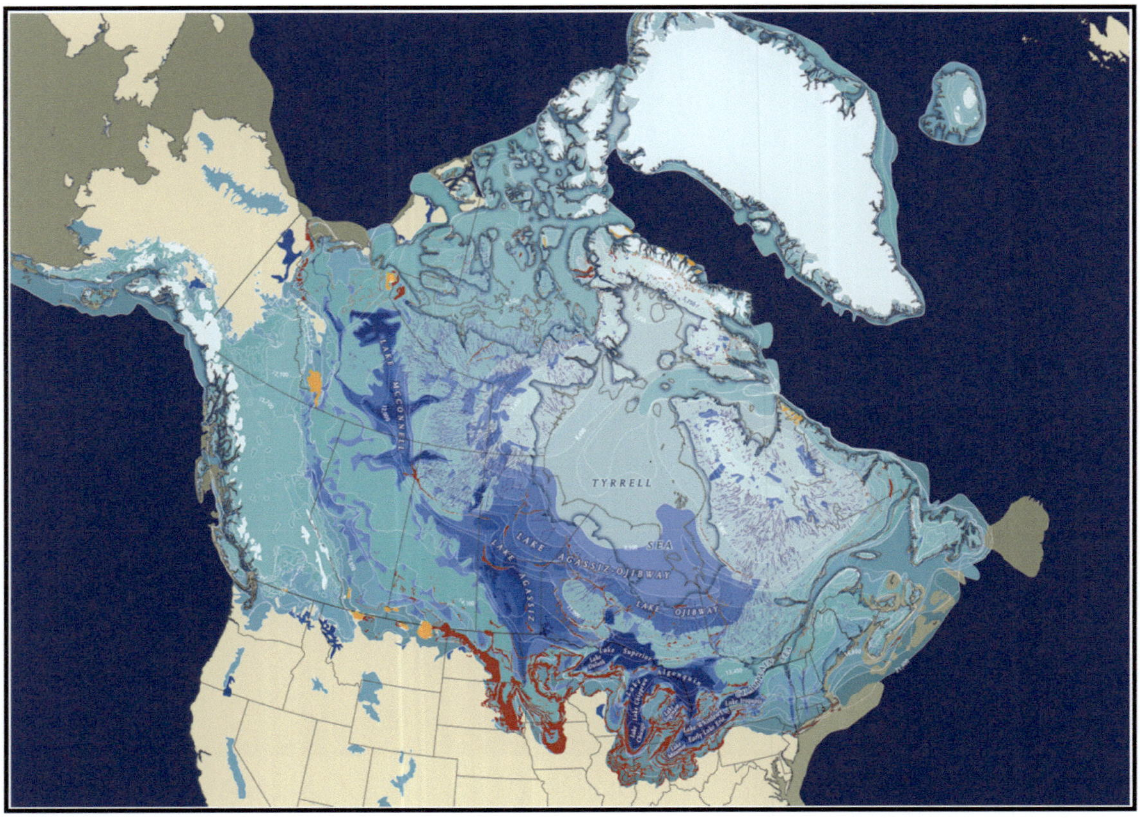

Fig. 14.23 Extent of glaciation in North America (credit: Chris Brackley/Canadian Geographic)

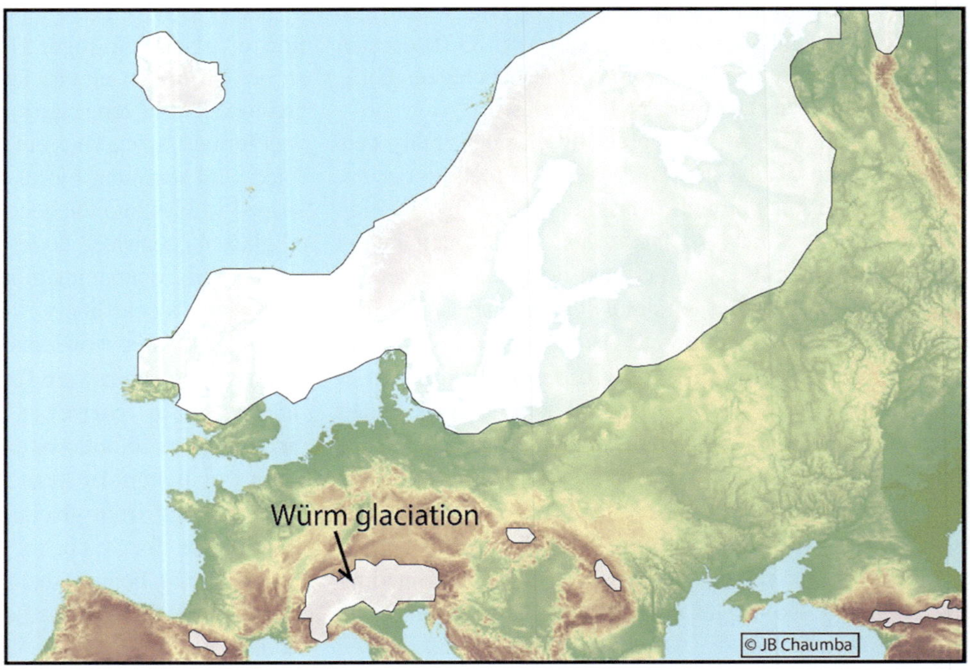

Fig. 14.24 Map showing areas covered by glaciers in Europe (Weichselian glaciation) during the last ice age. In the Alpine region, the glaciation is referred to as the Würm glaciation (credit: USGS)

During glacial periods, temperatures are colder, and the planet is drier and dustier than during interglacial periods. The last ice age began ~ 115,000 and ended ~ 11,700 years ago. It was the most recent glacial period within the Quaternary glaciation. During the last ice age, large portions of North America (Fig. 14.23), Europe (Figs. 14.24 and 14.25), Asia, and even South America were covered by glaciers. Global temperatures during this

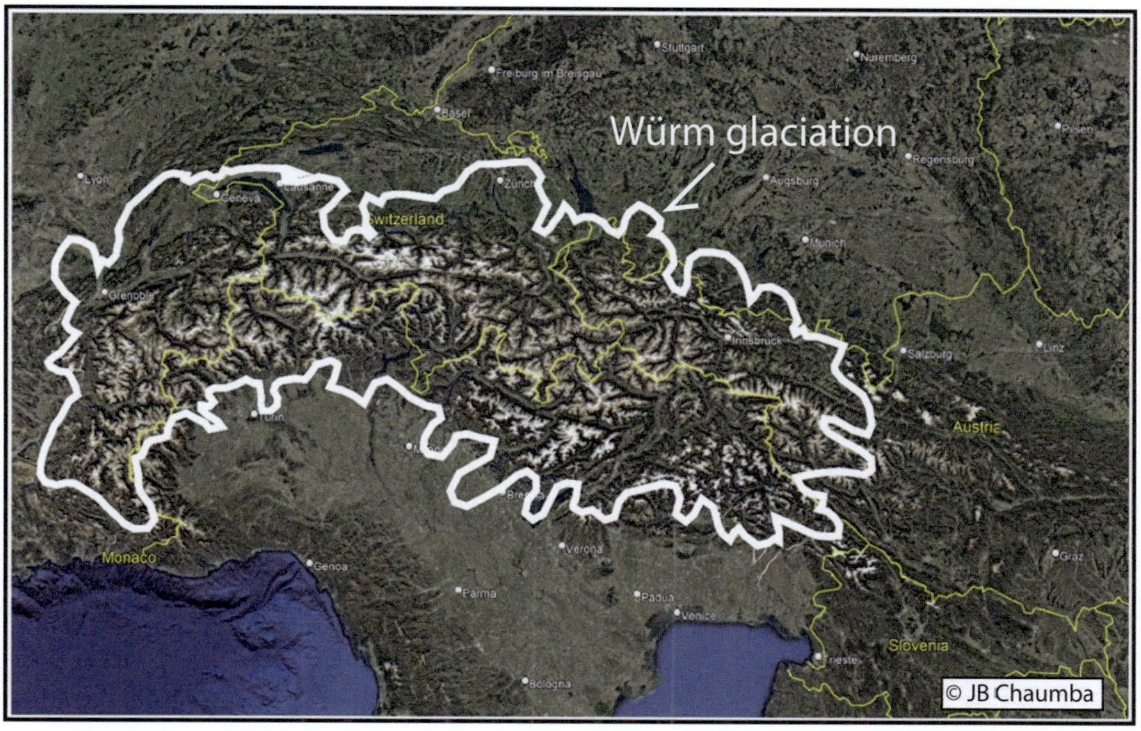

Fig. 14.25 Map showing the extent of Würm glaciation

period were about 11°F (6 °C) lower than those we experience today. Several plants and animals moved closer to the equator. Since the end of the last ice age, our planet has been in an interglacial period called the Holocene. What caused the end of the last ice age is not known. However, storms may have released huge amounts of water vapor which acted as a greenhouse gas. In addition, sea level drop may have cut off a current that normally diluted the salty waters of the Indian Ocean leading the Indian Ocean to precipitate salty water into the Atlantic Ocean. The currents and temperatures of the Atlantic Ocean may have been altered.

Glaciers covered much of Canada and some portions of the northern United States during the last ice age. These glaciers underwent thickening and thinning processes for thousands of years. At the height of the last glaciation, massive ice sheets spread across what is now Canada and northerly parts of the United States. A variety of landforms and the landscape of Canada and northern parts of the United States that we're all now familiar with were also created during this time. Today, the affected portions of North America are still rebounding from the weight of the ice sheets when they were at their maximum extent and thickness. Sea levels also fluctuate in response to the thickness of glaciers; the sea level drops when the glaciers are thick, and when glaciers are thin, sea level rises. Remnants from the last ice age include eskers, drumlins, moraines, pluvial lakes, among others. Earth started to warm approximately 19,000 years ago and the ice started to melt causes glaciers to retreat. Champlain and Tyrrell seas, for example, are remnants of "proglacial" lakes, formed during the time of glacial advance.

This map of Europe as it must have looked like during its last glaciation is shown in Figs. 14.24 and 14.25, approximately 20,000 to 70,000 years before present. This glaciation in northern Europe is called Weichselian Glaciation (Fig. 14.24), and in the Alpine Region it is called the Würm Glaciation (Fig. 14.25).

Dramatic cycles of glacial expansion and retreat occurred during the Pleistocene epoch, where large ice sheets covered vast areas of the Northern Hemisphere. The glacial advances and retreats are interpreted to have been driven by variations in Earth's orbital parameters (Fig. 14.23), resulting in significant changes in global climate and sea levels. Various methods such as analyses of ice core, cores from ocean sediments, and glacial landforms are utilized in the reconstruction of the extent and dynamics of ice sheets throughout the Pleistocene epoch.

Investigations of Pleistocene ice sheet are useful for a variety of reasons. Repeated cycles of expansion and retreating (interglacial periods) of glaciation (ice sheet expansion) which show evidence of multiple glacial advances and retreats over millions of years. Investigations of glacial landforms such as erratics, drumlins, and moraines on land gives insights into the extent as well the movement of past ice sheets. Additionally, studies of deep-sea sediment cores permit reconstructions of past ocean conditions and changes of volume of ice through analyzing oxygen isotope ratios in marine organisms. In addition, studies of ice cores from both Greenland and Antarctica yield important information about atmospheric conditions in the past, such as concentrations of greenhouse gas through analyses of trapped air bubbles within the layers of ice.

During glacial periods, the **Laurentide ice sheet** covered extensive sections of North America whereas the **Fennoscandian ice sheet** covered Northern Europe, both described above. While relatively stable throughout the Pleistocene, fluctuations in the **Antarctic ice sheet** had a role in climate change on a global scale. Variations in the orbital parameters of Earth are referred to as **Milankovitch cycles**. These cycles are the primary drivers of glacial-interglacial cycles through changing the amount of solar radiation that is received at different latitudes. **Mid-Pleistocene Transition**, or simply MPT, is a vital shift in glacial cycles that is observed ~ 1.2 million years ago. The dominant period of the glacial cycle changed from 41,000 years to 100,000 years.

Milankovitch cycles are a series of changes in the Earth's orbital characteristics that impact the climate over thousands of years. Milanković are named after the Serbian geophysicist and astronomer Milutin Milanković. The cycles (Fig. 14.26) are thought to be a consequence of variations in three astronomical parameters in the Earth's: i. **obliquity**—the angle of axis of the Earth's tilt relative to its orbital plane (Fig. 14.26a, ii. **eccentricity**—the shape of Earth's orbit (Fig. 14.26b), and iii. **precession** or wobble—the direction which the Earth's axis of rotation points (Fig. 14.26c). Milanković cycles of 26,000 years are thought to be responsible for variations in climate on Earth, which include ice ages and interglacial periods. The cycles also affect the distribution of solar radiation that is received by the Earth's surface. Milankovitch cycles are inferred to have caused the planet's climate to swing for millions of years between ice ages and warmer periods. The tilt of the Earth's axis, for example, varies between 22.1 and 24.5 degrees over a cycle that averages around 40,000 years. The axis of the Earth's rotation is presently tilted at 23.4 degrees, which is approximately halfway between the extremes. This angle is decreasing very slowly in a cycle spanning approximately 41,000 years. The angle was last at its maximum tilt ~ 10,000 years ago and it is inferred to reach its minimum tilt in the next ~ 10,000 years. The seasons are more severe, with warmer summers and colder winters, when the tilt is greater. Snow and ice can be caused to build up in high latitudes, which can reflect more of the sun's energy back into space, causing further cooling. With decrease in obliquity, seasons on Earth become progressively milder, which leads to increasingly warmer winters and cooler summers. Based on Milanković's findings, thousands of years of colder summers cause ice ages. Over time, the cooler summers will gradually permit more snow and ice at high latitudes to accumulate into large ice sheets. With increases in ice cover, more of the Sun's energy gets reflected ack into space, which leads to further cooling of temperatures on Earth.

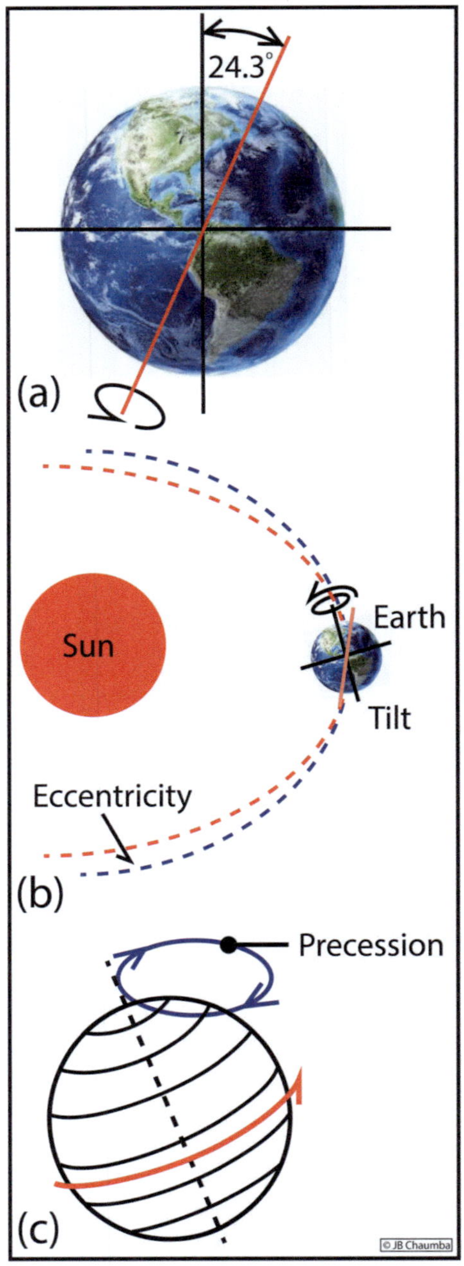

Fig. 14.26 Diagram illustrating Milutin Milanković's cycles of obliquity (**a**), eccentricity (**b**), and precession (**c**)

It must be pointed out that this evidence of obliquity, eccentricity, and precession on which Milutin Milanković's cycles hinge on, has been shown by other scientists to be flawed. Milanković's cycle predicts, for example, that ice ages occur at opposite times in the Northern and Southern Hemispheres. For example, when the northern hemisphere is experiencing cold summers, and thus heading towards an ice age, the southern hemisphere would be experiencing warm summers, heading away from an ice age. New research, however, has shown that both hemispheres simultaneously experienced the last major ice age at a peak ~ 20,000 years ago. This means that some long-term climate predictions based on Milanković's cycles may be more inaccurate than realized in the scientific community.

Glossary

Abrasion wind erosion process in desert landscapes whereby windblown sand particles collide with, and wear away, rock surfaces by sandblasting them forming smooth facets on the rocks

Alpine glacier also called a mountain glacier, is a glacier that forms in the mountains and confined by the terrain of the area

Antarctic ice sheet the largest mass of ice on Earth which covers almost all of the Antarctic continent

Arête a sharp-edged ridge of rock separating two valleys or glaciers

Attrition process whereby wind-carried sand particles collide with each other during transportation, breaking apart into smaller particles, and rounding them

Bad livestock practices raising farm animals using methods that are harmful to their welfare such as excessive confinement, painful procedures without proper anesthetic, inadequate nutrition, and practices that negatively impact the environment

Barchan a crescent-shaped sand dune in sandy deserts formed by wind that predominantly blows from one direction and is characterized by its convex side which faces the wind and its crescent horns pointing downwind

Barchanoid crescent-shaped dune that composed of alternating barchan and linguoid segments

Climate change long-term shifts in temperatures and weather patterns caused by natural or human activities

Cold ocean current movement of cold water in the ocean from cold higher latitudes near the poles towards the equator thereby bringing cool water to warmer regions

Continental glacier also called an ice sheet, is a large and continuous mass of ice covering a huge portion of a continent or an island

Corries *see cirque*

Cirque a bowl-shaped depression with steep walls and a relatively flat floor that was carved into a mountain by glacial activity at high elevations

Cirque glacier a glacier that forms in a cirque, a small, bowl-shaped depression in a mountain sidewall of a valley or mountain

Creep gradual downward movement of soil or rock on a slope

Crop rotation practice of planting different crops in the same area of land in a sequence over time

Deflation from the Latin word *deflare*, meaning "to blow away," is a type of wind erosion occurring when wind removes fine-grained soil particles of sediment from flat and dry areas

Deflation hollow a depression, or hollow, in the ground formed by wind erosion due to wind removing fine-grained sediment and leaves behind a hollow that is stabilized by a layer of heavier particles too heavy for the wind to lift

Deforestation clearing of forests and reduction in their size by humans

Demoiselles a rock pillar composed of more resistant harder rock towering above the surrounding softer rock which easily erodes away leaving a pillar-like rock produced by selective weathering

Desert a large arid area of land that receives little rainfall of less than 250 mm (10 inches) on average per year to none and has little vegetation

Desertification the process by which vegetation in arid and semi-arid regions decreases until it eventually disappears

Desert pavement a surface occurring in deserts comprised of closely packed rock fragments

Drumlin an elongated glacial deposit that forms an oval-shaped hill composed of sand, gravel, and glacial till

Drylands areas that experience significant scarcity of water with precipitation that is balanced by evapotranspiration

Dust storm weather phenomenon whereby strong winds pick up loose dust and soil from the ground in arid regions after periods of drought to form a large cloud of airborne particles that significantly reduces visibility and travel over a wide area

Eccentricity degree to which path of Earth's orbit around the Sun deviates from a perfect circle and essentially measures how elliptical the orbit of Earth is

Esker a long, narrow, and winding ridge of stratified till that was deposited as a glacier

Facets smooth and flat surfaces that are cut on a fragment of rock due to the action of windblown sand or water

Fennoscandian ice sheet an extensive ice sheet encompassing the Scandinavian Peninsula and extended into western Russia. It originated from the Scandinavian mountain range covering much of Northern Europe during the last glacial period

Fjord a narrow, elongated, and deep sea or drain of lake that composed of steep land on three sides formed by glaciers carving U-shaped valleys into bedrock followed by melting of glaciers and flooding of the valleys by seawater

Fog thick cloud composed of tiny water droplets or ice crystals suspended in the air that forms close to the ground and often seen on cold mornings or near bodies of water

Glacial erratic a rock that was transported by a glacier and deposited in a new location far away from its source

Glacial striation scratches called striations which occur where rocks transported by glaciers come into contact with other rocks or with bedrock and scratch the rocks and bedrock

Glacial till a poorly sorted sediment deposited directly by a glacier

Glaciation process whereby glaciers form across the Earth's surface and slowly move and melt, leaving behind unique landforms

Glacier large mass of ice, snow, rock, sediment that forms on land and slowly moves downslope due to gravity

Ground moraine a layer of till deposited on the ground by a retreating glacier

Hanging valley side valley that is at a higher level than the main valley it feeds into, and it is typically characterized by a waterfall

Horn a pointed, or pyramid-shaped, mountain peak formed by glacial erosion

Ice cap a large and dome-shaped ice mass of less than 50,000 square kilometers in area covering a high-latitude region

Ice field a large area of ice composed of many interconnected glaciers covering a mountainous region. Ice fields are smaller than both ice caps and ice sheets

Ice sheet a huge mass of glacial ice covering at least 51,800 km^2 (20,000 mi^2) of land

Improper agricultural practices practices that reduce soil porosity, increase soil alkalinity and salinity, disturb natural composition of nutrients in the soil, increase soil erosion, among others

Inselberg an isolated, steep-sided and resistant body of rock or hill rising above a flat plain which appears as an "island mountain" from the surrounding land being eroded away

Interior desert a desert occurring in the center of a continent and far from the coast formed by lack of winds loaded with moisture that reaches the continent interior

Inappropriate irrigation irrigation practice not optimized for a specific environment which results in inefficient water usage, reduced crop yields, and potential damage to soil

Kame an irregularly shaped hill or mound composed of sand, gravel, and till formed when meltwater streams carry material from a glacier and deposit it when the glacier melts

Kettle lake body of water produced when a block of ice breaks off from a glacier and melts leaving behind a depression that fills up with water

Kettle lake *see kettle lake*

Lateral moraine linear mound of debris that is deposited along the sides of a glacier and composed of material eroded from the valley walls by the glacier

Laurentide ice sheet principal glacial covering North America about 2,600,000 to 11,700 years ago during the Pleistocene Epoch

Loess wind-blown sediment composed of silt-sized particles covering more than 10% of the surface of our planet

Longitudinal dune a long, elongated dune that forms parallel to the prevailing wind direction and form in areas with a limited sand supply and consistent strong winds that blow from different directions over time.

Mid-Pleistocene transition a period of time when the glacial-interglacial cycles shifted in duration from 41,000 years to 100,000 years and it occurred between 1.25 and 0.85 million years ago

Milankovitch cycles periodic changes in orbital characteristics of a planet that control how much sunlight is received and hence affect its climate and its suitability for life over periods of hundred thousands of years

Moraine poorly sorted debris that is eroded, transported, and deposited by a glacier

Nunatak mountain peak that is isolated and once projected through an Alpine glacier or a continental ice sheet

Obliquity the angle between the plane of its orbit around the Sun and the Earth's equator

Outwash plain a flat and broad area of land that is formed in front of a glacier and composed of sediments deposited by glacial meltwater

Overgrazing excessive grazing which leads to damage of grassland

Parabolic dune a U-shaped dune composed of elongated "arms" pointing upwind and are formed by wind blowing sand out of a depression and gets trapped by vegetation at the edges to form crescent-like shapes with tips facing into the wind

Paternoster lake a series of small glacial lakes connected by a single stream in a formerly glaciated valley, and resembling the rosary beads based on their similar appearance when strung together and produced by a retreating glacier in U-shaped valleys

Pediplain a flat and broad rock surface in an arid environment that formed from merging of several pediments, or gently sloping bedrock surfaces, formed by erosion

Piedmont glacier glacier that forms when a valley glacier flows out of a confined or mountain valley and spreads out onto a flatter area to a fan-shaped and large lobe of ice and expands laterally in the process

Precession slow rotation of the stars around the ecliptic axis which lasts approximately 26,000 years for a single revolution

Poor irrigation management practice that leads to wasteful or inefficient use of water during irrigation with consequences of uneven distribution of water across fields, overwatering, poor timing of irrigation, and failure to monitor soil moisture levels with consequences of reduced crop yields water pollution and soil salinization

Pyramidal peak *see horn*

Ribbon lake long, narrow lake created in a valley carved out by a glacier in glaciated mountainous regions, and it fills the trough left behind by glacial erosion producing a characteristic "ribbon-like" appearance

Rock glacier a tongue-shaped, slow-moving landform resembling an alpine glacier and is covered by a thick layer of debris and rock

Rock pedestal a naturally occurring rock with a shape and name that resembles a mushroom

Sand dune mound of sand that is formed by deposition of wind-blown sand over time

Saltation movement of relatively small pebbles or particles which are too big to be carried in suspension by the air or water but are transported through bounce along the desert surface or bottom of a riverbed

Star dunes dunes composed of three or more arms that form in areas with winds blowing from multiple directions per year

Serac block of glacial ice on a glacier that is formed where crevasses intersect

Steppe a dry but grassy plain with a temperate climate and located between the tropics and the polar regions

Steppe a dry grassland with a temperate climate characterized by short grasses and few trees

Suspension a process whereby very fine particles like dust and silt are lifted into the atmosphere and transported by wind

Terminal moraine poorly sorted sediments deposited at the edge, or terminus, of a glacier and mark its furthest or maximum advance

Yardang an elongated ridge-like landform carved by wind erosion in desert environments with a steep face that faces the prevailing wind direction formed by wind erosion of softer rock layers leaving behind more resistant rock ridges

Tidewater glacier a glacier that originates on land and ends in the ocean where it breaks off into the sea through calving

Valley glacier also known as alpine or mountain, it is a huge mass of ice that moves down a valley, often beginning in a cirque at the head of valley or in a plateau ice cap

Ventifacts a rock or stone which was shaped by windblown sand or ice crystals to produce distinctive features such as facets or polishing due to abrasion

Wind bridge a rock formation, typically created by wind abrasion, with a resistant rock layer remaining intact while surrounding softer rocks erode away, creating an arch-like structure resembling a bridge, commonly with openings, or windows, on either side

Wind window an area of consistent and favorable winds where the direction and strength of the wind permits for optimal movement within a particular zone, or window, of conditions

U-shaped valley flat-bottomed and steep-walled valley that is formed by glaciers through widening of floors and over-steepening walls of previously V-shaped valleys carved by rivers

Yardang a wind-eroded hill encountered in deserts formed by wind abrasion sculpturing soft rock into parallel ridges and furrows with the long axis parallel to the prevailing wind direction

Zeugen a table-like resistant rock formation in desert regions formed by wind erosion, where alternating layers of hard and soft rock are exposed, with the harder rock forming a cap rock producing distinctive ridged appearance from differential erosion

Ocean Floor and the Water It Contains 15

Abstract

After a discussion of Earth's oceans and seas and of major features of its ocean floor to begin this chapter, a description of the composition of seawater aided by a table and a pie chart follows. The chemical composition and variation of seawater temperature and density versus depth in both low and high latitudes in both low and high latitudes is discussed. Thermocline and pycnocline are described and illustrated using diagrams. Ocean zones of epipelagic (Sunlight Zone), Mesopelagic (Twilight Zone), Bathypelagic (Midnight Zone), Abyssopelagic (Abyssal Zone), and Hadalpelagic (Trenches) are described next, as are ocean layers of surface (mixed) zone, transition zone, and deep zone, which are also illustrated diagrammatically. The geometry of ocean currents, ocean gyres, and the thermohaline circulation are all illustrated by means of diagrams. Also discussed with the aid of diagrams are marine sediments, the formation of ocean waves, coastal landforms, coastal erosion, and stabilization of shores.

Learning Goals

Studying this chapter will enable you to:

- Recognize Earth's oceans and their distribution.
- Distinguish between Earth's oceans and seas.
- Understand the distribution and formation of the major features of the ocean floor.
- Define seawater and its composition.
- Describe temperature and density variations of seawater with depth.
- Understand processes affecting salinity of seawater.
- Recognize Earth's ocean layers.
- Describe the formation and distribution of ocean floor sediments.
- Describe the formation of ocean waves.
- Recognize coastal landforms and discuss how they form.
- List and describe some methods utilized in stabilizing shorelines.

15.1 Introduction and Earth's Oceans

The **hydrosphere**, which includes all water on our planet, covers ~71% of the Earth's surface. **Freshwater** constitutes only approximately 2.5% of the water on Earth. Oceans on Earth (Fig. 15.1) are one continuous body of water that contains ~97% of our water, the only one in our solar system that is known to contain liquid water. Volumetrically, the oceans contain about 1.35 billion cubic kilometers (324 million cubic miles) of water. This lion's share of water on our planet is saltwater which covers ~71% of its surface. The oceans are critical in that they make all life on Earth possible. Additionally, oceans also make our planet appear blue (the blue planet) to an observer from space. Scientists who specialize in studying the oceans, called oceanographers, divide seawater on Earth into five oceans which are, in order of decreasing size, the Pacific, Atlantic, Indian, Southern, and Arctic Oceans (Table 15.1; Fig. 15.1). The Pacific, Atlantic, and Indian Oceans merge into icy waters around Antarctica. The Pacific Ocean is the largest as well as the ocean with the deepest ocean basin of the five oceans (Table 15.1; Fig. 15.1). The Pacific Ocean has an average depth of 4,282 m (~14,040 ft) and it plunges to the deepest part of our planet of 10,935 m (~35,876 ft) (Table 15.1) in the Challenger Deep located in the Mariana Trench. The most recently named ocean, the Southern Ocean (Table 15.1; Fig. 15.1), was only recently (in 2021) officially recognized by the National Geographic.

Supplementary Information The online version contains supplementary material available at https://doi.org/10.1007/978-3-032-20231-4_15.

J. B. Chaumba, *Introductory Physical Geology*, Springer Textbooks in Earth Sciences, Geography and Environment,
https://doi.org/10.1007/978-3-032-20231-4_15

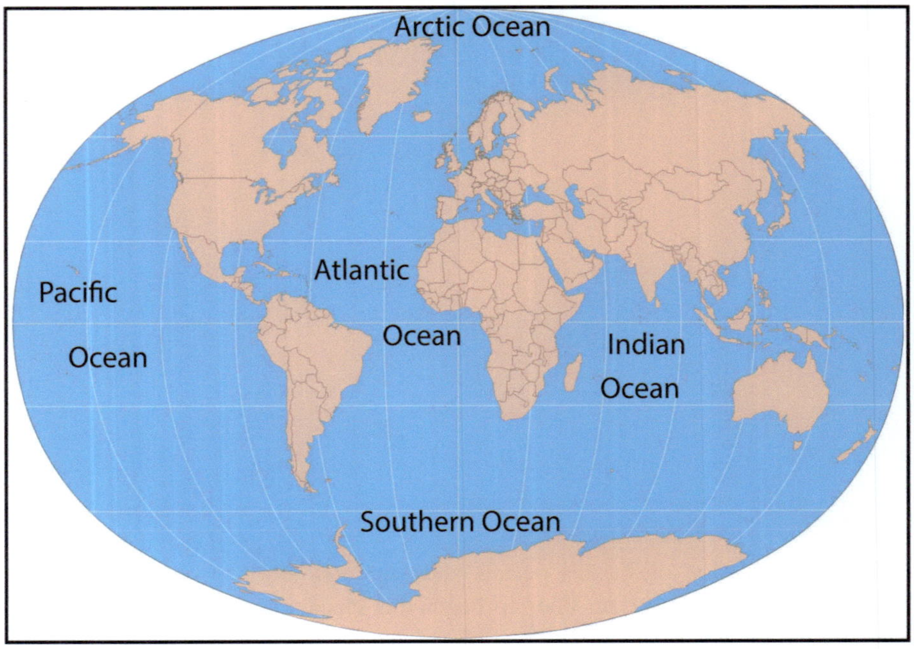

Fig. 15.1 World map showing the location of the oceans. Base map credit: Free World Maps (credit: https://www.freeworldmaps.net/printable/blank-printable-world-map-hd.jpg)

Earth's oceans are mostly defined by the continents that surround them, and they are separated by arbitrary boundaries. The Southern Ocean, for example, is separated from the southern parts of the other oceans by the 60°S latitude. The 60°S latitude is also the approximate position of the Antarctic Circumpolar Current. Oceans play a significant role in the hydrologic cycle (Chap. 12). Oceans are also significant to life on Earth as they influence our planet's weather, temperature, and the supply of goods and food. A large portion of our planet's oceans is still not mapped, however.

With an area covering 15,558,000 square kilometers, or ~6,007,000 square miles, the Arctic Ocean is the smallest of the world's five oceans and accounts for only 4.3% of the Earth's oceans.

15.2 Earth's Seas

The word "**sea**" is often erroneously used to refer to the ocean. A sea, however, is a subdivision of the ocean that is partially or entirely enclosed by land. The Baltic, Bering, Black, Caribbean, Coral, Mediterranean, North, Red, and Yellow Seas are examples of seas. Unlike inland lakes that can be composed of salty or freshwater, there is a constant exchange of water between an ocean and a sea. Seas, thus, are divisions of the ocean that are enclosed or partially enclosed by land. Inland salt lakes like the Caspian, Dead, and quickly disappearing Aral are erroneously referred to as "seas" since they don't exchange waters with any ocean and are entirely enclosed by land. Seas, therefore, are all salty or saline. Some seas are called bays. The Bay of Bengal, between Bangladesh, India, Indonesia, and Myanmar, is an example. Some lakes are also referred to seas, such as the Caspian Sea, which lies within the borders of Azerbaijan, Iran, Kazakhstan, Russia, and Turkmenistan. More than 50 seas are recognized on Earth (Fig. 15.2).

Three major subdivisions of seas are recognized. These are i. enclosed, ii. partly enclosed seas, and iii. hypersaline lakes. Almost all enclosed seas are connected with the open ocean by narrow waterways called straits and they extend into continents. Seas of this type include the Baltic Sea and the Mediterranean Sea. Since these seas are almost landlocked, only a small range of tides are experienced.

Table 15.1 Earth's oceans, and their sizes, average depths, and deepest points

Ocean	Area in km^2 (mi^2)	Average depth m (ft)	Deepest point and depth in m (ft)
Pacific	168,723,000 (65,144,314)	4282 (14,040)	Challenger Deep 10,935 (35,876)
Atlantic	85,133,000 (32,870,035)	3338 (10,932)	Milwaukee Deep 8,380 (27,493)
Indian	70,560,000 (27,243,368)	3,741 (12,274	Java/Sunda 7,450 (24,442)
Southern	21,960,000 (8,478,803)	3,270 (10,728)	South Sandwich trench 7,432 (24,383)
Arctic	15,558,000 (6,006,977)	1,038 (3,406)	Molloy Hole 5,550 (18,210)

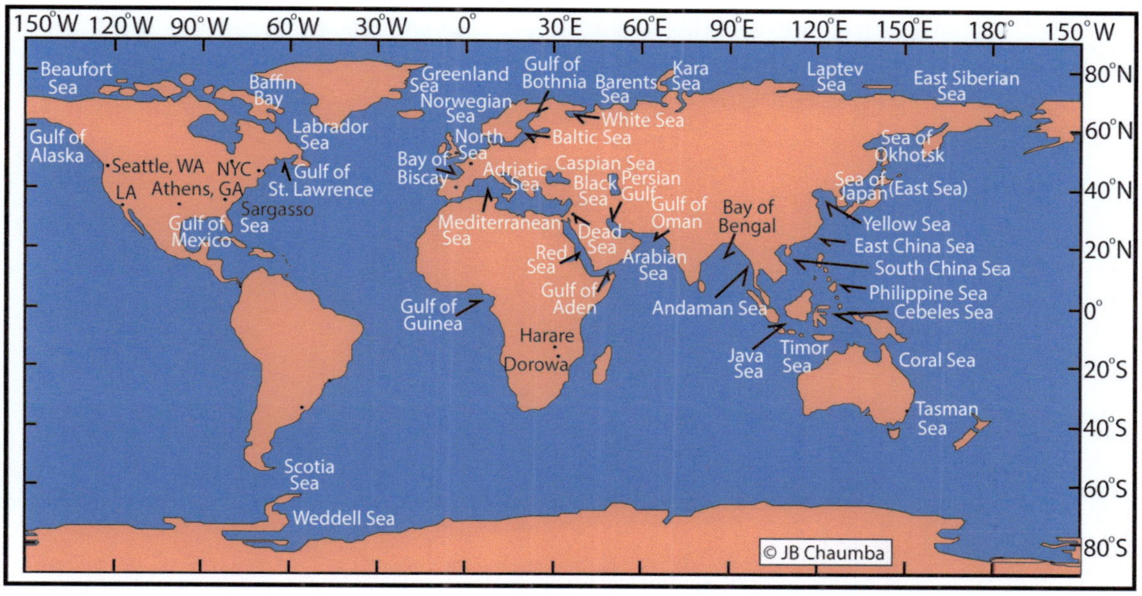

Fig. 15.2 Map showing the seas of the world

Some of the enclosed seas, however, virtually have no tides. Although the Red Sea is connected by a narrow strait, Bab-el-Mandeb Strait, it is much saltier than the Indian Ocean. The organisms that live in the Red Sea adapted to salty water. Approximately 10% of fish found in the Red Sea are unique to this sea.

Partially enclosed seas are comparable to the open ocean, particularly regarding the circulation of their waters. A wide opening, for example, links the Weddell Sea of Antarctica to the ocean. The South China Sea is connected with the ocean by passages between islands. Organisms of the South China Sea are similar across the South Pacific Ocean just organisms in the Weddell Sea are similar to those encountered in the Atlantic Ocean. However, large range of tides of oceans are comparatively smaller in seas such as those, respectively, in the Pacific Ocean and the South China Sea. Some saltwater bodies of seas are comparable to lakes since they were part of ancient oceans or seas. The access to larger bodies of water of some prehistoric oceans and seas were blocked by tectonic shifts leaving them bound by land. Some seas are hypersaline due to their very high salt concentrations. Some seas are called hypersaline if these saltwater bodies are blocked lakes, due to the extremely high salt content in their waters. An example of a landlocked sea encountered between Europe and Asia is the Caspian Sea. Since the Volga River empties its water into the Caspian Sea in the north, the northern part of the sea in Russia and Kazakhstan is almost comparable to that of freshwater lakes. The chemistry of the water in southern part of the Caspian Sea encountered in Iran and Turkmenistan, however, is saline.

The Dead Sea, which receives water from the Jordan River, but no rivers flow out of it, is another landlocked sea. The Jordan River has been managed for the benefit of agriculture and industry. Canals and dams for irrigation and power that were constructed on this river result in less water reaching the Dead Sea. Consequently, this sea is shrinking at an extremely high rate of ~1 m (3.3 feet) a year. The North Sea, shallow sea in the Atlantic Ocean, the Baltic Sea in the North Atlantic Ocean, the South China Sea, a marginal sea of the western Pacific Ocean, the Black Sea, a large inland sea in southeastern Europe, and the Persian Gulf, a shallow marginal sea of the Indian Ocean, as all seas that are encountered where land and ocean meet. These seas are all located on the margins of oceans and are partially enclosed by land.

15.3 Major Features of the Ocean Floor

Our planet was formed about 4.6 billion years ago, as discussed in Chap. 1. Earth's ocean waters are thought to have been sourced from rocks inside the newly forming Earth. As the molten material of our proto planet cooled, it released water vapor and other gases into the atmosphere. Eventually, the water vapor condensed and covered the crust with a primitive ocean. Hot gases from the Earth's interior are interpreted to continue to produce new water at the bottom of the ocean. In Chap. 6, we briefly discussed the morphology of our planet's ocean floor. Mapping of Earth's ocean floor began in the 1920s, with scientists using instruments called echo sounders which measure water depths using sound waves. Echo sounders use sonar technology. Sonar is an acronym for **SO**und **N**avigation **A**nd **R**anging. The sonar showed that the ocean floor has dramatic physical features, including huge mountains, deep canyons, steep cliffs, and wide plains.

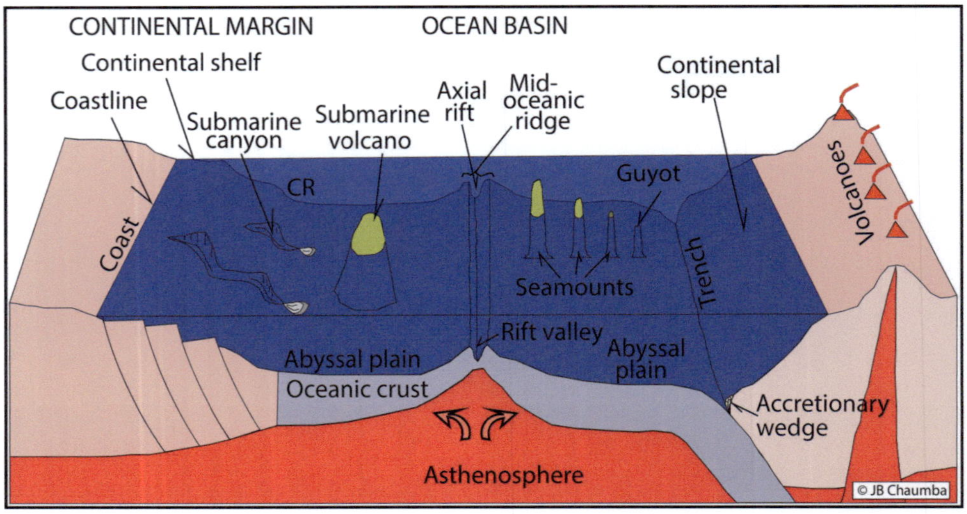

Fig. 15.3 Diagram showing some major features of the ocean floor. CR stands for continental rise

The ocean floor has been subdivided into several different areas. As one moves from the continent toward the ocean floor, the continental shelf is encountered first. The continental shelf dips very shallowly toward the ocean such that it is nearly flat. The continental shelf can be viewed as an underwater extension of a continent. The width of the continental shelf varies widely, ranging from less than 1 km to wider than 1,500 km, with an average width around 65 km. Continental shelves have the widest widths in the Arctic Ocean, whereas the narrowest widths are encountered along South America's Pacific coast. In general, the continental shelf is widest along low-lying coasts and narrowest along mountainous coasts. The transition zone between the edge of a continent and the deep ocean floor is called a **continental margin** (Fig. 15.3). One can envisage a continental margin as the boundary between the continental crust and the oceanic crust.

Two types of continental margins can be distinguished. A **passive continental margin** is one in which the transition from continental to oceanic crust lacks a plate boundary. At this boundary, a roughly wide and flat coastal plain is characterized by a broad continental shelf. Examples of passive continental margins are the east coast of North America and almost the entire coast of Africa. The other type, called an **active continental margin**, is one where the coast coincides with a plate boundary and it is characterized by interaction of at least two plates. Subduction, earthquakes, volcanic activity, and mountain building are associated with an active continental margin. The east coast of Japan, California, and the U.S. Pacific Northwest as well as southern Alaska are examples of active continental margins. Active margins tend to have steeper and narrower continental shelves and slopes.

Sediments that are sourced from the nearby continent cover the continental shelf. Still other sediments on the continental shelf are sourced from far away, then transported and deposited by rivers. Yet other sediments are trapped by lakes and natural dams. In rivers that originate or flow through formerly glaciated parts of our planet, **relict sediments** are those that originated from the last glacial period, or Last Ice Age, when the oceans receded and exposed the continental shelf.

The **continental slope** is encountered at the outer margin of the continental shelf where the land drops off sharply seaward (Fig. 15.3). Here, the slope of the continental shelf descends almost to the bottom of the ocean. The slope of the continental shelf then decreases to merge with the continental rise. The continental rise descends to the deep ocean floor, which is called the **abyssal plain** (Figs. 15.3 and 15.4). An abyssal plain is flat region with a gentle slope formed from accumulation of sediment over long periods of time. Abyssal plains are broad, flat areas that lie at depths of about 4,000–6,000 m (13,123–19,680 feet). Abyssal plains cover 30% of the ocean floor and are the flattest feature on Earth. They are covered by fine-grained sediments like clay and silt. Pelagic sediments, the remains of small ocean organisms, also drift down from upper layers of the ocean. Scattered across abyssal plains are abyssal hills and underwater volcanic peaks called **seamounts** (Fig. 15.3).

An **ocean trench** is a narrow, long, and steep depression that is encountered on the ocean floor (Figs. 15.3 and 15.4). An ocean trench which can reach depths in excess of 6,000 m (>19,600 feet) form the deepest parts of the ocean and include the deepest part of the ocean such as the Challenger Deep (Fig. 15.4). Challenger Deep is located in the Mariana Trench in the Pacific Ocean, not far from the island of Guam. Ocean trenches are formed as a result of plate collision at convergent plate margins.

The mid-ocean ridge, which rises from the abyssal plains of all the three major oceans, is a huge but largely undersea mountain chain. The mid-ocean ridge stretches over 64,000 km (40,000 miles) circling our planet. At mid-ocean ridges, an **axial rift**, or a rift valley (Fig. 15.3), is

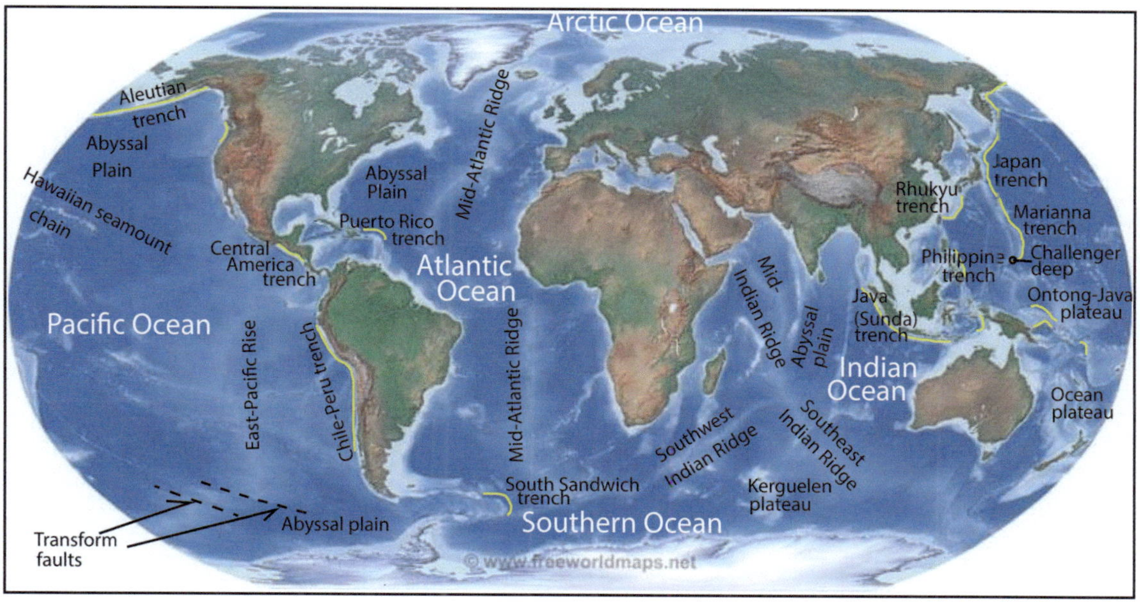

Fig. 15.4 Map of the world showing selected major features of the ocean floor. The darker the ocean floor color, the deeper the water. Base map credit: FreeWorldMaps.net: https://www.freeworldmaps.net/

encountered. The rift valley is the valley, or central fracture, situated directly along the crest of the ridge. Here, tectonic plates are moving apart which causes decompression accompanied by melting with the newly formed magma rising and erupting on the ocean floor. The rift valley, thus, marks the zone where formation of new oceanic crust is occurring through seafloor spreading. The rift valley is the most active volcanic zone on the mid-ocean ridge. Hydrothermal vents are a common occurrence along rift valleys, and so are earthquakes which are typically of small magnitude. The Mid-Atlantic Ridge is one portion of the mid-ocean ridge encountered in the middle of the Atlantic Ocean (Fig. 15.4). In the Pacific Ocean, however, the ridge is located on the eastern part of the ocean, hence it is referred to as the East Pacific Rise (Fig. 15.4).

A deep V-shaped valley with steep and narrow walls encountered in the ocean floor connecting the continental shelf to the abyssal plain is termed a **submarine canyon** (Fig. 15.3). Plate tectonics and ancient currents are inferred to have been responsible for the formation of submarine canyons. Submarine canyons are for the exchange surface and deep-sea waters through ocean currents pushing nutrients cold water rich upward through the canyon. Dense waters loaded with sediments flow through submarine canyons, which erode the continental slope in the process, ultimately depositing the sediments on the abyssal plain. The longest submarine canyon, measuring about 500 km (310 miles) long, is the Bering Canyon near the Aleutian Islands.

A volcano that erupts beneath the surface of the ocean, with an underwater volcanic vent whereby magma erupts on the ocean floor, is called a **submarine volcano**. A submarine volcano is located deep on the ocean floor, typically near divergent plate boundaries. Along mid-ocean ridges are where the bulk of submarine volcanoes are encountered. With continued magmatism at slow plate movement rates, some submarine volcanoes develop to form islands over time. Submarine eruptions are often less explosive than land-based volcanoes due to the immense pressure of the overlying ocean water. Rapid cooling underwater lava contributes to the formation of seamounts and **guyots** (Fig. 15.3). Seamounts are underwater mountains that rise from the seafloor to at least 1,000 m (3,300 feet) above the surrounding seafloor. Seamounts can be isolated or constitute part of a mountain chain. When seamounts rise above sea level, they form volcanic islands, such as the Hawaiian Island chains. Guyots are flat-topped underwater volcanic mountains that rise from the ocean floor (Fig. 15.3). Guyots once rose above sea level, and they form when volcanoes grow large enough to reach the ocean's surface, but then get eroded away chiefly by waves and weathering, as well as by wind and rainfall. The coast (Fig. 15.3) is the margin between land and sea or ocean and it includes both the land next to the ocean and the water near the shore. The coastline is where land meets water formed by tides and waves.

15.4 Seawater: Its Composition and Significance

The oceans and seas are composed of **seawater**, which is a mixture of water, dissolved salts, and other substances. The

Table 15.2 Major ion composition of sweater in mg/L

	Typical seawater	Eastern Mediterranean	Arabian Gulf at Kuwait	Red Sea at Jeddah, Saudi Arabia
Chloride (Cl^-)	18.98	21.2	23	22.219
Sodium (Na^+)	10.556	11.8	15.85	14.255
Sulfate ($SO_4{}^{2-}$)	2.649	2.95	3.2	3.078
Magnesium (Mg^{2+})	1.262	1.403	1.765	742
Calcium (Ca^{2+})	400	423	500	225
Potassium (K^+)	380	463	460	210
Bicarbonate ($HCO_3{}^-$)	140	–	142	146
Strontium (Sr^{2+})	13	–	–	–
Bromide (Br^-)	65	155	80	72
Borate ($BO_3{}^{3-}$)	26	72	–	–
Fluoride (F^-)	1	–	–	–
Silicate ($SiO_3{}^{2-}$)	1	–	1,5	–
Iodide (I^-)	0.05	2	–	–
Total dissolved solids (TDS)	34.483	38.6	45	41

Source Magazine—Water Condition and Purification, January 2005

chemical composition of seawater is comprised of approximately 96.5% water and 3.5% dissolved salts. Sodium and chloride ions account for the bulk of the dissolved salts (Table 15.2; Fig. 15.5). Additionally, seawater is also composed of smaller amounts of other substances such as organic and inorganic materials, as well as particulates as shown in Table 15.2 and Fig. 15.5. The density of seawater is higher than that of freshwater due to the higher concentrations of dissolved salts which are responsible for increasing the mass of saltwater more in comparison to the volume. Seawater is not neutral as its pH is approximately 8, which makes it slightly alkaline. The freezing point of seawater is lower than the temperature of pure water due to the dissolved salts it contains. Submarine volcanic eruptions, chemical weathering of rocks on the coasts, as well as chemical reactions which involve volcanic matter are the

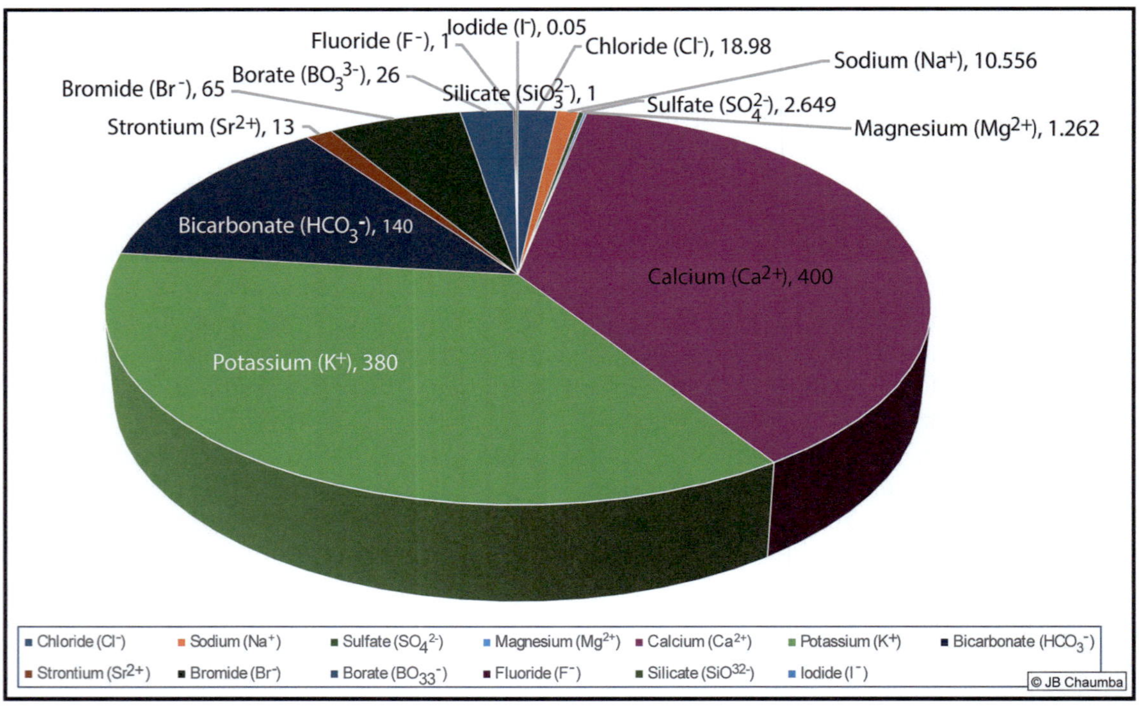

Fig. 15.5 Pie chart showing the composition of seawater. *Data source* Magazine—Water Condition and purification, January 2005

sources of salts in seawater. The amount of dissolved salts in any body of water is called **salinity**. It is measured in grams of salt per 1,000 g of water. Seawater has an average salinity of approximately 3.5%, which means that every kilogram of seawater contains approximately 35 g of dissolved salts. This makes seawater a prized source of valuable chemical elements from which many are commercially exploited.

Weathering and erosion of rocks on continents is the major source of salts in seawater. Rainwater and groundwater dissolve minerals that make up the rocks on the continents and transport them to the ocean via rivers and streams. This process is called weathering, and it is the primary source of ocean salinity. Weathering rates are higher in hot and humid climates.

The country with the world's largest desalination facilities is Saudi Arabia. Desalination facilities in this country include the Shoaiba and Al Jubail complexes. With about 70% of its freshwater demand sourced from desalination plants, Israel is the leading country in desalination technology. The Soreq Plant in Israel is the largest seawater reverse-osmosis membrane plant in the world. United Arab Emirates hosts one of the world's largest power and desalination plants in the Dubai Electricity and Water Authority (DEWA) M-Station project, whereas Australia currently opened seven desalination plants from 2006 to 2012. In Spain, a staggering 765 desalination plants supply freshwater daily to approximately 8 million of its citizens. As of early 2025, there were approximately 16,000 operational desalination plants in 177 countries. Out of these plants, 360 treat seawater, whereas 405 treat brackish water. There are less dissolved salts in brackish water in comparison to those in seawater, but there are more dissolved salts in brackish water relative to those in freshwater.

There are currently 11,000 desalination plants in the world that produce 20 million cubic meters per day. These desalination plants treat seawater with an average of 40.000 ppm and brackish water with an average TDS of 10.000 ppm, respectively. The lion's share of these desalination plants (60%) is located in the Middle East and Western Asia. Recall from Chap. 14 that this region is dominated by arid and semi-arid conditions and, together with the high population density in this part of the world, necessitates installation of desalination plants to meet household water requirements. Other parts of the world with desalination plants are North America (11%), North Africa and Europe both with 7%, and South and Central America with 4%. The remaining 11% of desalination plants are in coastal or near-coastal regions throughout the world such as, for example, Australia, Ghana, and Kenya. The Unites States has almost 2000 desalination plants, based on early 2025 figures.

15.5 Processes Affecting Salinity of Seawater

In low latitudes, that is, in equatorial regions, temperature decreases significantly with depth (Fig. 15.6). In low latitudes, also referred to as equatorial regions, surface water is much warmer than the deep water (Fig. 15.6), while in high latitudes (near the poles), there is little change in temperature with depth, as both surface and deep water are relatively cold; essentially, at low latitudes, there is a pronounced temperature gradient with depth, whereas at high latitudes, the temperature remains fairly constant throughout the water column.

In low latitudes, surface water is warm and the temperature versus diagrams is characterized by a sharp temperature

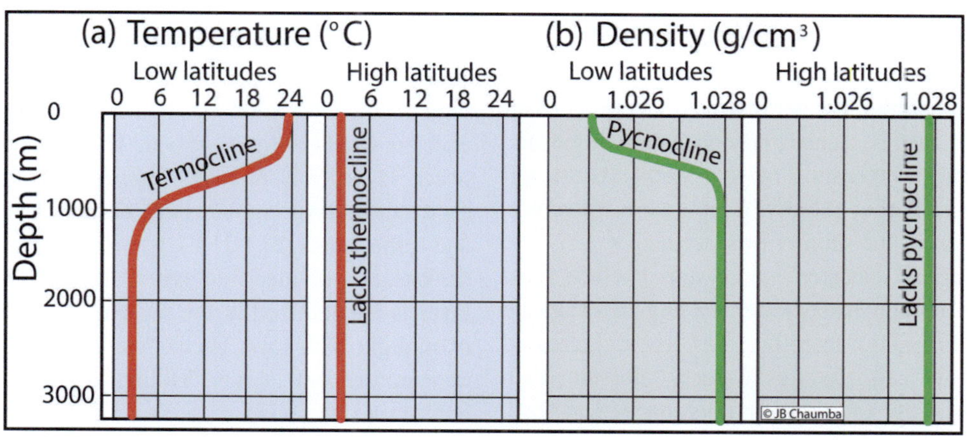

Fig. 15.6 Graphs showing the variation of depth versus seawater temperature in both low and high latitudes (**a**) and of ocean depth versus density in both low and high latitudes. Note that surface seawater temperatures are very warm at low latitudes at depths of $\leq$1000 m but cold throughout the entire water column at high latitudes (**a**). Shown in (**b**) is the variation of depth versus density of seawater in both low and high latitudes. Note the rapid increase in density of seawater with depth at shallow depths of less than 1000 m in low latitudes. At high latitudes, seawater density is high through the entire ocean column

decline as depth increases displaying a well-defined thermocline (Fig. 15.6). High latitudes are characterized by cold surface water which displays no change in the thermocline, lacks temperature change with depth (Fig. 15.6). The sun's rays strike the Earth more directly at low latitudes, which cause warmer surface waters. As the warm water sinks deeper, it cools and forms a temperature gradient, producing the thermocline. Global ocean currents are responsible for redistributing heat from the equator toward the poles. The ocean currents influence temperature variations at different latitudes as the heat is redistributed poleward.

On a graph of variation of density ocean water with depth, a trend of increasing density with increasing depth is displayed, with the significant change encountered in a relatively shallow layer near the surface termed the pycnocline. The pycnocline is where temperature and salinity decrease rapidly, which leads to a sharp increase in density. Below the pycnocline, the increase in density gradient becomes more gradual down the ocean floor where the highest density water at the ocean floor is registered. The lowest density is recorded by the ocean surface layer caused by both warmer ocean temperatures and low salinity. The pycnocline is a distinct zone of rapid increase in density with depth. The change coincides with the thermocline. Beneath the pycnocline, relatively uniform density occurs in the deep layer which increases gradually with depth due to increasing pressure. Higher salinity leads to increased density and temperature being the chief factor influencing density. The colder the water is, the denser it becomes. As depth increases, pressure increases too, which contributes slightly to higher density at greater depths.

Precipitation, evaporation, formation, and melting of sea ice, together with freshwater input from rivers, are the primary processes affecting the salinity of seawater. Precipitation affects salinity through addition of freshwater to the ocean which dilutes the concentration of salt in seawater. Precipitation, thus, decreases salinity. The level of salinity in any given location in the ocean is determined by the balance that occurs between water that is lost through evaporation and water that is gained through precipitation and river runoff. During evaporation of seawater, salts are left behind, resulting in higher salinity. Evaporation, therefore, leads to an increase in the salinity of seawater.

During freezing of seawater, ice crystals exclude most salt, salt crystals remain in seawater leading to increased salinity values. Freezing, therefore, results in increased seawater salinity. As ice melts, freshwater is released which helps lower the salinity of seawater. Melting of ice, thus, decreases the salinity of ocean water. When freshwater flowing from rivers is emptied into the ocean, the salinity of seawater is greatly reduced, particularly along coastlines. Ocean currents are capable of transporting water with varying levels of salinity to different locations, changing the salinities of the waters at their new locations. Wind can affect seawater salinity through mixing of different volumes of water. Mixing of different masses of water results in waters with net salinities that are distinct from the starting salinities of the mixed waters. Temperature influences seawater salinity through its effect on rates of evaporation. With increase in temperature, as in equatorial regions more water evaporates which leaves behind seawater that is higher in concentration of salt. Consequently, seawater salinity also increases. Conversely, colder temperatures, in contradistinction, lead to reduced evaporation and hence lower salinity. In cold regions that characterize high latitudes, less water evaporates which leads to lower seawater salinities. This is caused by cold seawater forming sea ice upon freezing which leaves the salt behind in seawater. The increased salt concentrations in seawater make it saltier. The opposite also holds: freshwater is released upon melting of sea ice which dilutes, or lowers, seawater salinity.

15.6 Ocean Layers

On a vertical cross section of the layers of ocean, starting from the surface of the ocean downward to the deepest point, distinct zones are encountered. The topmost layer is called the epipelagic (Sunlight Zone), Mesopelagic (Twilight Zone), Bathypelagic (Midnight Zone), Abyssopelagic (Abyssal Zone), and Hadalpelagic (Trenches), each characterized by decreasing light levels and distinct marine life adaptations as depth increases; with the Epipelagic being the shallowest, well-mixed layer, and the Hadalpelagic representing the deepest ocean trenches.

The natural separation of the ocean's water into distinct horizontal layers caused primarily by differences in the density of water caused by variations in temperature and salinity is called ocean layering. Warmer and less salty water floats above colder and saltier water. Distinct zones with distinct characteristics with increasing depth and differences in levels of light formed. The major factor that drives ocean layering is density. Temperature and salinity influence ocean layering whereby denser water that is colder and saltier sinks (Fig. 15.7). From the surface to the bottom, the ocean is divided into several layers. The surface mixed layer is encountered at the top and, thus, receives the bulk of sunlight. This layer is characterized by relatively uniform temperature and salinity. Mixing in this layer is caused by both waves and wind.

The zone occurring below the surface mixed zone is called the **thermocline** or transition zone (Fig. 15.7). In the transition zone, temperature rapidly decreases with increasing depth (Fig. 15.6). The last layer, called the deep zone (Fig. 15.7), or the deep ocean layer is character-

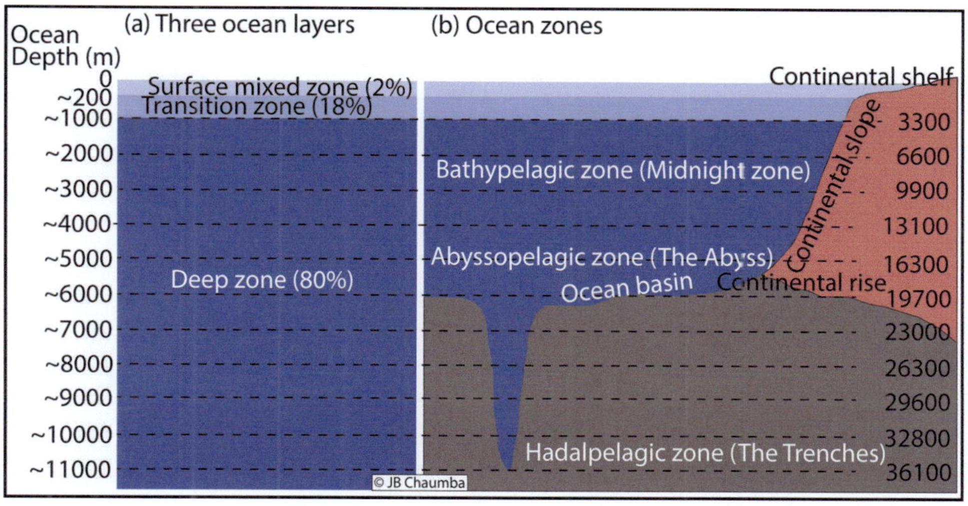

Fig. 15.7 a Diagram showing the three main layers in ocean water based on differences in their densities due to variations in salinity and temperatures. Shown in (**a**) are the various zones in the ocean with increasing depth from the surface of the ocean

ized by minimal mixing. The deep zone is a stable but cold layer encountered below the thermocline (Fig. 15.7). These different ocean layers support distinct ecosystems in which organisms have adapted living in, depending on temperature, nutrients, as well the amount of available light.

15.7 Geometry of Ocean Currents

Oceanic currents denote the movement of water from one location to another in the world's oceans. The significance of ocean currents is that they play a critical role in regulating our planet's climate. Water contained in the oceans has currents that display a range of patterns (Fig. 15.8). Ocean currents can be classified as warm or cold based on the direction in which the water moves as well as the effect they have on the overlying air. Currents that transport warm water from the tropics toward the poles and release energy into the air are called **warm currents**. Warm currents are typically encountered along the east coasts of continents, such as the Agulhas current in Africa and the Gulf Stream in North America (Fig. 15.8). **Cold currents**, on the other hand, transport cold water from high latitudes toward the equator (Fig. 15.8). Cold currents are typically found along

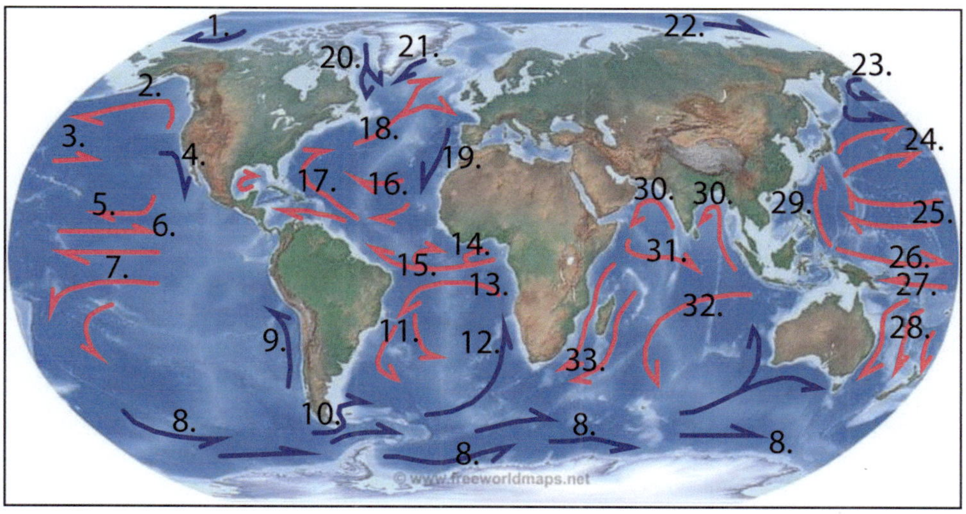

Fig. 15.8 Map showing the main ocean currents of the world: 1. Bring current, 2. Alaska current, 3. North Pacific drift, 4. California current, 5. Pacific Equatorial current, 6. Pacific Equatorial countercurrent, 7. South Pacific Equatorial current, 8. West wind drift, 9. Humboldt current, 10. Falkland current, 11. Brazil current, 12. Benguela current, 13. South Atlantic Equatorial current, 14. Guinea current, 15. Atlantic Equatorial current, 16. North Atlantic Equatorial current, 17. Gulf stream, 18. North Atlantic Drift, 19. Canary current, 20. Labrador current, 21. East Greenland current, 22. Bering current, 23. Oyashio current, 24. North Pacific Drift, 25. North Pacific Equatorial current, 26. Pacific Equatorial countercurrent, 27. South Pacific Equatorial current, 28. East Australian current, 29. Kuroshio current, 30. North Indian Equatorial current, 31. Indian Equatorial countercurrent, 32. South Indian Equatorial current, and 33. Agulhas current. Base map credit: FreeWorldMaps.net: https://www.freeworldmaps.net/

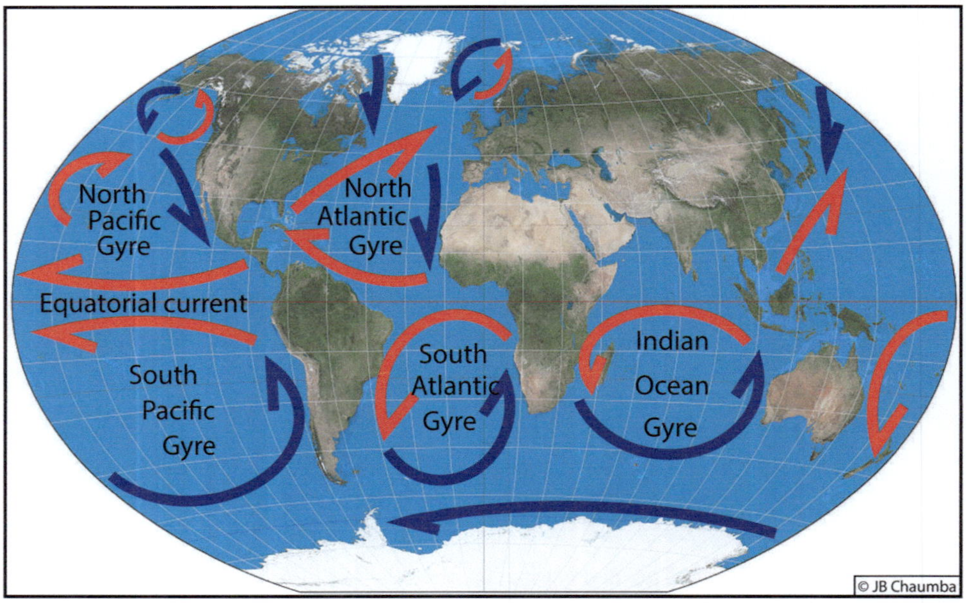

Fig. 15.9 Map of the world showing gyres. Base map credit: National Geographic

the west coasts of continents, and they absorb energy that is received in the tropics. Cold currents cool the air above. The coastline of California, for example, experiences a current that runs north to south, which transports cool water and keeps coastal temperatures relatively cool (Fig. 15.8). Ocean currents are either caused by wind or by differences in density and salinity of ocean water. Since warmer water is less dense in comparison to colder water, it pushes through the colder water and creates currents. Surface currents and deep ocean currents are two main types of ocean currents. These patterns include thermohaline circulation, gyres, as well as upwelling. Ocean currents act in a manner akin to a conveyor belt by transporting warm water, together with precipitation, from the equator toward the poles and then cold water from the poles back to the tropics. Ocean currents play a critical role in regulating global climate, and they also help to counteract the uneven distribution of solar radiation that reaches the Earth's surface. The band of low pressure that circles the Earth near the equator where the trade winds of the Northern and Southern Hemispheres meet is called the **intertropical convergence zone** (ITCZ). The ITCZ is also referred to as the "**doldrums**" due to the equatorial region's rising air caused by rising air to produce calm surface conditions. It is not static, but shifts based on the season, which affects wet and dry seasons in equatorial regions. This low-pressure band and clouds occur near the planet's equator at the confluence of the northeast from the northern hemisphere and the southeast trade winds from the Southern Hemispheres. It is a convergence of warm and moist air which rises, cools, and releases moisture resulting in frequent, intense showers of rainfall as well as thunderstorms. The ITCZ is also known as. A large system of rotating ocean currents which produce a circular pattern of movement of water that is driven by both the Earth's rotation and wind patterns in an ocean is called a **gyre**. Gyres form large but slow-moving "whirlpools" across the surface of the ocean. Five major gyres are recognized in Earth's oceans: the North Atlantic, South Atlantic, North Pacific, South Pacific, and Indian Ocean gyres (Fig. 15.9). Gyres are chiefly caused by the Coriolis Effect which deflects currents as a consequence of the Earth's rotation. The chief role of gyres lies in the distribution of heat, nutrients, as well as marine life across all the oceans. Additionally, weather and climate are also determined by the movement of ocean water, together with the movement of air. Tidal currents are generated by tides. Tides are lengthy but slow waves that are caused by gravitational pulls of both the sun and moon. Tides are part of ocean currents as well. In each gyre, a strong western boundary current and a weaker eastern boundary current are encountered. The Gulf Stream, for example, forms the western boundary current of the North Atlantic Gyre. The Central American Gyre often produces hurricanes or tropical cyclones, of which some develop into hurricanes that typically impact North America.

Gyres also collect and transport varying amounts of marine debris or pollution that is released in coastal areas. The Great Pacific Garbage Patch (GPGP) in the North Pacific Ocean between Hawaii and California and composed of plastic and other floating trash is a good example. GPGP is comprised of pieces of plastic, fishing nets, and other debris. The debris is sourced from the Pacific Rim areas in Asia, North and South America. It is formed in the North Pacific Subtropical Convergence Zone where warm water from the South Pacific interacts with cooler water from the north. Gyres can impact the climate of coastal regions. For

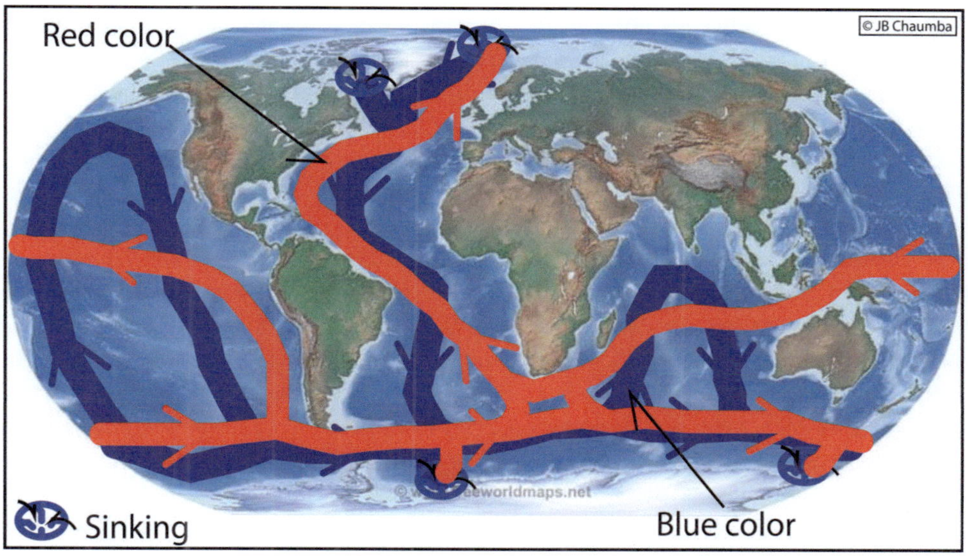

Fig. 15.10 Map of the world showing the thermohaline circulation. Note that cold and dense ocean water sinks at high latitudes and is circulated into all of Earth's oceans. Base map credit: FreeWorldMaps.net: https://www.freeworldmaps.net/

example, the Gulf Stream in the North Atlantic warms the shores of Great Britain and other parts of Northern Europe.

Ocean water is also moved by a large-scale ocean current system termed the **thermohaline circulation** or the thermohaline conveyor belt (Fig. 15.10) that moves water, nutrients, heat, as well as other materials around Earth's oceans. Thermohaline circulation is also referred to as the conveyor belt of the oceans. Earth's oceans cover ~71% of the surface, and the oceans can retain heat for longer periods than the atmosphere. Consequently, oceans are capable of redistributing heat around the globe more efficiently compared to winds. Thermohaline circulation is characterized by waters which typically flow clockwise in the northern hemisphere but counterclockwise in the southern hemisphere. In the Northern Hemisphere, the Coriolis effect deflects major surface ocean currents to the right, whereas the deflection is to the left in the Southern Hemisphere. The Coriolis force is an invisible force that appears to deflect winds, and it is perpendicular to the object's axis. The Coriolis force is applicable to motion on rotating objects. The mass of an object and its rate of rotation determine this force. For our planet, the spin on its axis is from west to east.

Different processes drive ocean currents, such as the speed and direction of wind, Coriolis force, and the influence of different continents. Differences in the density water, which are caused by variations in temperature and salinity, or haline, act as the engine driving thermohaline circulation. As ocean water gets saltier, it becomes denser which causes it to sink near the poles (Fig. 15.10). Furthermore, as sea ice forms in polar regions, salt is left behind which makes surrounding water saltier and heavier, triggering sinking of the water. Thermohaline circulation begins in the polar regions with cold ocean water sinking to the bottom of the ocean. Cold, dense sinking ocean water replaces deep cold water with surface water that is warmer and less dense and traveling toward higher latitudes where it eventually sinks. Upwelling in the thermohaline circulation occurs in the Southern Ocean particularly near Antarctica. Here, cold and dense water formed from formation of sea ice sinks and rises gradually and moves northward. Although winds drive surface ocean currents, ocean currents flow thousands of meters deep in the ocean.

The warm Atlantic Ocean current originates in the Gulf of Mexico, called the Gulf Stream, and drifts all the way to Europe via the northern parts of the Atlantic Ocean. The climate of the eastern United States as well as that of Western Europe is influenced by the Gulf Stream. The Antarctic Circumpolar Current is circumpolar since there is no landmass that connects Antarctica, hence warm ocean waters are kept away from this continent. The nutrient-enriched California Current originates in the Arctic Ocean to the north and travels southward through the Bering Sea in the western part of the North Pacific Ocean. The warm Western boundary current's flow migrates from the tropics to temperate latitudes.

15.8 Marine Sediments

Sediment is a term used for fine-grained particles of rocks and other materials that sink to the ocean floor where they eventually accumulate and settle. Marine or seafloor sediments are deposits of insoluble material derived from various sources. These sediments can be subdivided into clastic, siliceous, biogenous, pelagic, chemical, and cosmogenic sediments. Sediments that are derived from weathering are called clastic sediments. Clastic sediments are also referred to as terrigenous sediments, and they are derived from land. Clastic sediments accumulate on the continental shelf, continental rise, and abyssal plain where they are deposited. **Siliceous oozes** form a type of biogenic pelagic sediment that is encountered on the deep ocean floor. Siliceous oozes are dominantly composed of the silica-based skeletons of microscopic marine organisms like radiolarians and diatoms. Sediments that are produced from the remains of marine organisms like plankton, mollusks, and corals are called **biogenous sediments**. Biogenous sediments are enriched in organic compounds, particularly those composed of both calcium carbonate and silica. **Pelagic sediments** are restricted to the large ocean basins, with red clay and globigerina ooze as the predominant types of deposits. Pelagic sediments are fine-grained deposits that accumulate in deep water on the ocean floor, far from land and composed of clay-size siliciclastic sediment, volcanic ash, and meteoric dust. Approximately three-quarters of the seafloor is covered by pelagic sediments. The three main types of pelagic sediments are calcareous oozes, siliceous oozes, and red clays. **Chemical sediments** form from precipitation of minerals out of seawater and the precipitates are called evaporites. Sediments which originate from outer space such as cosmic dust form **cosmogenous sediments**.

When dissolved materials in ocean water precipitate out, **hydrogenous sediments** are formed. **Manganese nodules** (Fig. 15.11.) are widely on the floor of the world's oceans. Manganese nodules are concretions of the size of potato dominantly composed of manganese as well as iron, nickel, and copper. Hydrogeneous sediments, which include manganese nodules, are formed when metal compounds dissolved in the water column crystallize directly from water via different chemical reactions with deposition occurring around different nuclei. These nodules are inferred to form from accumulation of chemicals onto a nucleus of manganese. Manganese nodules vary in size and are typically the size of marbles to tennis balls. Chemicals that make up these nodules originate either in seawater or within water that is trapped between the sediment grains occurring beneath the floor of oceans to format pellets of metals near the surface of the ocean floor. The source of the manganese and iron, as well as algae and bacteria in the nodules, is from eruption of submarine volcanoes. Growth of the manganese nodules, which grow at very slow rates of only a few millimeters per million years, ceases once the nodules are buried beneath at least a few centimeters of sediment.

Evaporites form when seawater evaporates, leaving behind dissolved materials that precipitate into halite. Green clays, which contain high proportions of iron and potassium, form grains of the size of sand. Gypsum forms concentrates of seawater in areas experiencing evaporation of freshwater. Hydrogenous calcite sediments form aragonite crystals that can be used to rebuild reefs and remove pollutants. **Turbidity currents**, which are fast, high-density downhill basal flows, with thicknesses of several meters thick, transport huge amounts of sediment downslope. The sediments transported by turbidity currents are composed mostly of sand and mud, making turbidity currents denser than seawater. Turbidity currents are a type of gravity flow encountered in oceans and some lakes. The high quantities of sediment suspended in the seawater contribute to the high densities of turbidity currents. These currents, travel distances of hundreds of kilometers (miles), have speeds that reach 20 m/s and often move down submarine canyons. Sediments that turbidity currents transport are from coastal regions to the deep-sea floor. These cause damage to, or destroy, any underwater infrastructure such as cables and pipelines.

15.9 Formation of Ocean Waves

Wind that blows across the surface of the water is chiefly responsible for the generation of ocean waves. Friction of

Fig. 15.11 **a** Image of extensive ferromanganese nodules field interspersed with wide patches of ripple-ornamented sediment from Dive 17 of the 2021 North Atlantic Stepping Stones expedition (credit: NOAA Ocean Exploration, 2021 North Atlantic Stepping Stones). **b** Nodules from the abyssal plain in the central Pacific Ocean. *Credit* NOAA

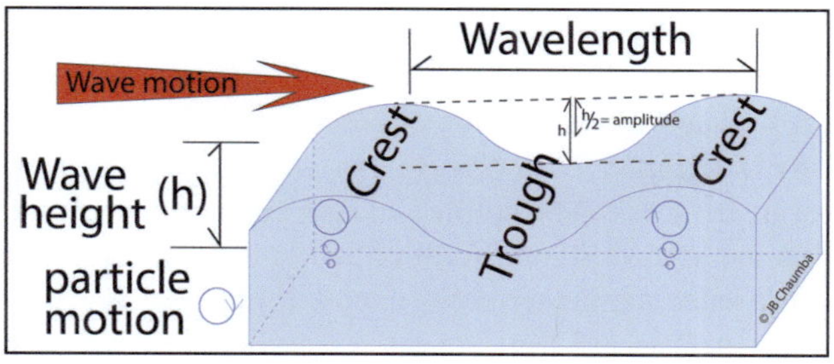

Fig. 15.12 Diagram illustrating a wave with associated features of crest, trough, wavelength, wave height, and particle motion

the wind as it blows over the ocean surface disturbs the water, thereby generating ocean waves (Fig. 15.12). These wind-driven waves are formed through the transferring of energy from the wind to the water. Consequently, the water moves in a circular motion resulting in the formation of crests and troughs that are observed on the surface of the ocean. The highest point of a wave is called the crest, whereas the lowest point is referred to as the trough.

The **crest** of a wave is its highest point, whereas its lowest point is called the **trough** (Fig. 15.12). **Wave height** is the total vertical distance between the crest and the trough, or the perpendicular distance between a line connecting crests of a wave and that connecting the troughs of a wave (Fig. 15.12). The **amplitude** is half of the wave height when it is measured from the resting position to either its crest or trough (Fig. 15.12).

Waves travel huge distances in the oceans. The motion of particles of water in circular paths as a wave passes through (Fig. 15.12), in which water molecules move up and down as well as slightly forward and backward, creates circular orbits as the energy of the wave itself travels across the water surface. The particles of water, however, don't travel with the wave but instead move in a circular pattern around their original position as the wave passes by (Fig. 15.12). Although water appears like it is moving with the wave, the particles of water themselves don't travel far. Instead, the water particles only move in circular paths around their original positions. The size of the water particles decreases its circular motion with depth (Fig. 15.12) as water particles near the surface produce the largest circular motion, whereas increasingly deeper water particles form increasingly smaller circles.

Fetch is the distance over which the wind blows consistently across the water surface. The size of waves is dependent on the fetch. The fetch area, wind speed, as well as the duration of wind blowing influence the size of the waves that are produced. The energy of a wave travels through the water in a circular motion whereas molecules of water don't travel long distances. Large and destructive waves are caused by disturbances that occur underwater, such as volcanic eruptions, landslides, and/or earthquakes.

Ocean waves cause erosion along coastlines through the sheer force of waves breaking against the coast wearing away the land. Rocks and sediment are removed by waves which slam against the land chiefly due to the energy of waves. In addition, currents and tides also erode the shoreline. The energy of waves as they are crashing against cliffs dislodges rocks and causes cracks to form through **hydraulic action**. Breaking waves erode the coast due to particles of sediment that are suspended in ocean water which also dislodge rocks. Ocean water also erodes the coast through rock particles transported by waves which grind against any rock faces through **abrasion**. Additionally, particles of rocks that are transported by waves collide with each other and further break up rocks into smaller pieces through a process called **attrition**.

If a wave has enough energy to break at its peak, it is referred to as a **breaker** (Fig. 15.13). Breakers form upon waves approaching the shore when the bottoms of the waves touch the floor of the ocean, which causes the fronts of the waves to slow down. The wavelengths of waves decrease as the waves slow down, resulting in the height of the waves increasing to form breakers. Waves become unstable and form a breaker when their height exceeds one-seventh of their wavelength. The bottoms of the waves slow down before their tops which causes the wave crests to get ahead

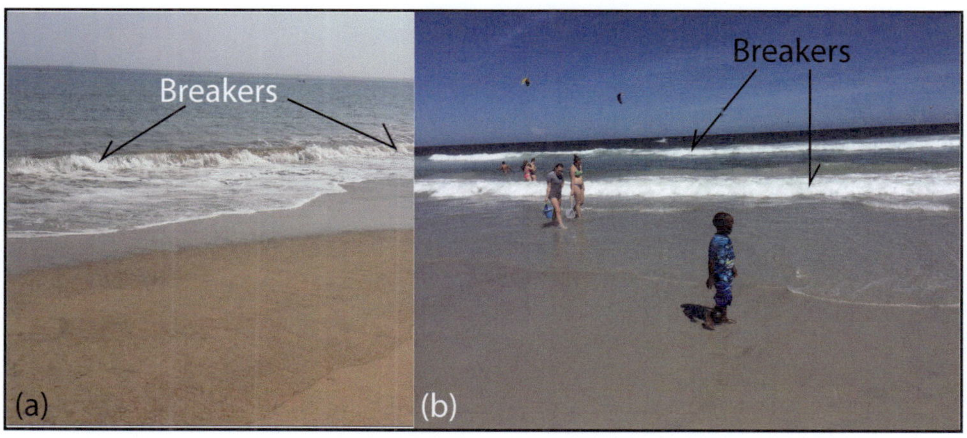

Fig. 15.13 **a** and **b** Breakers at Long Beach, Oak Island, North Carolina, USA

of the rest of the waves. Consequently, the wave crests fall forward as breakers (Fig. 15.13).

15.10 Coastal Landforms

Wave erosion creates landforms which are unique to coastal regions. Vertical cliffs, which are created by constant erosion at the base of a shoreline, leads to the formation of **wave-cut platforms**. A wave-cut platform (Fig. 15.14a) is a relatively flat and rocky surface at the base of a cliff that is formed by the abrasion of waves. It is formed by waves that crash repeatedly against the base of sea cliff, through processes abrasion and hydraulic action. With time a notch at the cliff base is formed. With deepening of the notch, the cliff becomes unstable until it finally collapses. Continuation of this process causes the cliff to retreat landward. As erosion occurs, waves transport away the eroded material from the cliff, producing a relatively flat surface at the base of the retreating. A wave-cut platform is uplifted to create the **marine terrace** (Fig. 15.14a). A marine terrace is a broad and flat shelf often encountered along the coastlines of the islands. Marine terraces typically surround the central part of islands, and they are also common along the coastline of California, USA (Fig. 15.14a). As uplift occurs, some marine terraces can be uplifted and get exposed further away from the new coastline. North of San Francisco, California, such a paleo marine terrace is exposed (Fig. 15.15).

Sea arches (15.14b) are natural formations that occur when waves erode rock over time, creating a hole in a headland. They form through waves eroding coastal rocks to form sea caves that eventually connect below a headland creating sea arches. Sea arches can be likened to holes that were formed by wave erosion through sea cliffs. With continued wave erosion, sea arches eventually collapse and leave behind sea stacks.

Sea cliffs (Fig. 15.16) have steep slopes or vertical drops along a coastline that are formed by the erosion of waves. Waves break down and remove rock from the cliff face, to form a notch that undercuts the ground above it. The ground becomes unstable and collapses, and the process repeats itself, causing the cliff to slowly retreat inland.

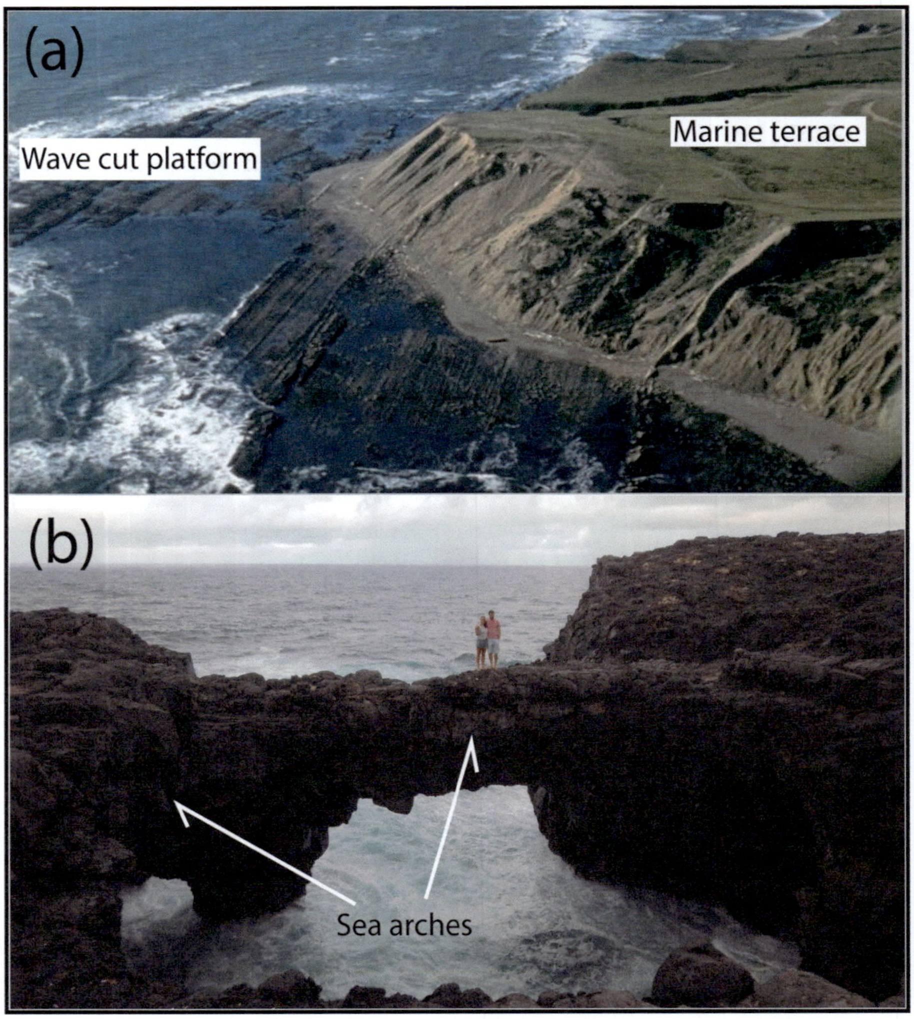

Fig. 15.14 **a** Photograph of wave-cut platform and marine terrace, Bolinas Point near San Francisco along the California coast, USA (Credit: J. S. Shelton). **b** Photograph of sea arches formed from erosion pf the basaltic southern coast of Mauritius by ocean swells. *Credit* D. L. Reid

Fig. 15.15 Photograph of paleo marine terrace from terrace exposed north of San Francisco, California, USA. *Credit* D. T. Nelson

Fig. 15.16 Photographs of sea cliffs. **a** Sea from Mediterranean with villas on top (Credit: Andreas Geissler via https://www.pexels.com/photo/stunning-clifftop-villa-overlooking-mediterranean-sea-29874112/), **b** Coastal scene with modern cliffside architecture, San Francisco, California (credit: Robert So via https://www.pexels.com/photo/coastal-scene-with-modern-cliffside-architecture-29867243/). **c** Roughly linear sea cliffs in California, USA. *Credit* D. T. Nelson

Waves break down and remove rock from the cliff face, creating a notch that undercuts the ground above it. The ground becomes unstable and collapses, and the process repeats itself, causing the cliff to slowly retreat inland. Sea cliffs are determined by the hardness of rocks. Hard cliffs are characterized by rock faces that can tower >300 m high. Cliffs composed of soft rocks have varied appearances. Older soft cliffs, such as those encountered on the eastern coats of Southeastern USA and east coast of Mozambique and South Africa, have marshy areas, scrub, and grassland. Recent cliffs, such as those on the west coast of the Unites States, often have some plants in seepages or plants that grow on bare ground.

A **promontory** is high land or rock that projects into a body of water. A good example of a promontory is Cape Point in South Africa (Fig. 15.17a). Cape Point is a prominent mass of land that projects into the South Atlantic Ocean and it marks southeast corner of the Cape Peninsula in Western Cape Province of South Africa. A **sea stack** is a rock formation that resembles a pillar, and it is detached from the coast (Fig. 15.17b). Sea stacks are spectacular features of rocky coasts. They are created by different rates of erosion, which is caused by the varied resistance of bedrock.

Higher waves that are powerful are capable of more coastal erosion. The rock types that comprise a shoreline also determine the features encountered along the coast. A beach is an accumulation of sediments deposited by waves and located along the **coastline** (Fig. 15.18a, b). A coastline is the boundary between the sea and land. Rocky coastlines

Fig. 15.17 Shown in (**a**) is a promontory that makes Cape Point, South Africa (credit: Jean-Baptiste D., https://unsplash.com/photos/a-view-of-the-ocean-from-a-cliff-UgTA_mzTP3M). **b** Sea cliff and sea stacks on the California coast, USA. *Credit* D. T. Nelson

Fig. 15.18 Photograph of beach along the Pacific Ocean, west of Highway 1 north of San Francisco, California, USA. *Credit* D. T. Nelson

are more susceptible to abrasion in comparison to sandy beaches. Furthermore, where storms are severe, high-energy waves develop which lead to extensive coastal erosion. A beach is a narrow strip of land with a gentle slope along the shoreline of a body of water such as an ocean, lake, or river and typically composed of loose sand, pebbles, and gravel or any mixture of these (Fig. 15.18). A **beach** is formed by the accumulation of sediment carried by waves and currents.

We discussed desert dunes in Chap. 14. Dunes also form along some coasts due to waves and winds picking up sand from the coast and shore, which it then blows landward (Fig. 15.19a, b). The windward side of coastal dunes, therefore, is on the ocean side. When you go for a swim or to relax at the beach, you will be between the beach dunes and the coastline. A narrow coastal landform that projects from the coastline into the ocean and composed of sand or gravel is referred to as a **spit**. A depositional landform, a spit is inferred to form through longshore drift whereby sediments are deposited by ocean currents typically at the mouth of estuaries or bays. Spits generally form where longshore currents that transport sediments along a coast and deposit them where the current changes direction such as around headlands. Thus, spits tend to form in areas where the reduction in energy occurs at the mouth of a river where it empties its water into the ocean or sea. An example of a spit is at the mouth of the Russian River in Jenner in California which forms during the spring (Fig. 15.19c). Spits are often made of sand or shingle and are formed by longshore drift. This process involves waves moving sediment along the shore, which eventually collects at the end to form a spit. Spits tend to form where there is an abrupt change in the direction of the coastline. Spits are dynamic landforms that change size and shape with time in response to erosion and deposition.

A **tombolo** is a narrow piece of land such as a bar or a spit that connects an island to the mainland. The word tombolo is derived from the Italian word tombolo, which means "cushion" or "pillow." Tombolo, one or more sandbars or spits that connect an island to the mainland. Tombolos are encountered where the coastline is composed of rocks, such as in California and the northeastern United States (Fig. 15.20).

A **baymouth bar** is formed when a spit reaches across the opening of a bay, i.e., when a spit closes off a bay. Baymouth bars are encountered along low coasts. States that border the Gulf of Mexico have tidal inlets which are associated with baymouth bars. Examples of baymouth bars include Stinson Beach in California and the Rodeo Beach

Fig. 15.19. **a** View of dunes to the right and Fernandina Beach to the left, Florida, looking south with the Atlantic Ocean to the left. **b** View looking north with dune to the left and Oak Island beach to the right, Oak Island Beach, North Carolina. **c** Russian River spit, Goat Rock Beach, California, USA. *Credit* D. T. Nelson

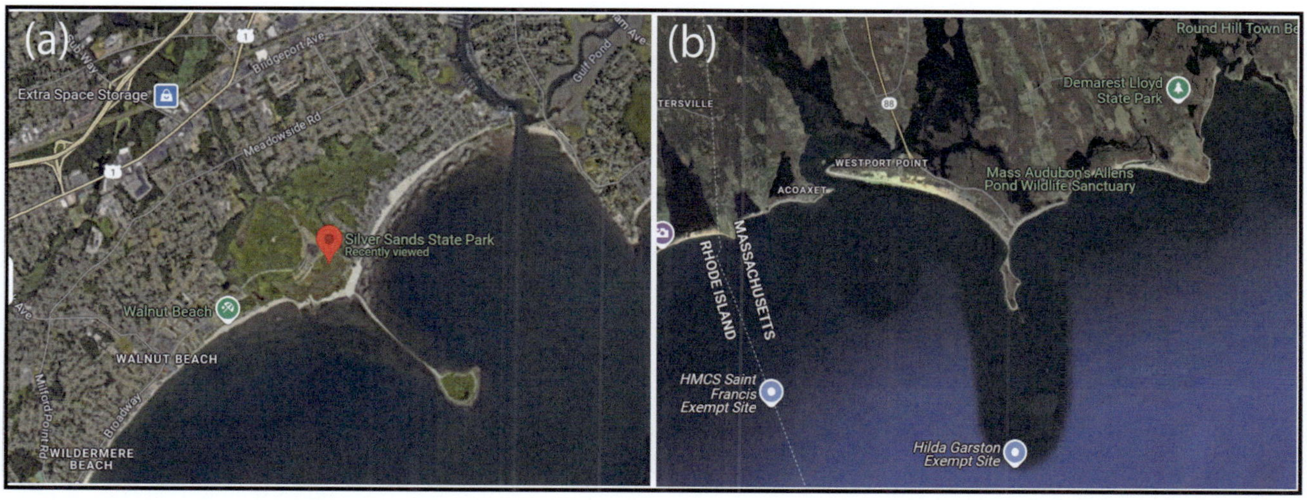

Fig. 15.20 Tombolo connecting **a** Eaquahaug Island to Silver Sands State Park in Connecticut, USA and **b** Gooseberry Island to Horseneck Beach State Reservation Campground in Massachusetts, USA. *Images credit* Google Earth

which separates Rodeo Lagoon from the Pacific Ocean, also in California, USA. Port Bay, a sandbar that blocks the mouth of a bay in New York, USA, is another example of a baymouth bar. Another example of baymouth bars includes the Bay Mouth Bar off Alligator Point in Florida shown in Fig. 15.21.

Lagoons (Fig. 15.21a) are shallow and protected bodies of water behind baymouth bars and barrier islands. A **bay** (Fig. 15.21a) is a body of water partially surrounded by land. A bay is usually smaller and less enclosed than a gulf. The water is connected to a larger water body of water such as a lake or an ocean. Bays are widespread on coasts. Examples include Richards Bay in KwaZulu-Natal, South Africa, Walvis Bay in Namibia, Wineglass Bay in Tasmania, Australia, the Bay of Bengal in India, Hangzhou Bay in

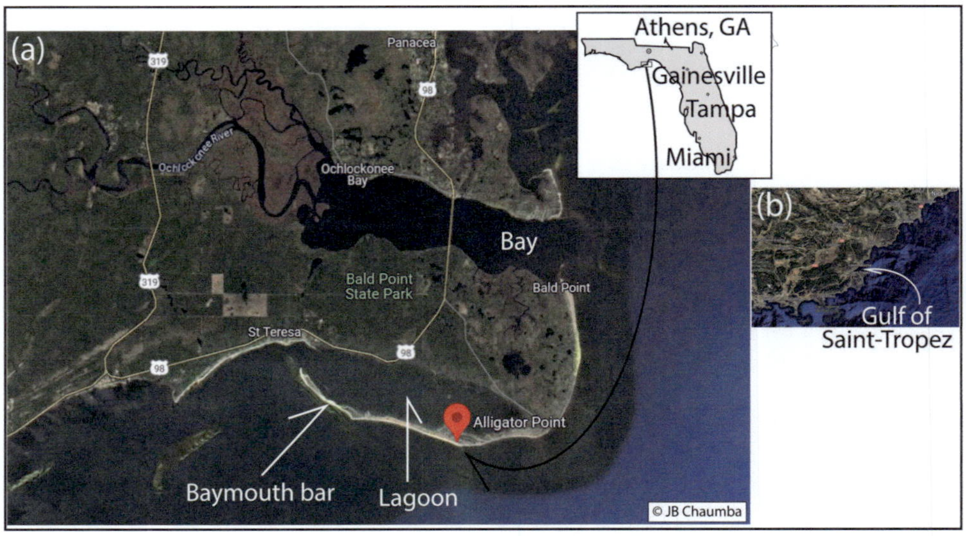

Fig. 15.21 **a** Google Earth map of the Alligator Point, Florida, USA area showing baymouth bar and lagoon, and Ochlockonee Bay. **b** Google Earth map showing Gulf of Saint-Tropez, Southeastern France

Fig. 15.22 **a** and **b** Photographs from Long Beach, Oak Island in North Carolina, USA, showing longshore current or longshore drift—where ocean waves are striking the shoreline at an angle

China, and the Bay of Naples in Italy. Shown in Fig. 15.21b is the Gulf of Saint-Tropez in France.

Refraction of ocean waves occurs when ocean waves change direction as they move from deep to shallow water leading to bending and moving parallel to the shoreline through **longshore drift** (Fig. 15.22). Longshore drift is movement of material along coasts through waves approaching the shore at an angle but recede directly away from it. Since the different parts of the waves interact at different times with the bottom of the ocean, waves approach the coast at an angle, slowing down at different rates (Fig. 15.22). The waves, thus, bend or refract as they slowdown in shallower ocean water causing the waves to approach to shore at a small angle as shown in Fig. 15.22. Water is pushed water along the shore as each one arrives, to form longshore drift in the areas where waves break, referred to as the breaker zone or **surf zone**.

15.11 Coastal Erosion and Stabilization of Shores

The process by which local wave activity, rise in sea level, flooding of the coast break down or and transport rocks, soils, and/or sands away of along the coast is referred to as **coastal erosion**. Storms, high tide, strong waves, and rise in sea level due to global warming all affect coastlines. According to the United States Climate Resilience Toolkit, coastal erosion in the United States is responsible for a loss of about half a billion dollars annually in coastal property. Coastal erosion occurs either through natural processes, such as waves, tides, currents, and winds, or through human activities such as dredging, installation of seawalls, groins, and jetties.

Coastal erosion can cause problems such as extensive damage property and infrastructure, shoreline migration, undermining dunes, shrinking of both beaches and acreage for recreational use, as well as loss of habitat for wildlife and plants. To continue the lives humans are presently enjoying in coastal regions, it is important for us to reduce the risks of living on coastlines.

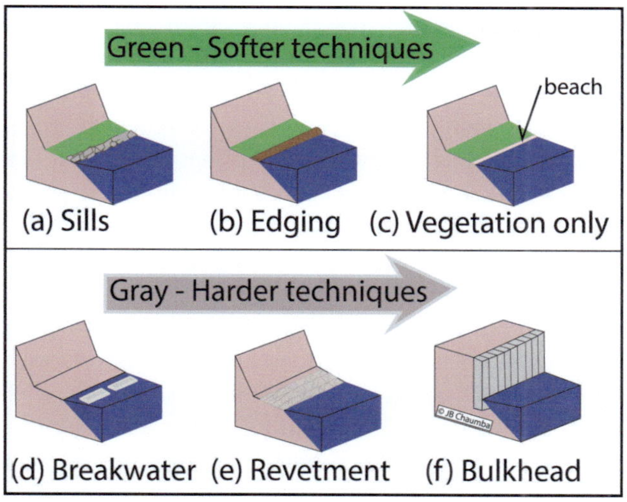

Fig. 15.23 Diagram illustrating some green-softer techniques of **a** sills that are comprised of rock gabion baskets and/or rocks with vegetation, **b** edging where a structure is installed to protect against erosion of shoreline by hold toe of existing slope or existing slope with vegetation, **c** vegetation only with its roots holding soil in place and providing buffer to areas upslope by braking small waves. Gray—harder techniques involve structures offshore such as d breakwater, parallel to coast, which break force of waves thereby promoting accretion of sediments. Groins (not shown) are perpendicular to the beach. **e** Revetment on shoreline slope with where hardened slope pre-exist. A bulkhead **f** consists of vertical retaining wall that is parallel to shoreline which holds soil in place

Reducing risks to coastal systems can be achieved through implementation of strategies that preserve natural habitats and installation of man-made structural features (Fig. 15.23). Natural-based strategies are natural features achieved through natural processes such as marshes and dunes. An example of a nature-based feature created by humans to mimic nature is a living shoreline. Modifications in management practices, regulatory policy, and public policy are non-structural measures. Examples of non-structural measures are flood preparedness planning, flood proofing of structures, and implementation of flood warning systems. Cost, reliability, objectives, and the desired level of risk reduction determine the types of risk reduction measure implemented. Examples of structural measures include breakwaters, groins, and sea walls. Structural measures act to reduce coastal risks through decreasing flooding, erosion of shoreline, and damage from waves.

Living shoreline is a technique of stabilizing coastal areas from erosion through utilizing natural materials such as vegetation, sand, and rocks to protect shorelines. This approach creates protected shorelines such that they mimic natural ecosystems hence they are called living shorelines. Living shorelines, thus, utilize natural materials for protecting coastal areas from erosion. Planting native coastal vegetation which includes grasses, shrubs, and trees, along shorelines, referred to as **soft** or "**green techniques**" (Fig. 15.23a–c), helps in stabilizing sediments and soil through anchoring their roots which reduces the erosive impact of waves and storms. Green technologies also provide habitat for wildlife.

Systems Approach to Geomorphic Engineering (SAGE) encourages a hybrid engineering approach that integrates soft, or green, natural and nature-based measures (Fig. 15.23). Hard, or gray, structural stabilization approaches involve use of man-made protective structures to lessen or control erosion. Plantings with structural techniques such as bulkheads or breakwaters (Fig. 15.23). Walls that are erected to prevent the sea from eroding or encroaching on land an area of land, called seawalls, are also utilized.

Green techniques include vegetation. Vegetation-only techniques involve roots of vegetation binding soil in place to reduce erosion. A buffer is also provided by vegetation to upland areas which also breaks small waves. Vegetation-only techniques are effective for low wave energy environments. Vegetation-only techniques work best with Native plants and act to slow down transfer of ocean water inland while promoting infiltration of storm water. Rock structures that are places parallel to the shoreline to form a natural buffer so as to protect habitats from erosion are called **sills** (Fig. 15.23a). Vegetation can thrive on the land behind the sills (Fig. 15.23a). Sills prevent erosion as well as reduce the energy of waves.

Edging (Fig. 15.23b) is a structure that holds the toe of existing or vegetated slope in place. Edging protects against shoreline erosion, and it is suitable for most areas with the exception of high wave energy environments. A **vegetation-only** approach relies only on plant roots to hold sediments in place, which helps to prevent coastal erosion in the areas where only vegetation is planted (Fig. 15.23c). Vegetation-only approach is the greenest approach that can be utilized to control coastal erosion. By absorbing water, plants also reduce coastal erosion caused by runoff. Vegetation also breaks the impact of raindrops or wave-splash, and physically slows down and disperses the speed of water flow from storm surges.

A method of protecting shorelines where hard structures such as seawalls are not constructed continuously along the entire coastline, but leaves spaces or gaps between them to allow natural sediment to move and improve connectivity of habitat for marine life is referred to as a **gapped approach** to coastal erosion. Greater tidal exchange, habitat connectivity, and better waterfront access are permitted by a gaped approach. A gaped approach is suitable for most areas with the exception of high wave energy environments.

Green-softer techniques include living shoreline and beach nourishment and vegetation on dune. Beach nourishment only involves sand that is sourced away from source to a beach that is eroding widens the shoreline and causes it to migrate seaward. This approach works best for low-lying oceanfront areas, and it also expands the beach area. Beach nourishment has lower environmental impact in comparison to hard structures. Additionally, beach nourishment only provides habitats for beach organisms. Continued sand supply for renourishment of the beach is a disadvantage, together with lack of high water protection and possible impacts to transportation of regional sediment are disadvantages of beach nourishment only.

Beach nourishment and vegetation on dune provide a buffer to protect inland areas from waves, erosion, and flooding that helps to anchor sand. This method, which utilizes sand with vegetation, is applicable for low-lying oceanfront regions equipped with existing sources of sediment and sand. Beach nourishment strengthens dunes with rocky core as well as geotextile tubes. Beach nourishment expands usable area of the beach, and its environmental impacts are low. Beach nourishment provides habitat and ecosystem services. Any vegetation increases resilience to storm events as well as strengthening dunes. Beach nourishment, however, relies on continued sand supplies for renourishment and lacks high water protection. Beach nourishment can impact transport of sediments on a regional scale and is only suitable for limited situations.

Gray, or **hard techniques** (Fig. 15.23d–f), is also utilized to control beach erosion. Hard techniques involve construction of artificial structures to minimize coastal erosion. In coastal environments, hard stabilization structures are typically utilized to counteract erosion in areas that are deficient in supply of sediment, or to reduce accumulation of sediments in areas such as inlets that are rich in sedi-

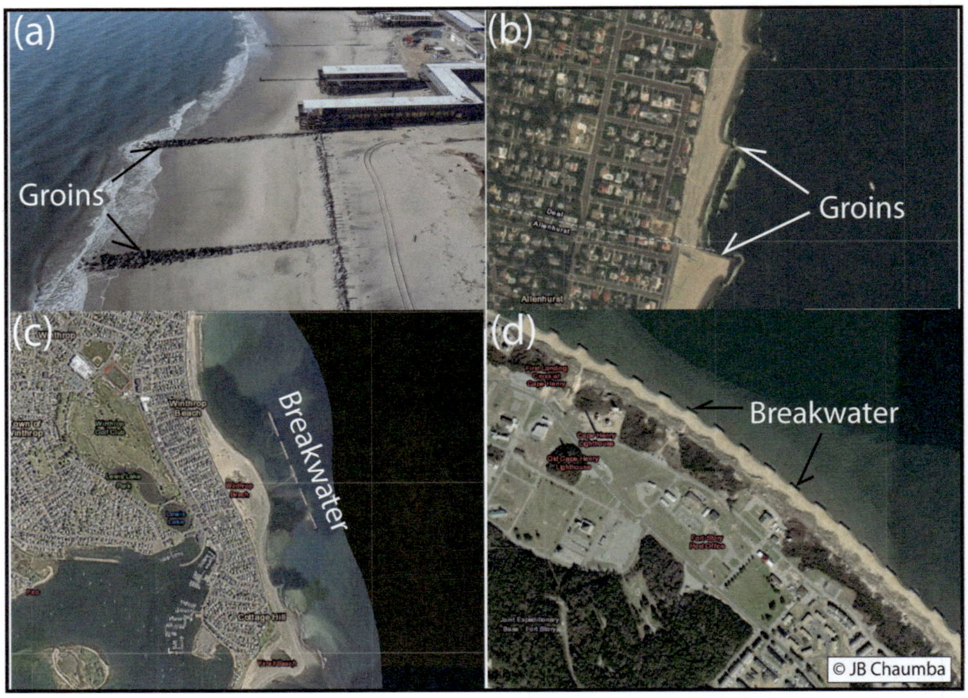

Fig. 15.24 **a** Groins along the Ocean City, New Jersey, USA shoreline, taken March 15, 2012, **b** Groins from Allenhurst, New Jersey, USA. **c** Breakwaters in Winthrop, Massachusetts, USA, and **d** Breakwaters in Fort Story, Virginia Beach, Virginia, USA. *Images credit* USGS: EarthExplorer—https://earthexplorer.usgs.gov/

Fig. 15.25 Aerial photograph of part of the Mississippi coast showing a living shoreline project where both natural infrastructure and artificial breakwater materials are utilized to reduce erosion of the shoreline *Credit* NOAA

ments. Unfortunately, Anthropogenic modifications through installation of seawalls, groins, jetties, breakwaters, and **revetments** can hugely alter natural shoreline processes.

Breakwaters (Figs. 15.23d, 15.24c, d and 15.25) are offshore structures, constructed parallel to the coast, which break waves, encourage accretion of sediments, and reduce the force of wave action. Breakwaters can be attached to the shore, fixed to the ocean floor, or floating. Additionally, breakwaters can either be segmented or continuous. A gaped approach allows connectivity of habitat, greater tidal exchange, and better waterfront access. Suitable for most areas except high wave energy environments often in conjunction with marinas. Breakwaters can be fabric bags filled with grout, or wood, rock, armor stone, precast concrete blocks, or rock. Breakwaters reduce both the height and force, and height of waves. They also stabilize wetlands and can function as reefs. The disadvantages of breakwaters are that in deep water they are expensive and can act as hazard to navigation and require more land area. Furthermore, breakwaters lack high water protection and reduce circulation of water. Sloping structures constructed parallel to shoreline on the beach in order to dissipate and reduce forces of waves attacking a boundary between the sea and the land are called revetments (Fig. 15.23e). Revetments reduce coastal erosion through absorbing and dissipating the energy of waves and offer an extremely effective and quick solution for reducing shoreline erosion. A **bulkhead** (Fig. 15.23e) is a vertical structure that is installed parallel to the coastline to reduce coastal erosion. Bulkheads, however, are instead now inferred to encourage erosion since they only deflect the energy of waves to other areas along the shoreline. Bulkhead is intended to hold soil in place as well as promote a stable shoreline. Bulkhead is suitable for high energy settings as well as sites that have pre-existing hardened shoreline structures. Finally, bulkhead accommodates working water fronts, for example, docking for ferries and ships.

Finally, **seawalls**, structures made mostly of concrete, and constructed parallel to the shoreline, protect the inland area from coastal erosion through deflecting the force of waves away from the coastline. Seawalls can either be vertical or with sloped walls. On one side of the wall, soil has the same elevation as that of water on the other. Since seawalls are designed to withstand the full force of waves and storm surges, they are typically very massive structures which are expensive to erect. Seawalls are appropriate for areas that are highly vulnerable to storm surges as well as wave forces.

Groins (Fig. 15.24a, b) are structures that are perpendicular to the coast that project from the shoreline. Groins act to intercept flow of water and movement of sand parallel to the shoreline to prevent beach erosion and break waves. The advantages of groins are that they retain sand placed on beach and they are suitable for coordination with beach nourishment. The disadvantages of groins lack high water protection, erosion of adjacent sites, and they can be detrimental to shoreline ecosystems.

In summary, living shorelines play a role in reducing coastal erosion. They also improve resiliency through utilizing the full array of coastal risk reduction techniques. Implementation of only hard or gray structures may negatively impact coastal habitats. Gray structures have higher upfront costs, and they are also visually obstructive. According to NOAA, natural infrastructure serves as effective solutions for minimizing flooding along coastal areas, runoff, and erosion. Shoreline structures are utilized to restrict transport of sediment through longshore drift, maintain updrift beaches, or lessen migration of inlet. Shoreline structures, however, may destabilize the coastal system as well as disrupt natural sediment regimes.

Glossary

Abrasion physical erosion whereby water, wind, or ice transport particles that scrape, grind, and wear away rock and land surfaces through friction

Abyssal plain located at the base of a continental rise, it is characterized by a gentle slope and accumulation of encompassing a large, flat area on the deep ocean floor.

Active continental margin coastal region where mountain building, earthquakes, and volcanic activity occur due to movement of plates.

Amplitude maximum displacement, or distance of a wave, from its equilibrium or resting position.

Attrition process erosional where rocks and sediment transported by water, wind, or ice collide against and knock each other, breaking into smaller, smoother, and more rounded particles with passage of time.

Bay broad body of water at coast curving inward and partially surrounded by land that also connects directly to a larger body of water such as a lake or an ocean.

Baymouth bar coastal landform composed of a sand and/or gravel barrier extending across the mouth of a bay or estuary which partially or entirely blocks its connection to the ocean. It is formed by deposition of sediment transported by currents and longshore drift resulting in growth of a spit across the bay entrance to form the sheltered area in the bay.

Beach narrow, transitional area encountered between a body of water and land, and it is characterized by continuous erosion, transport, and deposition of unconsolidated sediment.

Biogenous sediments sediments sourced from the skeletal remains of dead organisms such as bones, teeth, and shells that are resistant to decay and accumulate on the seafloor largely as microscopic tests from organisms like plankton or as macroscopic fragments from large animals.

Breaker wave that becomes too steep such that its crest collapses and topples forward and often transforming into white foam as it collapses onto the shore.

Breakwater structure constructed offshore shore and parallel to coastline on which waves break, thereby fostering sediment accretion between the beach and the breakwater and reducing energy of waves striking the beach.

Bulkhead vertical walls, often made of materials like timber, steel, or concrete structures, constructed along shorelines for retaining and/or prevention of erosion of the land by holding back soil from water bodies. The structures serve as barriers against wave action and shifting soil thereby stabilizing the land and preventing property loss due to erosion, floods, or tides.

Chemical sediments sediments that form as minerals precipitation out of seawater to form evaporites.

Coastal erosion gradual wearing away and removal of rocks, sediments, and land from coastline due to action of waves, tides, currents, wind, and storms.

Cosmogenous sediments sediments originating from outer space such as cosmic dust.

Crest highest point or peak of a wave that represents the maximum displacement from wave's undisturbed position.

Coast margin of continent encountered along an ocean or sea.

Coastline dynamic and constantly changing boundary line where land and the ocean or sea meet.

Continental margin edge of continent occurring underwater composed of the relatively shallow continental shelf whereby the continental edge merges with the deep ocean floor, the steeper continental slope, as well as the continental rise.

Continental slope steep, oceanward sloping area of the continental margin connecting the continental shelf to the ocean floor and forms the transition zone where continental crust becomes thinner and merges into the thinner and denser oceanic crust.

Doldrums *see* intertropical convergence zone.

Edging method utilizing materials placed along the base or toe of a shoreline to prevent beach erosion, specifically in areas where wave energy is low to moderate, such as in marshes and estuarine shorelines. The edging structure shields the shoreline's edge as well as encouraging the establishment of vegetation for further stabilization.

Evaporates sedimentary mineral deposit that forms when a saline water body such as a sea or lake evaporates, concentrating dissolved salts which precipitate out of the solution and crystallize to form anhydrite, gypsum, and halite.

Freshwater water containing very low levels of dissolved salts and other solids of less than 500 parts per million (ppm) occurring in lakes, rivers, aquifers, and glaciers and it is critical for life on Earth including for humans with most of it is locked up in glaciers.

Fetch undisturbed distance over which wind blows across a water body in a constant direction. The longer the fetch, the more energy the wind can transfer to the water leading to formation of larger and more powerful waves up to a certain limit.

Gapped approach construction of a series of segmented and detached breakwaters offshore that are parallel to the shoreline in order to protect the beach from erosion while permitting movement of both controlled water and sediment.

Green techniques also referred to as soft engineering or living shorelines, methods for beach stabilization that utilize natural materials and ecological principles to reduce erosion and are therefore considered sustainable as they work with natural coastal processes to protect the shoreline while enhancing or preserving the local ecosystem.

Groins a structure perpendicular to shore and typically made of stone or concrete, groins trap sand from longshore current and develop widening beach on the updrift side where sand accumulates, whereas erosion and loss of beach are promoted on the downdrift side.

Guyot flat-topped underwater volcanic mountain characterized by a flat summit that can exceed 10 km (6 miles) in diameter and once rose above sea level but has since submerged due to ocean wave erosion.

Hard structure solid, man-made coastal engineering structure to prevent coastal erosion and designed to withstand and deflect wave energy and currents rather than absorb them.

Hydraulic action physical erosion due to the sheer force of moving water such as ocean waves or rivers that dislodges and transports rock particles through exerting pressure on joints and cracks.

Hydrosphere the total amount of water on Earth which includes water occurring on the surface, as groundwater, in the atmosphere as vapor, liquid, and solid forms, as well as water dissolved in substances.

Intertropical convergence zone band of low pressure encircling the Earth near the equator where the trade winds of the Northern and Southern Hemispheres meet.

Lagoon typically brackish shallow body of relatively calm and quiet water separated from larger body of water by a natural barrier such as barrier islands, coral reefs, or sandbars.

Living shoreline nature-based method to stabilize beach that utilizes natural materials such as sand, rock, and plants to reduce erosion while restoring and enhancing shoreline habitats.

Longshore drift also termed littoral drift is the repeated sediment transportation along a shoreline due to waves action where waves approach the beach at an angle, to form a zigzag movement where the swash (water moving up the beach) carries sediment diagonally and the backwash (water moving back to the sea) pulls it perpendicular to coastline back down the slope. This movement of sand gradually shifts the sediment parallel to the coastline.

Manganese nodules rock concretions encountered on ocean floor which are composed of concentric layers of manganese and iron oxides and contain cobalt, copper, manganese, and nickel. They form slowly around a central core from sediment pore water and dissolved metals in seawater.

Marine terrace elevated, flat-lying bench of land that previously formed a shoreline, it forms through wave erosion and deposition of sediments at or near sea level, which either is later uplifted or has an in sea level, preserving it as a former coastline above current high tide line.

Mid-ocean ridge extensive mountain chain occurring at divergent plate boundaries where new oceanic crust is formed.

Ocean trench steep canyon-like features encountered where an oceanic slab is subducted beneath either another, but older, oceanic plate, or a continental plate and exceeds 6000 m (19,600 ft) in depth.

Onlap sequence phenomenon whereby younger rock layers progressively extend further over an erosion surface on inclined and older rock due to a relative rise in sea level, a process termed marine transgression.

Passive continental margin a wide and gently sloping area of transition between continental and oceanic crust occurs and it lacks tectonic activity.

Pelagic sediments fine-grained deposits occurring on deep ocean floor away from continental landmasses primarily comprised of microscopic shells from plankton, volcanic ash, wind-blown dust, and calcareous oozes, siliceous oozes, and red clays.

Promontory raised mass of land projecting into a body of water or a lowland.

Relict sediments sediments that originated from the last glacial period, or the Last Ice Age.

Revetment sloping structures constructed along the base of cliffs, on embankments or shorelines, or in front of sea walls for the purpose of absorbing and dissipating energy of waves thereby reducing coastal erosion.

Sea subdivision of the ocean that is partially or wholly enclosed by land.

Sea arch a bridge-like opening or tunnel formed through rocky headlands by the erosive action of waves leading to the formation of sea caves on either side of the headland which enlarge and eventually merge to form an arch-shaped structure.

Sea cliff vertical or near-vertical rock face formed by waves erosion. Steep coastal landforms are formed as waves attack weaknesses bases of shorelines, eroding a notch that leads to the collapsing of the unsupported cliff face above.

Sea stack forms from continued erosion and widening of a sea arch until the rock can no longer support the roof of the sea arch and collapses into the sea leaving behind pillar of rock called a sea stack.

Seawall type of coastal defense from coastal erosion that is constructed where the ocean or sea and associated coastal processes impact directly upon the landforms of the coast.

Seamount an underwater volcanic mountain that rises at least 1,000 m (3,300 feet) above the surrounding seafloor.

Salinity amount of dissolved salts present in water.

Sea a huge body of salty water covering a sizable portion of the earth.

Seawater water in oceans and seas which is characterized by a huge concentration of dissolved salts.

Siliceous oozes biogenic sediment occurring on the ocean floor mostly composed of the silica-based skeletons of microscopic marine organisms such as diatoms and radiolarians form in areas of ocean upwelling and high surface water productivity.

Sills submerged and shore-parallel structure mostly constructed of rock or wood that acts as a low-lying breakwater in reducing energy of waves to protect wetlands or create a foundation for new marsh vegetation to be planted behind it. The sill prevents erosion at the water's edge by dissipating waves which allow marsh grasses to establish and grow, further stabilizing the shoreline.

Soft techniques environmentally friendly approach utilizing natural materials and processes such as vegetation, sand replenishment, rocks, logs, and other organic materials to mimic natural features in order to lessen beach erosion and protect shorelines, slow wave energy, and anchor sediment.

Spit narrow, elongated landform of sand or sediment formed by the deposition of material transported by longshore drift extending from the coast into a body of water.

Surf zone land near a shoreline where ocean waves break as waves approach shallow water depth and it encompasses zones between the beach and where waves begin to break.

Submarine canyon a deep valley with steep sides in the ocean floor connecting the continental shelf to the abyssal plain.

Submarine volcano a submarine fissure or opening on the Earth's crust through which magmas erupt.

Thermocline temperature gradient in a large body of water such as a lake denoted by a layer above and below which the water is at different temperatures due to temperature differences associated with depth. Thermocline in the ocean divides the upper mixed layer from the calm deep water beneath.

Thermohaline circulation large-scale movement of ocean water propelled by differences in density ascribes to variations in both salinity (haline) and temperature (thermo). It is also referred to as the ocean conveyor belt due to its involvement of dense and deep water that descends and circulates globally transporting heat, nutrients, and gases throughout the oceans of the world before it eventually rises.

Tombolo sand bar of sand or rounded particles that joins an island to the mainland.

Trough minimum displacement or the lowest point of a wave.

Turbidity current gravity-induced swift-flowing underwater density currents composed of sediment-laden water flowing downslope triggered by slumping sediments and earthquakes among others. As the currents flow downslope, they erode the seafloor forming submarine canyons and as the velocity eventually decreases, graded beds of sand and mud, called turbidites, are deposited.

Vegetation only also called living shoreline, or soft armoring, is a method of exclusively using plants to reduce erosion and protect shorelines through use of natural, stabilizing properties of vegetation.

Warm current warm ocean current such as the Gulf Stream in the North Atlantic that flows poleward away from the equatorial region occurring on western margins of ocean basins.

Wave height vertical distance from the crest to the trough of a wave.

Wave-cut platform a flat and/or gently sloping rock surface formed at the base of a cliff through the erosive action of waves.

16 Geologic Resources

Abstract

After defining earth resources, the classification of economic resources using a flow chart follows. Recommended stages to be followed in the development of a mineral deposit are discussed and illustrated. Critical minerals are defined and discussed. Several types of ore deposits and their formation are described with the aid of diagrams. These deposits include magmatic ore deposits, hydrothermal ore deposits, secondary enrichment deposits, orogenic gold deposits, volcanogenic massive sulfide (VMS) deposits, sedimentary-exhalative (SEDEX) deposits, placer deposits, and residual deposits. Other ore deposits similarly discussed include placer deposits and residual deposits. Non-metallic mineral resources such as aggregates are also described. Also described are energy resources such as fossil or biomass, nuclear, hydroelectric, wind, solar, geothermal, and wave energy. Renewable energy such as solar, wind, hydroelectricity, and geothermal energy is also discussed. Non-renewable energy resources of coal, oil, natural gas, and petroleum, as well as metals, and nuclear fuels such as uranium are also described. Diagrams are utilized to describe common oil and gas traps of anticline, fault, stratigraphic, and salt dome traps. Hydraulic fracturing (fracking), oil and gas extraction both onshore and offshore are also discussed.

Learning Goals

Studying this chapter will enable you to:

- Define Earth resources.
- Define and differentiate metallic, non-metallic minerals, and rock resources.
- Differentiate between a reserve base and a reserve.
- List any three examples of mineral deposits and briefly describe how each is formed.
- List the stages involved in the development of a mineral deposit.
- Distinguish between measured, indicated, demonstrated, and inferred reserves.
- Distinguish between hydrothermal fluids and hydrothermal ore deposits.
- Describe how disseminated deposits form.
- Define SEDEX deposits and volcanogenic massive sulfide deposits.
- List the types of non-metallic mineral deposits.
- Define renewable and non-renewable resources.
- Differentiate renewable and non-renewable resources.
- Explain how fossil energy resources form.
- List and describe fossil energy resources of coal, oil, natural gas, oil sands, and nuclear energy.
- Define hydraulic fracturing and its importance in the oil and gas industry.
- List the various types of renewable energy resources.
- Describe how renewable energy resources form.
- Provide examples of renewable energy resources.
- Distinguish between minerals and/or rocks resources and energy resources.
- List and describe renewable sources of hydropower, solar, wind, and geothermal energy.

Supplementary Information The online version contains supplementary material available at https://doi.org/10.1007/978-3-032-20231-4_16.

16.1 Introduction

Geologic resources can be defined as naturally occurring resource such as rocks, minerals, and water that are used by human beings as well as other living things. Some minerals (see Chap. 2) find use in various applications in several industries. In developed countries such as the

J. B. Chaumba, *Introductory Physical Geology*, Springer Textbooks in Earth Sciences, Geography and Environment,
https://doi.org/10.1007/978-3-032-20231-4_16

United States, for example, almost all students in middle, high school, college, and even some elementary school now own a smartphone or have access to one. Minerals such as rare earth elements, copper, silica, and lithium are extensively used to make a wide variety of goods such as smartphones, computers, automobiles, and even in buildings. Limestone, when combined with sand, gravel, and other building materials, is used extensively in the manufacturing of concrete and asphalt. Some questions one may ask are: What are some examples of geologic resources? How are geologic resources formed and where do they occur? Are geologic resources infinite and do they occur in every country or state/province? In this chapter, we will discuss some of the answers to these questions, and more. Geologic resources include metals such as aluminum, copper, and iron, as well as non-metals like gypsum, clay, and salt. Earth resources are generally classified into metallic, non-metallic, and energy resources. "*Geologic resources*" is used here as a general term used to describe the concentrations of valuable earth materials of minerals and rocks of economic and/or industrial importance.

This chapter deals with the distribution, characteristics, and origin of some of Earth's geologic resources, equipping the reader to apply the acquired knowledge in understanding how some resources may have been formed and gain knowledge on profitability of extraction of these resources. Metallic mineral deposits are substantial concentrations of metals containing high concentrations of metals such as copper, aluminum, gold, and iron, among others, which can be extracted at a profit. Three types of mineral resources are generally recognized: 1. metallic mineral resources (e.g., gold, platinum, and copper), 2. non-metallic mineral and rock resources (e.g., barite, salt, limestone, sand, gravel, and other building materials), and 3. energy resources (e.g., coal, petroleum, and uranium) (Fig. 16.1). It must be borne in mind that geologic resources occur in finite amounts and are therefore non-renewable as the processes that lead to their formation take millions to even billions of years. As geologic resources are exploited, new deposits are also explored for and discovered, or their substitutes are found if prevailing living standards are to be improved upon. Consequently, there appears to be need for resource geologists, and to understand how geologic resources form and successfully discover new deposits, if geologists are to be indispensable in a world of rapidly advancing technology.

16.2 Mineral and Rock Resources

A **mineral resource**, which is a geologic resource, can be defined as a mineral deposit containing useful concentrations of valuable mineral(s) that may or may not be extracted at a profit. Mineral resources (Figs. 16.1, 16.2 and 16.3) are mineral deposit which consist of concentrations that may or may not be economic to extract. Technology process, economic needs, and the market price are dependent on whether or not, and when the mineral or rock becomes a raw material. The mineral copper is extensively used in the manufacture of coins as well as for making wires and pipes. Gold, silver, platinum and other glittery,

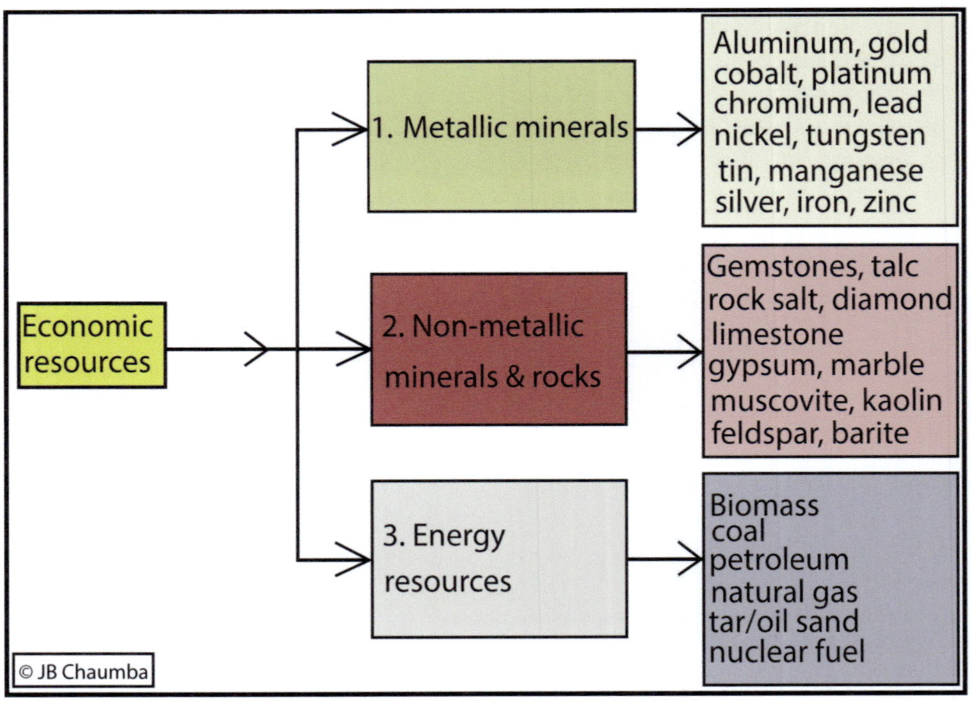

Fig. 16.1 Diagram illustrating the various subdivisions of economic resources of metallic minerals, non-metallic minerals and rocks, as well as energy resources, and examples

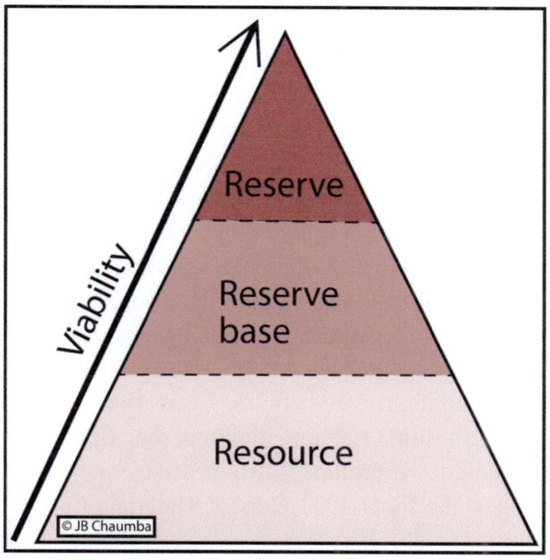

Fig. 16.2 Diagram illustrating resource and reserve base. The boundaries between them are dynamic, hence they are shown as dashed lines

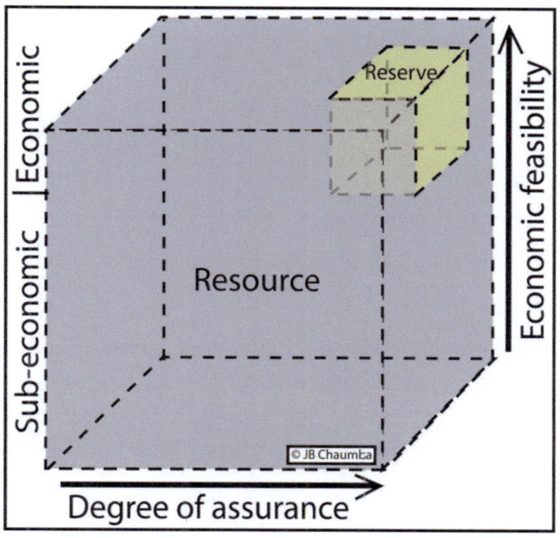

Fig. 16.3 Diagram illustrating resources and reserves. Reserves, which are economic, have the highest degrees of both assurance and economic feasibility

or shiny, minerals, for example, find use as jewelry. Silicon, which is obtained from quartz, is extensively used in the computer industry. Almost every product produced in industries is comprised of minerals, underlying the importance of mineral resources to society. Since the bulk of elements, with the exception of those comprising the silicon tetrahedron, occur in tiny concentrations in the Earth's crust, an element becomes valuable after natural processes concentrate that element by some orders of magnitude. Elements that constitute the least proportions in the Earth's crust such as gold and platinum, for example, must be concentrated by several orders of magnitude for them to be economically exploitable. Additionally, the profitability of an element is subject to change due to factors such as an increase in their prices due to increased demand which makes them less affordable and likely leading to the search for substitutes. In still other cases, some deposits considered unprofitable can become profitable overnight due to demand, and vice versa. Improvements in technology often result in previously unprofitable ores being made extractable at lower costs than would have previously been possible.

The part of an identified resource where a usable mineral and/or energy commodity can be extracted at a profit at the time of making the decision is referred to as a **reserve** (Fig. 16.3). The segment of resource that is regarded to be both economically and technologically feasible to extract at the time of determination is referred to as the **reserve base** (Fig. 16.3). Thus, a reserve base can be thought of as a subset of identified resources that may be potentially mined at the time of determination. A reserve of some type of largely metallic, but also non-metallic, mineral commodity is called **ore**. Minerals occurring in a form that can be easily extracted, as well as in high concentrations such as hematite and galena, are referred to as **ore minerals**. Some non-metallic minerals are sometimes also referred to as ore. Ore, thus, is generally applicable to reserves of metallic mineral commodities although it is also sometimes applied to non-metallic commodities. Reserves can also be categorized on the basis of inferred, measured, and indicated resources. Identified economic resources are called reserves as well as identified subeconomic resources. The distinction between a resource and a reserve is important to note: a mineral resource is broader as it encompasses all identified deposits whereas a reserve is economically viable to extract at the time the determination is made.

Deposits that have their quality and quantity determined within a <20% margin of error are referred to as **measured**. Economic materials whose quantity and quality have both been estimated in part from analyses and measurements and in part from geologic inferences that are reasonable are **indicated**. **Demonstrated** is a collective term applied to the total amount of materials in both measured and indicated resources, whereas materials that have been identified but unexplored deposits whose quality and quantity have been estimated from geologic projections are **inferred**. An aggregate is a rock or material consisting of minerals that is independently used as filler in asphalt, cement, plaster, as well as other materials.

With further investigation of a resource, reserves, which are more viable than resources (Fig. 16.2), may be delineated. Reserves, therefore, constitute relatively smaller fractions of a resource (Fig. 16.2). Reserves represent higher degrees of both assurances and economic feasibility (Fig. 16.2). The boundaries between resources and reserve base are dynamic, and hence they are shown in Fig. 16.2 as dashed lines.

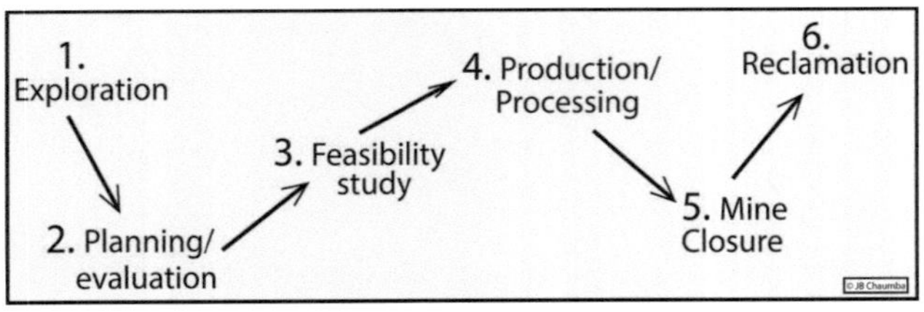

Fig. 16.4 Example of stages in the development of a mineral deposit

A reserve is generally a smaller proportion of any known resource. Reserves, which are economic, have the highest degrees of both assurance and economic feasibility, whereas resources are subeconomic and have both low degrees of assurance and economic feasibility. Upon exploration and eventual discovery of a mineral resource, evaluation of the resource is undertaken which provides information on both quantity and quality of economic minerals and its geologic characteristics. Mine development and exploitation can follow if it is economic to extract the reserve. Reserves generally comprise relatively small proportions of the resource (Fig. 16.3). A reserve may even get smaller with continued evaluation of the resource. Additionally, during exploitation, reserves may diminish with time. However, additional exploration may also lead to the identification of more reserves from the identified resources.

Stages that are typically encountered in the development of a mineral deposit are shown in Fig. 16.4. Exploration is, in general, the first stage, and this is usually done after examination of several factors including the political climate, the marketability of the commodity of interest, the regional geology of an area, manpower, and several other factors. After exhaustion of the commodity being mined, closure of the mine is followed by reclamation of the affected area to try and restore it as close as possible to the condition it was in before mining operations began. Most countries now have stringent laws in place concerning reclamation of areas affected by mining operations.

Some minerals, termed **critical minerals**, are necessary for the world economy, national security, and top-notch educational system, among other applications. Critical minerals are also useful in a wide variety of applications such as medical devices, electronics, sophisticated defense systems, and renewable energy technologies, for example. Based on the Energy Act of 2020, the definition of a critical mineral, according to the U.S. Geological Survey, is any mineral, element, substance, or material that is designated as critical by the Secretary of the Interior acting through its director.

Any disruption in the supply chain of critical minerals, which are largely concentrated in a few countries, ensures that security for their availability is prioritized for industries worldwide and their governments. Critical minerals are also important for the overall health of the global society. Critical minerals are also essential in the transition to clean energy and emerging technologies. In the United States the critical minerals list, which is determined by the Secretary of the Interior, acting through the director of the U.S. Geological Survey, include 50 minerals of aluminum, antimony, arsenic, barite, beryllium, bismuth, cerium, cesium, chromium, cobalt, dysprosium, erbium, europium, fluorspar, gadolinium, gallium, germanium, graphite, hafnium, holmium, indium, iridium, lanthanum, lithium, lutetium, magnesium, manganese, neodymium, nickel, niobium, palladium, platinum, praseodymium, rhodium, rubidium, ruthenium, samarium, scandium, tantalum, tellurium, terbium, thulium, tin, titanium, tungsten, vanadium, ytterbium, yttrium, zinc, and zirconium, according to the Final 2023 Critical Materials List.

16.3 Types of Ore Deposits and Their Formation

Ore minerals, also referred to as economic minerals, occur almost everywhere. For example, it is estimated that >70 million tons of gold occur in the ocean. This gold in the ocean, however, because the extremely high cost of extracting this gold is currently impossible to meet due to its occurrence in trace concentrations in seawater. To be exploited, minerals have to occur in concentrations that are high enough to form deposits that can be economically mined. A mineral deposit containing one or more minerals that can be extracted profitably is called an **ore**. Many minerals are commonly found together (e.g., quartz and gold; molybdenum, tin and tungsten; copper, lead and zinc; platinum and palladium). Because various geologic processes can create local enrichments of minerals, mineral deposits can be classified according to the concentration process that formed them. A wide variety of ore deposit types is recognized. Five basic types of mineral deposits are: **i. magmatic**, **ii. hydrothermal**, **iii. sedimentary**, and **iv. residual and secondary enrichment** deposits.

16.3.1 Magmatic Ore Deposits

Magmatic ore deposits, also referred to as orthomagmatic ore deposits, are formed either within igneous rocks when ore minerals precipitate from a magma or along contacts of magmas and country rocks. Felsic igneous rocks and mafic/ultramafic igneous rocks, for example, act as host, respectively, to magmatic deposits of tin and nickel. One of the reasons for these differences is that the tectonic environment in which magmas are generated lead to differences in resulting chemical compositions of magmas. The world's deepest mine, Mponeng Gold Mine in the Witwatersrand gold field, South Africa, extracts gold from a depth of >4 km (2.5 miles) below the Earth's surface with a temperature of 60 °C (140 °F). Sophisticated cooling systems, however, maintain air temperature of ~30 °C (86 °F) for the miners at such great depths. Chalcophile and siderophile elements such as nickel concentrate in mafic magmas, whereas lithophile elements, such as tin, are concentrated in felsic or alkaline magmas. Magmatic ore deposits, therefore, are formed within and are transported in magma. Deposits hosted by felsic igneous rocks, therefore, are more consequently concentrated in crustal rocks whereas deposits hosted by mafic rocks should be more concentrated in the mantle. However, due to its great depth and high temperatures, among other factors, deposits in the mantle cannot be easily extracted with present technology and these resources have to be extracted from mafic rocks in the crust. Where crustal rocks are mafic, exploration for chalcophile and siderophile elements is conducted in mafic rocks hosted deposits within the Earth's crust.

Platinum-group metals and chromite are examples of magmatic ore deposits. Chromium is also useful in chromium chemicals. Chromium has no known substitute in stainless steel. World production of chromium is dominated by South Africa, followed Turkey, Kazakhstan, and India (Fig. 16.5). Chromium is also of huge importance in human health, where it acts as a trace element that influences the metabolism of glucose and insulin. Chromium is also useful in dyes and pigments.

Looking to the future of this commodity, reserves for chromium are dominated by Zimbabwe and South Africa (Fig. 16.6). Türkiye, Kazakhstan, and India also contain huge reserves of chromium, as shown in the pie chart in Fig. 16.6.

16.3.2 Hydrothermal Ore Deposits

Hydrothermal ore deposits also form within igneous rocks or along their contacts in which ore minerals were transported in a melt and mineral precipitation occurred from hot aqueous solutions. These circulating hot aqueous solutions are referred to as hydrothermal fluids (Fig. 16.7), and the ore deposits formed from precipitation of these fluids are called hydrothermal ore deposits. There are various types of hydrothermal fluids involved in the formation and/or modification of ore deposits. Hydrothermal fluids are fluids that are sourced from magmas during both cooling and crystallization at various crustal depths (Fig. 16.7). These fluids are called juvenile or primitive, and they are responsible for circulation of fluids accompanied by scavenging and alteration of rocks through which they circulate. Other hydrothermal fluids originate from meteoric water, that is, from the atmosphere through precipitation, and also snowmelt, which subsequently infiltrates into the Earth's crust. Still other hydrothermal fluids are expelled from pores in sedimentary rocks during compaction of sediments, and still other fluids originate from dehydration reactions during metamorphism. Hydrothermal fluids are of huge importance in the formation of ore deposits as they can significantly enrich sparsely mineralized rocks into economi-

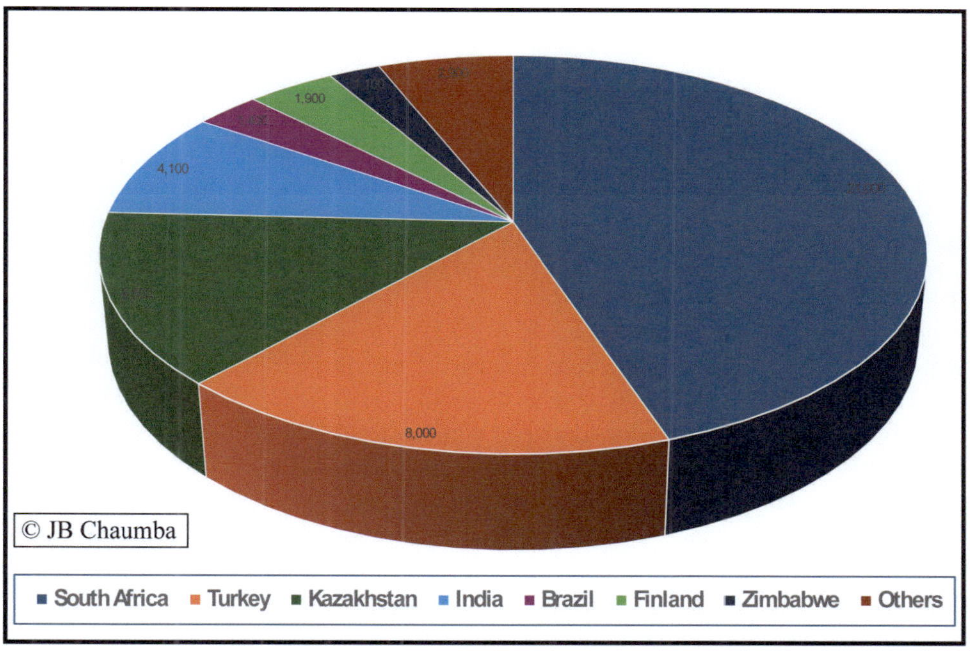

Fig. 16.5 Pie chart showing the world's leading producers of chromium

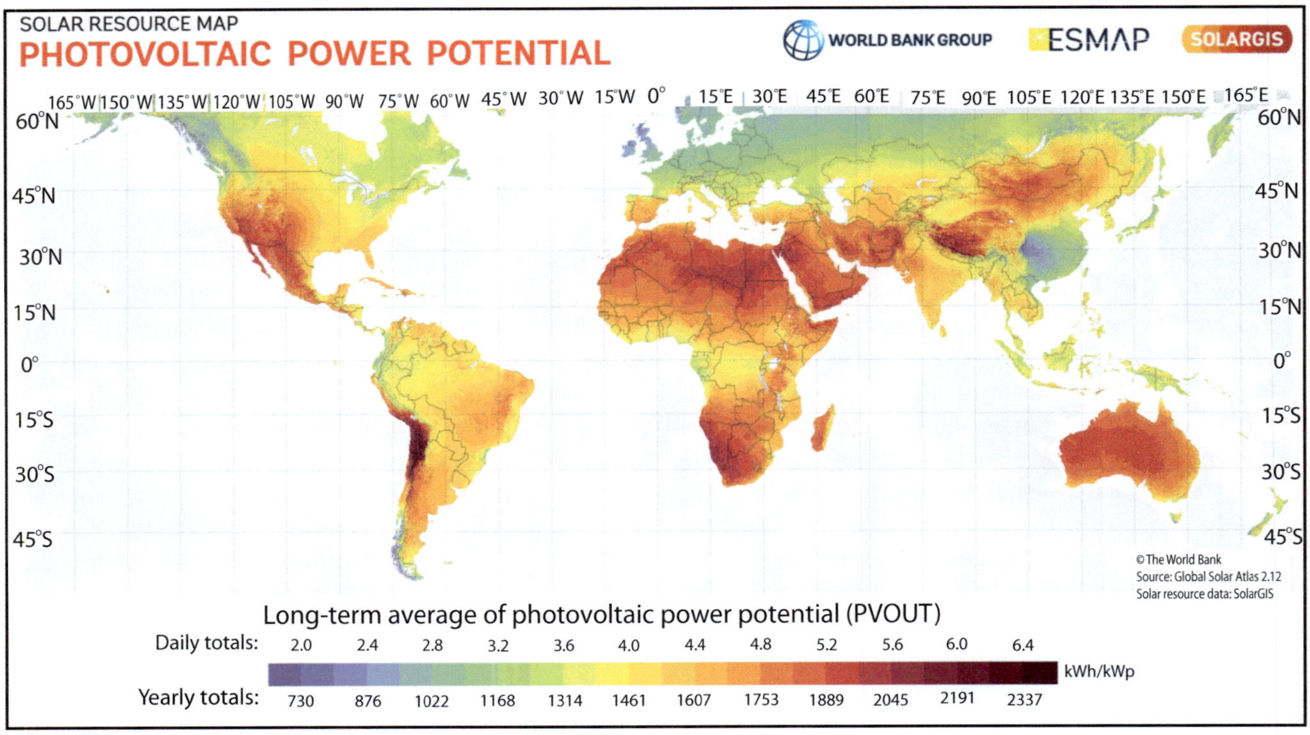

Fig. 16.6 Reserves of chromium are dominated by Zimbabwe, Kazakhstan, South Africa, and India

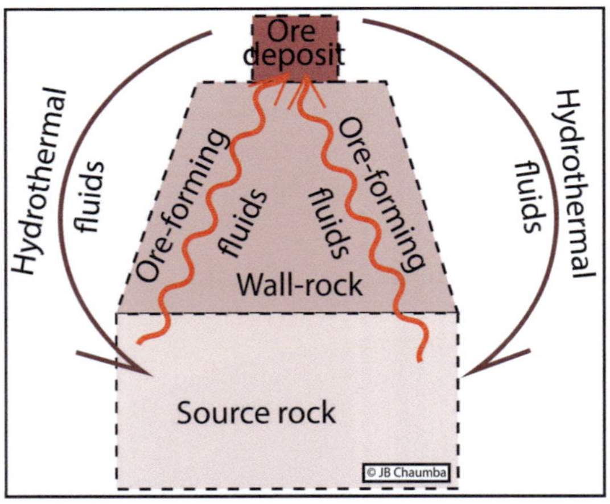

Fig. 16.7 Diagram illustrating hydrothermal fluids circulating from the hot source rock through wall rocks resulting in formation of an ore deposit at high crustal levels relative to the source rock

cally viable ore deposits through concentration of metals (Fig. 16.7).

Some mineral deposits are characterized by fine-grained ore minerals which occur dispersed within the host rock. Such deposits, which are of low-grade nature and relatively of small size, contain ore minerals that are disseminated and are referred to as disseminated deposits. These deposits contain ore minerals that are scattered throughout the host rock, and they can be profitably extracted due to their large size by bulk mining techniques despite their low grades. Elements such as copper, molybdenum, and even diamonds can occur as disseminated deposits. Porphyry copper deposits are a well-known example of disseminated deposits, where copper and molybdenum are found in porphyritic intrusive rocks. Some of the Carlin-type gold deposits in Nevada, USA, the Cananea copper deposit in Mexico, and the Argyle diamond mine in Australia are examples of disseminated deposits. The Cananea copper deposit, located in Cananea State, Mexico and near the Arizona, USA border, is one of the world's largest open-pit copper mines.

Hydrothermal ore deposits are thought to form from depths of ~8 km deep in the crust to less than 1 km below the Earth's surface. Hydrothermal fluids, also referred to as aqueous solutions, are interpreted to be responsible for the formation of, or ascribed to modify, some magmatic ore deposits, particularly in deposits that are relatively close to the Earth's surface. Magmatic-hydrothermal deposits are linked spatially and genetically at, or near, the surface of the Earth. Ore deposits in which the aqueous solutions were sourced from magma are called **magmatic-hydrothermal ore deposits**. Most ore deposits are inferred to have been formed due to concentration of hydrothermal fluids in the crust, or the ore deposits themselves were concentrated by hydrothermal fluids. Hydrothermal vein deposits tend to be associated with outcrops of quartz veins with a milky-white coloration. Where intrusions are formed, country rocks are heated together with any surrounding groundwater to generate hydrothermal fluids that transport and/or redistribute and concentrate metals from both the magma and enclosing rocks. The role of hydrothermal fluids is thought to have affected ore deposits occurring in both felsic and mafic rocks. Additionally, hydrothermal fluids are also interpreted to play an important role in sedimentary ore deposits formed at great depth in the lithosphere to those formed at shallow depositional settings.

Metamorphic rocks often contain ore deposits concentrated by fluids that dissolve, transport, and deposit metals during metamorphism. The fluids, of metamorphic origin, are responsible for the deposition of metals in different locations from their source regions. Hydrothermal fluids, therefore, are thought to play significant roles in the formation of ore deposits associated with igneous, sedimentary, as well as metamorphic rocks. Hydrothermal ore deposits are formed when valuable metals are dissolved from a primary deposit and redeposited in a more concentrated form, often near the surface. Groundwater, rich in dissolved minerals, percolates through existing ore deposits, dissolving metals, such as copper and silver, and redeposits the minerals in zones located closer to the surface. Still other hydrothermal ore deposits are formed due to dissolution of some metals from a primary deposit which are later redeposited in forms containing higher concentrations that are located closer to the Earth's surface. Percolating ground water, enriched in dissolved minerals, traverses through crustal rocks, dissolving elements such as copper and silver, and later redepositing them in shallower environments to form **secondary enrichment deposits**.

Several different mechanisms are inferred to be involved in the deposition of ore metals by hydrothermal fluids. Firstly, any change in pressure results in the precipitation of minerals from hydrothermal fluids. Secondly, changes in temperature resulting from cooling of hydrothermal fluids lead to precipitation of ore minerals due to the reduced capability of hydrothermal fluids to transport them. Thirdly, hydrothermal fluids chemically react with country rocks (Fig. 16.5), which alter their composition leading to precipitation of ore minerals. Several examples of hydrothermal ore deposits are known. They include: i. **orogenic gold deposits** which are important sources of gold and are often linked to metamorphism, ii. **volcanogenic massive sulfide (VMS) deposits** which occur in association with volcanism and are comparable to **sedimentary-exhalative (SEDEX) deposits**, and iii. SEDEX deposits are dissolved metal-laden hydrothermal fluids released into water environments such as the ocean or other water bodies from which metals precipitate out in the form of sulfide minerals. SEDEX deposits are discussed further under sedimentary ore deposits.

Volcanogenic massive sulfide deposits can be classified further into subtypes, such as the Besshi type, Cyprus type, Kuroko type, Noranda type, and Urals type, which were derived from areas where these deposits were mined extensive. Besshi-type VMS have simple mineralogy composed mostly of pyrite and chalcopyrite and occur in association with mafic volcanic rocks. Besshi-type VMS are stratiform and sheet-like ore bodies often encountered in ancient, metamorphosed settings. The Cyprus-type VMS deposits occur in ophiolites, are comprised of basalt, and are enriched in copper and zinc. Cyprus-type VMS deposits are lens- or mound-shaped, and they form largely at mid-ocean ridges or back-arc basins. These deposits are comprised of massive pyrite that is underlain by stringer zones enriched in copper.

Kuroko-type VMS, dominated by felsic rocks, often occur either as lens-shaped or mound-like accumulations of massive pyrite at mid-ocean ridges as well as back-arc basins and in ophiolites. Copper and zinc are enriched in Kuroko-type VMS deposits, and they are formed from hydrothermal fluids venting onto the seafloor where they deposit sulfide minerals. Noranda-type VMS, which are polymetallic nature, form within primitive arc settings in bimodal mafic volcanic sequences. Huge amounts of copper, zinc, gold, and silver, which are hosted in felsic volcanic rocks occurring dominantly in mafic volcanic environment, are a feature of Noranda-type VMS. The last class of VMS deposits, Urals type, occur in the Urals mountains in Russia, and are characterized by basalt-rhyolite as well as basalt-andesite-rhyolite rock types. The Urals-type VMS deposits form within volcanic arcs and act as host to huge amounts of copper, lead, zinc, gold, as well as silver.

16.3.3 Sedimentary Ore Deposits

Sedimentary ores include those formed primarily by gravitational settling of mineral particles at or near the surface of the earth, compacted, followed by cementation. Heavy minerals/elements such as native gold, diamonds, magnetite, and ilmenite. Deposits associated with sedimentary processes or rocks include several metallic and industrial mineral deposit types and commodities.

Deposits that are formed from the weathering and erosion of rocks that contain heavy minerals such as gold, PGEs, tin, or gemstones are referred to as **placer deposits**. Since these minerals are denser and relatively more resistant to weathering in comparison to surrounding materials, they tend to be transported away by agents of erosion such as water and wind to be deposited in regions where velocities are reduced. The deposited minerals form placer deposits. To explore for such minerals, therefore, one may target ancient depositional environments. Weathering, erosion, and transportation of some initially dispersed heavy materials such as gold, diamonds, and platinum by water may result in deposition of the heavy minerals along riverbeds and shorelines to form placer deposits. Other placer mineral ore deposits include monazite for rare earth elements as well as thorium, cassiterite for tin, both rutile and ilmenite for titanium. Other minerals such as magnetite, garnet, and zircon may also originate as placer deposits.

Certain types of ore deposits occur in association with certain plate boundaries as well as certain types of rocks. At subduction zone settings, magma generated by dehydration melting transports metals toward the Earth's surface forming mineral deposits such as epithermal deposits, skarns, and porphyry copper deposits. For example, platinum-group elements tend to occur in mafic–ultramafic igneous rocks such as layered intrusions. The bulk of igneous rocks on our planet occur at divergent and convergent plate boundaries. At divergent plate boundaries, rising magmas may generate massive sulfide deposits that are enriched in elements such as gold, copper, silver, and zinc that occur in association with hydrothermal vents. At convergent plate boundaries, associated uplifting often exposes mineral deposits generated through metamorphism and hydrothermal activity, as well as other mineral deposits that were elsewhere. Along continental rift zones, rifting may result in the formation of magmatic deposits such as platinum-group elements, chromite, together with associated base metal sulfides. Along transform plate margins such as those along divergent plate margins, faults act as conduits for fluids enriched in, and concentration gold, silver as well as other metals.

Although not directly related to plate boundaries, intracontinental settings often host ore deposits that are formed from processes such as magmatic intrusions and accompanying hydrothermal activity. The distribution of various types of mineral deposits is thought to reflect the processes that occurred during the formation of those deposits as well as how they were preserved. For example, the distribution and preservation of deposits in the early stages of our planet's history called the Archean and Palaeoproterozoic (see Chap. 10) were linked due to their preservation in continental crust that was linked to thick buoyant subcontinental lithospheric mantle but were decoupled by the Neoproterozoic and Phanerozoic; however, the distribution and preservation of the deposits were decoupled due to the evolution to thinner and increasingly dense subcontinental lithospheric mantle. This transition is thought to have marked a change from plate tectonics influenced by mantle plumes to the modern style of plate tectonics. This change is also inferred to have been linked with the nature and abundance of preserved mineral deposit types.

SEDEX deposits are formed in sedimentary basins on the seafloor, particularly along faults, where metal-laden fluids vent into seawater and deposit their dissolved minerals. SEDEX deposits have also been suggested to form through sedimentary fan model whereby ore-forming fluid of sedimentary-exhalative deposits is akin to turbidity currents. The fluid source has been suggested to have originated from the exhalative process which traversed syngenetic faults into marine basin to form sedimentary ore deposits shaped like fan. A pipe, central, and marginal facies form sedimentary fan-shaped ore bodies. Zonations of mineralization, trace elements, and other parameters are displayed by the fans. Pipe, central, and marginal facies are recognized, with the central and marginal facies mainly occurring as stratiform ore bodies of sedimentary origin. Increasing seawater components enter into the ore-forming process from the central facies to the marginal facies, producing changes in metallogenic environments.

Submarine hydrothermal activity is acidic (pH of 4–6) hot water of up to 400 °C that is reducing and enriched in metals and hydrogen sulfide (H_2S). The water gets heated by submarine magma chambers chiefly at submarine divergent plate margins and becomes less dense than surrounding water which causes it to rise toward the seafloor. The result is the triggering of submarine hydrothermal circulating systems which are responsible for economic concentration of metals as well as hydrothermal alteration of affected rocks. **Volcanogenic massive sulfide (VMS)** deposits (Fig. 16.8), also known as volcanic-associated, volcanic-hosted, or volcanosedimentary-hosted massive sulfide deposits, act as significant sources of elements such as Zn, Cu, Pb, Ag, and Au. Additionally, VMS deposits also host Sn, Se, Co, Mn, Cd, In, Te, Bi, Ge, and Ga. VMS deposits also occur as lenses of polymetallic massive sulfides near or on the seafloor in submarine volcanic environments. VMS deposits are classified depending on base metal or gold content or lithology or the lithology of the host rock. It is estimated that 350 known VMS deposits occur in Canada alone, and in excess of 800 deposits occur worldwide. VMS deposits occur in submarine volcanic terranes whose ages vary from the Paleo-Archean to the present, including those being formed

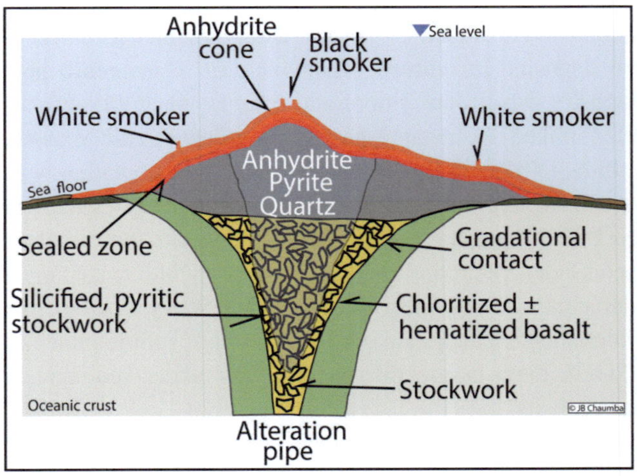

Fig. 16.8 Diagram illustrating some of the key features of a volcanogenic massive sulfide deposit

in modern seafloor environments. Black smokers are now interpreted to be present-day analogues of the formation of VMS deposits.

All types of VMS deposits are thought to form in extensional environments on divergent as well as at arc tectonic environments. VMS deposits are, thus, formed in submarine, volcanic-associated, extensional environments such as back-arc basins, or volcanic arcs, or ancient oceanic ridges. The bulk of preserved ancient VMS deposits preserved in the geological record are interpreted to have been generated in arc tectonic settings. The bulk of economic VMS mining districts are characterized by deposit clusters that were generated in calderas or rifts attributed to a single heat source triggering large-scale fluid circulation systems in the subseafloor. The subvolcanic intrusions are thought to supply metals to VMS hydrothermal systems.

16.3.4 Residual Deposits

Some mineral deposits, called **residual deposits**, form due to accumulations of valuable important minerals that are left behind after pre-existing rocks have weathered and eroded away. Consequently, such deposits are formed in situ, and they contain high concentrations of minerals that are relatively insoluble. Residual deposits tend to form in hot and humid climates such as tropical regions characterized by high weathering rates. Such climatic regions promote erosion of soluble components leaving behind high concentration of minerals that are more resistant to decomposition which constitute placer deposits. Laterite is a type of residual deposit rich in both iron and aluminum oxides occurring in tropical regions. If a laterite is enriched in nickel, it is referred to as a nickeliferous laterite. Weathered and oxidized zones exposed at the surface of sulfide ore deposits are called **gossans** (Fig. 16.9). They are formed from the breakdown of sulfide minerals, which results in a distinctive reddish-brown to black color caused by the presence of iron oxides and hydroxides such as limonite and hematite. In the zone of oxidation, sulfide minerals such as pyrite (FeS_2), which are unstable at the Earth's surface, react with oxygen and water to form gossan. Gossan has a reddish-brown to black caused by iron oxides and hydroxides. When sulfide deposits are exposed to the atmosphere, sulfide minerals oxidize to form sulfuric acid. Sulfuric acid dissolves metals and transports them downward, but leaving behind iron oxides and hydroxides. Sulfuric acid, which is a product of this reaction, further dissolves the surrounding rock together with other sulfide minerals. Iron released from the sulfides later precipitates as iron oxides and hydroxides to form gossan. Soluble minerals and other metals are leached away by acidic solutions, leading to concentration of metals at deeper levels to form an ore deposit. A boxwork or honeycomb structure is left behind as a consequence of dissolution of the sulfide minerals, and the boxwork structure reflects the original outline of the sulfide ore. The gossan, which occurs at the surface above sulfide ore deposits, is largely comprised of iron oxides such as limonite, goethite, hematite although also contains some secondary minerals like sulfates and carbonates. The significance of gossans in mineral exploration is that of an indicator of an ore body. The occurrence of gossans may be suggestive of a sulfide ore deposit in the subsurface. The importance of gossans is that they act as indicators for the occurrence, as well as the nature of underlying sulfide mineralization. In mineral exploration, the occurrence of gossans is a useful indicator of underlying sulfide mineralization. Metalliferous material that has not yet been enriched to become ore is referred to as **protore**.

Laterite is a reddish, iron-rich soil called left behind by intense weathering leaches away soluble materials in tropical climates. An aluminum ore, called bauxite, and nickel deposits commonly occur in laterites. Sedimentary deposits referred to as banded iron formations (BIFs), comprised of alternating layers of chert and iron oxides such as hematite and magnetite, occur in ancient rocks. Upon weathering, the iron oxides are concentrated, rendering BIFs important sources of iron ore. In other instances, weathering can dissolve important minerals from primary ore deposits, later redepositing the dissolved minerals at lower levels, and concentrating them to produce highgrade ore bodies. This process is referred to as supergene enrichment.

16.3.5 Role of Metamorphism, Weathering, and Erosion

Metamorphism of sedimentary rocks enriched in iron oxides results in the formation of banded iron formations (BIFs). BIFs are mined for their iron, and they represent good targets for exploration of host rocks of iron ore. Hydrothermal fluids triggered by intrusions often result in alteration of carbonate rocks such as limestone and marble which leads to deposition of ore minerals like copper, zinc, and tungsten. This process results in the formation of skarn deposits.

Weathering, erosion, transportation, and deposition can also result in the formation of mineral deposits caused by surface processes. Redistribution of minerals from these surface processes may lead to the formation of placer deposits in stream sediments as well as iron ore in banded iron formations. Bauxite in laterites also form as a product of surface processes. Certain minerals in rocks are subject to alteration, dissolution, or decomposition caused by chemical reactions with water, oxygen, and acids. Elements are then released into solution through this process. In addition, weathering of aluminum-rich minerals in rocks can result in the formation of deposits of bauxite.

Physical weathering of rocks occurs through their breaking down into smaller fragments via the processes of abrasion, freeze–thaw action, or exfoliation. These smaller fragments can then be transported by erosion and deposited to form placer deposits. Weathered material, including minerals, is also transported away by wind, water, and ice and eventually deposited to form placer deposits. Flowing water erodes, transports, and finally deposits heavy minerals, such as gold and diamonds, and some metal ores to form placer deposits. Placer deposits tend to be concentrated either along stream beds or along coastlines due to their high densities, leading to the formation of placer deposits. The mouths of major rivers that drain source areas of these minerals, such as in paleochannels of the Orange River in Southern Africa, formed placer deposits of diamonds where it empties into the Atlantic Ocean at Oranjemund in Namibia, near the border of Namibia and South Africa.

In orogenic zones, metamorphism which accompanies orogenic episodes may be associated with formation of certain mineral deposits. During metamorphism, the composition and migration of metamorphic fluids play a crucial in transporting, concentration, as well as deposition of valuable minerals. Hydrothermal alteration is accompanied by metamorphic fluids which can be enriched in dissolved minerals. These hydrothermal fluids permeate through rocks, depositing certain minerals such as antimony, copper, mercury, zinc, lead, and even gold in fractures and veins. Some types of metamorphic rocks such as serpentinites are formed during metamorphism, and they host chrysotile asbestos. Zvishavane and Mashava chrysotile asbestos deposits in Zimbabwe, as well as similar deposits hosted in Archean greenstone belts in Zimbabwe, South Africa, and Canada, among others, act as host to huge deposits of chrysotile asbestos. Metamorphic fluids transport, concentrate, and deposit gold to produce economic deposits, particular gold occurring in association with arsenopyrite. Volcanogenic massive sulfide deposits occurring in metamorphosed volcanic and sedimentary rocks often contain zinc and copper, among other metals. Garnet, emerald, and various types of corundum (ruby and sapphire), such as those in the Blue Ridge of western North Carolina, USA, are hosted by metamorphic rocks. Finally, minerals such as kyanite, sillimanite, graphite, garnet, pyrophyllite, wollastonite, asbestos, talc, and mica commonly occur in association with metamorphic rocks.

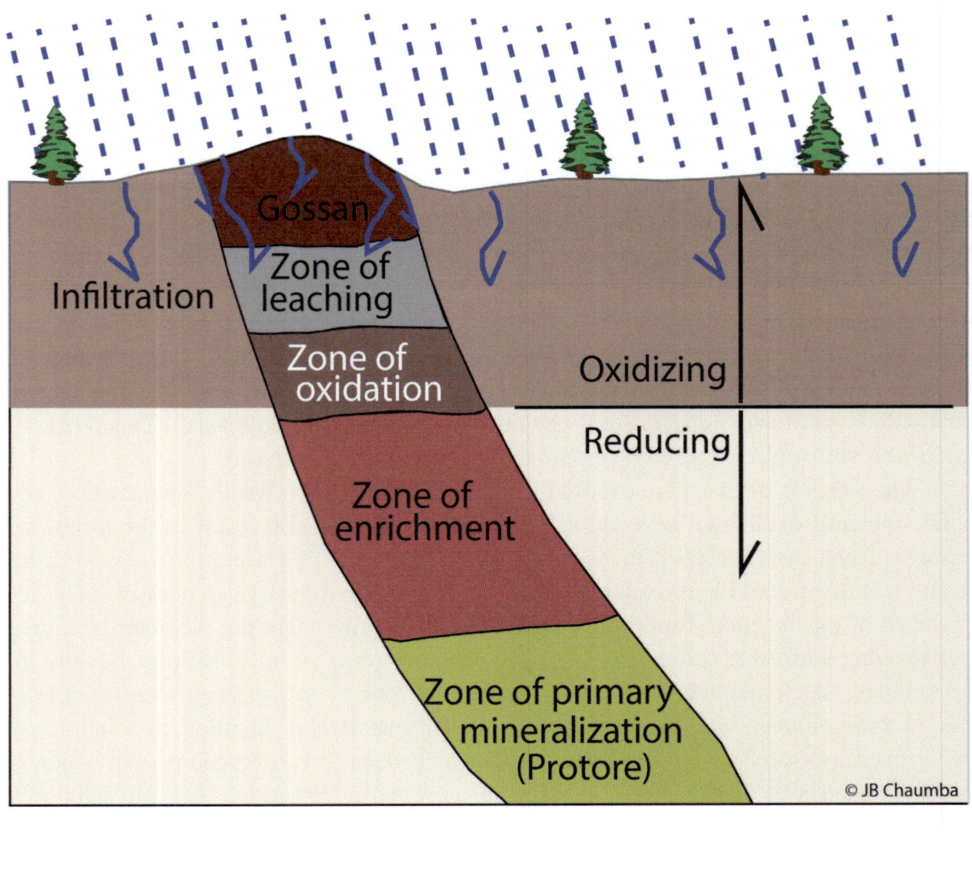

Fig. 16.9 Diagram illustrating the formation of gossan in the oxidizing zone and associated leaching and oxidation zones. Reducing environments accompanied by the zones of enrichments and zone of primary mineralization occur beneath the zone of oxidation

16.4 Non-metallic Mineral Resources

Non-metallic mineral resources (Fig. 16.1) can be defined as geologic resources that are exploited for commercial value but are neither utilized as fuels nor as sources of metals. These non-metallic mineral resources are indispensable in various types of industries such as in agriculture, manufacturing, and construction. Sand, gravel, salt, limestone, clay, gypsum, phosphate, crushed rock, sulfur, and potassium salts are examples of non-metallic mineral resources. Materials often used for building, such as sand, gravel, or crushed stone, are called **aggregates**. Water is the most extracted natural resource by humans, followed by sand.

At one sand and gravel operation, B V Hedrick Gravel & Sand Company, a sands and gravel company near Lilesville in southeastern North Carolina, USA, has been in operation since 1924. This operation mines unique high-purity quartzite gravel and sand products that are used in decorative, metallurgical, exposed aggregate concrete products, golf and turfgrass, concrete, and other specialty applications (Fig. 16.10a–d). Quartz grains of different compositions and varieties dominate the mineralogy of the particles (Fig. 16.10a–d). Sand products at this operation are used for various applications including white mortar sand, extra white mortar sand, volleyball sand, and C-33 concrete sand.

Sand and gravel (Figs. 16.10a–d and 16.11a) are the preferred resources for landscaping and they are also widely used in recreation. Furthermore, sand and gravel are the resources of choice at golf courses, lakeside beaches, as well as at volleyball courts. An aggregate is a loosely compacted mass formed from mixing fragments or particles. To help reduce both erosion along coasts and infiltration of water in certain areas, aggregates are widely used. Finally, sand and gravel are also used widely as golf course and turfgrass quartzite sand. At golf courses, the grade of sand matters as premium bunker sand (USGA 0.25 to 1.0 mm) and natural bunker sand (USGA 0.1 to 1.0 mm) are widely utilized. Further uses for sand and gravel occur in the form of sand trap sand (2.5 mm), washed fairway sand, utility topdressing sand, and rootzone sand.

Non-metallic mineral resources lack luster and are not malleable. In addition, non-metallic mineral resources are poor electrical conductors and poor thermal conductors. However, non-metallic mineral resources possess diverse applications as they are used as construction materials (cement and concrete) as well as in industrial products such as ceramics and glass. Additionally, non-metallic mineral resources are widely useful in industries such as manufacturing and agriculture. Building materials are examples of non-metallic mineral resources. Sand, gravel, clay, limestone, and dimension stone such as granite, marble, and slate

Fig. 16.10 Sands from a company in North Carolina, USA, showing piles of mined aggregate (**a**), a plant processing the aggregates by different grain sizes (**b**), a photograph of some of the different particle sizes mined by the company (**c**), and packed containers of the various grain sizes sold by the company (**d**)

Fig. 16.11 **a** Photograph of sand and gravel (credit: Ohio Department of Natural Resources), **b** phosphate (credit: ICL-Group), **c** fertilizer held in hands (credit: Farm Sahel), and **d** barite (credit: Carles Millan via Energy Virgina)

find use in construction as well as in infrastructure development. Another non-metallic resource used widely is phosphate (PO_4) (Fig. 16.11b) and it is essential to life on earth. Phosphate is a natural source of phosphorus, and it typically occurs as calcium phosphate ($Ca_3(PO_4)_2$) in apatite. Phosphorus is part of several products that are useful in modern life. No substitute exists for phosphorus in human, animal, or plant nutrition, and, consequently, in the production of food. A single phosphate-based compound is useful for different applications in a wide variety of industries such as agriculture, pharmaceutical, personal care, construction, and chemical manufacturing, among others.

Gypsum, a hydrated calcium sulfate, is a soft industrial mineral useful in wallboard and plaster. Sulfur finds use in fertilizers and acids. Substances that are added to soil for enhancing the growth of plant through providing essential nutrients are called fertilizers. Fertilizers (Fig. 16.11c) are critical for replenishing nutrients that plants absorb from the soil, ensuring increased healthy yields. Salt, another industrial mineral, is useful in food preservation as well in de-icing. Chemical and fertilizer minerals such as phosphate rock are essential ingredients in fertilizers. Gemstones such as ruby, emeralds diamonds, and other gemstones are useful in both jewelry and industrial applications such as abrasives. Other non-metallic minerals such as barite (Fig. 16.11d), graphite, mica, and asbestos play significant roles in a wide range of different industrial processes. Approximately 72% of baryte is used worldwide in oil and gas exploration as a weighting agent for drilling fluids to suppress high formation pressures and prevent blowouts. The non-metallic mineral industries, thus, function as a major component of the domestic non-fuel mineral industries. We are familiar with the importance of salt in our everyday lives through foodstuffs. Sodium chloride, in particular, plays a significant role for our bodies to function. Table salt helps transmit nerve impulses and it is crucial for the functioning of our nerves and muscles to function. Non-metallic resources play a critical role in manufacturing goods, supporting agricultural productivity, and building infrastructure. These resources are essential in the production of glass, cement, ceramics, and lime products.

The process of verifying data collected from a distance, such as geophysical surveys or satellite imagery through field observations and measuring conditions at the actual mining site, is called ground truthing. Ground truthing is important as it confirms, or disproves, the accuracy of data obtained through remote methods. Finally, ground truthing ensures the validity of surface geological features and interpretations of the subsurface geology. Ground truthing is also applied in the prospecting of energy resources that we discuss next.

16.5 Energy Resources

Energy can be defined as the capacity to do work, permitting things to move or change. Energy can also be converted from one form to another. Several types of energy occur, such as thermal energy (energy of heat), potential energy (stored energy), kinetic energy (energy of motion), chemical energy (energy in the bonds of molecules released during chemical reactions such as burning fossil fuels or food), electrical energy (energy of moving electric charges, like in lightning strikes or batteries), and nuclear energy (energy stored in the nucleus of an atom that is released during nuclear reactions such as nuclear power plants), for example.

Energy can be converted from one form to another. For example, water in a river that flows downhill, turns a turbine connected to a generator, which then produces electricity at a hydroelectric power plant. Chemical energy stored in gasoline is converted to kinetic energy in a vehicle. According to the law of conservation of energy, energy can neither be created nor destroyed, but it is only transformed from one form to another.

Every country consumes energy, and the amount of energy consumed by countries varies widely (Table 16.1). For example, consumption of energy in the United States in 2024 was approximately 94 quadrillion British thermal units (or quads), an increase of 1% from 2023. This consumption was largely (82%) accounted for by petroleum, natural gas, and coal whereas the remaining 18% was accounted for by nuclear energy as well as solar, wind, hydropower, and geothermal energy. Petroleum remained the most-consumed fuel, and nuclear energy consumption surpassed coal for the first time.

The capacity to do work is called energy. Anything that possesses the capability to power life, move objects, produce heat, or produce electricity is referred to as an **energy resource** (Fig. 16.1). Natural sources of power that are useful as fuels, in electricity generation, heating and cooling, or power various activities, for example, are called energy resources. A total of seven types of energy resources are recognized: i. fossil or biomass, ii. nuclear, iii. hydroelectric, iv. wind, v. solar, vi. geothermal, and vii. wave energy.

Energy has several forms such as motion, heat, electrical, gravitational, light, and chemical energy, among others. Energy is grouped into either kinetic (working) or potential (stored) energy. Kinetic energy can be converted to potential energy, and vice versa. Consumption of food, for example, contains chemical energy which is stored in a person's body to be later converted to kinetic energy when the person plays or works. Similarly, chemical energy stored in coal or natural gas and the kinetic energy of running water can be converted to electrical energy. The electrical energy can then be converted to light and heat Energy can be subdivided into renewable or non-renewable energy resources. Energy resources that are not depleted with usage, or can be renewed in relatively short spans of time of up to decades, are called **renewable energy** sources.

Table 16.1 List of the top 25 countries for 2023 by electric energy consumption in terawatt-hours (TWh) and per capita figures are in megawatt-hours (MWh)

Rank	Country	Consumption (TWh per annum)	Per capita (MWh per annum)
1	China	9,443.07	6.64
2	United States	4,272.91	12.44
3	India	1,956.55	1.36
4	Russia	1,162.44	7.99
5	Japan	1,013.26	8.15
6	Brazil	725.1	3.43
7	South Korea	617.92	11.94
8	Canada	605.52	15.41
9	Germany	505.76	5.98
10	France	467.81	7.04
11	Saudi Arabia	422.83	12.71
12	Iran	380.28	4.2
13	Mexico	357.75	2.76
14	Indonesia	351.49	1.25
15	Turkey	323.8	3.71
16	Italy	312.72	5.26
17	United Kingdom	309.22	4.5
18	Taiwan	282.14	12.1
19	Vietnam	278.61	2.78
20	Australia	273.11	10.33
21	Spain	265	5.53
22	South Africa	231.79	3.67
23	Thailand	224.04	3.13
24	Egypt	218.54	1.91
25	Malaysia	186.66	5.31
-	World	29,664.60	3.67

Source Ember-Energy.org

16.5.1 Renewable Energy

Renewable energy, also known as green energy, is energy that is generated from renewable resources that are replenished on human timescales, or derived from natural sources that are replenished at rates that are higher than they are consumed. Solar, wind, hydroelectricity, and geothermal energies are considered to be the cleanest sources of energy since they produce negligible amounts of greenhouse gases when they are in operation. Consequently, these sources of energy are also referred to as "low-carbon" or "clean" energy. Since the availability of uranium is limited, nuclear power is also a low-carbon energy source, it is not considered renewable due to the limited availability of uranium. Renewable energy resources, such as solar, wind, hydroelectric power, or simply hydropower, and geothermal energy, are replenished naturally. Although renewable energy constitutes a smaller percentage of global energy consumption of approximately 13–15%, the share of renewable energy in electricity production is higher. Worldwide, sources of renewable energy (hydropower, wind, solar, biomass, and geothermal) account for approximately 30% of global production of electricity.

The Grand Coulee Dam, a hydroelectric power plant that was constructed on the Columbia River in Washington State, is the largest one in the United States. It has a large capacity for generating electricity of more than 6,809 megawatts. The combination of the vast size of the dam as well as huge water supply makes the Grand Coulee Dam the United States' largest electricity generator by capacity. It produces more than 21 billion kilowatt-hours of electricity annually, enough to supply ~2 million households with power. Like other dams, the Grand Coulee plays a critical role in flood

Fig. 16.12 Photographs of some renewable energies. **a** Grand Coulee Dam, on the Columbia River west of Spokane, Washington (credit: US Bureau of Reclamation), **b** wind turbines in Taft, Texas, USA, **c** geothermal energy plant (credit: IPSGroupBV), and **d** biomass (corn stalks) in Crowley, Texas, USA

control, provides a reliable source of power, water for irrigation, and regulates streamflow which regulates migration of fish. The three largest electricity-producing dams in the world are Three Gorges Dam in Hubei, China, with an instantaneous generating capacity at 22,500 **Megawatt** (MW) of power, followed by the Baihetan Dam, also in China, with a capacity of 16,000 MW, and the Itaipu Dam in Paraguay and Brazil with a power capacity of 14,000 MW.

Hydropower (Fig. 16.12a) is a type of renewable energy source that utilizes the kinetic energy of flowing water to generate electricity. The flowing water in rivers or streams spin turbines that are connected to generators, which then generate electricity. Hydropower is driven by the water cycle, which in turn is driven by the sun, and hence it is a renewable source of energy. **Wind energy**, a type of renewable energy, relies on the kinetic energy of air in motion to generate electricity through wind turbines (Fig. 16.12b). A combination of unequal heating of the Earth's surface by the sun, irregularities in the Earth's surface, and Earth's rotation around the sun generates wind. Differential heating of Earth's surface leads to different areas heating up at different rates which generates areas of high and low pressures resulting in air in motion, called wind. Propeller-like blades on wind turbines then capture the wind's energy, ultimately generating electricity which is then transmitted to consumers.

Energy that is generated by geothermal energy is called **geothermal energy** (Fig. 16.12c), and it is utilized for either electricity generation or heating purposes. Although the number of countries utilizing geothermal energy can fluctuate, it is estimated that geothermal electricity is currently generated in 26 countries, whereas geothermal heating is utilized in 70 countries. In late 2025, engineers were building the hottest geothermal energy plant on Earth on the slopes of Newberry Volcano in Oregon, United States. Newberry Volcano is one of the largest active volcanoes in the United States, as well as one of the most hazardous ones. Geothermal production in late 2025 accounted for less than 1 percent of the total production of electricity worldwide. According to the International Energy Agency (IEA), however, this figure is projected to increase to 8 percent by 2050 if scientists can tap into superhot rock, coupled with other advances in technology. The IEA estimates that geothermal energy tapping superhot temperatures could generate 150 times more electricity than the world uses at present.

Pioneering projects during the past few decades began to produce energy from hot dry rocks through cracking it followed by pumping in water to generate steam. This approach is akin to hydraulic fracturing, or fracking, which is utilized in the oil and gas industry. In France, Japan, Germany, and Switzerland among others, geothermal energy is being utilized, whereas in Nevada and Utah pilot geothermal energy projects already exist.

The Newberry Volcano project, part of Mazama Energy, is one of the hottest geothermal sites in the world with temperatures that have already reached 331 °C (629 °F) with the goal of cranking up the temperature to more than 398 °C (750 °F). Geothermal energy has an advantage of low risks of water pollution since geothermal plants recycle the same water which is sealed and isolated in wells by passing it through reservoirs located at depths that are deeper than the depths of common groundwater.

Energy from plants and animals that is derived from organic matter is called **biomass energy** (Fig. 16.12d). Biomass is solar energy that is stored within, or recently living, organisms released when they are used as fuel. Biomass is either burned to produce heat, or it is converted into other forms of energy such as electricity or biofuels. Biomass comes from various sources, such as agricultural crops and waste, wood and forestry residues, animal manure, and sewage. Organic components of garbage also contribute to biomass. Organic matter, which is burned to produce heat that can be used directly for heating, or for generating steam that powers turbines which produce electricity. Biomass is low cost, and a renewable source of energy as organic matter is replenished through plant growth. Use of biomass energy is considered carbon neutral since the carbon released during combustion is offset by the carbon which plants absorb during photosynthesis resulting in zero net carbon emissions. Biomass energy is generally considered to be the most environmentally friendly alternative to using fossil fuels.

Technologies in use include dry steam power stations, flash steam power stations, and binary cycle power stations. In 2023, sources of renewable energy accounted for 9% of the total energy consumed in the United States. This is much lower than countries such as Iceland and Norway which generate more than 70% of their energy from renewables, largely from hydropower and geothermal resources. Fifty percent of Sweden's energy is derived from renewable sources.

Radiant energy from the sun's electromagnetic radiation, in particular, its heat and light, is the source of **solar energy**. Solar energy is harnessed to produce heat, electricity, or other forms of energy. Solar energy is renewable, clean, and sustainable. Our planet receives more solar energy than it currently uses. Solar energy is abundant.

When it comes to solar power, solar cells and panels often get all the attention. However, there exists a different kind of solar power generator referred to as a **solar thermoelectric generator** (STEG). The efficiency of STEG was recently increased by a factor of 15. STEG utilizes a principle called Seebeck effect whereby a difference in temperature exists across any two different conductors leading to the generation of voltage. Thus, STEGs are devices where one side of STEGs is cold whereas the other one is hot, resulting in electricity flowing through semiconduc-

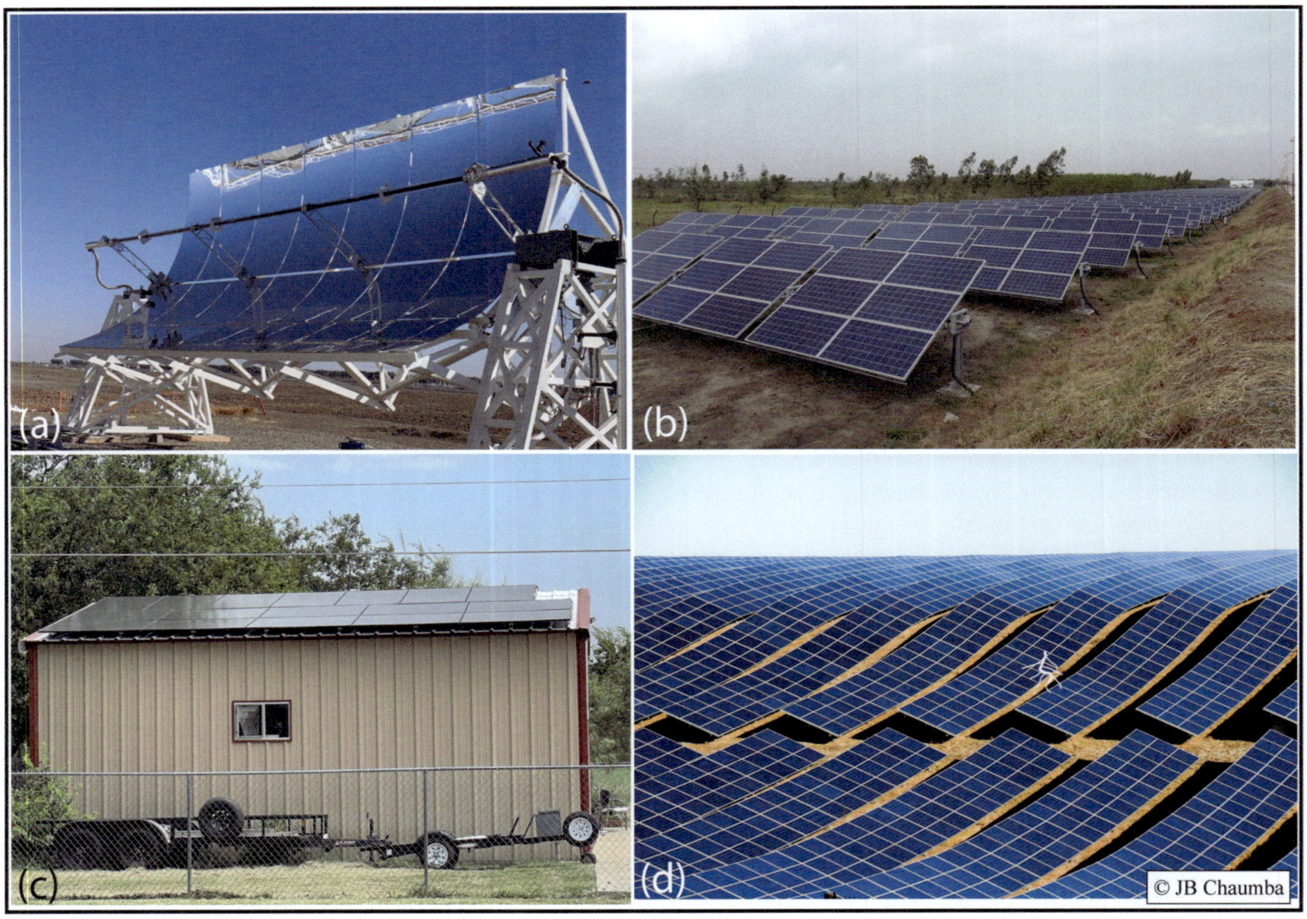

Fig. 16.13 Photographs of some common types of solar panels: **a** parabolic troughs, **b** solar tiles, **c** solar panels on roof of a workshop in Crowley, Texas, USA, and **d** thin-film solar panels. *Credit* American Solar Energy Society

tors placed between them. In STEGs, there are no moving parts hence they are solid-state devices. Although the Sun's energy heats up the hot side, it can also utilize any kind of thermal energy to maintain a high temperature. Traditionally, STEGs have been able to convert only less than one percent of sunlight into electricity. In comparison, perovskite/silicon solar cells exceed 30% in power conversion. Consequently, STEGs appear to be unlikely to be viable sources of clean power anytime soon.

Radiant energy, a renewable source of energy from the sun, can also be harnessed and converted into usable forms of energy such as heat and electricity. Radiant energy is captured using various technologies. **Photovoltaic (PV) cells** capture solar energy for electricity generation, whereas solar thermal systems capture solar energy for heating. Photovoltaic systems are also referred to as solar panels. In addition to its abundance, solar energy is clean and doesn't contribute much to global warming. Solar panels, composed of PV cells, typically made of semiconductor materials such as silicon, convert energy directly into electricity. **Concentrating Solar Power**, or CSP, are panels that utilize mirrors to focus sunlight onto receivers resulting in the generation of heat. The heat then drives turbines which generate electricity. Parabolic troughs (Fig. 16.13a) are examples of CSPs.

Solar Thermal Systems, or STS, make use of solar energy to heat fluids which are then be used for generating electricity through turbines, heating, and cooling. Solar energy is commonly used to generate electricity through solar panels typically located on top of roofs and at large solar farms. Solar energy also finds extensive use can power homes, businesses, and entire communities as well as being integrated into heating and cooling systems in buildings. Solar thermal systems also provide hot water used in industries or for domestic use and are also utilized in powering electronic devices, cooking, and even in solar ponds for chemical production. The advantages of solar energy are that it is clean and a renewable form of energy. Additionally, solar energy is cost-effective since its cost has decreased substantially in recent years. Solar energy is abundant for entire continents such as Africa and countries like Australia. Solar energy is abundant in a number of other regions such as most of the southern and southwestern parts of the United States. The disadvantages of solar energy are

Table 16.2 Table showing the six main types of solar panels

Solar panel type	Approximate lifespan (years)	Efficiency (%)	Key negative	Key positive
Monocrystalline	30–40	20–25	Highest efficiency	More expensive than some types
Perovskite	25–35	29–31	Highest efficiency	Unavailable yet
Transparent	25–35	1	Blends in with windows	Low efficiency
Solar tiles	25–30	10–20	Blends in with roofs	Expensive
Polycrystalline	25–30	18–21	Cheaper than monocrystalline	Discontinued
Thin film	10–20	17–19	Most flexible option	Low power output

that it is intermittent since its production is dependent on sunlight, weather conditions, and time of day both affect solar energy. The initial investment in solar panel systems is often huge, although with time this is becoming less of a barrier. Finally, large solar farms are required to power large buildings which take significant land areas which can be utilized for agriculture, industries, and housing, among others.

Solar panels come in several types. There are presently six main types of solar panels (Table 16.2) which are i. perovskite, ii. monocrystalline, iii. transparent, iv. solar tiles (Fig. 16.13b, c), v. polycrystalline, and vi. thin film (Fig. 16.13d). All of these solar panels utilize daylight to generate electricity, that is, they are all **photovoltaic panels**. These panels are all categorized based on the type of material they are made of. Solar panels with both the best efficiency and longest lifespan are monocrystalline solar panels. The most popular type of solar panel in the United Sates is monocrystalline solar panel due to its sleek black appearance, long lifespan, and high efficiency. It must be noted here that in the United States, the traditional mounted solar panels remain more widespread although use of solar tiles and shingles is also increasing. In the United Kingdom, for example, the monocrystalline solar panel is the preferred choice for residential rooftop installations. No new polycrystalline solar panels are still available for installation from top installers in the USA. Thin-film panels are a good, low-cost option for mobile homes. It must be borne in mind that solar panels still continue to produce electricity lasting their proposed lifespans as shown in Table 16.2 but at gradually reduced rates.

Building-integrated photovoltaic (BIPV) products that combine the functions of shingles or traditional roofing tiles with generation of solar energy are referred to as solar tiles or solar shingles (Fig. 16.14a, b). In addition to electricity generation from sunlight, some solar tiles are designed to function and resemble common standard roofing materials such as slate which allows for a seamless integration with the design of buildings. Rather than simply being installed on top of a roof, solar tiles are also integrated into the roofing system to replace traditional roofing materials such as tiles or shingles. Solar panels are designed to be aesthetically pleasing by mimicking appearances of traditional roofing materials. This makes solar panels more discreet options for some homeowners that are concerned about the appearance of solar panels.

According to the American Solar Energy Society, thin film is the future of the solar industry. Thin-film solar panels are very economical as they require less material. Additionally, thin-film solar panels don't contain toxic components, generate less waste, and they are very easy to manufacture. Their disadvantages, however, are that they have the lowest lifespan of only 10–20 years, and they also have a lower power output.

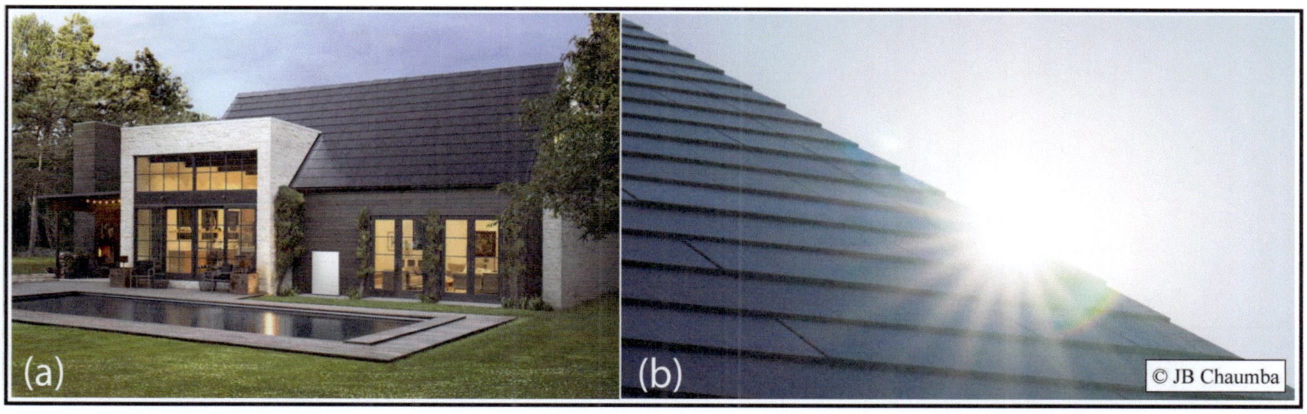

Fig. 16.14 **a** and **b** Photographs of Tesla Solar Roof. *Credit* https://www.tesla.com/solarroof

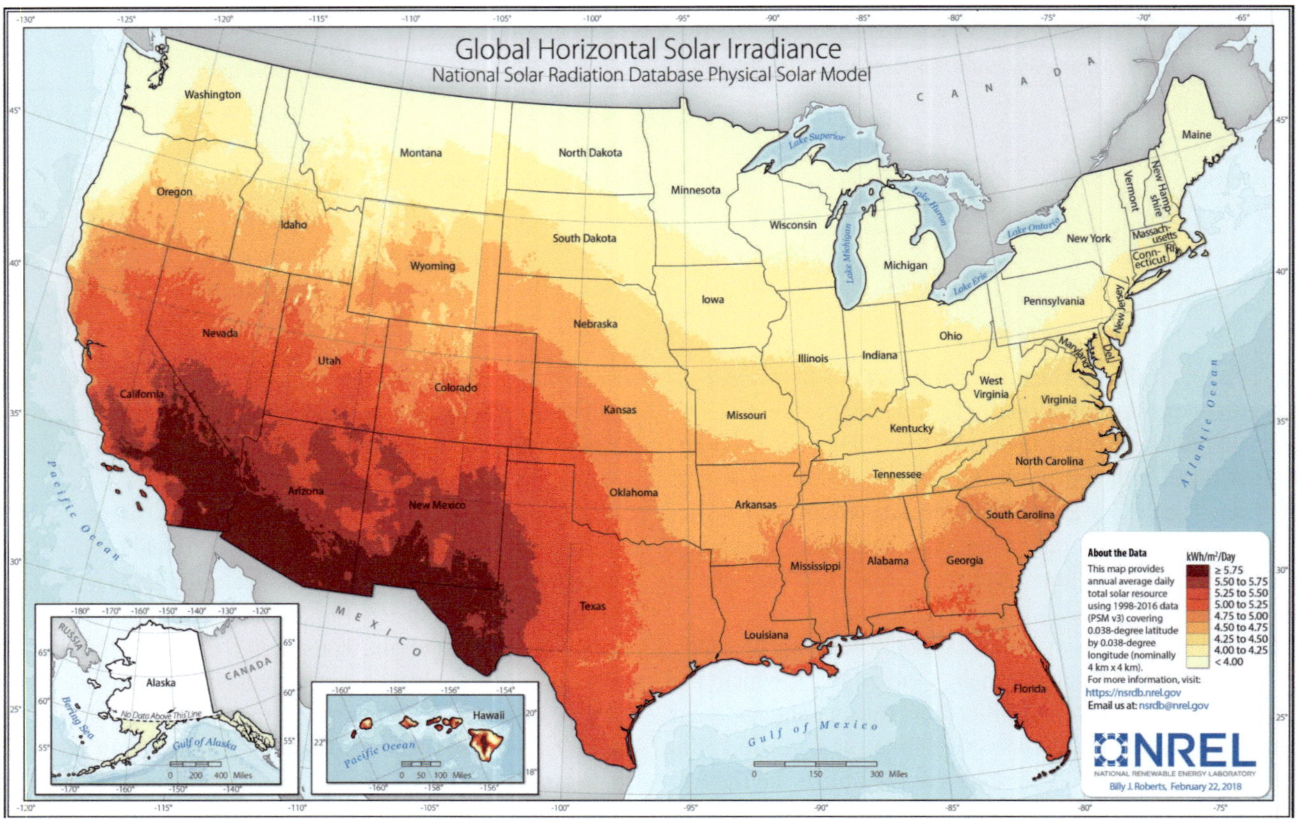

Fig. 16.15 Horizontal solar irradiance map of the United States showing the highest vales in the southwestern part of the country followed by the southeastern part of the country

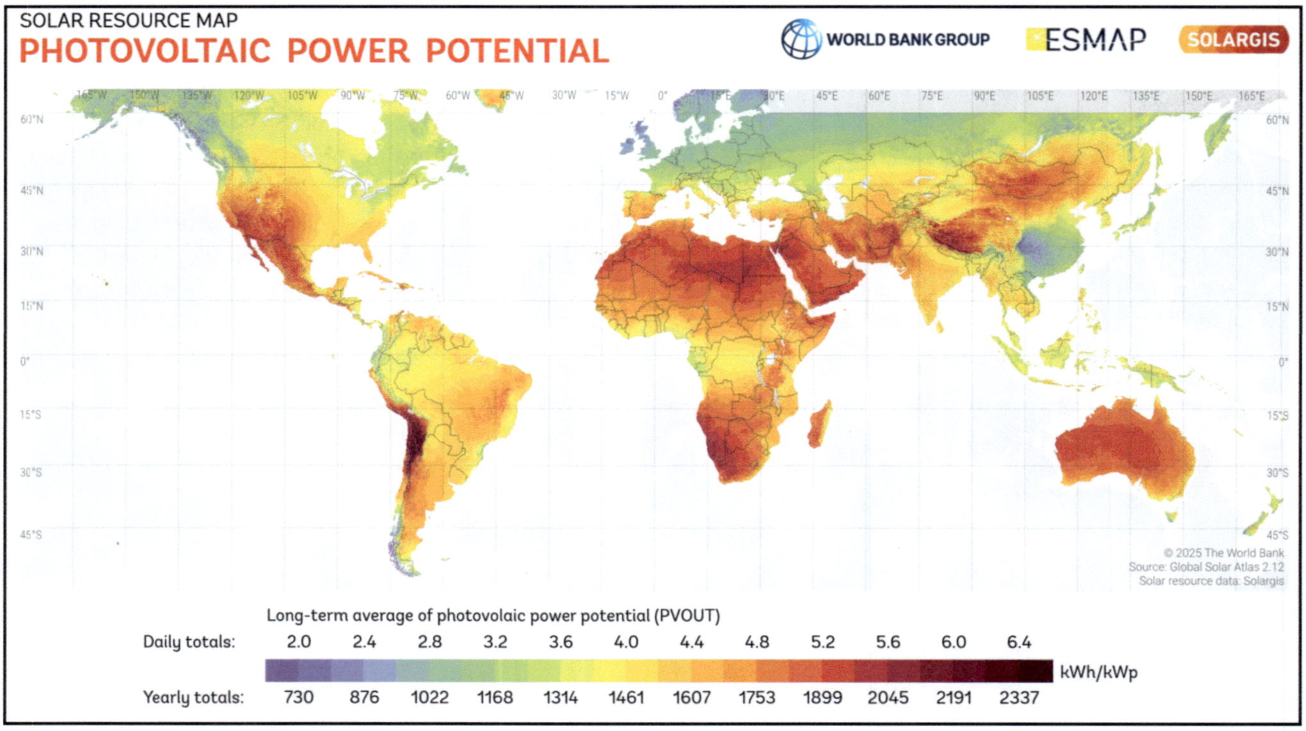

Fig. 16.16 Photovoltaic power potential of the world showing the highest vales in Australia, Africa, the Arabian Peninsula, Central America, the United States and South America as well as portions of southern Europe

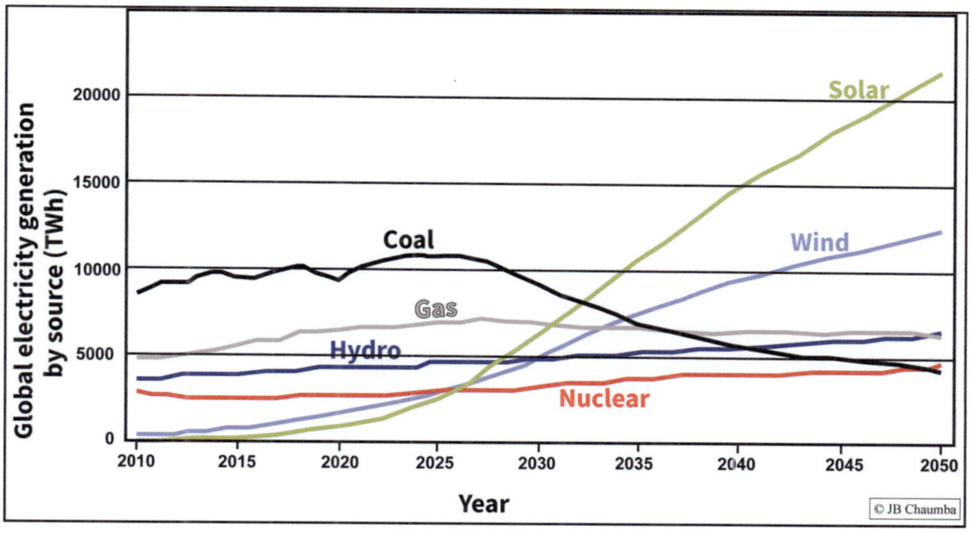

Fig. 16.17 Diagram showing the amount of global electricity generation by source in Terawatt-hour (TWh) versus year since 2020. Solar energy is predicted to be the dominant source of global electricity generation (After World Energy Outlook, 2024)

In the U.S., monocrystalline solar panels are bulk of solar panels that are installed, particularly for residential use, due their high efficiency. For the utility-scale market, thin-film panels, especially thin-film cadmium telluride (CdTe) panels, are the most common since they are preferred in large projects. In Texas, monocrystalline and polycrystalline panels are the most common solar panels. Although they cost more, monocrystalline panels are favored for their higher efficiency and performance in high-temperature Texas climates, despite their higher cost. Texans, for example, should also consider solar panels with a low-temperature coefficient for better efficiency in the heat, and robust construction to withstand severe weather such as hail and strong winds. Solar tiles can now produce significant amounts of electricity, with excess power either being stored in batteries or fed back into the grid.

Historically, coal has been the dominant form of electricity generation for the world as shown in Fig. 16.17. However, in approximately 2032, coal is projected to lose its pole position to solar energy (Fig. 16.17). Looking to the future, or around 2050, in order of increasing importance, global electricity generation is forecast to be dominated by solar, wind, hydro, gas, nuclear, and coal, as shown in this figure.

16.5.2 Non-renewable Energy

Non-renewable energy resources are energy resources that cannot be replenished at a rate comparable to their consumption and they form over relatively long periods of time such as millions of years. Coal, oil, natural gas, and petroleum took thousands of years to form from the buried remains of ancient sea plants and animals that lived millions of years ago, which is why these energy sources are also referred to as **fossil fuels**. Coal (Fig. 16.18a), a brownish black to black sedimentary rock with a high amount of carbon and hydrocarbons that make it combustible, mostly occurs in rocks of the Carboniferous Period (~359 to 299 million years ago) belonging to the Paleozoic Era. Coal contains high amounts of carbon and hydrocarbons which make it combustible through utilizing the energy stored by plants that thrived in swampy forests some hundreds of millions of years ago. Coal is classified as a non-renewable energy source since it takes millions of years to form.

Another important fossil fuel is oil (Fig. 16.18b), which forms from remains of minute ancient marine organisms such as algae and plankton over millions of years under specific conditions. When algae and plankton die, they sink to the bottom of oceans where they accumulate within sedimentary beds. These layers become buried at increased depth over time, which subjects the organic matter to higher temperatures and pressures, transforming the organic matter into hydrocarbons, which are the major components of crude oil. The organic matter is later transformed into the waxy kerogen through diagenesis in the absence of oxygen. With continued increase in temperature and pressure, kerogen breaks down into hydrocarbons through the process of catagenesis. Finally, the hydrocarbons migrate through porous layers of rock, and some get trapped in reservoirs to form oil deposits.

Metals, and nuclear fuels such as uranium, are other examples of non-renewable energy resources. Non-renewable energy sources account for most U.S., and many other countries, energy consumption. Nuclear energy is produced from uranium, when atoms are split through nuclear fission generating heat and, eventually, electricity. Uranium is thought to have been formed >4 billion years ago when stars formed. Uranium occurs throughout the earth's crust, but most of it is too expensive, or too difficult, to mine and process into fuel for nuclear power plants.

Fig. 16.18 **a** Coal from Hwange coal field, Hwange Colliery Company, Zimbabwe, credit: newZWire, **b** photograph of motor oil in container and lid, **c** Plant Vogtle in Waynesboro, Georgia, USA (credit: Southern Nuclear), and **d** photograph of ground pipes connecting to underground natural gas storage container (credit: US Department of Transportation)

These energy sources are called non-renewable because their supplies are limited to what we can mine or extract from the earth. Most of the petroleum products consumed in the United States are made from crude oil, but petroleum liquids can also be made from natural gas and coal. Renewable and non-renewable energy sources can be used as primary energy sources to produce useful energy, such as heat, or they can be used to produce secondary energy sources such as electricity and hydrogen.

Major sources of renewable energy are hydropower from flowing water, solar energy from the Sun, geothermal energy from the interior of our planet, wind energy caused by differential heating of our planet, and biomass from vegetation. Renewable energy sources are replenished by natural processes due to several factors. For example, the Sun's rays strike our planet daily, causing vegetation to grow. Wind also blows daily on our planet, similarly with evaporation occurring daily which leads to precipitation. Precipitation and groundwater eventually result in rivers flowing daily, which results in our planet's renewable energy. Wave energy is a type of renewable energy that utilizes the movement of the ocean's waves to generate electricity. For areas in the interior of continents far away from coastlines, access to wave energy is limited due to prohibitive costs of making it accessible to continental interior regions.

In an effort to reduce the amount of carbon emissions which contribute to global warming, incentives are now being offered by some countries to increase the use of renewable energy resources, particularly in electricity generation. In the United States, incentives are geared to lower the cost of transition to renewable energy through grants, loans, tax credits, and rebates, among others. The goal is to promote use of renewable energy and make it more affordable.

Nuclear energy, an important powerful source of energy through generation of energy through nuclear fission reactions. However, concerns are raised concerning safety and disposal of nuclear waste. Other sources of energy are ocean tides and waves. Plant Vogtle (Fig. 16.18c), operated by Southern Company subsidiary Southern Nuclear, includes the only new commercial nuclear units built in the U.S. during the past 30 years. Vogtle-3 began commercial operations on July 31, 2023, and Voglte-4 followed suit on April 29, 2024. Most underground gas storage facilities take advantage of depleted oil and gas fields. Other underground storage facilities utilize depleted aquifers as well as salt formations to store natural gas (Fig. 16.18d). Non-renewable energy resources, or fossil fuels, are finite and are not replenished at rates comparable to their consumption. At present, fossil fuels are currently the major sources of energy despite them being the biggest contributor to the emission of greenhouse gases.

A raw form of energy occurring in nature that has not been converted into another form is called a primary energy source. A raw form of energy is the initial source of energy that is harnessed and used directly, or it can be converted into other forms of energy. Fossil fuels and renewable sources are examples of primary energy sources. Examples of primary sources of energy include energy that is stored within the nucleus of atoms, nuclear energy, and fossil energy. A form of energy that is generated through the conversion of a primary energy source is referred to as a secondary energy source. A secondary energy source, thus, is energy that has been transformed, or processed, from its original state into a form that is more usable or transportable. Gasoline in containers or tanks and electricity via power lines are examples of secondary energy sources. Transformation of primary energy source to secondary energy source occurs when crude oil is refined into gasoline or when coal is burned to produce electricity.

Recall that we discussed the structure of an atom in Chap. 2. Energy that is stored within the nucleus of an atom is called nuclear energy. Nuclear energy is released either through combining atoms, called **nuclear fusion**, or through splitting of atoms, called **nuclear fission**. A reaction whereby the nucleus of heavy atoms such as uranium is split into two or more smaller nuclei releasing huge amounts of energy in the process is termed nuclear fission. Power plants currently utilize nuclear energy to heat power plants through nuclear fission to produce steam. The steam in turn spins turbines which generate electricity. Nuclear fission releases energy. Nuclear fusion involves combining light atoms such as hydrogen to form heavier ones. Currently, there is no energy produced from nuclear fusion as the technology to generate electricity from nuclear fusion is still being investigated.

Fossil fuels, such as coal, petroleum, and natural gas, are sources of energy formed from mixtures of compounds comprised of fossilized plant and animal remnants source that were formed in the Earth's crust from decayed organic material. Fossil fuels, which encompass both liquid and gaseous fuels, fall within the field of petroleum geology. Fossil fuels are important in the generation of electricity in which the mechanism providing the initial force, or mechanical energy, to spin the generator's rotor and produce electricity is an internal combustion engine or turbines. The turbines are rotated either by high-pressure steam that is produced in a boiler or by hot exhaust gases produced from the burning of fossil fuels.

Although we discussed the formation of coal in Chap. 4, in this section, we will briefly focus on its energy content. Four main types, or ranks, for coal based on both the i. types and amounts of carbon contained in the coal and ii. on the amount of heat energy the coal produces are 1. anthracite, 2. bituminous, 3. subbituminous, and 4. lignite. The rank of a deposit of coal is dependent on the pressure and heat to which the plants (coal) were subjected.

The highest grade of coal recognized, the one with the highest heating value of all ranks of coal is anthracite. Anthracite is characterized by 86%–97% carbon but it only accounts for a paltry <1% of all the coal that was mined in the United States in 2022. Anthracite mining in the United States is confined to northeastern Pennsylvania. Anthracite in the United States is mainly used by the metal industries.

Bituminous coal, the second highest grade of coal, contains 45–86% carbon and it is the most abundant rank of coal occurring in the United States. In 2022, bituminous coal accounted for about 46% of total U.S. coal production. Bituminous coal is used to generate electricity and is an important fuel and raw material for making coking coal for the iron and steel industry. Bituminous coal was produced in at least 16 states in 2022, but five states accounted for about 78% of total bituminous production. States with the highest production of bituminous coal in the Unites States in 2022 were West Virginia—31%, Illinois—14%, Pennsylvania—14%, Kentucky—11%, and Indiana—9%.

With a lower heating value than bituminous coal, subbituminous coal contains 35–45% carbon. In the United States, most subbituminous coal is at least 100 million years old. Subbituminous coal accounted for ~46% of total U.S. coal production in 2022. The leading U.S. subbituminous-producing states in 2022 and their percentage share of total production were Wyoming—89%, Montana—8%, New Mexico—2%, Colorado—2%, and Alaska <1%. A map of US showing the approximate coal deposits and their ages is shown in Fig. 16.19. A map of most of the major coal deposits throughout the world is shown in Fig. 16.20. From this map, it is apparent that northern hemisphere continents, and Australia in the southern hemisphere, are richly endowed with coal deposits in comparison to southern hemisphere continents.

Lignite, with the lowest energy content of all coal ranks, contains 25–35% carbon. Deposits of lignite coal are relatively young and consequently have not been subjected to relatively high heat or pressure. Lignite, therefore, has high moisture contents and is crumbly, accounting for the low heating value of all types of coal. The five highest lignite-producing states and their percentages were North Dakota—56%, Texas—36%, Mississippi—7%, Louisiana—1%, and Montana <1%. At the Great Plains Synfuels Plant in North Dakota, lignite is converted to synthetic natural gas that is sent to consumers in the eastern United States via pipelines of natural gas. Since 2008, production of all types of coal in the U.S. has declined.

Any plant that utilizes coal, petroleum, or gas as its source of energy is a fossil fuel plant. A fossil fuel steam-

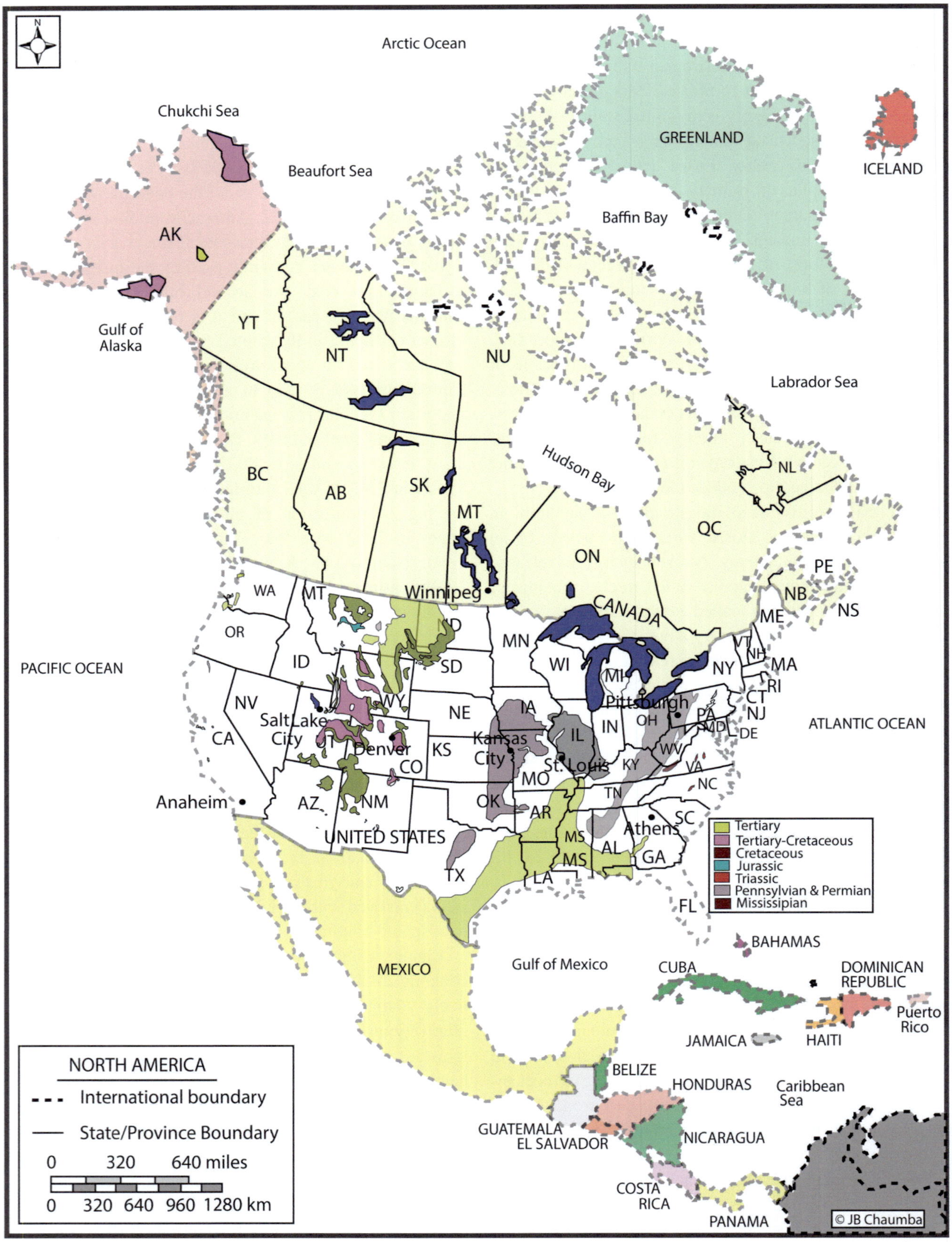

Fig. 16.19 Map of Central and North America showing the approximate location of US coal deposits. The distribution of major coal deposits on a world scale is shown in Fig. 16.20

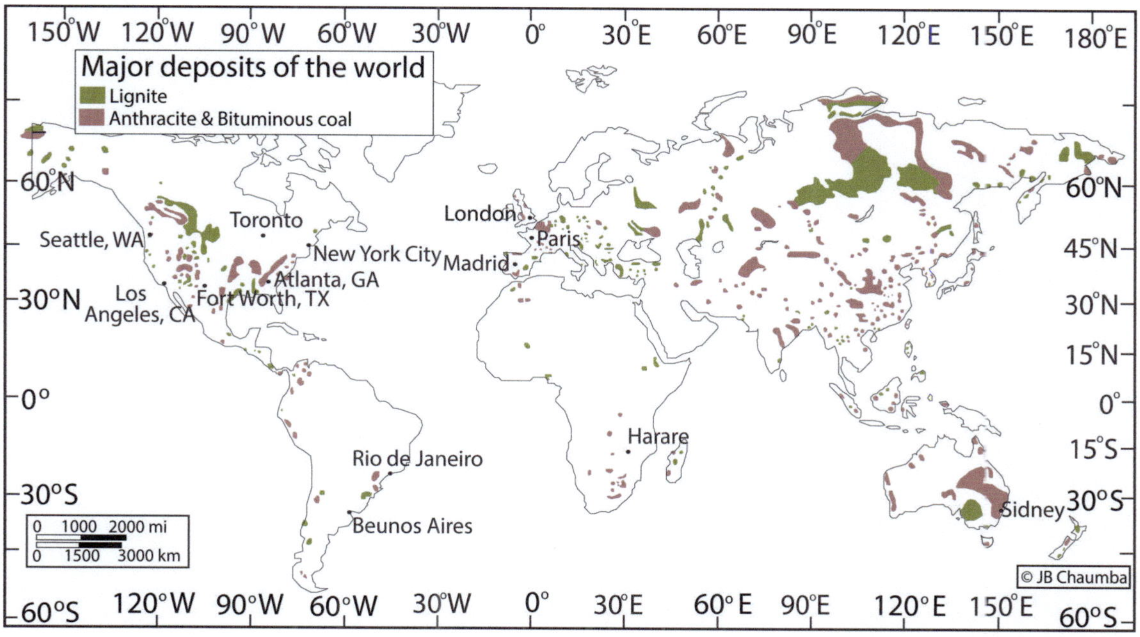

Fig. 16.20 Map of world, except for Antarctica, showing the approximate distribution of major coal deposits. Australia and the northern hemisphere are richly endowed in deposits of coal

electric power plant is an electricity generation plant in which the prime mover is a turbine that pis rotated by high-pressure steam produced in a boiler by heat from burning fossil fuels. Energy formed from ancient plants and animals which died and then gradually buried by beds of coal, oil, and natural gas is referred to as fossil energy. Fossil energy is a **non-renewable resource**, that is, the sources of energy are not replenished naturally on human timescales. Once fossil fuels are mined and used up, they are depleted and won't be available for future use.

The largest nuclear power plant electricity generator in the United States is Plant Vogtle located in Waynesboro, Georgia. This nuclear power plant is comprised of large, gigawatt-scale light water reactors which include some of the newest units in Vogtle-3 and Vogtle-4. These newest units utilize the advanced passive pressurized water reactor design. A total operating fleet of 94 reactors power homes and businesses in 28 states. Passive safety describes reactors with the ability of advanced reactors or newest designs that can shut down and remove excess heat without human intervention. In the event of a nuclear plant loses power, though highly unlikely, passive safety systems keep reactors safe and cool the core to prevent fuel from damage. Gravity, or the natural circulation of coolant, moves heat away from the reactor core without the need for external power sources, action of operator, or pumps, also termed "walk-away safe." "Advanced reactors" refer to a broad suite of designs that come in an incredible variety of shapes, sizes, materials, and fuel types. All Enhanced Safety of Advanced Reactors have in common is the ability to achieve safety, enhanced efficiency, and versatility over designs of conventional reactor. The simpler design of Enhanced Safety of Advanced Reactors compared to conventional reactors means less maintenance, fewer components, and fewer workers are required. Enhanced Safety of Advanced Reactors are designed to self-adjust and be fail-safe with passive safety systems thereby preventing the possibility of over-heating.

The disadvantages of nuclear energy relate to the likelihood of accidents which could result in high fatalities, as well as the extremely high costs of building nuclear facilities. Construction of nuclear power plants requires significant upfront investments in construction, safety systems, and eventual decommissioning. In addition, the high costs associated with potential for malfunctions such as meltdowns are major concerns. Nuclear power plants require huge amounts of water for cooling, which exerts strain on available water resources, especially in regions with scare water supply such as deserts. Some people are wary of nuclear power plants due to safety concerns. Consequently, citing and construction of new plants are often difficult undertakings.

Uranium, the element used in nuclear power plants, is a non-renewable resource although it is relatively abundant. Despite stringent safety measures in existence at nuclear power plants, accidents such as the Chernobyl in Russia and the Fukushima in Japan disasters are proof of the potential for catastrophic consequences, such as environmental contamination on a long-term as well as leaks in radiation. Finally, radioactive wastes produced by nuclear power plants remain hazardous for hundreds of years. Safe and

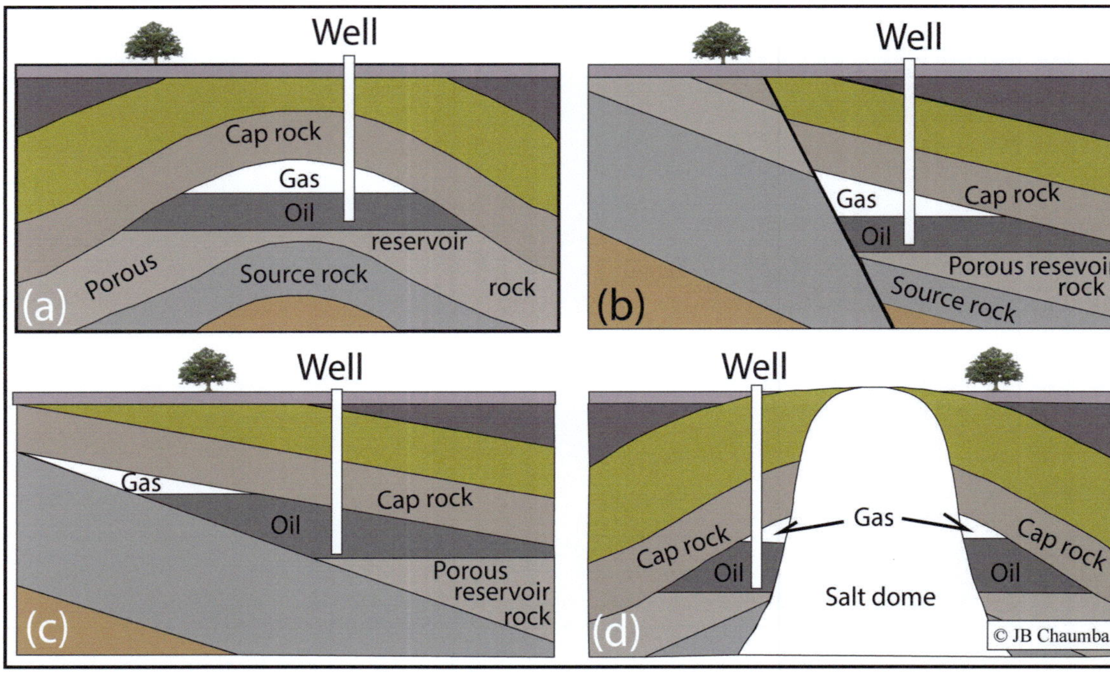

Fig. 16.21 Diagram showing some common oil and gas traps of anticline (**a**), fault (**b**), stratigraphic (pinchout) (**c**), and salt dome (**d**) traps

permanent disposal of radioactive wastes are expensive and pose unique technical challenges.

Hydrocarbons, which are compounds of hydrogen and carbon naturally occurring in sedimentary rocks in the Earth's crust, make up crude oil and natural gas. Hydrocarbons are created by the compression of the remains of plants and animals in sedimentary rocks such as sandstone, limestone, and shale that lived in ancient oceans and seas. The remains of plants and animals such as plankton, algae, and other organic materials decayed and were integrated into the sedimentary rocks. The organic material eventually transformed into oil and gas after being subjected to specific temperatures and pressure ranges deep in the Earth's crust.

Oil and gas, both being less dense than water, migrate through porous sedimentary rock to migrate and accumulate, or be "trapped" and preserved, in various settings called traps (Fig. 16.21). For petroleum to accumulate, a porous and permeable reservoir rock such as sandstone or limestone is required, where it can accumulate and be stored. An impermeable layer of rock, called a cap rock (Fig. 16.21), is required to prevent the oil from escaping to the surface. Consequently, where cap rocks were missing, huge volumes of oil must have escaped to the existing surface at the time the oil and gas deposits were formed. An anticline trap (Fig. 16.21a) is a structure where oil and natural gas accumulate due in folded rock layers that form an arch-like shape, with an impermeable layer above, a cap, preventing hydrocarbons from escaping.

A fault trap (Fig. 16.21b) is a structural trap whereby a fault forms a barrier that prevents hydrocarbons from escaping through a fault juxtaposing the porous reservoir rock against an impermeable rock layer which prevents hydrocarbons within the reservoir rock. A stratigraphic trap is one in which hydrocarbons are prevented from escaping due to variations in the lithology of the layers resulting in impermeable layers trapping the oil and gas (Fig. 16.21c). Deposits of salt, largely from seas that evaporated, are buried deep underground. The salt deposits, due to their lower density, pierce through overlying sedimentary beds as they gradually migrate upward deforming surrounding beds to forming folds and faults. Oil and gas deposits are often trapped adjacent to the salt domes (Fig. 16.21d).

The combination of source rock, reservoir rock, and cap rock, together with specific geological formations, creates traps where oil can accumulate in commercially viable quantities. When hydrocarbons are trapped beneath less-porous cap rock, an oil and gas reservoir is formed (Fig. 16.21). Reservoirs of oil and gas represent sources of crude oil and natural gas. Petroleum, commonly referred to as 'oil', and natural gas A are produced from tiny plant sand algae that have been accumulated in seas and lakes millions of years ago. When subjected to heat and pressure, the organic material reacts to form oil and/or natural gas, or simply gas. Gasoline, heating oil, propane, and kerosene are examples of petroleum products. Finally, some types of combinations of any two or more of these four traps can occur to form what are referred to as hybrid traps. For example, the initial configuration of a reservoir can be that of a fault-controlled structural trap, but the caprock can be composed of the draping deposition of mudstones during the process where oceans are formed on the sites of continents by vertical mixing of crust and upper mantle called oceanization.

Mixtures of sand, water, clay, and bitumen are called oil sands, or tar sands, or bituminous sands. The mixture is typically between 1 and 20% bitumen. Bitumen is the heaviest hydrocarbon mixtures that occur in crude petroleum. Bitumen, a viscous semisolid, is extremely viscous such that traditional oil extraction techniques do not work. Bitumen often occurs as solid in cold weather. To mine bitumen, the one removes the overburden which then permits the tar sand to be extracted in a comparable manner to the open-pit mining methods employed in surface mining of coal. Bitumen may be extracted alternatively by utilizing in situ methods such as steam injection to change the properties of the bitumen.

A type of soil that is characterized by a high proportion of particles of sand is called soil sand or sandy soil. Soil sands, containing particles larger than silt and clay, are responsible for the loose, coarse texture of sandy soils. Furthermore, soil sands quickly drain water and nutrients away as they have high contents of greater than 85% particles of sand-sized by mass making them prone to drought. Sandy soils are also characterized by coarse-grained particles of sand which are responsible for poor retention of water. Soils sands have low nutrient retention and have difficulty in retaining essential plant nutrients due to their large pore spaces.

Oil sands are concentrated in Alberta, Canada, and Venezuela although they occur in other countries. In Alberta, Canada, huge oil sand reserves of 1.7–2.5 trillion barrels of oil occur in the oil sands, the world's largest single reserve of oil. However, only 15% of this vast reserve, representing ~75% of the petroleum reserves in North America, is currently recoverable. On account of Alberta, Canada's oil sand reserves, the country now occupies the second largest proven oil reserves in the world tally of global petroleum reserves after Saudi Arabia. Sandy soils commonly occur in areas such as deserts and coastal regions.

Hydraulic fracturing, or fracking, is a technique utilized to induce wells in the extraction of oil and natural gas from deeply buried shale and other low-permeability reservoirs. Fluids, chiefly water, sand, and chemicals, are

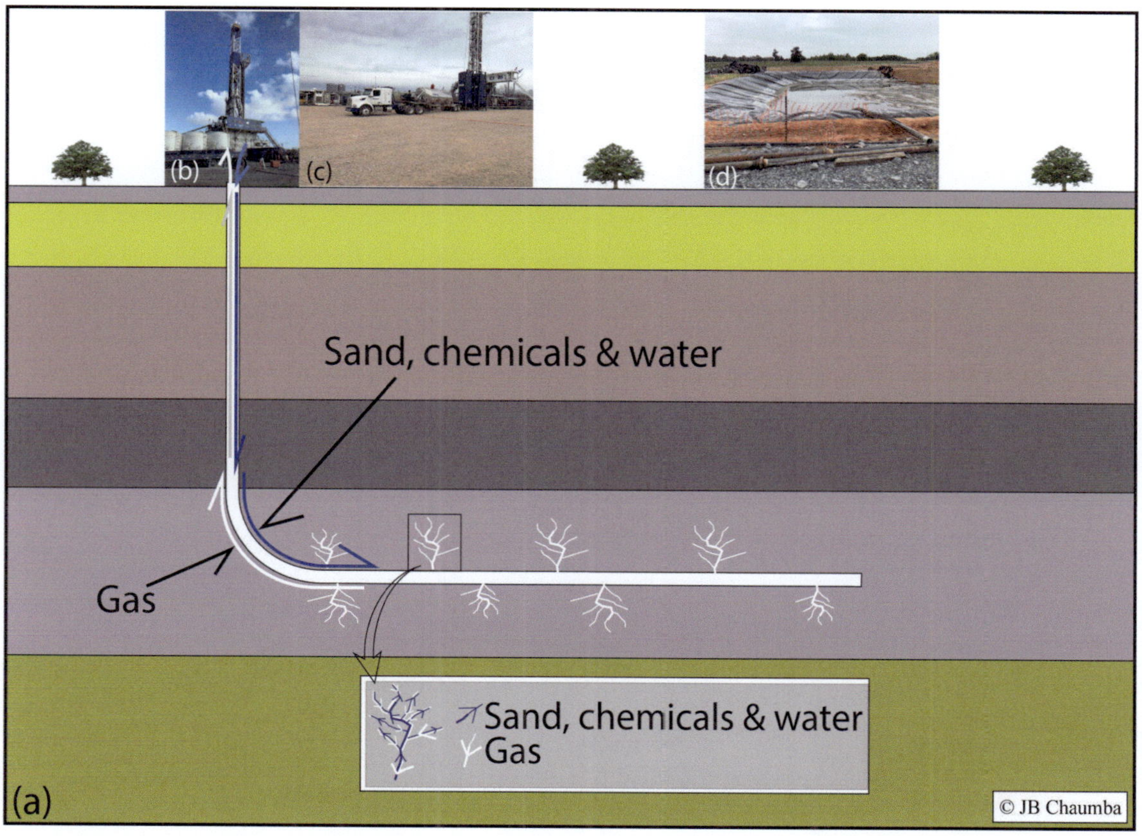

Fig. 16.22 **a** Diagram of illustrating fracking. **b** and **c** are photographs of drill rigs and containers with water and chemicals utilized during drilling. After drilling is completed, materials such as sand, water, and chemicals used pumped into well to fracture the shale (blue arrow to the right) containing gas pumped to the surface (white arrow to the left). **d** is a photograph of a containment tank for drilling wastes. All photographs from south of Coushatta, Louisiana, USA, by the author

injected at high pressure into a well to generate fractures in the rock along which oil and gas can flow more freely. After a well is drilled, a high-pressure mixture of water, sand, and chemicals is injected into the well to fracture the rock and enhance recovery of trapped oil and gas.

A well is first drilled vertically into the shale hosting the hydrocarbons, and then later horizontally into a shale formation (Fig. 16.22). This is followed by perforating holes in the well casing which permit the fracking fluid to escape into the shale (Fig. 16.22). A high-pressure mixture of sand, water, as well as chemicals is then pumped into the well through fracking leading to fracturing of the shale. Oil and gas flow through cracks in the shale generated by fracking (Fig. 16.22a–c). Sand (Fig. 16.22a) plays a crucial role in propping open the fractures, which prevents them from closing. Finally, the oil and gas flow back up the well where they are harnessed (Fig. 16.22a, b). Although fracking has enabled previously unreachable energy reserves to be harnessed, contamination of groundwater, air pollution, and inducement of relatively small earthquakes are serious problems that may never be addressed.

A **pump jack** (Fig. 16.23a, b) is an instrument that is used for lifting oil wells artificially. This instrument is composed of horse-shaped head that moves up and down to provide power for a sucker rod that extends into the well. The pump jack extracts oil when natural pressure of the well is insufficient to force it to the Earth's surface. The energy is provided by various sources such as counterweights, combustion engines, prime movers, gear boxes, and electrical motors. Thus, electrical or gas engines provide the energy needed to pump the oil or hydrocarbon to the surface.

Drilling wells into sedimentary rocks deposited above crystalline rocks of the oceanic crust for the purpose of extracting oil and natural gas from is called **offshore drilling**. Highly specialized equipment and techniques are involved in drilling in marine environments due to harsh conditions. Consequently, offshore drilling deals with unique safety and environmental problems. During offshore drilling operations, drilling rigs, which are typically floating platforms or **drillships** or fixed platforms, are utilized to drill into the sedimentary rocks of the seafloor.

Huge offshore drilling platforms are structures from where oil and gas are extracted from their host rocks beneath the seabed underlying deepwater sea waters. To extract oil and gas from deeper areas, floating platforms such as semi-submersibles are used whereas drillship fixed platforms are used for exploiting these resources in shallower waters. Fixed platforms, which are typically used in shallower waters, are anchored to the seabed through steel or concrete legs. Offshore drilling platforms are equipped with facilities for drilling, production, and sometimes **processing**. Most platforms also contain living quarters for the crew.

Fig. 16.23 Some photographs of pump jacks from **a** the Gulf of Mexico in Louisiana, USA Courtesy of Adrian Arismendez and **b** near Hearne in South Texas, USA. *Credit* Francis Nwachukwu

A wide variety of floating platforms exist. Platforms that utilize pontoons and columns to float, which providing stability in deeper waters, are called semi-submersibles. Self-propelled vessels that are equipped with drilling equipment and are typically used in ultra-deepwater locations are known as drillships. Platforms with legs that can be lowered to the seabed, jacking up the platform above the water, are called jack-up rigs. Slender and tall structures that are capable of flexing with the movement of the water and are called compliant towers. They are used in deeper waters. Platforms that are anchored to the seabed with tension tendons are called tension-leg platforms. Tension-leg platforms are used for deepwater drilling locations. Floating, Production, Storage, and Offloading (FPSO) vessels are a combination of production facilities. Drill ships (Fig. 16.24) and semi-submersibles provide flexibility and cost-effectiveness for accessing reserves that are challenging. Floating platforms play a critical role in offshore drilling through their provision of movable stable bases for drilling and oil and gas production in deep ocean waters. Additionally, floating platforms also permit operations in areas either with no pipeline infrastructure or in remote locations through serving as storage, offloading system and as a mobile production. Platforms use dynamic positioning buoyancy and anchoring to steady drilling equipment, housing for crews, storage, and processing facilities. FPSOs, drill ships, and semi-submersibles provide flexibility and cost-effectiveness for accessing reserves that are challenging.

Production facilities are equipment used in the processing and preparation of the extracted hydrocarbons to be transported. Additionally, most platforms are equipped with accommodation for the platform crew referred to as living quarters. Offshore drilling operations have the potential to cause environmental impacts of destruction of habitat: construction and operation of platforms can damage sensitive marine habitats. In addition, oil leaks and spills, together with routine discharges, may cause both water and air pollution. The downside of drilling operations is that they disturb marine life through noise pollution. Furthermore, during drilling operations and when drilling is completed, wastes that are generated need to be properly managed to prevent contamination.

Floating platforms (Fig. 16.24) such as drill ships or semi-submersibles find use in deeper waters whereas fixed platforms (Fig. 16.25) have preference in shallower waters. Jack-up rigs are characterized by legs that are lowered to the seabed in order to drill in moderate depths. To maintain well pressure, drilling mud is pumped down to cool the bit as well as transport away rock cuttings. Wells are drilled through the seabed and casings (steel pipes) are then installed and cemented for purposes of well stabilization. To prevent uncontrolled releases of oil or gas, a blowout preventer is installed. Production equipment is installed on the seafloor or platform after a successful well is drilled and intersects oil or gas. The oil and gas are then transported to shore through pipelines or shuttle tankers. Offshore drilling, however, poses great risks to marine life and ecosystems

Fig. 16.24 Pictures of offshore drill ship and rig in the Gulf of America (**a**) and a close-up of an offshore drill rig (**b**). *Credit* Mark Gross

Fig. 16.25 **a** Photograph of drilling rig on platform, Russia and **b** photograph of a drilling platform, location unknown

through potential oil spills and disruption of habitat. To protect workers from hazards, such as explosions, fires, and malfunctions of equipment, safety measures in offshore drilling operations are strictly enforced.

A wide variety of contaminants like hydrocarbons, heavy metals, and naturally occurring radioactive materials (NORM) form part of drilling fluids. **Drilling fluid** is a specialized fluid that is used during drilling in the oil and gas industry to facilitate drilling as well as to maintain wellbore stability. After drilling is completed, the drilling fluids, also known as drilling mud (Fig. 16.22d), are typically managed through a combination of methods including recycling, disposal, or treatment. The specific approach chosen depends on the type of fluid used and local regulations. Recycling and reuse are increasingly becoming more common, particularly with water-based fluids, to help in the reduction of both environmental impact and operational costs. Water-based drilling fluids are processed to recover valuable oil and water and are subsequently reused in drilling or other wells, as well as in fracking.

Drilling fluids play a critical role in cooling both the drill bit and drill string during drilling, as well as reducing friction and wear. Drilling fluids also improve drilling efficiency. Drilling fluids also maintain hydrostatic pressure, prevent walls of the wellbore from collapsing, and also prevent any unwanted fluids such as water or gas from entering the wellbore. In addition, trilling fluids transport samples of cuttings, or fragments of the rock being drilled, to the surface where they are analyzed to provide information about the lithology being drilled.

There are some instances when it is not either feasible or cost-effective, or both to disposed of drilling fluids. When this happens, drilling fluids are either buried in the ground through injection wells or are transported to landfills that are designated for drilling fluids where they are treated. Chemical, physical, or biological treatment methods are used to neutralize or remove contaminants before disposal or reuse of drilling fluids. State and federal regulations govern regulations pertaining to the handling and disposal of drilling fluids. Specific rules and regulations applicable in the handling and disposal of these fluids often vary depending on location. Rock fragments generated during drilling, called drill cuttings, form part of drilling wastes. These rock fragments are typically removed from the drilling fluids and treated separately. The rock fragments are either disposed of, or treated, for beneficial use.

Glossary

Aggregate granular materials such as crushed stone, sand, gravel, and slag and utilized primarily in construction. They are the main components of building materials, such as asphalt, concrete, mortar, and foundations, providing stability, strength, and structure to infrastructure and buildings.

Biomass energy renewable energy derived from organic materials such as plants or municipal, agricultural, and animal waste from the stored the solar energy which is released through burning or decomposition in generation of electricity, heat, and biofuels such as biogas and ethanol.

Concentrating solar power or CSP utilizes mirrors to reflect and concentrate sunlight onto receivers with the concentrated sunlight energy which is then used to heat high-temperature fluid in the receivers.

Critical minerals minerals that are essential to a nation's economic or national security and possess a supply chain that is vulnerable to disruption, and it also serves a vital function in manufacturing products where its absence poses significant consequences.

Disseminated ore deposits high tonnage, low concentration ore deposits where ore minerals are scattered as fine-grained particles and/or veinlets throughout the host rock. The vast sizes of these deposits make them economically viable sources for exploitation and are the source of most of the world's copper, diamonds, molybdenum, silver, and tin, among others.

Drillship vessel that is self-propelled and designed to drill in deep ocean waters for oil and gas and maintaining its location over the subsea wellhead through use of dynamic positioning systems rather than anchors.

Earth resources naturally occurring materials such as fossil fuels, minerals, rocks, water, and air, among others, which are useful in manufacturing of goods for use by human beings.

Energy resource natural source from which energy for electricity, power, and other uses by humans can be generated.

Fossil fuels energy sources that are formed from remains of ancient plants and animals over millions of years and they are non-renewable.

Geothermal energy heat generated from the interior of planet Earth and used either for direct heating and cooling or in electricity generation.

Gossan highly leached and oxidized zones occurring at the surface or upper parts of sulfide ore bodies with reddish-brown coloration from iron oxides such as hematite and goethite produced by leaching and oxidation of underlying sulfide minerals by water and air. Soluble minerals are transported away to leaving behind insoluble oxides of iron and manganese.

Geologic resources solid, liquid, and gas elements and compounds which are obtained from the surface and within the Earth's crust with high enough concentrations to be extracted at a profit.

Hydraulic fracturing process whereby large volumes of water, sand, and chemicals are injected underground at high pressure to fracture rocks for extracting oil and gas.

Hydropower Energy source that uses the kinetic energy of moving water to generate electricity, typically by diverting a river's natural flow with a dam to create a reservoir. Water released from the reservoir spins turbines, which then activate a generator to produce electricity. This system relies on the natural water cycle and is a flexible and reliable source of power.

Hydrothermal ore deposits minerals that form from mineral-rich hot water circulating through the Earth's crust and dissolves elements, concentrates them, and later precipitates them through either replacing the rock or within rock fissures.

Magmatic-hydrothermal ore deposits formed due to metal enriched magmatic fluids sourced from cooling magma migrate through the Earth's crust depositing copper, nickel, platinum, and gold in veins, fractures, stockworks, and cavities during pressure release and cooling.

Magmatic ore deposits mineral deposits such as platinum-group elements, chromium, and nickel that form in association with igneous rocks.

Megawatt a unit of power which measures the rate at which energy is being generated or consumed at a given moment, and it is equivalent to one million watts (1,000,000 W) or one thousand kilowatts (1,000 kW).

Mineral resource naturally occurring inorganic minerals in high enough concentrations that can be economically extracted.

Non-renewable energy resources energy resources existing in finite amounts that cannot be replenished at rates that are comparable to those at which they are consumed.

Non-renewable resources finite natural resource that cannot be replenished at rates comparable to those at which consumption occurs and will be eventually depleted.

Nuclear fission nuclear reaction whereby the nucleus of heavy atoms splits into two or more smaller nuclei and other particles accompanied by large amounts of energy being released together with more neutrons.

Nuclear fusion combining of light atomic nuclei to form a single, heavier nucleus accompanied by release of huge amounts of energy.

Solar energy light and heat energy sourced from the Sun's radiation harnessed to produce electricity, heat buildings, and even power chemical reactions. Essentially, solar energy is the radiant energy emitted by the Sun, which can be utilized for various purposes on Earth.

Offshore drilling drilling into the rock formations beneath the ocean or seafloor to extract oil and gas.

Ore minerals minerals containing high of metals in forms that can be extractable.

Photovoltaic (PV) cells devices capable of converting solar energy directly into electricity through photons from sunlight being utilized to excite electrons in semiconductor materials to generate an electric current. PV cells are also called solar cells.

Photovoltaic panels or solar panels—are devices that convert sunlight directly into electricity.

Placer deposits accumulations of weather-resistant dense minerals such as gold, platinum, and gemstones through weathering, transportation, and deposition by both gravity and water, and subsequently concentrated on riverbeds, beaches, and in any locations where the stream velocity decreases.

Processing stages followed in benefaction of ore from mining to the finished products through various operations which liberate the targeted minerals or commodities to desired concentrations. In energy resources it refers to extraction of coal, uranium, petroleum, and sunlight to their conversion into electricity, gasoline, or heat.

Protore metalliferous material that has not yet been enriched to form ore.

Pump jack equipment utilized in oil wells for artificially pumping oil from the ground.

Renewable energy energy sourced from natural resources such as water, sunlight, wind, and geothermal heat that are replenished naturally on a human timescale and be depleted.

Renewable energy source a virtually inexhaustible natural resource capable of replenishing itself at a rate faster than it is consumed.

Reserve part of an identified resource from which a mineral or energy commodity can be extracted at a profit at the time of determination.

Reserve base sum of both measured and indicated resource, demonstrated, from which estimation of reserves is conducted.

Residual deposits mineral concentrations form in situ from weathering and erosion of pre-existing rocks, leaving behind insoluble, concentrated ore materials. As soluble components of rock are leached away, less soluble minerals such as iron oxides (laterites) or aluminum silicates (bauxite) are left behind.

Secondary enrichment deposits also called supergene enrichment deposits—are ore deposits where metals are dissolved from oxidized rocks near the surface by circulating groundwater and are then redeposited and concentrated below the water table. Thus, gangue material dissolved and transported was leaving behind concentrations of copper, silver, and gold, to produce high-grade ore compared to the original low-grade primary deposit.

Sedimentary-exhalative (SEDEX) deposits ore bodies formed when hot metal-laden hydrothermal fluids are vent and precipitate lead, zinc, silver, with or without copper, gold, and barite on the seafloor in deep sedimentary basins. Dewatering and metamorphism of sediments, rather than from volcanic activity, is the source of the brines. They are the main source of lead and zinc.

Sedimentary ore deposits concentrations of mineral deposits due to sedimentary processes occurring in sedimentary rocks.

Solar thermal systems or STS—capture radiant heat from the sun to produce thermal energy which is then utilized to generate electricity, heating water, and space heating.

Volcanogenic massive sulfide deposits or VMS deposits—are accumulations of metallic minerals, rich in sulfides, which form at or near the seafloor during volcanic activity. VMS deposits are significant sources of copper, zinc, lead, gold, and silver that are formed when mineral-rich, superheated seawater that is heated by nearby magma rises to the seafloor and mixes with cold seawater. The rapid temperature changes generated cause dissolved metals to precipitate out and accumulate, forming a massive sulfide cap on seafloor and an underlying feeder zone of mineralized veins in the volcanic rock.

Wind energy renewable energy that harnesses wind power for electricity generation through wind turbines. It is sometimes referred to as wind power.

Textbook Summary

Physical Geology is an introductory-level geology course which most non-science majors probably prefer to take to satisfy their science requirements. Consequently, Physical Geology tends to have the highest enrollment of science introduction courses at the six universities, ranging from research to teaching universities, where I have taught since graduation in 2009. This text has numerous activities such as multiple choice questions, Discussion questions, and short answers. Additionally, it has activities based on videos that explain various aspects of Physical Geology in greater detail and students then answer accompanying questions.

This textbook is well suited the traditional face-to-face, hybrid, and ONLINE delivery methods. As the Instructor of record I have used Blackboard, CANVAS, and D2L, and this textbook is well suited for all these packages and more. After having graduated from five different Universities in three countries, I accumulated a wealth of information on writing, teaching, and research. This textbook utilizes geologic examples from throughout the world such that it can conveniently be adopted by higher education institutions internationally, which is the author's goal.

J. B. Chaumba, *Introductory Physical Geology*, Springer Textbooks in Earth Sciences, Geography and Environment,
https://doi.org/10.1007/978-3-032-20231-4

MIX
Papier aus verantwortungsvollen Quellen
Paper from responsible sources
FSC® C105338

If you have any concerns about our products,
you can contact us on
ProductSafety@springernature.com

In case Publisher is established outside the EU,
the EU authorized representative is:
Springer Nature Customer Service Center GmbH
Europaplatz 3, 69115 Heidelberg, Germany

Printed by Libri Plureos GmbH
in Hamburg, Germany